Root Methods: A Handbook

Springer
Berlin
Heidelberg
New York
Barcelona
Hong Kong
London
Milan
Paris
Singapore
Tokyo

A.L. Smit · A.G. Bengough · C. Engels
M. van Noordwijk · S. Pellerin · S.C. van de Geijn
Editors

Root Methods

A Handbook

With 108 Figures, 4 in Color, and 39 Tables

Springer

Dr. Albert L. Smit
Plant Research International
(Wageningen UR)
Postbus 16, 6700 AA Wageningen,
The Netherlands

Dr. Meine van Noordwijk
ICRAF, JI CIFOR
PO Box 161, Situ Gede
Sindang Barang Bogor 16680
16001 Bogor, Indonesia

Dr. A. Glyn Bengough
Scottish Crop Research Institute (SCRI)
Invergowie
Dundee, DD2 5DA, United Kingdom

Dr. Sylvain Pellerin
INRA
B.P. 81, 71 Avenue Edouard Bourleaux
33883 Villenave d'Ornon, France

Prof. Dr. Christof Engels
University of Bayreuth
Institute of Geosciences
Department of Agroecology
95440 Bayreuth, Germany

Dr. Siebe C. van de Geijn
Plant Research International
(Wageningen UR)
Postbus 16, 6700 AA Wageningen,
The Netherlands

Cover illustration: Maize root system (see Fig. 4.3a in Chap. 4 by L. Pagès et al.)

This book is the result of a project financed by the European Union: Concerted Action AIR3-CT 93-0994: The dynamics of rooting patterns in relation to nutrients and water in soils. Development, standardisation and documentation of methodologies. The contents of the book is the sole responsibility of the Editors and Authors and does not represent the views of the Commission or its services.

ISBN 3-540-66728-8 Springer-Verlag Berlin Heidelberg New York

Library of Congress Cataloging-in-Publication Data
Root methods: a handbook / A.L. Smit . . . [et al.], (eds.). p. cm. Includes bibliographical references. ISBN 3-540-66728-8 (hc : alk. paper) 1. Roots (Botany) I. Smit, A. L. (Albert L.), 1950–
QK644.R656 2000 575.5′4—dc21

Production: PRO EDIT GmbH, Heidelberg, Germany
Cover design: Design & Production GmbH, Heidelberg, Germany
Typesetting: Best-set Typesetter Ltd., Hong Kong

Printed on acid free paper SPIN 10688779 31/3130 5 4 3 2 1 0

Preface

Why a New Handbook on Root Methods?

Uptake of water and nutrients is a key process in agricultural and natural ecosystems. Plant roots play a vital role in the supply of resources for growth, and yet we have a relatively poor understanding of how they function in the natural soil environment. Dr. Böhm began the Preface of his standard text on root research methods (Böhm 1979) with the comment:

"Root research under natural field conditions is a step-child of science. The reason for this is primarily methodological. The known methods are tedious, time-consuming and the accuracy of their results is often not very great".

Although improvements have been made recently for some methods, working on roots is still tedious and time-consuming. Compared to some other disciplines, a root researcher requires a considerable budget for his or her work. It is for this reason that in root research careful planning, the choice of appropriate methods and also a clear idea of *which* root characteristic should be observed in relation to the goal of the research question are fundamental. Consequently, it is recommended to make use of "root experience" developed elsewhere as much as possible.

A logical choice is then to produce a handbook in which root methods in current use are made available to a broader public. Since Böhm's book several new methods have been developed: e.g. the use of mini-rhizotrons (already mentioned in his book) has increased considerably and several (time-saving) new methods of observing roots using mini-rhizotrons have been developed. However, many older methods have also been revised and are described in this book.

Scientific Contents

At a time when researchers are often forced to reduce their budgets, root research is under pressure because of its time-consuming and expensive character. This book enables the researcher to make use of the expertise and knowledge available worldwide as much as possible, as it gives a broad overview of all available methods.

Not only sophisticated new methods (image analysis, CT scans etc.) are described, but also time-saving "quick and (hopefully not too) dirty tricks" to quantify root growth and root distribution in the soil profile.

Each chapter of the book was written by authorities on the subject. Virtually all methods which can be used to expose "the hidden half" of the plant (in reference to a well-known book on root growth and function, Waisel et al. 1996) are described.

In the introductory chapter the importance of roots in ecology and crop production is explained (why and what to measure). The following two chapters deal with methods to measure *root system properties* (anatomy, root hairs, mycorrhizae) and to measure the effect of *environmental interactions* (mechanical stress, water, oxygen, temperature) on root growth and functioning, under both laboratory and field conditions.

A number of chapters focus on the assessment of root distribution and root dynamics under field conditions. Advice is given on sampling strategy and statistics, before treating in detail auger sampling, in-growth cores, pinboard methods, trench profile techniques and core break methods. Also, the use of transparent interfaces in a soil profile (mini-rhizotrons and root windows) is explained in detail. One chapter is devoted to processing the data from root observations to analyse (fine) root longevity, an important aspect in ecosystem processes. Relatively new methods described in the book include modelling (describing root growth and architecture as well as functioning), the use of root image analysis (IA), computer-assisted tomography (CAT) and magnetic resonance imaging (MRI).

Finally, techniques for the experimental determination of functioning of roots are described in the second part of the book, e.g. a review on the use of isotope techniques in root research and methods to assess nutrient acquisition, water uptake and plant anchorage.

We hope that this book will be a guide for research workers in many areas of the plant and soil sciences, agriculture, forestry and horticulture, to the benefit of production systems and natural ecosystems.

Acknowledgement. The production of this handbook was possible as part of a project sponsored by the European Union (Concerted Action AIR3-CT93-0994 –"RootAction"). The objectives of RootAction were to establish a representative network of research centres where plant root studies are done in relation to nutrient and water dynamics of the plant/crop/soil system. In addition, the objective of RootAction was to standardise and update methods used currently in root research. At an early stage of the Concerted Action it was considered important to produce a book in which the work could be made available to the scientific community worldwide.

More people than the mentioned authors contributed to this book. Several participants of RootAction, who are specialists on a certain topic, shared their expertise with the authors. They are gratefully acknowledged. Furthermore, to guarantee scientific quality, the individual chapters were reviewed by at least two independent reviewers. Special thanks are extended to these reviewers (listed on the pages IX, X). They not only thoroughly reviewed the chapters but, in many cases, also made valuable suggestions to the authors.

We also thank many colleagues and editors of journals for permission to use photographs and figures. Many thanks also to Springer-Verlag (Germany) for excellent and efficient support in the final stages of the production of this book.

Finally, we hope that this handbook will not only be found in libraries of institutes and universities, but will also have a permanent place on the desks of many scientists working on roots. This would be an indication that the book is used as it was intended: as a handbook!

On behalf of the Editors, A.L. Smit
Wageningen, April 2000

References

Böhm W (1979) Methods of studying root systems. Ecological Studies 33. Springer Berlin Heidelberg New York

Waisel Y, Amram E, Kafkafi U (eds) (1996) Plant Roots. Marcel Dekker, New Yok

Reviewers

We would like to gratefully acknowledge the kind help of the following referees.

The Editors

Dr. J. Alegre, ICRAF-PERU, Lima, Peru
Dr. R.R. Allmares, Southern Experiment Station, University of Minnesota, Waseca, USA
Dr. P.W. Barlow, Department of Agricultural Sciences, Bristol, UK
Dr. G.M. Berntson, Harvard University, Cambridge, USA
Dr. W.L. Bland, University of Wisconisn, Department of Soil Sciences, Madison, USA
Dr. T. Boutton, Department of Rangeland Ecology, Texas A & M University, Texas, USA
Dr. L. Bruckler, INRA, Unité de Science du Sol, Avignon Cédex 9, France
Dr. D.T. Clarkson, IACR-Long Ashton Research Station, Department of Agricultural Sciences, University of Bristol, Bristol, UK
Dr. Mike Coutts, c/o Forestry Commission Northern Research Station, Penicuik Midlothian, UK
Dr. W.J. Cram, Department of Biological and Nutritional Sciences, University of Newcastle upon Tyne, Newcastle upon Tyne, UK
Dr. P.R. Darrah, University of Oxford, Department of Plant Sciences, Oxford, UK
Dr. L.A. Dawson, Macaulay Land Use Research Institute, Aberdeen, UK
Dr. M.C. Drew, Texas A&M University, College Station, Texas, USA
Dr. W.R. Eason, Institute of Grassland and Environmental Research, Aberystwyth, Dyfed, Wales, UK
Dr. A.H. Fitter, Department of Biology, University of York, York, UK
Dr. O. Gasparikova Institute for Botany, Slovak Academy of Science, Bratislava, Slovak Republik
Dr. J. Graham, School of Agriculture, Food and Environment, Cranfield University, Silsoe Bedfordshire, UK
Dr. T. Heaton, NERC Isotope Geosciences Lab, Nottingham, UK
Dr. T.C. Kaspar, USDA/ARS National Soil Tilth Lab, Ames, USA
Dr. E.A. Kirkby, The University of Leeds, Department of Biology Leeds, UK

Dr. M.B. Kirkham, Department of Agronomy, Kansas State University, Manhattan, Kansas, USA

Dr. F. Lafolie, INRA, Station de Science du Sol, Montfavet, France

Dr. H. Majdi, The Swedish University of Agricultural Sciences, Department of Ecology and Environmental Research, Uppsala, Sweden

Dr. H. Persson, The Swedish University of Agricultural Science, Department of Ecology and Environmental Research, Section of Soil Ecology, Uppsala, Sweden

Dr. H. Rogers, National Soil Dynamics Lab, USDA-ARS, Auburn, USA

Dr. M. Smith, Institute of Hydrology, Wallingford, UK

Dr. P. Stamp, Institut für Pflanzenwissenschaften ETH Zürich, Switzerland

Dr. Judy Tisdall, School of Agriculture, La Trobe University, Victoria, Australia

Dr. E.W. Tollner, Biological and Agricultural Engineering Department, Athens, USA

Dr. W.B. Voorhees, North Central Soil Conservation Research Laboratory, Morris, Minnesota, USA

Contents

Principal Authors

The following list contains the addresses of those authors who had the main responsibility for the chapters. The addresses of all contributors to a chapter can be found on the first page of the contribution.

Chapter 1: Prof. Dr. D. Atkinson, The Scottish Agricultural College (SAC), West Mains Road, Edinburgh, EG9 3JG, UK, tel. +44 131 5354004, fax +44 131 5354340, e-mail: d.atkinson@ed.sac.ac.uk

Chapter 2: Dr. Tara S. Gahoonia, The Royal Veterinary and Agricultural University, Department of Agricultural Sciences, Plant Nutrition and Soil Fertility Laboratory, Thorvaldsensvej 40, 1871 Frederiksberg C, Copenhagen, Denmark, tel. +45 35283497, fax +45 35283460, e-mail: TSG@KVL.DK

Chapter 3: Dr. W.R. Whalley, Silsoe Research Institute, Wrest Park, Silsoe, Bedford, MK45 4HS, UK, tel. +44 1525860000, fax +44 1525860156, e-mail: richard.whalley@bbsrc.ac.uk

Chapter 4: Dr. L. Pagès, INRA-Centre d'Avignon, Écophysiologie et Horticulture Site Agroparc, Domaine St. Paul, 84914 Avignon Cédex 9, France, tel. +33 0432722431, fax +33 0432722432, e-mail: loic.pages@avignon.inra.fr

Chapter 5: Dr. A.G. Bengough, Scottish Crop Research Institute (SCRI), Invergrowie Dundee, DD2 5DA, UK, tel. +44 1382 562731/ext 2527, fax +44 1382 562426, e-mail: G.Bengough@scri.sari.ac.uk

Chapter 6: Prof. Dr. Maria do Rosário G. Oliveira, Universidade de Évora, Apartado 94, 7002-554 Évora, Portugal, tel. +351 266 760800, fax +351 266711163, e-mail: mrol@uevora.pt

Chapter 7: Dr. M. van Noordwijk, ICRAF, Jl CIFOR, P.O. Box 161, Situ Gede, Sindang Barang Bogor 16680, 16001 Bogor, Indonesia, tel. +62 251 625 415, fax +62 251 625 416, e-mail: m.van-noordwijk@cgiar.org

Chapter 8: Dr. A.L. Smit, Plant Research International (Wageningen UR), Postbus 16, 6700AA Wageningen, The Netherlands, tel. +31 317 475877, fax +31 317 423110, e-mail: A.L.Smit@plant.wag-ur.nl

Chapter 9: Dr. J.E. Hooker, School of Applied Sciences, University of Glamorgan, Pontypridd, Mid-Glamorgan, CF37 1DL, UK, tel. +44 1433 482453, e-mail: jhooker@glam.ac.uk

Chapter 10: Dr. W. Richner, ETH Zürich, Institute of Plant Sciences, ETH Zentrum, LFW A4, 8092 Zürich, Switzerland, tel. +41 1 632 4237, fax +41 1 632 1143, e-mail: walter.richner@ipw.agrl.ethz.ch

Chapter 11: Dr. S. Asseng, CSIRO Division of Plant Industry, Private Bag PO, Wembley, WA 6014, Australia, tel. +61 8 93336615, fax +61 8 93878991, e-mail: s.asseng@ccmar.csiro.au

Chapter 12: Prof. Dr. D. Robinson, Department of Plant and Soil Science, University of Aberdeen, Aberdeen, AB 24 3UU, UK, tel. +44 1224 273 662, fax +44 1224 272 703, e-mail: david.robinson@abdn.ac.uk

Chapter 13: Prof. Dr. Ch. Engels, University of Bayreuth, Institut für Geowissenschaften, Abt. Agrarökologie, 95440 Bayreuth, Germany, tel. +49 921/552292, fax +49 921/552315, e-mail: christof.engels@uni-bayreuth.de

Chapter 14: Dr. J.E. Fernández, Inst. de Recur. Natur. y Agrobiol. de Sevilla, Avenida de Reina Mercedes, No. 10, Aptdo 1052, 41080 Sevilla, Spain, tel. +34 954 624711, fax +34 954 624002, e-mail: jefer@irnase.csic.es

Chapter 15: Dr. P. de Willigen, Alterra, Green World Research, Postbus 47, 6700 AA Wageningen, The Netherlands, tel. +31 317 474204, fax +31 317 419000, e-mail: p.dewilligen@alterra.wag-ur.nl

Chapter 16: Dr. S. Pellerin, INRA, B.P. 81, 71 Avenue Edouard Bourleaux, 33883 Villenave d'Ornon Cédex, France, tel. +33 556843051, fax +33 556843054, e-mail: pellerin@bordeaux.inra.fr

CHAPTER 1

Root Characteristics: Why and What to Measure

D. Atkinson

SAC, West Mains Road, Edinburgh, EH9 3JG, UK

CONTENTS

A.L. Smit et al. (Eds.), Root Methods

1.1 Introduction

A volume on available methods for the measurement of roots must begin by questioning why such a volume is needed and thus with questions such as "Why measure roots?" "What root parameters need to be measured?" and "What is the functional, physiological, or ecological significance of a particular morphological assessment?" To provide some perspective to this volume, these key questions are discussed here. Measurements of root properties are made and used by those involved in quantifying and managing both agricultural and natural ecosystems. It is important therefore that methods and definitions can be used in relation to the full range of species found in a complete range of habitats.

1.2 Why Study Roots?

A study of roots in the field is probably justified only when there is reason to believe that the amount of below-ground material is likely to be statistically and functionally different to that which might be predicted by the allocation of a fixed amount of photosynthate to a below-ground compartment using an allometric model, or where there is a need to achieve basic understanding of the system (Atkinson 1996). In many situations, however, measurements will be needed because available data will be inadequate to allow carbon partitioning or root mass to be estimated. Many studies which have included measurements of roots, have either been indecisive, i.e. have not produced a clear functional link between the root measurement and the problem under investigation (were the root measurements really necessary?) or demonstrate that the root information could as easily have been generated from a model. Prior to beginning any study of roots, it is important to ask why it is being carried out. Despite this proviso, the main reasons for studying roots are:

Ecological Significance. In many situations, but especially in relation to natural vegetation, there is little basic relevant information e.g. amount of root, distribution of root weight with depth. This information is needed to answer questions, such as "Why do particular plant species grow in the places they do?"

Resource Capture. Roots represent the principal means whereby plants extract resources such as nutrients and water from soil. Current expenditure on irrigation systems and fertilisers attest to the importance of nutrient and water to crop production. An understanding of roots will help to eliminate wastage and adverse environmental effects.

Soil Microbes. The root system represents the major pathway for the flow of carbon to the soil and to soil organisms, especially those in the rhizosphere. As rhizosphere organisms are responsible for many key processes, such as N immobilisation, NH_4 oxidation, denitrification, and root nodulation, the supply of resources to the soil is potentially critical to the evaluation of soil carbon budgets. In addition, there is an increasing body of information on the effects of plant species on soil microbial composition. This has gained additional emphasis as a consequence of the debate about the impact of raised atmospheric levels of CO_2 and nitrous oxide.

Resource Allocation. Information on the relative allocation of resources below-ground, above-ground, and to different types of root and mycorrhizas tells us about the coupling of the plant to its environment.

Plant Interactions. Roots represent one of the key means whereby plants of the same and other species interact. These interactions are now being seen, both in relation to temperate and tropical multi-crop systems, as means of improving the efficiency of resource use.

Soil Structure. The root and its associated microflora have a major effect upon soil structure and the stability of aggregates. The input of organic matter to the soil which they represent will influence key soil properties such as cation exchange capacity.

Anchorage. Roots are essential for plant stability and anchorage. While this is particularly important for tree crops, it has significant economic implications for many field crops, e.g. cereals.

Root Products. Roots may be used as an energy source in tropical production systems. They may also be a source of pharmaceutical compounds or of food additives and flavourings.

Basic Biological Information. To obtain basic information on a part of the plant which consumes a significant proportion of total resources and which is of physiological and developmental interest in its own right.

These topics are reviewed briefly to provided a background to the later discussion of methods.

1.2.1 Ecological Significance

The ecological significance of roots is poorly understood (Atkinson 1991). Although observation of most habitats or groups of habitats indicate that they have species which are both characteristic of and different between habitats, we are still unaware of the cycle of interactions between soil characteristics and the amount of root. It is clear that in most plants roots exist to absorb nutrients and water and to anchor the plant to its substrate. "Roots are natural selection's design solution to the problems of obtaining these resources from a heterogeneous, porous, semi-compressible medium containing solid, liquid and gaseous phases" (Robinson 1991). Against this basic requirement, plants have evolved a series of strategies to solve the problem of maintaining an upright stance; at least for their reproductive structures, and to capture sufficient assimilates to reproduce and transfer their genes to the next generation. This variety of strategies allows for different genotypes to co-exist and for the acquisition of resources from different soils. These adaptations occur at several levels (Robinson 1991) and are illustrated in Fig. 1.1. Although activity at subcellular, cellular, organ and whole plant levels must have a molecular basis, it seems that such higher level functions as morphological variation probably account for much of the variation in the response to soil factors. Atkinson (1990a) reviewed variation in morphological and physiological traits and found that while a morphological feature such as root length density could vary between species and habitats by a factor of 10^3, physiological attributes normally varied only by around 10^1. This emphasises the need for the types of values which are measured in field studies of roots.

Developments in soil microbiology (see also Sect. 1.2.2), especially in the use of molecular methods to identify organisms and their effects, indicate the association between specific microbes and individual species. This suggests that roots may be critical to the establishment of microbial populations in soil which then influence factors such as soil structure and nutrient transformations, which go on to influence vegetation development. This and other ecological processes suggest that in natural ecosystems a range of measurements, which are wider than commonly made in crop situations, will be necessary. In a crop situation, plants principally interact with other plants with whom they share the whole or most of their genotype. All individuals are relatively similar and so the purpose of studies tends to be the prediction of the use of soil water or the extraction of nutrients, which will then inform either a fertiliser or irrigation programme. In natural ecosystems, a range of genotypes will be present. A range of strategies will be exhibited with the possibility of particular characteristics leading to effective resource acquisition in a limiting situation. As roots are not effective in radiation capture, or the production of reproductive structures, there will be many situations where resource allocated to the root system, even when needed to absorb water and nutrients, represents a reduction in total productivity. The challenge of having sufficient plasticity to cope with variations in environmental resources, is illustrated in Fig. 1.2.

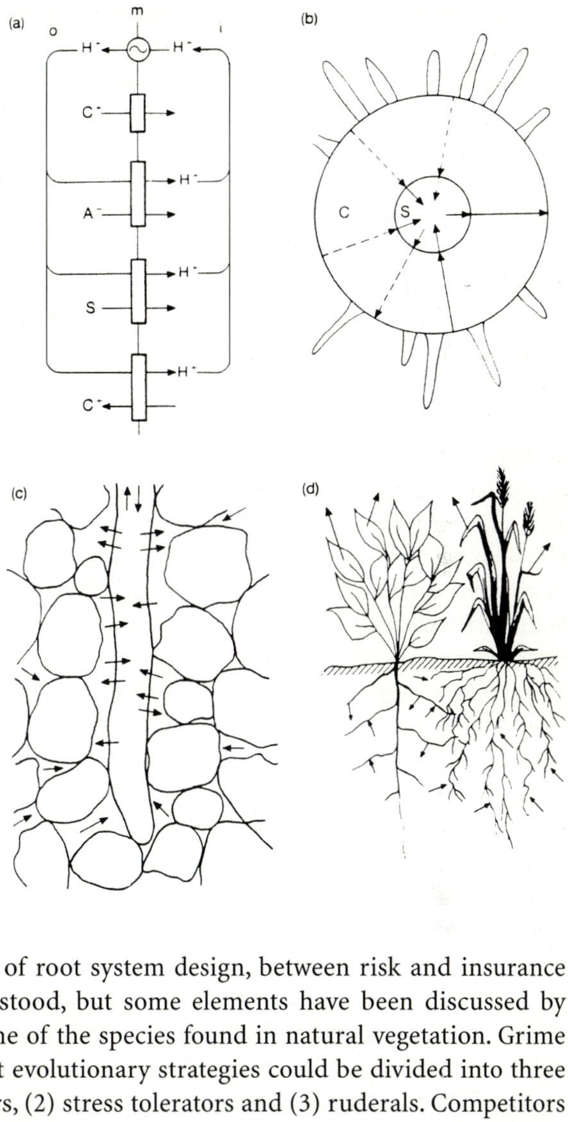

Fig. 1.1. Hierarchies of nutrient and water flux in roots and root systems. **a** Transmembrane fluxes of cations (C^-), anions (A^-) and uncharged solutes (S) and their relation to active proton efflux (modified from Smith and Smith 1986): *m* membrane, *o* outside, *i* inside. **b** Symplastic (*solid arrows*) and apoplastic (*broken arrows*) fluxes within and between cells in a transverse section of a root cortex (C) and stele (S). **c** Fluxes in the soil solution to and from a single root. **d** Fluxes to and from root systems and within plants growing in a community. (After Robinson 1991)

The balance, in terms of root system design, between risk and insurance strategies is poorly understood, but some elements have been discussed by Grime et al. (1991) for some of the species found in natural vegetation. Grime (1974) suggested that plant evolutionary strategies could be divided into three trait groups (1) competitors, (2) stress tolerators and (3) ruderals. Competitors are associated with rapid growth rates and therefore depend upon the ability to sustain high rates of resource capture. Morphological plasticity is important here. These species have high rates of reinvestment of the captured resources in new roots. This type of investment must result in access to significant edaphic resources. Stress tolerators depend upon their capacity to both capture and to retain scarce resources. This tends to mean that roots will be long lived with plasticity expressed by physiological changes. Ruderal species exhibit the premature development of reproductive structures. In their vegetative phase,

Resource plentiful habitats	Optimal Functional Design	Habitats with acute resource limitations
←	———	→
Low C allocation (R:S)		High C allocation (R:S)
Risk strategy		Insured strategy
Ephemeral structures		Perennial structures
Limited microflora		Abundant microflora

Fig. 1.2. The range of possible variation in root system optimal function design

adaptive foraging for nutrients will be important. The complexity of the above attributes indicates the need for a varied range of methods for the study of natural ecosystems.

1.2.2 Resource Capture

In cultivated situations, but rarely in more natural situations, nutrient supply and to a lesser extent water supply and crop health are, within limits, guaranteed. Breeding programmes for crops have therefore aimed at maximising partitioning into grain and, as such, have gone for a higher risk scenario (Fig. 1.2). In agricultural species, the ability to absorb nutrients from very fertile soils can be substantial. Spring barley is able to absorb 80% of its total nitrogen content of 170 $kg\,ha^{-1}$ in a single month, i.e. at a rate of around 4.5 $kg\,ha^{-1}\,day^{-1}$. This uptake will occur via the roots; although, in all situations, uptake will depend upon both the supply of available nutrients in the rooting medium and upon the root system. The importance of the root system will, as a consequence, vary. Where nutrients are abundant and soil conditions are such as to make them effectively available, then beyond the need for a minimum length of root, root system characteristics are likely to be unimportant. As nutrient supply reduces in total, or in the intensity of supply, or as it becomes diluted through a large volume of soil, as it becomes more varied and complex in terms of its chemistry or where spatial distribution is particularly complex, then the root system and its various properties become critically important. Concerns about the losses of easily available nutrients and about the environmental impact of sustainable agricultural systems are resulting in an increase in more extensive systems of production and in systems, e.g. crop rotations and organic farming systems, where nutrients are applied in a more measured way and more complex sources are used (Atkinson 1990b).

Our understanding of the impact of individual root and root system properties on nutrient capture is limited, so that it is easier for us to list root system parameters than to specify the functional and ecological significance of these parameters. The linkages and interdependence of properties means that a correlation between a property and function may not indicate an unambiguous functional role for that characteristic, but merely that it is a good surrogate for a root or root system key characteristic. During the 1970s, considerable efforts were made to relate nutrient supply to root length (Nye and Tinker 1978). The ability of plants to absorb nutrients, in a range of glasshouse trials, could be substantially explained in terms of the intensity of nutrient supply from the soil and the length of root present in the soil. This led to the assumption that in relation to nutrient supply, contact between root length and the soil was the key element in root system design. The correlation between root length and root mass, itself a complex character, means that true interpretation is far from simple. In addition, it is clear that evolutionary success has given importance to other characteristics. A high specific root length (SRL, length per unit mass) represents an effective use of resources to maximise soil contact. However, SRL varies between species (Atkinson 1990a). Recorded values for SRL are as low as $5\,m\,g^{-1}$ for apple and as high as $750\,m\,g^{-1}$ in *Lolium perenne*. In general, SRL seems lower in dicotyledenous species than in monocotyledonous species. The amount of root in the soil, root length density (RLD), also varies in the same way as seems evident for SRL. These properties are also influenced by soil microbes and will be affected by the ecological strategies discussed above. Until the interaction of these factors is wholly understood, there will be a need for a suite of measurements to characterise root system performance.

1.2.3 Soil Microbes

Soil microbial characteristics must relate to the morphological and physiological properties of the root system. As indicated in Section 1.2.2, a number of morphological characteristics, which seem counter-intuitive to the simple exploitation of the soil by the development of a maximum root length are known to be of significant value. Characteristics such as a large root diameter and the development of an extensive cortex (Atkinson 1989) will promote interchanges and associations between roots and microbes. In the rhizosphere, interactions occur with soil bacteria, fungal symbionts, such as arbuscular mycorrhizal fungi (AMF), fungal pathogens and the wide range of bacteria and protozoa involved in key nutrient transformations such as N mineralisation and NH_4 oxidation. The flow of assimilates to these organisms may be as high as 50% (Coleman and Crossley 1996), while the range of activities carried out by soil microbes is such that an understanding of carbon flow via roots is critical to the understanding of processes as diverse as the stabilisation of soil aggregates to nitrogen mineralisation (Coleman and Crossley 1996). In addition to their role as a microbial substrate, roots have direct effects on nutrient uptake (Atkinson 1990a).

Recent studies have highlighted the importance of feedback between processes occurring at vegetation, individual plant, root and molecular levels. A key question in relation to soil microbes is the extent to which the amount of microbial material and the species present are a function of the plant species present or of independent soil microbial factors of which the flow of carbon from roots is but one. Assessments of the likely impact of an increasing concentration of CO_2 in the atmosphere on plant growth and biological activity in the soil, is a dominant theme in many current research programmes. A number of studies e.g. Curtis et al. (1995) have indicated that increased CO_2 can lead to increased root growth and to altered soil microbial activity. The importance of the soil component within the global carbon budget emphasises the need for "root" inputs to be firmly based. Advances in molecular genetics are indicating the extent of biodiversity in soil microbes (Coleman and Crossley 1996). The implication of apparent "excess" species diversity in soil (functional redundancy), clearly indicates the possibility of substantial plasticity in relation to ecosystem functioning. Measurement of root features, such as length and diameter, will be important to the resolution of this problem. These factors also relate to the sensitivity of roots to soil-borne plant diseases (Campbell 1989). Research to date has shown there are interactions between microbes and the host which control disease. AMF infection, which is dependent upon root properties, has been shown to influence the level of infection due to pathogens such as *Phytophthora fragarae.*

1.2.4 Resource Allocation

There is no clear consensus on the functional significance of individual morphological and kinetic parameters of roots. Atkinson (1990a) reviewed the potential impact of changes in root diameter, root length, root branching, root distribution and a range of physiological properties. He concluded that the level of variation exhibited in morphological properties was greater than that found in physiological properties.

A number of relative parameters indicate changes in the adaptation of the plant to the environment in which it finds itself. Of these, the root/shoot ratio is perhaps the most commonly measured. A high value of R:S indicates the movement of substantial amounts of photosynthate below-ground which may relate to the genotype of the species or to the extent to which growth is being limited by a restricted supply of soil resources, e.g. nitrogen. Although the amounts of root and shoot are functionally related and can only vary within certain limits, a range of studies indicate that for all species a certain minimum proportion of root is needed to supply a given quantity of leaf or shoot (Atkinson and Fogel 1997). Hellriegel (1883) formulated his fundamental law of agriculture, a consequence of which is that every limitation to root growth leads to reduced shoot growth. Brouwer (1962) proposed the concept of a functional shoot-root equilibrium and showed that, in situations of good supply of water and mineral nutrients, small root systems can allow maximum shoot production (the risk strategy of Fig. 1.2). Large root systems may result in an increase in the efficiency of water and nutrient extraction, especially when nutrient supply is patchy in either chemical, spatial or temporal terms (the insurance strategy of Fig. 1.2). This type of variation is shown in Fig. 1.3 for young trees of *Betula pendula* (Lavender 1992) supplied with low or high levels of nitrogen. With low rates of N, shoot growth was limited but constant, while the root-shoot ratio varied between 0.45 and 0.85. In contrast, with a high N supply, shoot growth was substantial, around 5 times that with low N but varied from 3.5 to 6.5 × mean low N growth. The R : S ratio in contrast varied only between 0.3 and 0.5, i.e. to a much smaller extent than under low N conditions. Under conditions of low N supply, available resources were used to increase the size of the root

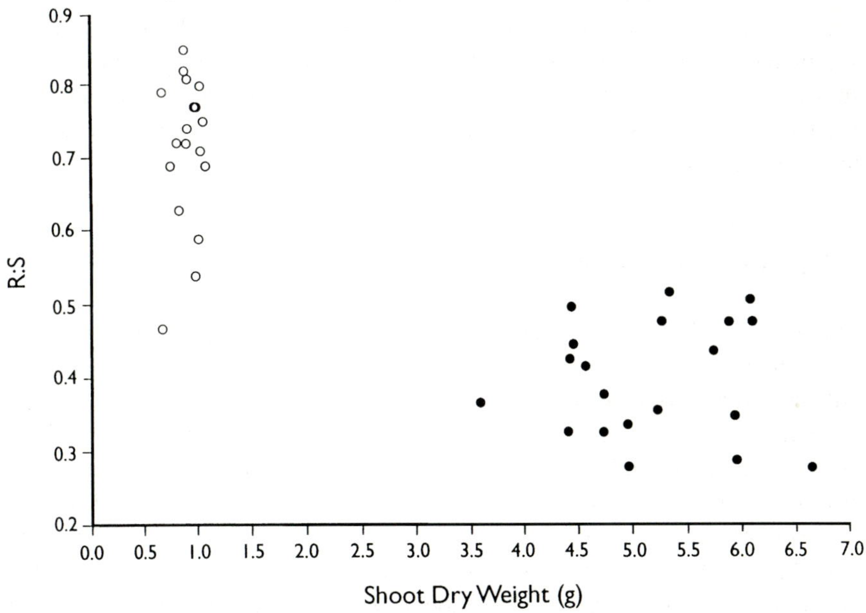

Fig. 1.3. The relationship between R : S and shoot dry weight (g) for four *Betula pendula* clones at two levels of nitrogen addition. *Closed circles* represent high and *open circles* low levels of nitrogen addition. Each point represents one plant. (After Lavender 1992)

system, while under conditions of good N supply, shoot growth increased in association with small increases in root. Comparable data for other species would allow the analysis of comparative strategies.

The relationship between root growth and soil condition is complex (Gowing et al. 1993). However, it is clear that the root system is non-homogeneous in its response to drying soil. Responses appear to be under hormonal control. Roots seem to be able to monitor water availability in soil and regulate the whole plant's response. Signals generated by roots near the soil surface move to both deeper roots and to the shoot. Field measurement of root parameters will not indicate the mechanisms of co-ordination, but describing effects may allow mechanisms to be identified.

1.2.5 Plant Interactions

The existence of different communities of plants on different soils, some of which have species in common, indicates that below-ground interactions influence both growth and survival. A number of studies have shown the importance of root-root interactions, both in studies of populations of a single species (Atkinson et al. 1976) and in mixtures of species (Atkinson 1983; Caldwell 1987). These studies have shown that when plants of the same or of different species interact, a whole range of root properties are changed in a major way. Competition between apple trees (Atkinson 1978, 1985) influenced radial spread, distribution with depth, total root length (Fig. 1.4), root length density, the development of woody roots, and root survival. Competition between apple trees and grass swards also influenced many of these but changed root system branching, the periodicity of new growth, AMF infection and root activity (Atkinson 1983). To date, studies of inter-specific competition between roots has been limited by our ability to identify the roots of the different species. Molecular methods seem likely to aid this, so allowing a functional base for studies of interaction.

1.2.6 Soil Structure

Soil structure influences the growth of roots (Passioura 1991; Tardieu 1994; Atkinson and Mackie-Dawson 1991). The whole "soil cultivation" industry depends upon this fact. Although species respond to different extents, a relatively mild mechanical stress reduced root elongation in wheat by 47% and in maize by 68% (Goss et al. 1989). Conversely, roots and their associated organisms can also influence the structure of soils. Miller and Jastrow (1992) have shown that roots and mycorrhizal hyphae "as a result of a series of mechanisms" are involved in the creation of water-stable soil aggregates. The formation of these

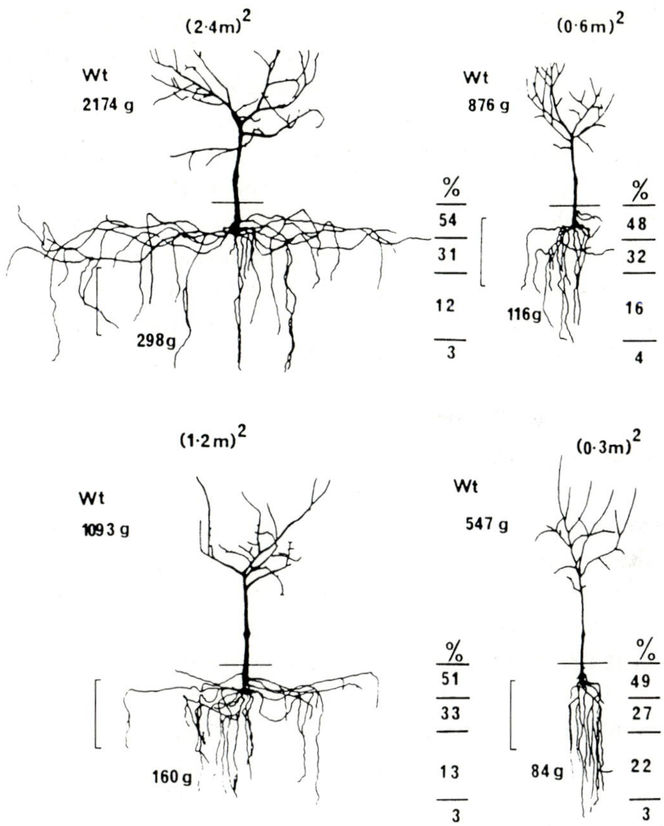

Fig. 1.4. Golden Delicious/M9. The effect of spacing on the form of the root system, the weight of root and shoot and the % distribution of root weight with depth. (After Atkinson 1978)

aggregates is an important prelude to soil stabilisation and the creation of a nutrient reserve. As a consequence, roots and mycorrhizae are important in processes such as the link between the restoration of vegetation and the reestablishment of normal soil processes which are critical to the formation of soil structure and the redevelopment of nutrient cycles. Miller and Jastrow (1992) found that the length of fibrous roots in a prairie restoration chronosequence gave a 0.69 correlation with water-stable aggregates greater than 2 mm, whilst the length of extra-matricula fungi hyphae gave a correlation of 0.88 with water-stable aggregates. Through the process of growth, development and death, roots and their associated micro-organisms create a network of biopores (Smucker et al. 1993). These pores, when vertically oriented, as a significant proportion are, allow the preferential flow of nitrate and some pesticides to ground water. This links to the growing concern in many countries around potable water quality. Biopores can also act as preferential pathways for the roots of following crops.

1.2.7 Anchorage

Anchorage is a critical feature for all trees and many other woody perennials. Listing the features of a root system which provide good anchorage (Coutts 1983) remains far from definitive. McKay and Coutts (1989) identified a series of components of which the weight of roots to be pulled from the soil, the resistance of roots pulled from the soil (related to root system diameter), and the soil area exploited by the root system were important. Root system properties influencing the anchorage of herbaceous plants have been discussed by Ennos et al. (1993). The bonding between root and soil is important in resisting vertical forces such as those occurring during grazing, while the production of a rigid vertical structure extending to a distance below the soil surface and with horizontal members at this depth also seemed significant in this situation. These architectural properties are distinct from those reviewed in Section 1.2.1. Many of the properties needed for anchorage will differ from those which relate to nutrient acquisition. A number of key properties need to be measured at a substantial geographic scale.

1.2.8 Root Products

An early publication (Evelyn 1662), indicates a means of recovering roots for use as an energy source, firewood (Fig. 1.5). Currently, there is interest in roots

Fig. 1.5. A Discourse. Recovering roots for use as firewood. (After Evelyn 1662)

as a source of secondary metabolites which can be extracted and used as pharmaceuticals, as food additives and as oils for aromatherapy. Storage roots are also important for the production of raw materials such as sugar and as vegetable crops. As food and other compounds differ between different roots, at different times during the season, development of these uses requires knowledge of phenology and demography in addition to data on root mass.

1.2.9 Basic Biological Information

Most field studies of roots and root systems are carried out for applied or strategic reasons. Information from field studies, e.g. total root length, root length density and specific root length are needed to validate the output from basic models and for use in detailed studies of individual root properties. The parameters needed for process orientated models of crop growth have recently been reviewed by Van Noordwijk and Van de Geijn (1996). They identified a need for parameters related to root-soil geometry, soils and plants. The plant parameters included both state parameters and physiologically linked characteristics such as hydraulic conductance and the minimum effective nutrient concentration at the root surface for uptake to occur. The investment in time to make many of these measurements asks difficult questions about minimum data sets and the ability to estimate particular root parameters from other properties. In addition to their role in resource acquisition, roots have a role in producing hormones such as ABA and in storing carbohydrates and proteins (Hennerty et al. 1980).

1.3 What Should Be Measured?

In the absence of absolute and unique links between particular root properties and functional objectives, there will be a need to measure a number of parameters to obtain a full picture. In addition, establishing functional and allometric links between different parameters will be important for the use of data in a wide range of circumstances. In many situations, available time and equipment will limit the measurements made and so the ability to estimate other parameters will be important. In addition, in other situations, only rough estimates of root presence or function may be needed. There is a need, therefore, to reconcile root measurement with function, to assess the extent of our knowledge on the interrelationships of root parameters and to ask what information exists on what can be obtained from rapid root system assessments. It is important to begin by defining key root system parameters which are then related to functional significance and to methods of determination (Table 1.1). These

Table 1.1. The definition of key root system parameters and their usual units

Parameter (alternatives; unit)	Definition	Function/significance	Usual means of measurement
Root length (m)	Length of all root members present.	Total root system size. Potential for absorption of nutrients or water from soil. Indicator of baseline soil microbial, especially AMF, activity in soil and of microbial functioning, e.g. organic phosphorus catabolism. Basis for most functional calculations. Large temporal variation.	Monolith methods – soil coring, needle boards, etc. Rhizotron and mini-rhizotron methods can be used but rely on assumptions (Atkinson 1985). Profile wall methods (Atkinson and Mackie-Dawson 1991) can be used when limited precision is required.
Root weight (mass, standing crop; g)	Oven-dry weight of the total root system (and attached micro-organisms). Corrections may need to be made for attached soil minerals.	Total root system size. Standing crop of below-ground material. Amount of assimilate moved below-ground.	Monolith methods, especially soil coring, but may involve partial or total excavations for woody perennials.
Root volume (ml)	The space occupied by the root system. (Frequently calculated from mean root diameter and length or assumed to be equal to the fresh weight of the root system).		
Root number	The total number of individual roots ($\Sigma 1°,2°,3°,4°,5°$).	Hormone production (e.g. cytokinin) potential, meristematic activity, presence in volume of soil.	Counts of samples obtained by soil coring, profile walls, rhizotron or mini-rhizotron methods. Root tips equal to 50% of all links regardless of system topology. Specifying whether total number or number of tips is important.
Root : shoot ratio (dimensionless)	The ratio of the dry weight of root to the dry weight of leaves + stem (branches).	Relative allocation strategy. Used against a bench-mark of plant nutrient status.	Estimated from root and shoot weight.

Root length density (intensity; cm cm^{-3} or ml ml^{-1})	The length of root present in a unit volume of soil to or at a specified depth. This is commonly identified as L_V.	Probable limitations to soil nutrient and water exploitation.	Calculated from root length and soil volume to which measure applies.
Cumulative root density (mm cm^{-2} soil)	The length of root present under a unit area of soil surface to a specified depth. This is commonly identified as L_A.		
Specific root length (SRL; m g^{-1})	The length of root associated with a unit of dry weight. (This is not always a good indicator of diameter for which it is sometimes used.)	Within-root system allocation strategy, relative importance of soil exploitation.	Calculated from root length and root weight.
Root fresh weight: dry weight (dimensionless)	The ratio of the wet and dry weights of the root system. (FW/DW-1 has been used to give an estimate of volume.)		
Root density (m ml^{-1})	The length of root associated with a unit volume of root tissue. (Together with SRL can be used to estimate root diameter.)		
Root radius (diameter; mm)	The radius (diameter) of an average individual root; usually assumed to be a plain cylinder. (Frequently calculated from length and volume or FW.)	Potential for mycorrhizal development, regulation of water stress, potential for apoplast-symplast exchange, growth potential, influences and responds to soil physical condition.	Measured directly or calculated from root length and volume.
Root of primary structure (white root)	A root with an intact cortex and where the external layer is usually the epidermis. (This is the normal state for the roots of annual species.)		
Root of secondary structure (woody, brown root)	A root with at least some woody tissues and where the external layer is usually a bark cell. (Many of the roots of perennial species are in this category.)	For perennial species the amount of investment in root system infrastructure.	Measured in terms of weight or length.

Table 1.1. *(cont.)*

Parameter (alternatives; unit)	Definition	Function/significance	Usual means of measurement
Root system branching pattern		Relative intensity of soil exploitation, mean root longevity, soil exploitation strategy.	Measured lengths or number of roots of different orders.
Primary root	The main root axis or a root which emerges from stem hypocotyl tissue and so includes both seedling (tap, radical seminal) roots and adventitious (nodal basal) roots (1°: first order).		
Secondary root	A root emerging from a primary root (2°). These are sometimes described as a first order lateral.		
Tertiary root	A root (second order lateral) emerging from a secondary root (3°) and bearing quaternary roots (4°) which may in turn bear quinary roots (5°). (This is the usual limit to branching.)		
Root longevity (d)	The length of time for which an individual root either is physically present or alive in the soil. Under soil conditions where decomposition is slow, defining whether longevity relates to present or alive is critical. This is most commonly given as a mean value or as a range for a population. This may not necessarily relate to the length of time for which the root is functional.	Potential for rapid adjustment in root length. Plasticity in relation to reduction in length. Flux of carbon to rhizosphere, nutrient release. Indicator of soil exploitation strategy.	Measured on cohorts, identified using rhizotrons or minirhizotrons.

Root cohort	A group or population of roots which are initiated (or become obvious) during a short, defined period of time.		
Root production ($kg\,ha^{-1}\,yr^{-1}$)	An estimate of the total amount of root produced during a given time period for a given land area (commonly a growing season or year) and estimated either from the sum of new root production on a weekly basis or from the difference between live root and dead root (necromass) amounts at the beginning and end of the assessment period.	Overall potential for soil exploitation, over specific time periods. The ability to increase root system length. The ability to exploit the soil.	Sequential soil core estimates or a combination of baseline standing crop and rhizotron or mini-rhizotron measurements.
Root growth rate ($mm\,day^{-1}$ or $g\,week^{-1}$)	The rate of increase in the length of an individual root or the short-term gain in mass.		
Mycorrhizal infection (AMF; %)	The degree of infection of a root with, usually, arbuscular mycorrhizal fungi. Infection is usually scored on a presence rather than intensity basis with scores based upon the proportion of 2 mm-long root sections containing evidence (hyphae, arbuscles, vesicles) of fungal presence.	Carbon allocation strategy, soil health, real effective surface for nutrient uptake.	Estimation from stained root samples.
Vertical distribution			
-Rooting depth (m)	The depth of the deepest root found on a plant root system.	Physical stability or anchorage, depth of soil exploited, potential for resource use and the estimation of available nutrient and water resources.	Soil core samples. Profile walls. Diagnostic pits for coarse estimates.

Table 1.1. *(cont.)*

Parameter (alternatives; unit)	Definition	Function/significance	Usual means of measurement
-Effective rooting depth (m)	A semi quantitative estimate of the use of the depth of soil available to the plant. The depth within which either 90% of roots are found or for which evidence is available for the presence of root activity, e.g by the depletion of soil water. Where significant capillary rise occurs this can be greater than rooting depth. The value of this parameter can be considered conceptually flawed because it will vary during a single season and between seasons depending upon the demands placed upon soil resources, e.g under very dry conditions survival may make the effective functioning of a small number of roots at depth critical and so modify the interpretation of the definition.		
Horizontal Distribution			
-Root system diameter (m)	A semi-quantitative estimate of the area of soil under which the root system is found or which, at a particular point in time, is influenced by the activities of the root system and its associated organisms.	Stability of anchorage. Potential for interaction with other species. Estimation of available soil nutrients and water.	Profile walls, partial or total excavation.

Root activity (varied)	Temporal changes in the amount of a soil resource or evidence of a material being moved into the above-ground part of the plant. This may also include evidence of hormone production or of storage and mobilisation.	Root system functioning, uptake per unit amount of root, k_m, k_{min}, exudation of physiologically active molecules, nutrient storage potential.	Depletion of soil water, or uptake of isotopes from soil.
Root uptake ability (k_m or k_{min}; μm)	The Michaelis-Menten constant for uptake of a specific nutrient at 50% of the maximum rate of nutrient uptake ("optimal" rate) by a unit length of root. This is calculated as the minimum soil solution nutrient concentration (c_{min}) from which a nutrient can be absorbed.		
Viable (functional) root length (m)	The length of root present which is able to carry out resource capture or physiological activity under field conditions. This is difficult to measure and so is frequently assumed to be equal to total (unsuberized) root length.		
Phenology (periodicity)	The period(s) during the year when root production or growth is occurring.	Functional plasticity, ability to complement activities of other species, potential to use temporarily available resources.	Rhizotron and mini-rhizotron methods.
Rooting volume (m^3)	A semi-quantitive estimate of the total volume of soil exploited by the plant root system.		

indicate that the various direct measures and estimated root parameters will have a different functional significance. The measurement of a number of the critical attributes of roots e.g. hormone production, require those measurements to be made under the controlled and commonly homogeneous conditions normally found only under laboratory conditions. These laboratory measurements need to be linked with field measurements to allow the full significance of the laboratory measurement to be interpreted (Atkinson and Fogel 1997). Physiological and other more detailed features of roots, often measured in the laboratory, but interpreted in field context or used in a field based model (Van Noordwijk and Van de Geijn 1996) are detailed in Table 1.2. The key questions in root methodology will thus be jointly advanced by novel means of measuring root and root system parameters and by increased understanding of the relationship between key morphological and physiological features of the root system and root functioning. This will link with issues relating to scale and effects based at the individual root and population levels. Decisions on what to measure, thus depend upon:

Table 1.2. Physiological parameters associated with root systems and their significance and estimation

Parameter	Function/significance	Usual means of measurement
Root growth	Development of the root system	Direct observation of change in root length
Root viability	Ability of root to function; usually in resource acquisition	Biochemical tests of functioning
Root hair density	Exploration of soil adjacent to root	Observation and estimation of number on sample length
Root respiration	Contribution to soil respiration and plant carbon budget	Direct on roots in cuvette or microcosm or by calculation from soil volume
Root exudation	Release of secondary metabolites from roots	Analytical measurement on root-associated soil or from media in solution culture
Root hydraulic conductance	Ability of water to move within a root	Pressure chamber or heat (thermal) pulse
Minimum useable nutrient concentration	Absorbing power of root, supplying power of soil or solution	Experiment with range of dilutions in solution culture

1. The availability of methods,
2. Our ability to interconnect values from different methods,
3. Understanding of the physiological significance of morphological change and the functional relationship between these traits,
4. The scale at which measurements need to be made and are required to be applied,
5. The importance of factors working at the population rather than individual root level, and
6. The minimum data set of values needed for the production of models involving root parameters.

These factors are discussed in the following sections.

1.3.1 Available Methods

Harper et al. (1991), reviewing the problems of analysing root behaviour, commented "more energy may have been spent on developing technologies than on studying roots. Each method so far developed has serious shortcomings . . . No one technique, of the many so far developed, informs us about all aspects of root growth and structure". They divided available methods into two groups. The first containing whole plant excavations, profile wall methods, pinboards, soil cores, the use of isotope applications to soil, resin embedding techniques and NMR. This group was able to provide information on root distribution and on the mass of the standing crop. The second group included rhizotrons, minirhizotrons and in-growth bags. This group allowed the assessment of changes with time. Although methods do exist for the assessment of mass and temporal changes, there are fewer methods available to measure either three-dimensional architecture or demographic parameters. Since 1991, however, considerable progress has been made in both of these areas. The use of NMR and tomography, together with the combination of information from soil sections (Krebs 1995) has developed our ability to see soil and the roots within it on a 3-D basis. Similarly, methods for viewing and establishing cohorts of roots have allowed progress in establishing values for root longevity and the demography of root populations. These are discussed further in Section 1.6.1.

Other analyses of our needs in relation to methods have been carried out. Böhm (1979) described a whole range of methods which could be used to measure roots in the field. Atkinson and Mackie-Dawson (1991) reviewed methods of root measurement which could be used to relate roots to the physical properties of soils. They concluded "There is no single method of root measurement applicable for all situations . . . The principal factors influencing the choice of methods are likely to be the availability of equipment and facilities, the crop and or soil to be investigated and the type of root system effect of interest". When measurements are made on the above-ground parts of plants, a range of features are recorded. Similarly, with the root system, a range of measurements are required so as to be able to characterise factors such as the rate and type of root growth, the standing crop and root activity. Mackie-Dawson and Atkinson (1991), who also reviewed a range of methods, concluded "The main criterion influencing the selection of a method is probably whether information is needed on changes with time or whether spatial data on distribution at one moment in time will be adequate. Where time is involved, an observational method will be indicated. Where spatial distribution is the key issue a direct sampling method will be the best". This dichotomy in objectives and methods still seems to exist and these two groups appear to be the most difficult to reconcile and to interconvert (Atkinson and Fogel 1997). Wherever roots growing in soil are sampled, in addition to the measurement of the quantity of root present, there is also a need to determine the soil volume with which this quantity of roots is associated. In extractive methods, this is simply measuring the volume of the core containing the roots, although allowances are needed for compression during sampling if a complete depth section is taken and later subdivided. For samples derived from observation methods, estimation of the depth of field represents a significant element in the scaling process (Atkinson and Fogel 1997).

1.4 The Interconversion of Values

Data on root systems is expensive to acquire and so there is a need to make the best use of available information including deriving new, or related values from existing parameters. The major interconversions of data which seem to occur are: (1) the estimation of information of one type from measurements of a different type, e.g. root length from root weight, (2) the estimation of information of the same type but for a different spatial or temporal situation.

1.4.1 Methodological Interconversions

Table 1.3 shows some of the more commonly attempted interconversions. Of all of those listed, the interconversion of weight and length is the most common. This interconversion is used to calculate length values in studies of nutrient inflow from fresh weight which can be estimated non destructively from volume, and as a means of relating root length values, assessed by observation methods such as the mini-rhizotron technique, to root weight values obtained by soil coring in method comparison studies and in attempts to derive total

Table 1.3. Interconversion of data assessed from a range of root measurements

Measurements (units)	Parameter to which converted	Method
Dry weight (g)	Length (m)	Regression
Fresh weight (g)	Volume (ml)	Calculation
Length (m), volume (ml)	Diameter (mm)	Calculation
Length (m), diameter (mm)	Volume (ml), surface area (cm^2)	Calculation
Length (m)	Number (–)	Regression
Length (m), soil volume (cm^3)	Root length density (cm^{-2})	Calculation
Length at depth a. (m)	Length at depth b. (m)	Regression
Length at time a. (m)	Length at time b. (m)	Regression
Profile wall counts (–)	Length (m)	Calculation
Length (m), weight (g)	Specific root length ($m\,g^{-1}$)	Calculation

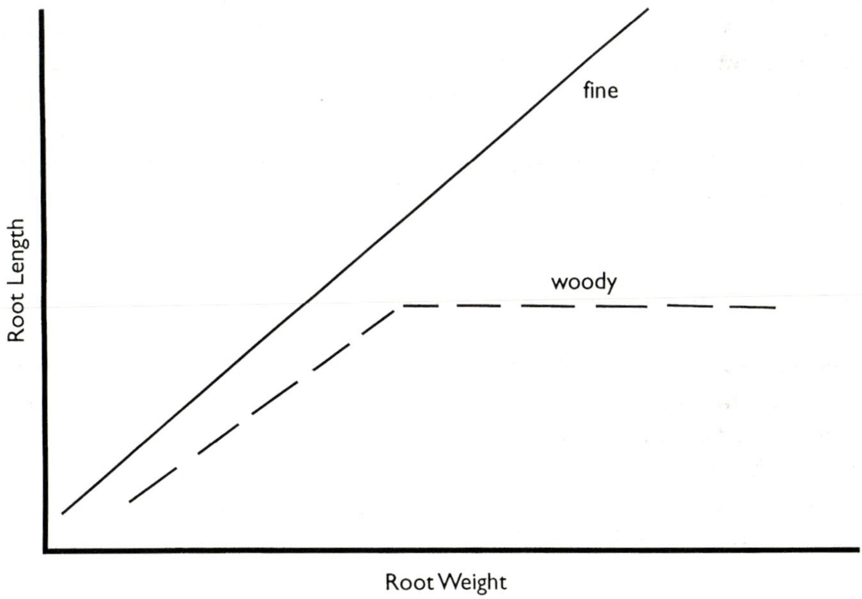

Fig. 1.6. Variation in root length with increase in root weight for fine and woody roots of *Betula pendula*. Data after Lavender (1992). The length scale for woody roots has been increased by a factor of 30

carbon budgets. For *Betula pendula*, Lavender (1992), found that for fine roots the relationship between length and weight was good but it was poorer for woody roots (Fig. 1.6). Similarly, the relationship between root length assessed using the mini-rhizotron method and root length assessed using other methods, although significant, was limited (Mackie-Dawson and Atkinson

1991). Methodological interconversions can also be used for indicators of root system activity. For example, the combination of measurements of water content and water potential allows estimates to be made of water flow. The combination of data derived from two methods is often essential. Depletion of soil water, often used as an indicator of root system activity, requires the use of a range of instruments e.g. tensiometers plus soil psychrometers so as to cover the entire range of soil water potentials which are of interest. Similarly, estimation of root length present at different periods during the year can be made using a combination of methods such as the profile wall plus rhizotron (Atkinson 1986).

1.4.2 Spatial and Temporal Interconversions

Although predictions of the size of the root system, in terms of weight, can be made from small numbers of infrequent measurements, many studies require detailed information about the distribution of root weight or length with depth (see Table 1.1 for rationale). Difficulties in understanding the vertical growth of root systems has meant that compromises between the level of sampling at any one point in time, and the frequency at which field sites are sampled, have to be made. This has often resulted in inadequate amounts of data being acquired for each of a number of time periods.

Plant breeding programmes rarely involve any assessments of root system development; most measurements are made on relatively few occasions. The issues discussed above therefore influence apparent minimum requirements for breeding programmes. For both physiological and crop breeding purposes, there is a clear need to establish whether the depth distribution of a root system can be evaluated from a basis of selected measurements and whether measurements made at one point in time, on a root system, can be used to estimate the size and distribution of the root system at a later point in time. A case study, carried out using 24 different varieties of spring barley, illustrates the possibilities of predicting spatial and temporal data from a given data set. Consideration of these points is important in the context of the use of data from any root measurement exercise and in respect of minimum data sets required from particular field experiments.

1.5 Functional Relationships Between Traits

Although root growth at various depths are on some occasions correlated, and although some future properties can be predicted, by and large the case study shows that where a detailed understanding of root growth and development is

BOX 1.1. Case Study

A study using barley assessed the extent to which the length of root present at a greater depth could be estimated from the length present at more superficial depths (Table 1.4) and the relationship between the length of root present on different occasions (Table 1.5; Atkinson 1989). The amount of root present at any one depth was always correlated with that at the next (greater) depth interval, i.e. root length at 15 cm depth showed a 0.46 correlation with that at 25 cm depth. For zones where root growth was most active at the time of the measurements, correlations were both greater and significant for a larger number of greater depths. This suggests that potential exists to assess root growth at greater depths

Table 1.4. Correlations between root growth at a number of depths for 33-day-old spring barley grown in transparent growth units, with root growth at other depths. NS = non-significant

Root growth at depth (cm)	Root growth at depth (cm)						
	15	25	35	45	55	65	75
5	0.32	NS	0.36	NS	NS	NS	NS
15	–	0.46	0.41	0.36	NS	NS	NS
25	–	–	0.56	0.33	NS	NS	NS
35	–	–	–	0.74	0.61	0.54	0.49
45	–	–	–	–	0.76	0.57	NS
55	–	–	–	–	–	0.72	0.42
65	–	–	–	–	–	–	0.76

Table 1.5. Correlation between root growth at a number of depths for 33-day-old spring barley grown in transparent growth units with root growth at the same depth, but on earlier dates (d)

Depth (cm)	Depth (cm)				
	day 19		day 26		
	15	25	15	35	45
35	0.55	0.73	0.36	–	–
45	0.48	0.65	NS	0.61	–
55	0.37	0.60	NS	0.58	0.75
65	NS	0.63	NS	0.59	0.74
75	NS	0.63	NS	NS	0.54

through its relationship with that at more superficial levels (Table 1.4). Similarly, growth on one date was correlated with that on later dates. The closer the dates and the nearer to the principal site of new growth, the better was the relationship. Growth at 15 cm depth on day 19 of the study was significantly related to that at 35 and 45 cm on day 33. This would seem to suggest the possibility of estimating growth on a later occasion at a greater depth from earlier measurements (Table 1.5).

required, and where firm data is required in relation to quantitative studies, then there is likely to be no alternative to adequately replicated measurements of root length and distribution made on a significant number of occasions. Where only general relationships are required, then predictive modelling is likely to be helpful and may avoid the need for large-scale field-based exercises.

The correlations discussed in Section 1.4.1 suggest the existence of functional links between properties. The correlations between root development at different depths in the soil and with time (Sect. 1.4.3) are evidence of clear physiological control during development. The existence of limits to R:S imply functional links between root and shoot development. The data in Fig. 1.6 suggest that in fine roots, an increase in assimilate allocation to the root system increases root length proportionally. With woody roots (Fig. 1.6) initially an increase in allocated resource (dry weight) increased the length of this type of root. However, beyond a certain root system size, 0.5 g in this study, increased assimilate, as evidenced by increased weight, did not result in an increased length, suggesting that assimilate is principally used for an increase in root diameter. McCully (1987) reviewed the development of maize roots in the field. She identified an information gap between laboratory and field studies, especially in relation to the state and condition of root tips which were usually absent from framework roots in the field. Under field conditions the relationships may be very different from those found in the laboratory. Gregory (1983, 1986) found that factors such as temperature could influence both the rate and form of development. Even for a given genotype, responses could be inconsistent between years. Effects on root growth tended to be different from other processes. Atkinson (1989, 1990a) assessed variation in a range of root characteristics. Root length, its persistence during a season, its distribution and activity all varied. The significance of both these parameters and the patterns of variation within them varied both within and between seasons as limiting components of soil resources varied. Consequently, the optimal root system needed to deliver these limiting resources changed: in addition, the ideal root systems for coping with limits to *P* and water supply are not the same. Improvements to our understanding of what root systems are really like in the field should help to bridge the gap between current field experiments and so allow the extrapolation of laboratory data (Atkinson and Fogel 1997).

1.6 Appropriate Scales for Measurement

Even for measurements made in field studies, the appropriate scale for determinations, and the limitations of scaling data up to field or landscape cause difficulties (Atkinson and Fogel 1997). The development of the mini-rhizotron method, and its use in a wide range of field studies, has drawn attention to a number of scaling issues. These have been identified by Hendrick and Pregitzer (1992b) and by Atkinson (1992). With the mini-rhizotron method, measurement of root length, root morphology or root survival are made using a window area of under 400 mm^2. This information must be converted from the 2-D basis on which it is measured to a 3-D basis to allow the estimation of length per unit volume. This estimate, based on a volume of 1–2 ml must be transformed to a volume representative of 10^4 m^3 for a 1 ha field. Although estimates with appropriate error terms can be obtained, there remains a need to express the functional variation and the difficulty of obtaining sufficiently precise estimates to generate useful field-scale estimates. In addition to the size of the variance term associated with field-scale estimates derived from small initial samples, there is also the problem of estimating the length or mass of a large root system e.g that of a tree from a limited number of small samples. In addition to spatial variation, there is likely to be pattern in the variation within the root system both with depth and distance from the trunk. Both actual values and variance are likely to vary along these gradients (Atkinson and Wilson 1980). Attempts to scale estimates of root system size from core samples must work with these gradients.

1.7 Individual and Population Measurements

Hendrick and Pregitzer (1992a) introduced the concept of the cohort, a group of roots, to form the basis of demographic assessments, for example of root longevity. The cohort approach has made the volume of information captured by the mini-rhizotron method, used to estimate values such as longevity, manageable. The cohort assessment gives a population value for root longevity. The population is composed of a series of individual roots. Variation in soils suggest most individual roots will experience a different set of physical, chemical and microbiological conditions to other roots and that the total response of the plant will be the sum of the activities of individual roots which may be separately controlled and which may respond in a different manner to a stress such as drought, (Gowing et al. 1993). Hooker et al. (1995), in a study of poplar roots, found that within a cohort of non-mycorrhizal roots 10% survived only 14 days and a further 40% only 42 days. The distinguishing features of roots surviving less than 14 days and greater than 42 days are not known but may be of functional significance. Infection with AMF increased the proportions, given above, to 20% and 50%. Again the distinguishing features of these roots is not known, but it seems important to understand at both individual and population levels. Different studies will require estimates to be based on either individuals or populations depending upon specific objectives. It will be important, however, and especially in studies of natural vegetation, to be sure whether values relate to single individuals or to populations.

1.8 Data for Modelling Studies

The difficulty of measuring roots has meant that their growth and production are not uncommonly estimated as a fixed fraction of net assimilation. The development of models would benefit from:

1. Improved allometric models to aid estimates of partition, and
2. The specification of a minimum data set for different types of models.

In many carbon allocation models, root values and variations between species for root values can have a significant impact. In the carbon storage model developed by Cannell et al. (1992) litter production represented by woody roots was around 50% of that due to branches. Production due to fine roots exceeded that for leaves. Parameterizing models with the right root characteristics is important. A range of root, shoot and soil parameters needed for process models of crop growth have been specified by Van Noordwjik and Van de Geijn (1996).

1.9 What Is the Functional Significance of Field Measurements?

Table 1.2 attempts to identify some of the more obvious links between root properties and function. These are expanded in Section 1.4.1 above. The clear answer to the question "What is the functional significance of, for example, large root diameter or a particular root length?" is, in general, that we do not know. We may, however, be able to assign value to features in particular situations e.g. to deep roots under conditions of drought or to a high root length under conditions of P deficiency. This is partially because the functional significance of even major attributes has not been definitively determined, but more so because plants have a range of root features which are designed to cope with more extreme conditions than they will or may normally experience during their life, i.e. there is a substantial functional excess capacitance built into root systems. This in-built functional redundancy, or insurance, clearly confuses the correspondence between a root property and the quantity of a function, e.g. NO_3 uptake. Field measurement of a wider range of root characteristics would give a more complete understanding of functional plasticity and the ranges of morphological and physiological variation possible and help the development of an understanding of relationships between environmental profiles and roots.

1.10 Future Trends

The concept of "fit for purpose" is now being widely used to ensure that facilities and equipment are designed for specific needs but without unnecessary features. Many of the measurements made on and of root systems are not used or are used in a qualitative (presence or absence) manner. Refining the question to be asked by clearly specifying the information required will allow the considerable effort required for any root study to be used to maximum effect, usually by increasing the number of measurements taken. The development of easily accessible modelling capabilities, computerised decision support systems and data bases to aid the design of sampling strategies should greatly enhance future root sampling exercises.

The capture of data, either from images of a mini rhizotron camera system or from systems designed to assess root length from washed root samples, will develop as a result of improved capacity for image analysis (Glasbey and Horgan 1995). Developments should improve our ability to extract data of increasing complexity from direct and video images and to distinguish roots at a range of development stages, including decaying, from a wide range of backgrounds, including soils. A restricted ability to document root development has, to date, aggravated a limited capacity to produce root systems with a desired set of characteristics on new, and other, cultivars. Developments in biotechnology now present the possibility of creating transgenic plants with specific genes controlling particular root or developmental features, or the ability, through the use of anti-sense genes of turning off particular aspects of gene expression. This will allow the production of material which can be used to test hypotheses on root functioning. The developments in rhizo-technology discussed here will be needed to allow these biotechnological developments to be realised.

References

Atkinson D (1978) The use of soil resources in high density planting systems. Acta Hortic 65: 75–90

Atkinson D (1983) The growth, activity and distribution of the fruit tree root system. Plant Soil 71: 23–36

Atkinson D (1985) Spatial and temporal aspects of root distribution as indicated by the use of a root observation laboratory. In: Fitter AH, Atkinson D, Read DJ, Usber MB (eds) Ecological interactions in soil. Blackwell, Oxford, pp 43–65

Atkinson D (1986) The nutrient requirements of fruit trees: some current considerations. Adv Plant Nutr 2: 93–128

Atkinson D (1989) Root growth and activity, current performance and future potential. Aspects Appl Biol 22: 1–13

Atkinson D (1990a) Influence of root system morphology and development on the need for fertilisers and the efficiency of use. In: Balligar VC (ed) Crops as enhancers of nutrient use. Academic Press, London, pp 411–451

Atkinson D (1990b) Land: agriculture resource or wildlife reserve: reorganisation in the food factory. Aberdeen Univ Rev 183: 218–225

Atkinson D (1991) Plant roots: an ecological perception. In: Atkinson D (ed) Plant root growth. Blackwell, Oxford, pp ix–x

Atkinson D (1992) How long is the life span of a root? Trends Ecol Evol 7: 173–174

Atkinson D (1996) Why study roots. Agrofor UK 7: 22–24

Atkinson D, Fogel R (1997) Roots: measurement, function and dry matter budgets. In: Van Gardingen PR Foody GM, Curran PJ. (eds) Scaling-up. Cambridge

Atkinson D, Mackie-Dawson LA (1991) Root growth: methods of measurement. In: Smith KA, Mullins CA (eds) Soil analysis: physical methods. Marcel Dekker, New York, pp 447–509

Atkinson D, Wilson SA (1980) The growth and distribution of fruit tree roots: some consequences for nutrition uptake. In: Atkinson D, Jackson JE, Sharples RO, Waller WM (eds) Mineral nutrition of fruit trees. Butterworth, London, pp 137–150

Atkinson D, Naylor D, Coldrick GA (1976) The effect of tree spacing on the apple root system. Hortic Res 16: 89–105

Böhm W (1979) Methods of studying root systems. Springer Verlag Berlin, Heidelberg New York

Brouwer R (1962) Nutritive influences on the distribution of dry matter in the plant. Netherlands Journal of agricultural Science 10: 399–408

Caldwell MM (1987) Competition between root systems in natural communities. In: Gregory PJ, Lake JV, Rose DA (eds) Root development and function. Cambridge University Press, Cambridge, pp 167–186

Campbell R (1989) Microbial interactions leading to biocontrol of root diseases – reality or myth. Aspects Appl Biol 22: 165–172

Cannell MGR, Dewar RC, Thornley JHM (1992) Carbon flux and storage in European forests. In: Teller A, Mathy P, Jeffers JNR (eds) Responses of forest ecosystems to environmental changes. Elsevier, London, pp 256–271

Coleman DC, Crossley DA (1996) Fundamentals of soil ecology. Academic Press, San Diego

Coutts MP (1983) Root architecture and tree stability. Plant Soil 71: 171–188

Curtis PS, Zak DR, Pregitzer KS, Teeri JA (1995) Above and below ground responses of *Populus grandidentata* to elevated CO_2 and soil N availability. Plant Soil 165: 45–54

Ennos AR, Crook MJ, Grimshaw (1993) A comparitive-study of the anchorage systems of Himalayan balsam (*Impatiens glandulifera*) and mature sunflower (*Helianthus annuus*). J Exp Bot 44: 133–146

Evelyn J (1662) A discourse of forest trees and the propagation of timber in His Majesty's dominions. Stobbard, London

Glasbey CA, Horgan GW (1995) Image analysis for the biological sciences. John Wiley, Chichester

Goss MJ, Barraclough PB, Powell BA (1989) The extent to which physical factors in the rooting zone limit crop growth. Aspects Appl Biol 22: 173–182

Gowing DJG, Davies WJ, Trejo CL, Jones HG (1993) Xylem-transported chemical signals and the regulation of plant growth and physiology. Philos Trans R Soc Lond Ser B Biol Sci 341: 41–47

Gregory PJ (1983) Response to temperature in a stand of pearl millet (*Pennisetum typhoides* S. and H.) III. Root development. J Exp Bot 34: 744–756

Gregory PJ (1986) Response to temperature in a stand of pearl millet (*Pennisetum typhoides* S. and H.) VIII. Root development. J Exp Bot 37: 379–388

Grime JP (1974) Vegetation classification by reference to strategies. Nature 250: 25–31

Grime JP, Cambo BD, Mackey JML, Crick JC (1991) Root plasticity, nitrogen capture and competitive ability. In: Atkinson D (ed) Plant root growth. Blackwell, Oxford, pp 381–398

Harper JL, Jones M, Sackville-Hamilton NR (1991) The evolution of roots and the problems of analysing their behaviour. In: Atkinson D (ed) Plant root growth. Blackwell, Oxford, pp 3–24

Hellriegel H (1883) Beiträge zu den naturwissenschaftlichen Grundlagen des Ackerbaus mit besonderer Berücksichtigung der agrikultur-chemischen Methode der Sandkultur: eine Auswahl von Versuchen ausgeführt an der Versuchstation Dahme. Braunschweig: Vieweg, 1883, 796 p

Hendrick RL, Pregitzer KS (1992a) The demography of fine roots in a northern hardwood forest. Ecology 73: 1094–1104

Hendrick RL, Pregitzer KS (1992b) Spatial variation in tree root distribution and growth associated with mini-rhizotrons. Plant Soil 143: 283–288

Hendrick RL, Pregitzer KA (1993) Patterns of fine root mortality in two sugar maple forests. Nature 361: 59–61

Hennerty MJ, O'Kennedy, BT, Titus JS (1980) Conservation and reutilization of bark proteins in apple trees. In: Atkinson D, Jackson JE, Sharples RO, Waller WM (eds) Mineral nutrition of fruit trees, Butterworth, London, pp 369–377

Hooker JE, Black KE, Perry RL, Atkinson D (1995) Arbuscular mycorrhizal fungi induced alteration to root longevity of poplar. Plant Soil 172: 327–329

Krebs (1995) Biogenetic soil fabrics: plant species, specific microfabrics of top soils and aspects of their genesis (German) PhD Thesis, University of Hohenheim, Hohenheim

Lavender, EA (1992) Genotypic variation in the root system of Betula pendula. PhD Thesis, University of Aberdeen, Aberdeen

Mackie-Dawson LA, Atkinson D (1991) Methodology for the study of roots in field experiments and the interpretation of results. In: Atkinson D (ed) Plant root growth. Scientific Publications, Oxford, pp 45–47

McCully ME (1987) Selective aspects of the structure and development of field-grown roots with special reference to maize. In: Gregory PJ, Lake JV, Rose DA (eds) Root development and function. Cambridge University Press, Cambridge, pp 53–70

McKay H, Coutts MP (1989) Limitations placed on forestry production by the root system. Aspects Appl Biol 22: 245–254

Miller RM, Jastrow JD (1992) The application of VA mycorrhizae to ecosystem restoration and reclamation. In: Allen MF (ed) Mycorrhizal functioning. Chapman and Hall, New York, pp 438–467

Nye PH, Tinker PB (1978) Solute movement in the soil root system. Blackwell, Oxford

Passioura JB (1991) Soil structure and plant-growth. Aust J Soil Res 29: 717–728

Robinson D (1991) Roots and resource fluxes in plant communities. In: Atkinson D (eds) Plant root growth. Blackwell, Oxford, pp 103–130

Smith FA, Smith SE (1986) Movement across membranes: physiology and biochemistry. In: Gianazzi Pearson V, Gianazzi S (eds) Proc Eur Symp on Mycorrhizae. INRA, Paris, pp 75–24

Smucker AJM, Richner W, Snow VO (1995) Bypass flow via root-induced macropores (RIMS) in subirrigated agriculture. In: Clean Water, Clean Environment for the 21st Century. Kansas City, MO. pp 255–258

Tardieu F (1994) Growth and functioning of roots and of root systems subjected to soil compaction – towards a system with multiple signaling. Soil Tillage Res 30: 217–243

Van Noordwijk M, Van de Geijn SC (1996) Root shoot and soil parameters required for process-orientated models of crop growth limited by water or nutrients. Plant Soil 185: 1–25

CHAPTER 2

Anatomy and Histology of Roots and Root-Soil Boundary

E. de Neergaard[1], O.B. Lyshede[2], T.S. Gahoonia[3]*, D. Care[4], and J.E. Hooker[5]

[1,2] The Royal Veterinary and Agricultural University, Department of Plant Biology, Thorvaldsensvej 40, 1871 Frederiksberg C, Copenhagen, Denmark

[3] The Royal Veterinary and Agricultural University, Department of Agricultural Sciences, Plant Nutrition and Soil Fertility Laboratory, Thorvaldsensvej 40, 1871 Frederiksberg C, Copenhagen, Denmark

[4] AgResearch, Grasslands Division, Ruakura Agriculture Research Centre, Private Bag 3123, Hamilton, New Zealand.

[5] School of Applied Sciences, University of Glamorgen, Pontypridd, Mid-Glamorgen CF37 1DL, UK

CONTENTS

* Principal author

A.L. Smit et al. (Eds.), Root Methods

2.1 Introduction

Growing roots undergo many anatomical and morphological changes, which influence their activity and nutrient uptake processes. Therefore, it is often necessary to obtain structural information on the inner (anatomy and histology) and outer (morphology) parts of roots. This chapter gives an overview of methods to obtain information on anatomical and histological as well as morphological (root hairs and mycorrhiza) properties of roots. The methods applied for the study of root anatomy do not, generally, differ from methods used for the study of plant stems and leaves. Methods can thus be found in general laboratory books and manuals (Johansen 1940; Sass 1961; Jensen 1962; Purvis et al. 1964; O'Brien and McCully 1981; Neergaard 1997). Before a root specimen and a thin section of root-soil boundary can be investigated under a microscope it has to pass along a chain of processes which include sampling, killing and fixing, embedding, sectioning, and staining. Details of these processes depend on whether light microscopy (LM), transmission electron microscopy (TEM), or scanning electron microscopy (SEM) is to be used. For LM and TEM, histochemical or immunological tests may be applied additionally if the purpose is to demonstrate the presence of certain compounds in cells or tissues. SEM deviates from the other two mentioned with the exception of initial fixation steps and will be treated in a separate section. Squash techniques for chromosome studies, most often carried out on root tips, are also dealt with separately. Information on the anatomy of roots can be sought in books on plant anatomy (Esau 1965, 1977; Guttenberg 1968; Mauseth 1988; Fahn 1990) and root physiology (Luxová and Ciamporová 1989).

2.2 Root Sampling

Sampling of roots for anatomical studies is more difficult and elaborate than for stems and leaves, because roots usually grow in soil. Adhering soil particles must be carefully removed by washing and careful brushing as they otherwise may interfere during preparation for microscopy. The cleaning problem may, in some cases, be avoided if plants are grown in a liquid nutrient medium or in a substrate from which the roots are easily removed. For anatomical studies, actively growing roots should be sampled. A soil-free white elongation zone immediately proximal to the root caps indicates actively growing roots (McCully 1995). When sampling the roots of target plants in fields, the roots of weeds may pose a problem. Roots for anatomical studies should be processed immediately or stored frozen in liquid nitrogen to avoid anatomical changes prior to further processing. Root segments can be frozen in liquid N_2 in cryo-vials for transportation and further processing. Time of root sampling is also important, if exudation of water and other substances from roots in situ is to be examined. Rhizosheaths adhering to roots may be wet from early morning collection and drier from midday and midafternoon collections (McCully 1995), because of transpiration fluctuations during the day.

For sampling of roots for mycorrhiza or root hair studies, it is necessary to retain their natural features. A sharp knife with a long blade can be used to take soil cores with roots to a depth of ca. 10 cm. To minimize damage to root hairs, hyphal connections between roots and ectomycorrhiza roots should not be pulled through soil. The soil cores can be wrapped in aluminium foil to prevent crumbling.

For the isolation of roots from soil cores for mycorrhizal or root hair studies, the soil cores should be kept soaked in water at least overnight. Ultrasound treatment (120 W, 47 kHz) in a water bath is useful for removing soil particles adhering to roots, especially for root hair studies (Gahoonia and Nielsen 1997) and perhaps also for mycorrhizal investigations. Moderate magnification (6 x), a black background and a lamp of daylight quality are helpful to get a real impression of the mycorrhiza colours from the beginning of the washing procedure. A white background masks the natural colour. Cleared mycorrhiza must not be stored in water for long periods because hyphae can continue to grow and colour changes can occur. Agerer (1991) and Brundrett et al. (1996) have reviewed the methods of root sampling for mycorrhizal studies.

2.3 General Methods for Studying Root Anatomy and Histology

2.3.1 Whole Mounts

Whole mounts of finer roots and rootlets may be studied directly with a light microscope at lower magnification (Fig. 2.1). Part of the root, e.g. the root tip, may be bathed in a drop of water, a clearing agent, or a suitable stain. It is thus possible to study the root zones comprising root cap, root meristem, elongation zone, root-hair zone, and branching zone where the formation of lateral roots occurs. Individual structures in the tissue may be viewed by focusing at different levels. Mounting in iodine (2 g KI + 1 g I_2 + 100 ml H_2O) may reveal the presence of statoliths in the root cap.

Staining of whole mount plant tissues can be done using appropriate stains and procedures (Sect. 2.3.6). Entire plant parts normally do not take up stain readily and, therefore, it is time-consuming to infiltrate plant tissue with

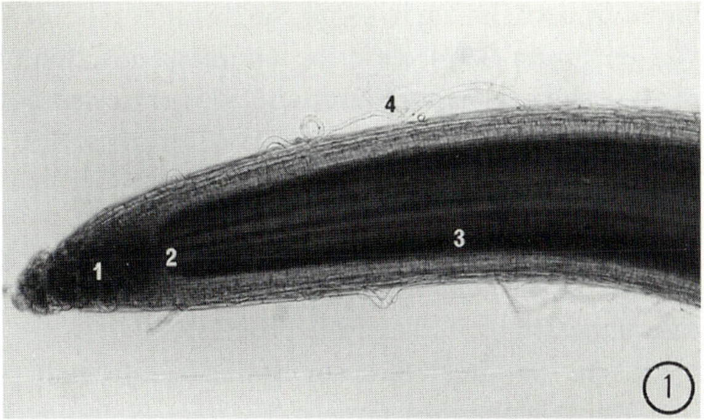

Fig. 2.1. *Lepidium sativum.* Root tip in whole mount. *1* root cap with statolith starch, *2* dividing zone, *3* elongation zone, *4* separated root cap cells alongside the surface. (155 x). (Photo by Ole B. Lyshede)

a staining solution. The use of ethanol, lactic acid, or lactoglycerol as solvents for the stain, will often facilitate infiltration. A brief heating of the specimen while immersed in the stain or a vacuum treatment speeds up the process. Stains dissolved in clearing agents or in the mounting medium may prove successful.

2.3.2 Tissue Clearing

Clearing of plant tissue in whole mounts facilitates observation of internal structures including infecting fungi. Clearing removes certain pigments (e.g. tannins and other polyphenols) and results in transparency of even rather thick tissue parts. Internal structures may thus be observed in the light microscope after mounting in a medium with a sufficiently high refraction index (e.g. glycerol).

Tissue clearing is a general consequence of chemical fixation. Ethanol (96%) is effective. It can be used cold, but works faster when the tissue is boiled in it for 5 min. Ethanol is coagulating and a treated specimen cannot be used afterwards for studies of fine cell structure. However, the procedure is excellent for removal of air due to the low surface tension of ethanol.

Cleared specimens for light microscopy are obtained if glycerol or chloral hydrate is used as the mounting medium. Tannins may be removed by treatment with mixtures of ethanol and acetic acid unless they have been fixed in aldehyde. NaOCl or H_2O_2 can be used for decolorization of polyphenolics.

Root tissue often requires fairly severe methods for clearing and must be treated with aggressive chemicals (e.g. alkalis), occasionally at elevated temperatures or for long periods. However, use of a strong alkali over very long periods leads to degradation of the cytoplasm and cellulose. This leaves little material for staining. Bevege (1968) recommends autoclaving in ethanol followed by autoclaving in 1–5% NaOH for studies of VA-mycorrhiza. O'Brien and von Teichman (1974) describe a number of different autoclaving procedures of plant tissue in ethanol. Gardner (1975) reviewed botanical clearing techniques.

2.3.2.1 Clearing Agents and Procedures

Ethanol and *methanol* are common clearing agents. Vacuum infiltration or careful heating speeds up the process.

Lactic acid in an 85% solution is traditionally used alone or as *lactophenol* (lactic acid + phenol + glycerol + water). Phenol is often replaced by glycerol in *lactoglycerol*, which can be prepared by mixing lactic acid + glycerol + water (v/v) in ratios 1 + 1 + 1 *or* 1 + 2 + 1 *or* 1 + 4 + 1.

Higher ratios of lactic acid (14 + 1 + 1) have proved especially useful for treatment of roots. Pure (100%) lactic acid is also recommended.

Lactophenol is made by replacing half the amount of glycerol with phenol. The use of phenol should be limited due to its harmful effects on human health. Lactoglycerol can be used together with stains. Aniline blue (= cotton blue), dissolved in lactoglycerol is a commonly used stain (0.5–1.0% solution).

The following clearing agents have proved useful on leaf tissue and should be tried on roots:

Lactoglycerol-ethanol (1 + 2) thoroughly mixed. Clearing is accelerated by heating in a boiling water bath for 3 min or when left in the solution overnight.

Acetic acid-ethanol. A mixture 1 + 1 (v/v) is generally used. Alternatively try this: treat with heated 96% ethanol and transfer the material to acetic acid +96% ethanol + water (1 + 4 + 5). The method may be combined with staining. Increased ratios of ethanol are reported in the literature, e.g. acetic acid and 50% ethanol (1 + 17) for clearing of leaf sheaths of wheat (Strausbaugh and Murrey 1989).

Clarke fixatives consist of mixtures of acetic acid and ethanol (for example 1 + 3). Clarke is used as a fixative of root tips prior to chromosome studies by squash techniques (Sect. 2.3.3). It may be used as a clearing agent, but is not suitable for cytological and anatomical investigations.

Acetic acid-glycerol. A 1 + 1 mixture is recommended for clearing very young roots.

Chloral hydrate is used in aqueous solution, alone or in combination with other chemicals (Table 2.1). Specimens can be left in the solution for 4–5 days or heating which removes air from the specimen can shorten clearing. *NB!* Chloral hydrate is very *injurious* to human health and should be used only if other methods fail and always in a fume hood using gloves.

Table 2.1. Chloral hydrate solutions

(Dhingra and Sinclair 1995):		**(Gerlach 1984):**	
Chloral hydrate	250 g	Chloral hydrate	160 g
Distilled water	100 ml	Distilled water	100 ml
		Glycerol	50 ml

Herr (1971) used the following procedure:
Lactic acid + chloral hydrate + phenol + clove oil + xylene
2 + 2 + 2 + 2 + 1 (by weight)

NB! Working with xylene should be done in a ventilation hood, and use of xylene substitutes (e.g. histoclear) is preferable.

Lactic acid-methanol-chloroform (1 + 1 + 1) is occasionally recommended as a clearing agent for entire leaves (24 h). By subsequent staining, fungal structures in the host tissue are demonstrated. The stain is dissolved in lactoglycerol, lactophenol, or 95% ethanol and boiled for 2 min. (For staining procedures see Sect. 2.3.6.) Chloral hydrate may be used as a final treatment (24 h) before observation. (NB: toluidine blue O is destained by chloral hydrate).

Chloroform can be used in a mixture with methanol (2 + 1, v/v) for clearing of leaf tissue. Kuck et al. (1981) used this solution (1–2 h at 20 °C) for pretreatment of rust-infected leaves before applying fluorescent stains for demonstration of fungal structures.

Alkalis. KOH or NaOH are commonly used for roots and lignified tissue. Woody tissue can, for example, be treated with 1*M* KOH e.g.:

1. Treat specimens with boiling 1*M* KOH. If necessary, infiltration can be facilitated by vacuum,
2. Neutralise with 1*M* HCl (briefly),
3. Mount with lactoglycerol or stain with staining reagents dissolved in lactoglycerol.

Roots of perennial plants for investigation of fungal infection can be treated with strong alkalis (S. Rosendahl, pers. comm.):

1. Immerse roots in 10% KOH at 70–90 °C for 30–60 min,
2. Add 1*M* HCl (98 ml 32% HCl + water until 1000 ml) for neutralising,
3. Stain with staining solution (e.g. trypan blue) in lactoglycerol at 70–90 °C.

An alternative method includes the following treatment between steps 1 and 2: 30 ml 10% H_2O_2 + 3 ml NH_4OH + 967 ml water for 40 min at 30 °C.

Vegetable oils. The commercial product Histoclear consists of citrus oil and vegetable oil. It is non-toxic and claimed to be an efficient clearing agent unless used on lipid-rich specimens. It is also used as a dewaxing agent.

2.3.3 Squash Techniques

Studies on chromosome number and morphology are very often carried out on root tips using squash methods (Box 2.1). Pre-treatment of the roots with colchicin, 8-hydroxyquinoline, or α-monobromnaphtalene for 3–4h will often make the chromosomes more distinct in the microscope (*NB!* Longer pre-treatment causes multiplication of the chromosomes). Sudden cooling of root tips is an alternative to the chemical treatment and may also be useful.

BOX 2.1. Squash technique by Aceto-orcein:

1. Pre-treatment with colchicin a.o. (if preferred);
2. 1–2mm of the root tips are excised and fixed in Clarke's fixative (absolute ethanol/acetic acid, 3:1) for 10min;
3. Heat root tips in 1M HCl at about 60°C till soft (5–10min);
4. Place root tips in a droplet of Aceto-orcein on a glass slide and mash with a needle;
5. Drop a large cover slip onto the mashed root tips and heat over a spirit flame, not exceeding 60°C. (Check with back of your hand; it should be very warm, but should not hurt);
6. Place two pieces of filter paper at opposite edges of the cover slip and hold them steadily with two fingers;
7. Squash the material by moderate tapping with the wooden end of a matchstick;
8. Replace filter papers with fresh on top of the cover slip; squeeze with your thumb;
9. Examine the material.

The preparation may keep for a few days if the cover slip is framed with nail polish.

Permanent slides can be made as described in Box 2.2.

BOX 2.2. Making Permanent Slides:

1. Freeze the squashed material with comprimated CO_2;
2. While frozen, separate slide and cover slip with a sharp knife and let both pieces fall into absolute ethanol for no more than 15s;
3. Transfer both to absolute ethanol/xylene (1:1) and on to xylene;
4. Make two permanent slides by mounting in Canada balsam or other suitable mounting medium.

Staining of chromosomes with Acetocarmine, Fuchsin, or other stains can be carried out with slightly deviating methods (Jensen 1962).

2.3.4 Tissue Fixation for Light- and Transmission Electron Microscopy

Observation of gross anatomical structxures can take place using whole mounts (see Sect. 2.3.1), or by hand sectioning of fresh tissue (see Sect. 2.3.5). However, before detailed studies of the fine structures of tissue can be done in LM or TEM it must be fixed, embedded, and sectioned. Prior to killing and fixation of the plant material it must be cut into appropriate pieces depending on the final investigation. For TEM at least one dimension should not exceed about 2 mm in order to facilitate a quick penetration of the fixative. The tissue compounds are immobilized and preserved by the fixation procedure and the purpose of the subsequent embedding is to stabilise and support all cell components at their original site during sectioning. Two fundamentally different methods of fixation are described: freeze fixation and chemical fixation.

2.3.4.1 Freeze Fixation

By freeze fixation the conservation of the tissue and immobilization of its components take place by low temperature and the supporting medium is the frozen water content of the tissue. Freeze fixation is generally faster than chemical fixation and the risk of chemical alterations of the cell components is much lower since all biological processes are rapidly arrested. Thus, changes of distribution of elements and loss of diffusible ions are kept at a minimum. As a consequence enzyme- and immunological properties of the tissue are relatively unaffected. The negative effects are mainly caused by the formation of ice crystals, which may damage cell components, especially the membranes. Liquid nitrogen is therefore extensively used because the freezing is rapid and strongly diminishes the formation and size of ice crystals. Most plant tissues, including roots, can be frozen directly from the living state.

Freeze fixation can be used prior to both light and electron microscopy. In the following, the first procedures described aim at light microscopical work. Freeze substitution and high pressure freezing can be used for both light and electron microscopy, depending on the embedding medium used.

General Freeze Methods for Light Microscopy. Most simple freezing of plant tissue takes place by infiltrating small specimens with a proper cryoprotective substance (Table 2.2) and freezing them on a freeze stage, preferably placed in a cryostat. This is a cooled chamber in which freeze fixation and sectioning

Table 2.2. Cryoprotective substances

Sucrose solution (saturated solution = sucrose + water (52 g + 30 ml, w/v)
Gum arabic (40% aqueous solution + a crystal of thymol)
Gum arabic + sucrose (40% gum arabic solution + saturated sucrose (5 + 3, v/v)
Glycerol
Glycerol-gelatin (= glycerol jelly)
Water agar
Gelatin
Polyethylenglycol
Polyvinylpyrrolidone
Dimethylsulphoxide

take place. The frozen tissue adheres to the freeze stage by the frozen cryoprotective substance and is supported by more cryoprotectant added gradually on all sides prior to sectioning. Knox (1970) recommends tissue infiltration before freezing with dimethylsulfoxide or glycerol. These chemicals can be used as anti-frost additives to gelatin used as supporting medium during freezing. Wood can be saturated with water before freezing, but most tissue types should be treated with cryoprotective substances.

A number of commercial products are available, marketed under various trade names.

Glycerol is a good cryoprotective substance, but diffuses very slowly into plant tissue. A precaution is to let the plant take up glycerol in the living state, but glycerol is osmotically active and there is a danger of plasmolysis unless applied in gradually increasing concentrations. When infiltrated gelatin is used as a cryoprotective substance, a graded series of gelatin solutions can be used for dehydration and as an embedding medium. Dehydration takes place by using 10%–40% (up to 70%) aqueous gelatin. After partial drying at 37 °C for some hours pieces are cut and dried in a vacuum desiccator.

2.3.4.2 Chemical Fixation

Warning: Chemical Fixatives Contain Ingredients that are Harmful to Human Health. Chemical fixation aims at protein fixation as a key point. The chemicals used for protein fixation are either *gel forming* (cross-linking fixatives) or *coagulating* (precipitating fixatives). The gel forming fixatives do not (or only slightly) denature the proteins. Coagulating fixatives act by irreversibly denaturing agents (Lyon 1991). Gel forming fixatives (formaldehyde, glutaraldehyde, croton aldehyde, acrolein, potassium dichromate and osmium tetroxide) act as stabilizers of the protein configuration by establishing bonds between polypeptides in the same protein and between neighbouring protein molecules. These fixatives are recommended for TEM. Coagulating fixatives (methanol, ethanol, chromium trioxide, mercury chloride, picric acid, hydrochloric acid, nitric acid and acetone) denature the proteins by opening the tertiary structure, displaying the reactive groups of the proteins. Coagulating fixation is in some cases an advantageous method for light microscopy provided that the protein network formed is not too coarse, but it cannot be used for TEM (Lyon 1991).

Rapid infiltration with the fixative is important. This may be obtained with small sizes of tissue, at least in one dimension; the use of vacuum will enhance the infiltration. The concentration of the fixative, its pH and osmolarity, and also the temperature at which it is applied are critical for a good fixation.

The concentration of the fixative is changed on its way through the tissue, and at the same time osmotic changes take place in the remaining living parts. At low temperature the autolytic cell processes are slower, but the diffusion time of the fixative is longer.

Commonly Used Fixatives for Light Microscopy. Formaldehyde has a high diffusion rate compared with other fixatives and therefore is a rapidly working fixative. Used alone as a formalin solution (1–8% in buffer) it is commonly used as fixative. Formalin (2–4% aqueous solution, Table 2.3) in combination with acetic acid and ethanol (FAA) is a widely used fixative for routine anatomical studies both for LM and SEM.

Other ratios are found in the literature and should be tried if unsatisfactory results are obtained. The dehydration process following FAA fixation starts with 70% ethanol. FPA (formalin-propionic acid-alcohol) is made by replacing acetic acid in FAA with the same amount of propionic acid and is considered a good fixative.

Glutaraldehyde (2–4% in phosphate-buffer) is used for both light and electron microscopy. It is used alone or in combination with other fixatives (Sect. 2.3.8). If cytological details are required, aldehyde fixation is strongly recommended; this is especially good for TEM. See O'Brien and McCully (1981) and Glauert (1975) for fixation methods.

Table 2.3. Formalin-acetic acid-alcohol (FAA)

Percentage formaldehyde:	2%	4%
70% ethanol	90 ml	85 ml
40% formalin	5 ml	10 ml
Glacial acetic acid	5 ml	5 ml

Clarke fixatives consist of mixtures of ethanol and acetic acid (for example 3 + 1). They can be used for studies of chromosomes (Sect. 2.3.3), but do not work well for tissue fixation.

CRAF fixatives are mixtures of chromium trioxide, acetic acid and formaldehyde in different ratios, depending on the tissue character. CRAF fixatives have similar properties to FAA (Sass 1961; O'Brien and McCully 1981).

Dehydration. Chemically fixed tissue must be dehydrated prior to embedding. Commonly used dehydration media are ethanol and acetone. Boyde and Maconnachie (1981) have demonstrated that minimum morphological changes of the specimen take place if dehydration starts with 70% ethanol or 70% acetone. It should be kept in mind that many cell constituents are removed from the tissue during the dehydration. A good fixation procedure is the only way to keep such losses at a minimum. Acetone should probably be preferred in most cases.

Propylene oxide is often used as an intermediate step between dehydration and resin infiltration, especially for TEM, but it is toxic and potentially carcinogenic and its use must be strongly limited.

Acetonitrile is recommended since it causes less extraction of cell lipids, including the phospholipids of membranes. It is miscible with water and with epoxy resins (Edwards et al. 1992).

Acidified dimethoxypropane (3 drops of 0.1 M HCl + 25 ml dimethoxypropane) is used as a dehydration agent in electron microscopy, followed by two changes of acetone (Carlson et al. 1991) or by a stepwise change directly to the embedding medium (gradually increasing the concentration to pure embedding medium).

Embedding. Tissue is embedded to stabilize the cell components during sectioning, thereby facilitating the production of thin uniform sections. Most simple is soaking in water (under vacuum) and embedding in ice or an aqueous cryoprotective substance (see text on freeze fixation in Sect. 2.3.4.1) using a freeze microtome or a cryostat. Tissue infiltration with embedding media is normally done at room temperature because the media are generally highly viscous and their diffusion into the tissue is slow at low temperatures. Cell walls, crystalline substances and storage structures such as starch in roots are difficult to infiltrate with embedding media and may create special problems. Starch may be removed with amylase prior to embedding for LM, but not for TEM (Lyshede 1977, 1979).

Paraffin wax or *different types of resins* are common embedding media. Wax embedding media include paraffin wax (Paraplast) and Steedman's wax. Paraplast is a commercial product, whereas Steedman's wax is produced by polyethylene glycol 400 distearate mixed with 1-hexadecanol, 9 + 1 (w/w; Norenberg and Barrett 1987).

Acrylic resins include the Lowicryls, the London Resins (LR White and LR Gold), and the methacrylates. The methacrylates and the London Resins are used for light microscopy.

All these are available as commercial products. *Epoxy resins* (Epon, Araldite and Spurr) are the most frequently used embedding media for transmission electron microscopy (TEM). The acrylic resins (Lowicryls and LR-resins) are used for special purposes for TEM, such as studies involving immuno-reactions.

Protocols for Fixation, Dehydration and Embedding of Tissues for Light Microscopy. Common procedures are suggested below with reference to only the most well-known fixatives, dehydration media and embedding media. The protocols should be modified to suit the user's special situation.

The duration of each single treatment step depends on size and character of the material. Since infiltration is based on diffusion, the degree of permeability of the tissue should be considered. Infiltration is regarded as complete when the root specimen sinks to the bottom of the infiltration vial. The suggested time course in the schemes will apply for tissue not exceeding 2 mm in at least one dimension, which means that no distance from a cut surface is more than 1 mm. In other cases all treatments should be extended.

BOX 2.3. Formaldehyde Fixation and Subsequent Embedding in Glycolmetacrylate:

1. Fixation with 2–4% formaldehyde in buffer solution (e.g. a 0.1 M phosphate buffer or a cacodylate buffer, pH 6.8), using vacuum infiltration until the tissue sinks to the bottom of the fixation vial. Leave the vials in the refrigerator overnight.
2. Wash with buffer three times in the course of 1 h;
3. 10% ethanol, 2 h;
4. 20% ethanol, 2 h;
5. 30% ethanol, 2 h;
6. 50% ethanol, 2 h (or overnight in refrigerator);
7. 70% ethanol, 2 h;
8. 70% ethanol, overnight;
9. 96% ethanol, 2 h;
10. 96% ethanol, overnight;
11. 96% ethanol + infiltration solution, 1 + 1, 2 h;
12. Infiltration solution (vacuum, if necessary) overnight in refrigerator;
13. Embedding in plastic or gelatine capsules.

BOX 2.4. FAA Fixation and Subsequent Embedding in Glycolmetacrylate:

1. Formalin-acetic acid-ethanol (FAA), using vacuum infiltration until the tissue sinks to the bottom of the fixation vial. Leave the vials in the refrigerator overnight.
2. 70% ethanol, 2 h;
3. 70% ethanol, overnight;
4. 96% ethanol, 2 h;
5. 96% ethanol, overnight;
6. 96% ethanol + infiltration solution, 1 + 1, 2 h;
7. Infiltration solution (vacuum, if necessary) overnight in refrigerator;
8. Embedding in plastic or gelatine capsules.

The dehydration procedures (in Boxes 2.3 and 2.4) apply if the embedding medium contains approximately 5% water. Otherwise the dehydration procedure should continue to 100% ethanol followed by 100% ethanol/infiltration solution (1 + 1, v/v).

Formaldehyde Fixation Followed by Dehydration with the Embedding Medium. For enzyme histochemistry Gerrits and Zuideveld (1983) suggest the following procedure (Boxes 2.5 and 2.6) based on dehydration with the embedding medium (metacrylate, in this case Technovit 7100) and with all steps at 4 °C:

BOX 2.5. Formaldehyde Fixation Followed by Dehydration with the Embedding Medium:

1. Fix in 4% formaldehyde in a 0.1 M phosphate or cacodylate buffer, pH 6.8, 2 h;
2. Wash with buffer, 18 h;
3. 70% Embedding medium 30 min;
4. 90% Embedding medium 30 min;
5. 100% Embedding medium 2 h;
6. 100% Embedding medium 12 h;
7. Embedding.

BOX 2.6. Formaldehyde/Histochoice Fixation Followed by Embedding in Paraffin Wax:

1. Fixation in 4% formaldehyde in buffer pH 6.8. 0.1 M (alternatively Histochoice). Use vacuum for infiltration;
2. Formaldehyde (alt. Histochoice) + absolute ethanol, 4 + 1;
3. 30% isopropanol;
4. 50% isopropanol;
5. 70% isopropanol;
6. 90% isopropanol;
7. 95% isopropanol;
8. 100% isopropanol, 3 times.;
9. Add paraffin wax at +60 °C, gradually increasing the concentration of wax to 100%;
10. Embed in paraffin wax (sets at room temperature);
11. After sectioning, the specimens are collected on glass slides specially treated for adherence. The wax is removed from the plant tissue by four times treatment with Histoclear, followed by five times isopropanol.

BOX 2.7. Formaldehyde Fixation Followed by Embedding with Steedman's Wax. Baluška et al. (1992) Recommended the Following Method for Fine Structure Studies of Primary Root Apices, Including Immuno-Analysis:

1. Fixation in 4% formaldehyde (buffer specified for protecting the structures of interest);
2. Dehydration in an ethanol series diluted in the same buffer as the fixative, until absolute ethanol;
3. Infiltration at +37 °C with a series of mixed absolute ethanol and Steedman's wax: 2 + 1, 1 + 1, 1 + 2 (v/v), 2 h in each step. (Steedman's wax: polyethylene glycol 400 distearate mixed with 1-hexadecanol, 9 + 1 (w/w);
4. Pure wax, changed three times under vacuum to remove ethanol;
5. Embedding. The wax sets at room temperature.
6. After sectioning, the specimens are collected on glass slides, pretreated for adherence. The wax is removed by ethanol.

2.3.5 Sectioning

2.3.5.1 *Hand Sections*

Fresh or fixed plant tissue can be sectioned using a razor blade. Small or thin plant parts can be supported between pieces of elderberry pith (*Sambucus nigra*) or different types of roots or tubers (carrot, potato) or other relatively soft, homogenous materials (soap, cork, foam plastic) to facilitate hand sectioning.

Lindauer (1972) gives an overview of sectioning methods. Treatment with ethanol is effective in removing the air from the plant tissue and furthermore it hardens soft tissues. On the other hand, glycerol-ethanol (1 + 1 (v/v) mixtures can soften hard tissues. Increasing the glycerol ratio augments the effect.

Woody roots may be sectioned directly in the fresh state by hand or after clamping in a sliding microtome. To soften the material (under vacuum, if necessary) it may be pre-treated by boiling in water or placed in a mixture of equal amounts of water, ethanol, and glycerol.

2.3.5.2 *Freshly Cut Tissues*

Tissue (living or formaldehyde fixed) is mounted with cyanocrylate (superglue) to slides and cut using a vibratome. This instrument produces 70-μm-thick sections. Staining or enzyme treatment can take place before microscopy. This method is used as a pre-treatment before confocal laser microscopy. The use of specific fluorescent stains (Blancaflor and Hasenstein 1993, 1997) can mark the object of interest.

2.3.5.3 *Embedded Tissues*

By embedding, the material is penetrated and surrounded by the embedding substance (e.g. ice, paraffin, epoxy resins, other plastics). Apart from ice, the material must be dehydrated prior to penetration of the embedding medium. The material is passed through an ascending concentration of the dehydrating agent, e.g. alcohol or acetone, followed by ascending concentrations of the embedding medium dissolved in the dehydrating agent and ending with pure medium. Finally the embedding medium is polymerized. For details of embedding methods, see Sass (1961), Glauert (1975), O'Brien and McCully (1981). Spurr's resin is especially popular embedding material for TEM (Spurr 1969).

Embedding of fresh tissue may be carried out in solid paraffin. This is a fairly rapid and simple method giving sufficient support to the tissue during the sectioning. The tissue is immersed in molten paraffin and then cooled down to a temperature slightly above the melting point. After a suitable period of infiltration the specimen is cooled down. If necessary the paraffin can be removed using xylene.

Paraffin infiltrated and embedded material is ready for sectioning immediately after the cooling of the paraffin. Material embedded in plastic can be sectioned shortly after polymerization. Hard or brittle blocks may be moistened on wet filter paper before sectioning. Warming in an incubator may harden soft plastic blocks.

The microtome is equipped with a holder(s) for glass knives or metal knives or adapters for disposable knives. Glass knives are produced on a special knife maker. They have a more perfect cutting edge, but for most light microscope work metal knives will provide sections of sufficient quality. In addition, they allow large sections and in this way one benefits from the advantages of many plastic embedding media in making large sections of less thickness. Metal knives are available as disposable knives or as knives which can be resharpened.

Cutting angle, cutting speed, humidity and temperature are critical factors during sectioning. Generally, most of the necessary information appears in the instruction sheets of the commercial embedding kits.

During sectioning, the material is compressed and the sections need to be stretched on a water surface, most conveniently on droplets or on a thin layer on the glass slide. Stretching is best obtained with pure, distilled water with maximum surface tension (Gerrits et al. 1987). Low concentrations of additives are occasionally required (for instance 1% acetone). As temperature affects surface tension the stretching and the subsequent water evaporation should preferably take place at room temperature, but often a heating plate is used to speed up the process.

2.3.6 Staining

Staining increases the contrast of biological material when observed under the microscope. Furthermore, depending on specificity, the staining performs a cyto-/histochemical analysis of the tissue components. There exists a wide range of stains for different purposes (Table 2.4). For TEM, staining with uranyl acetate followed by lead citrate is routinely used to increase contrast. Selected stains for special compounds may be used.

Paraffin-embedded material must have the paraffin removed and the tissue hydrated prior to staining. This is most often done through xylene (xylene/100%

Table 2.4. List of common stains

Stains for general anatomical/histological survey	*Stains for vitality*
Aniline blue	4′,6 Diamidino – 2 – phenylindole (DAPI)
Chatellier's solution	Neutral red
Fuchsin – safranin – astra blue (FSA)	Phenosafranine
Safranine-fast Green or	Trypan blue
Safranine-light Green	Evan's blue
Toluidine blue o	
Trypan blue	
Stains for localization of fungal structures in host tissue (mycorrhiza, root pathogens)	*Stains for lipids/suberin*
Aniline blue	Sudan III
Toluidine blue O	Sudan IV
Trypan blue	Sudan Black B
	Sudan Red 7B
Staining procedure for suberin, lignin and callose	*Staining method* for evaluation of tissue water contents
Berberine-aniline blue	Sulphorhodamine G

ethanol 1 + 1), followed by a graded ethanol series to water (ethanol 100%, 96%, 70%, 50%, 30%, water). Steedman's wax is removed from sections by ethanol. Plastic-embedded material may be stained when it adheres to the slide and is dry. Hand sections may be stained directly on glass slides prior to mounting or transferred through the stains in small sieves.

A number of pitfalls must be avoided in the interpretation of the stained material. Artefacts commonly occur for several reasons: the staining solution may vary from time to time, or between manufacturers. The active ingredient may only work in the predicted way when other substances are present or absent, under controlled pH, molarity, and temperature.

The quality of the staining agents is critical. Presence of impurities and the actual percentage of the active substance may vary from one batch to another. Proper controls will ensure that conclusions are drawn on a safe basis. Stained and unstained material should always be compared (positive control). Comparison of unfixed or cryofixed tissue with tissue treated with chemicals is recommended. It should be noted that different chemical fixatives affect the electrostatic charge of the resulting specimen and hence interfere with most staining reactions: Formaldehyde fixation of cytoplasm will result in an increased affinity to acid stains, whereas CrO_3 will increase the affinity to basic stains. On the other hand, chromosomes will show increased affinity to acid stains after treatment by either of these two fixatives (formaldehyde and CrO_3).

Other control measures include inhibition of the assumed process or extraction of the tissue component in question followed by comparison with an untreated specimen (negative control).

2.3.6.1 Stains

NB!!: Almost all stains are somewhat harmful to human health. Use a fume hood and take care when preparing the staining solution and when handling the stain. Waste should be deposed of in suitable waste containers. The most commonly used stains are listed in Table 2.4.

Staining for specific tissue components is relevant for more detailed studies. More Information concerning this can be found in Lillie (1977), Clark (1981), Gahan (1984) and Neergaard (1997).

2.3.6.2 Staining Procedures

Aniline Blue. Aqueous solution, 0.5–1.0%, alternatively dissolved in ethanol, glycerol or lactoglycerol. Glucans are positive.

As Leuco-Aniline Blue: the stain is dissolved in buffer pH 8.3 or higher, and the tissue observed using fluorescent light at 365–405 nm. Leuco-Aniline Blue is used for localisation of callose.

Berberine-Aniline Blue. Aqueous solution of berberine hemisulphate, 0.1% (w/v). Stain for 1 h, subsequently wash in water. After that, stain in aniline blue solution (0.5% aniline blue w/v, aqueous) 30 min and finally treat for several minutes with $FeCl_3$ in aqueous 50% glycerol, prepared by adding glycerol to filtered aqueous $FeCl_3$.

Mount in the last solution. Observe with fluorescence microscopy at 365 nm. Suberin is blue-white to blue, lignin is bright yellow, callose is blue-white, and Casparian strips in the endo- and exodermis are intense yellow (Brundrett et al. 1988).

Chatelliers Solution. Dissolve 1.1 g aniline sulphate in 70 ml distilled H_2O under heating. Filter this solution into a mixture of 60 ml lactic acid and 45 ml lactic acid saturated with Sudan III. Dissolve 1 g of KI in 10 ml distilled H_2O and add 10 ml 96% ethanol. Dissolve iodine (I_2) into this solution and pour into the lactic acid mixture. Add 6 ml conc. HCl and shake thoroughly.

Mount in this solution, add cover glass, and boil over flame. Lignin is yellow, cutin and suberin are red, starch is black, and protein is yellow. The lactic acid is a clearing agent. Preparations may keep for about a week.

DAPI (4,6-Diamidino-2-phenylindole). Used as a vital stain (0.002% in buffer) applied for 30 min in the dark, or for fixed material (0.01–0.05%) overnight at +4 °C, and observed at 365 nm indicates DNA, hence it can be used for studies of vitality. Phenols are positive, too.

Evan's Blue (3,7-Diamino-5-phenylphenazinium chloride) used as vital stain. Stain in a 0.05–0.25% aqueous solution for 5 min. Living cells remain unstained, dead cells become blue.

FSA (Fuchsin-Safranine-Astra Blue) is made up by mixing basic fuchsin (0.1 g) safranin O (0.4 g), astra blue (1.0 g) and acetic acid (20 ml). Subsequently water is added up to 1 l. Stain for 1–5 min. Cellulose is blue, lignin, cutin and some suberins, nuclei and plastids are red, whereas starch is unstained.

Neutral Red is a non-toxic stain used for studies of living cells, for example to observe plasmolysis. The stain is used as an aqueous solution (0.1%, w/v) and stains lipids in vacuoles and certain cell walls.

Phenosafranine is used to distinguish living from dead cells. A 0.1% aqueous solution, it is taken up by dead or abnormal cells immediately, while living cells remain unstained for at least 30 min.

Safranine-Fast Green/Safranine-Light Green. A mixture of 1.5 g safranine O and 0.5 g light green or fast green is dissolved in 1.25 ml conc. HCl + 75 ml distilled water + 120 ml 96% ethanol. Tissue sections, after being treated with ethanol, are stained for 5 min, followed by differentiation with 96% ethanol 3 × 30 s. Cytoplasm and cellulose walls become green, nuclei, chromosomes, as well as lignified, suberised, and cutinised walls become red. In the case of fungal infections, hyphae become red whereas host cells are green.

Sudan III and Sudan IV. Saturated solution in 96% ethanol. Stain in solution for few minutes. Lipids, cutin and suberin are red.

Sudan Black B. 1% Sudan black B is dissolved in 70% ethanol, the tissue is stained 1–1.5 h, subsequently washed in 70% ethanol (3 × 15 s), and mounted in 75% glycerol before microscopy. Lipids are positive.

Sudan Red 7B. The stain is dissolved (0.1% w/w) in polyethylene glycol 400-glycerol (1 + 9). Lipids are positive. It is especially good for demonstrating the presence of suberin.

Sulphorhodamine G (SRG). The specimen is treated 5 s in a 1 mM freshly prepared and filtered solution of sulphorhodamine G before sectioning. Hand sections are mounted in paraffin oil for fluorescence microscopy. Distribution of induced fluorescence corresponds to water content of the tissue. For details, see Watt et al. (1996).

Toluidine Blue O. The staining solution is prepared by dissolving 0.05% toluidine blue O in buffer, pH 4.4–6.8. Sections or whole tissues are stained for 5–10 min followed by differentiation using the same buffer until excess stain is removed. Cellulose and other glucans become red-violet depending on pH, lignified elements and other polyphenols become bluish-green. Fungal hyphae are often distinguished from the surrounding host tissue because of the difference in cytoplasmic pH.

Trypan Blue. A 0.01–0.2% aqueous solution is used for root staining. The method is often used for investigation of mycorrhizas. Suitable for whole mounts as well as sections. Also used as a stain for vitality, as living cells generally remain unstained.

2.3.7 Mounting

After staining, the sections are ready for mounting and microscopy. Initial observation is recommended using water (a drop of distilled water before the cover slip is placed). For permanent slides different embedding media are commercially available.

After staining with water-soluble stains, the use of non-aqueous media has the advantage of stabilizing the material without leaching the stain. The change from the hydrophilic staining phase to a hydrophobic mounting medium is done by drying or by dehydration through an ascending series of ethanol, ethanol-xylene (1 + 1) ending with 100% xylene before mounting. Often the mounting medium has to be hardened by heating in an incubator. Xylene-containing media must be stabilized in this way only under conditions with sufficient ventilation (fume hood or other arrangement), as the evaporating xylene is hazardous to health. For plastic-embedded sections drying is sufficient before mounting. Fluorescence microscopy requires a non-fluorescent mounting medium.

Aqua-mount contains phenol and is thus poisonous. Certain aqueous stains are destained by Aqua-mount. DPX and Canada balsam contain xylene (must be handled in well-ventilated places such as fume hoods). Specimens stained with zincchloride-iodine, Millon's reagent, Sudan IV and J-KJ should not be mounted in Canada balsam.

2.3.7.1 Mounting Media

Table 2.5. Mounting media

Aqueous mounting media	Non-aqueous mounting media
Water	Canada balsam
Glycerol	DPX
Glycerol-gelatine (= glycerol jelly)	Caedax
Aqua-mount	Eukitt
Polyvinylpyrrolidone (PVP)	Euparal
Aqueous saturated $CaCl_2$	
PVP + $CaCl_2$	*Non-fluorescent mounting media*
Polyvinyl alcohol (PVA)	Water
Shear's medium	Glycerol
Gelatin-sucrose	DPX
Malinol	Gurr's fluormount
Gelvatol	

Specimens of plant tissue are usually observed in distilled water or 70% ethanol or 0.1 *M* $CaCl_2$ or 0.25 *M* sucrose. If the mounting medium evaporates too rapidly, glycerol or lactoglycerol can be used instead. It should be kept in mind that these chemicals may cause swelling of the tissue.

Glycerol and Ethanol or Tween. This can be used as a mounting medium for specimens for immediate observation using 50% aqueous glycerol. In order to facilitate infiltration and remove air from the tissue, a small amount of ethanol or a drop of Tween 20 is added. Alternatively, an aqueous solution of Tween 20 (1 drop in 100 ml water) is often sufficient as a mounting medium.

Polyvinylalcohol (PVA). This is a commonly-used mounting medium (refractive index 1.39) for permanent mounts (Box 2.8). It can be used as a solvent for stains and as it hardens by heating, it can be used in a combined tissue treatment including staining, conservation and mounting.

BOX 2.8. Preparation of a Polyvinylalcohol Solution

Polyvinyl alcohol	1.66 g
Distilled water	10 ml
Lactic acid	10 ml
Glycerol	1 ml

The polyvinyl alcohol should be added very slowly to the water (magnetic stirrer) and it dissolves in 1–2 h. Lactic acid is then added (vigorous stirring is necessary), followed by glycerol.

Lactoglycerol. Another mounting medium for extended conservation is 50% glycerol or 20% lactic acid. The cover glass should be sealed with nail varnish.

Different types of mineral oil can be used if non-aqueous mounting media are preferred, for example on desiccated or hydrophobic surfaces. Mineral oil is useful especially for plant tissues containing high amounts of pectin substances or polysaccharide slime or with large water containing cells. *Warning*: mineral oil may affect flavonoid cell components.

Clearing, mounting, and staining may be combined in one procedure.

2.3.8 Procedures for Transmission Electron Microscopy

The preparation for transmission electron microscopy involves aldehyde fixation; dehydration using ethanol or acetone, and embedding in resins especially suitable for electron microscopy (epoxy resins or certain acrylic resins, such as the Lowicryls or the London Resins). Generally, the methods are similar to what is described for LM, but a number of modifications are necessary. The samples must be smaller. The values of pH and osmolarity of the solutions are much more critical. If *formaldehyde* is used as a fixative it should be freshly prepared from paraformaldehyde by slowly dissolving (in fume hood) the necessary amount of paraform in buffer. Use a magnetic stirrer during careful heating until the solution looks clear. Certain postfixation procedures (staining) are often required; a 1–2% osmium tetroxide solution is usually in the same buffer as for fixation. (*NB! OsO_4 is extremely hazardous to human health*). An ultramicrotome is needed for making sections in the range 60–70 nm. Preparation methods for TEM are described in Robards and Wilson (1993).

2.3.9 Procedures for Scanning Electron Microscopy

Scanning electron microscopy is used for studies of surfaces, including cut surfaces. The material may be chemically fixed or cryo fixed.

Chemical fixation may be done with either FAA or glutaraldehyde with or without postfixation with OsO_4. Chemical fixation should be followed by "critical point drying". The material is passed through an ascending acetone series and in the pressure chamber of the CPD-apparatus exchanged with liquid carbon dioxide. By elevating temperature the pressure is raised and when the critical point of CO_2 has been reached it passes from liquid to the gaseous phase.

The pressure is slowly released and the material coated with gold or gold/palladium in a sputter coater prior to observation.

For cryo fixation the material is immersed in nitrogen slush prior to transfer into the cooled microscope (−195 °C). After sublimation to remove condensed ice from the surface, the specimen is sputtered with gold prior to investigation. Both fixation methods are suitable for the study of root hair morphology and ultrastructure.

Cryoscanning electron microsocpy has been used to observe the contents of cortical intercellular space in roots (Watt et al. 1996) in studies concerning the influence of soil water contents on deposits in root tissue.

2.4 Root-Soil Boundary

2.4.1 Root-Soil Interface in Thin Sections

Roots modify biological activity, structure, water status and mineral constitution of the surrounding soil. These changes can be observed by *in situ* microscopic examination of the root-soil boundary. For this purpose, thin sections of soil with roots in place are needed (Jenny and Grossenbacher 1963; Lund and Beals 1965; Fitzpatrick 1990). Soil blocks for preparing for thin sections *in situ* can be obtained under field conditions by freezing the blocks with liquid nitrogen. Under laboratory conditions, plant roots can be grown in small tubes filled with soil. Undisturbed soil blocks are taken out of the tubes and are passed through procedures of fixation, dehydration, embedding and polymerisation in polyester resin. These procedures are described in Sections 2.3.4 and 2.3.5, but Tippkötter et al. (1986) provide very useful information on the procedures for preparation of thin sections for biological studies. Thin sections of desired thickness are prepared by grinding or by cutting the blocks with a diamond saw or microtomes, depending on the details to be observed under the microscope. Use of fluorescence microscopy and staining of the thin sections with acridine orange (1:1000 in 10% HCl) improve brightness and contrast between soil and plant material (Altemüller and van Vliet-Lanoe 1990). Blue light excitation exhibits cell walls of the root tissue as green or yellowish green. Soil material becomes orange to reddish, hence contrasts well with the root tissues. Microorganisms and fungi generally show green colours and are discernible in contact with soil particles. However, they may be invisible if they are attached to root cell walls. Trolldenier (1965) also used acridine orange in rhizosphere studies for producing good quality colour micrographs of bacteria on root hairs. Many of these thin section methods, however, provide mainly qualitative information, but they allow in situ observations of the processes occurring close to roots.

Watteau et al. (1996) described methods to study rhizosphere microsites in situ. The methods are based on electron energy Loss spectroscopy (Egerton 1986) and allow observations at the ultrastructure scale on:

1. Microaggregation,
2. Polysaccharides of bacterial or root origin,
3. Microporosity,
4. Organo-mineral associations (root/mineral, microorganism/mineral), and
5. Plant-microorganism interactions.

The limitation of these methods is the embedding of the sample in resin, because it may lead to element migration and collapsing of clay. Use of a carbon-rich resin may interfere with the detection of carbon in the sample. Cryo-ultramicrotomy may help to overcome these problems. Cryo-scanning electron microscopy also allows observation of undisturbed accumulation of water close to roots taken from the field (McCully 1995).

2.4.2 Root Hairs

Root hairs extend effective root surface area and thereby increase root-soil contact. The initiation and development of root hairs occurs due to cellular differentiation in the root epidermis. In the majority of plant species, any protoderm cell has the potential to form a root hair. In a few plant species, root hair formation occurs only in "trichoblasts". Peterson and Farquhar (1996) have recently reviewed the cytological and functional aspects of root hairs.

2.4.2.1 Quantification

Root hairs have typically been measured using a microscope and occular micrometer (Caradus 1979; Föhse and Jungk 1983; Lamont 1983; MacKay and Barber 1984). For root hair studies, the roots should be disturbed as little as possible. Root hairs are often measured on roots grown in nutrient solution culture, because it allows sampling of roots with minimum damage of root hairs. For studying root hairs of solution-grown roots, nutrient concentrations close to those found in the soil solution should be used. For soil- or field-grown roots, the roots bearing root hairs must be cleaned without damaging the root hairs. Ultrasound treatment is useful for cleaning roots for root hair studies (Gahoonia and Nielsen 1997; Box 2.9).

BOX 2.9. Sampling and Cleaning of Soil-Grown Roots for Root Hair Studies:

1. Take soil cores with intact roots using a sharp knife or auger,
2. Immerse the soil cores in water overnight in the dark at 5 °C,
3. Remove the floating roots carefully e.g. using a (kitchen) sieve,
4. Place the roots in an ultrasound water bath (e.g. Branson 5200) and subject them to an ultrasound treatment (120 W, 47 kHz) for 5–10 min. Intermittent treatment by switching off the ultrasound bath for few seconds and then switching it on again facilitates cleaning of roots.

The treatment with ultrasound removes the soil particles without damaging the root hairs and leaves roots with root hairs comparable to those grown in nutrient solution culture (Fig. 2.2), facilitating measurements of root hair parameters using image analysis as described below. Roots with longer and denser root hairs may have to be treated for longer with ultrasound.

Depending on the size of sampled root system, root hairs can be quantified on whole or sub-sampled root systems. To make sub samples, the root system is laid out in a film of water in a shallow plastic tray. The length of the root is measured from the root to shoot junction to the longest root tip, and a mark placed midway down the root. Excess water is blotted from the tray, and a ca. 1-cm strip of root is cut across the midway line using a scalpel and ruler. The top and bottom portions of the root are removed and placed into bags for dry weight determination. The 1-cm segments are re-floated in water, and ten segments are randomly chosen for root hair studies. The surplus segments are placed with the rest of the root, as are the ten segments after they have been analysed for root hair length and number.

Root hairs and root parameters can be measured on the roots grown in nutrient solution (Care 1995) or on roots washed out of the soil (Gahoonia and Nielsen 1997) using image analysis. A typical image analysis system for root hair studies consists of a microscope connected to a video camera, which is further interfaced with a computer containing appropriate software and cards. The quality of the optical system is critical for image clarity, as there is loss of image resolution from the microscope to the computer. The combination of microscope and image analysis package can be critical and should be evaluated beforehand, if possible. Though other combinations may be equally good, Care (1995) and Gahoonia et al. (1997) found that a Leica Wild M8 stereo microscope (Leica AG, Heerbrugg, Switzerland) to be the most appropriate when coupled with the MD30™ image analysis software package (Leading Edge Pty. Ltd., Australia). Each image requires a minimum of 0.2 Mb computer memory.

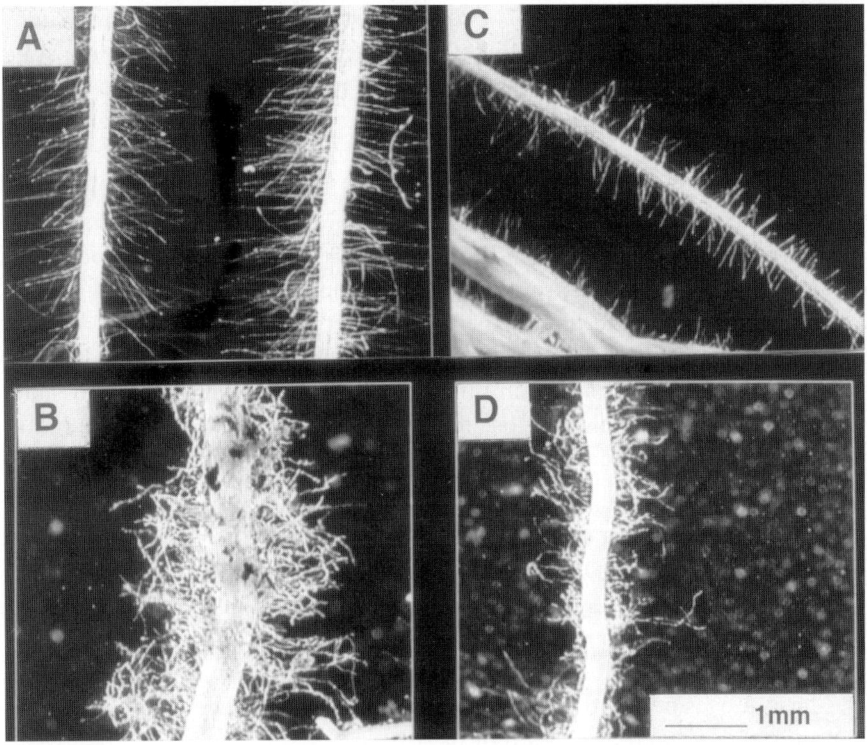

Fig. 2.2. Root hairs of two spring barley cultivars, Salka and Zita. **A** Root hairs on roots of Salka grown in nutrient solution culture. **B** Root hairs on roots of Salka grown in a field experiment. Root hairs were extracted from the soil using ultrasound treatment (120 W, 47 kHz). **C** Root hairs on roots of Zita grown in nutrient solution culture. **D** Root hairs on roots of Zita grown in a field experiment. Root hairs were extracted from the soil using ultrasound treatment (120 W, 47 kHz). (Photos by Tara S Gahoonia)

This should be kept in mind when planning experiments since capturing images of entire root systems may require a large data storage system.

To make root hairs clearly visible, appropriate magnification and illumination is needed. The best magnification must be determined for each particular case, and then all the root hair images captured at this magnification. The best illumination for root hair resolution is achieved by dark phase transmitted light. The image analysis system must be calibrated using a micrometer (usually supplied with the microscope) at the same magnification at which root hair images are captured. The calibration procedure is normally described in the accompanying user manuals.

To determine the length of the root hairs, an image is recalled on the computer screen and individual root hairs are selected and highlighted, using an "anchor and drop" process with the computer mouse. As each root hair is highlighted, its length in pixels or millimetres (depending upon the calibration scale chosen) and also the number of root hairs is recorded in a data file. All root hairs visible on each image are measured. The diameter of the root is also measured. The stored data files can be imported into an Excel spreadsheet.

To calculate the number of root hairs on the full root hair cylinder, it is necessary to determine the depth of field of the microscope (Box 2.10) and the diameter of the root because on thin roots a higher proportion of root hairs will be seen than on thicker roots. It should be noted that changing the microscope magnification and/or the aperture setting will alter the depth of field, hence all measurements should be done at the same settings that the depth of field is calculated at. To maintain the focus of as many root hairs as possible, the centre of focus should be close to the longitudinal axis of the root, and at the widest part of the root.

BOX 2.10. Measuring Depth of Field of a Microscope:

1. Place a vernier scale micrometer on the focus arm of the microscope,
2. Focus the microscope on root hairs at the magnification of interest,
3. Move the focus up or down and measure (on the vernier scale) how far it moves before the object is no longer in focus. The distance moved is equal to the depth of field.

The number of hairs in the full cylinder is calculated as:

$$\frac{\text{number of hairs observed on unit root} \times \text{diameter of root}}{\text{depth of the field}}$$

If a root map is done and the total root length measured (for methods see Chap. 10) and the length of root with root hairs is known, then the contribution of root hairs to total root surface area can be calculated. This is done by using the total length of root hairs (e.g. by multiplying mean root hair length by the total number of root hairs) in the equation for the surface area of a cylinder as $2\pi r l$, where r is the root hair radius and l is the total length of root hairs.

2.4.2.2 Vitality

Root hairs are often subjected to biotic and abiotic stresses that affect their vitality and activity. Various stains (Table 2.4) can be used to test the viability of root hairs. Oprisko et al. (1990) compared the effectiveness of some stains for assessing root hair viability and reported that Evan's blue (0.5 g/l) gave greatest precision and least variability. Evan's blue performs basically on the retention or loss of semi-permeable properties of the plasmalemma (Gaff and Okong'o-ogola 1971). Live root hairs remain colourless because of exclusion of Evan's blue and the dead root hairs are permeable and turn blue. To obtain dead root hairs (control), the root segments can be boiled in water for 10 min (Fischer et al. 1985).

2.4.2.3 *Nutrient Uptake via Root Hairs*

The role of root hairs in nutrient uptake can be assessed with microelectrodes. Recent work (Jones et al. 1995) using a vibrating microelectrode with *Limnobium* root hairs has shown that Ca^{2+} is taken up at the root hair tip, and K^+ is taken up over the whole root hair surface. This highlights the importance of root hair number for Ca^{2+} and root hair length for K^+ uptake.

Gahoonia and Nielsen (1998) described a direct method to measure uptake of soil phosphorus via root hairs. Only root hairs penetrated a tightly stretched nylon screen (53 μm) glued to the bottom of a PVC tube. The penetrating root hairs grew for 2 and 4 days in soil labelled with radioisotope phosphorus tracer ^{32}P (185 kBq g^{-1} dry soil) in another PVC tube. Transparent plastic rings of thickness ranging from 0.25 mm to 2.0 mm were inserted between the two PVC tubes. This provided slit width for microscopic observations in situ, which confirmed that only root hairs were growing into the ^{32}P labelled soil. The uptake of ^{32}P from soil via the root hairs was quantified by measuring activity of ^{32}P in the plant shoot (^{32}P uptake only via root hairs). In some cases, *no* rings were inserted between the two PVC tubes so that both root hairs and root surface could participate in ^{32}P uptake from the labelled soil (total ^{32}P uptake). This study provided direct evidence on the substantial participation of root hairs in uptake of phosphorus from soil.

2.4.2.3 *Root Hairs in Nodulation Process*

Root hairs are the primary sites of infection with *Rhizobium* in nitrogen fixation of legumes. The infection process follows curling and deformation of root hairs. The rate of root hair deformation can be used to compare the activities of different Nod factors (nodulation genes of *Rhizobium*). Heidstra et al. (1994) adapted the method of Bhuvaneswari and Solheim (1985) to assess root hair deformation induced by different Nod factors. To perform the assay, sterilised seeds are germinated for 2 days and transferred to Fåhraeus slides, modified as small trays, containing 1 ml of Fåhraeus medium (Fåhraeus 1957). All roots are examined under a microscope, and those with already deformed root hairs are discarded. The 1 ml medium is replaced by medium containing Nod factor and the slides are incubated at 22 °C. After 3 h (when maximum deformation occurs), the deformed root hairs are scored under a microscope as 0, 1 and 2, corresponding to 0–20%, 20–60% or >60% deformed root hairs respectively. The reliability of this assay lies in the fact that over 90% of the plants, when tested with different concentrations (ranging from 10^{-7} to 10^{-13} M) of a Nod factor, showed almost identical rates of root hair deformation. Based on these observations, Heidstra et al. (1994) suggested that by applying dilution series of Nod Factors, the root hair deformation could be used as a semi-quantitative assay of activity of Nod factors.

2.4.3 Mycorrhiza

Mycorrhizas are symbiotic associations between plant roots and certain fungi. They are almost universal, with most vascular plants forming a mycorrhizal symbiosis with a fungal partner. It is widely recognized that the association normally brings mutual benefits to partners; the plant receives mineral nutrients and the fungus receives carbon compounds derived from photosynthesis. Different types of association, involving different plant and fungal groups, have been identified and organised into the following four major categories:

1. Arbuscular mycorrhizas (AM) – the most widespread, involving many agriculturally and ecologically important plant species;
2. Ectomycorrhizas (EM) – involving most forest tree species;
3. Ericoid mycorrhizas – confined to plants in the *Ericales*;
4. Orchid mycorrhizas – confined to orchidaceous plants.

These divisions may be seen as arbitrary. For example, it is now evident that although such divisions are useful, in practice many variations exist (e.g. ectomycorrhizas can also be arbutoid, monotropoid and ectendo types: see Smith and Read [1997] for more details). Individual mycorrhizal associations thus need to be treated as specific cases and clearly characterized. It is therefore important, when making measurements of mycorrhizas, to identify the possible nature of the association and not to make assumptions based on any preconceptions. Therefore, it is the aim of the methods described here (which deal only with AM and EM) to provide guidance. Because each circumstance will be different they are not prescriptive protocols but should be seen as starting points, being revised where appropriate. Finally, it should be mentioned that these methods provide a single-point assessment of a mycorrhizal symbiosis and any assumptions made on the basis of data collected should take into account both spatial and temporal factors.

It is not possible to see root colonisation with vesicular-arbuscular mycorrhizae (VAM) fungi without staining and microscopic observation. However, the staining process does not require expensive instrumentation.

After washing and acidification of the root sample, fungal structures in the roots are stained with trypan blue, acid fuchsin, chiorazole black or other compounds (Brundrett et al. 1984; Grace and Stribley 1991). Staining of active fungal parts only may also be possible (Schaffer and Peterson 1993). After staining, in most cases the percentage of the total root length colonised by mycorrhizal fungi is determined (Giovannetti and Mosse 1980).

Spores of mycorrhizal fungi can be isolated from soil by a wet sieving and decanting technique (Box 2.12). Also, the length of the external mycorrhizal mycelium in the soil can be determined after washing the hyphae from soil and subsequent staining (Hamel et al. 1990). In the future, DNA-based methods are expected to facilitate characterisation and distribution of different fungal populations in the soil and in the roots.

Before designing a protocol to assess mycorrhizas it is important to establish the hypothesis being tested. Is the aim, for example, to establish the presence or absence of a particular type of mycorrhiza, the extent of the colonisation or perhaps the function? Each will require a different approach and procedure.

In this section, only a selection of the major methods is provided. Readers are encouraged to further read the papers referenced or for a more in-depth account of methods are referred to Ingleby et al. (1990), Norris et al. (1991a, b), Sieverding (1991) and Brundrett et al. (1996). Marschner (1995) and Smith and Read (1997) give a summary of the role of mycorrhizas in plant growth.

2.4.3.1 *Arbuscular Mycorrhizas*

Arbuscular mycorrhizal fungi (AMF) of the order *Glamales* colonise and form mycorrhizas in the roots of a wide variety of plants. They are common in both annuals and perennials and have a major agricultural and ecological significance. The mycorrhizas have internal fungal structures (hyphae and often arbuscles and vesicles) and external to the root, hyphae (often forming an extensive external mycelium) and spores growing from the external mycelium. In some cases spores can also be found growing within the root. Modifications to the morphology of the root system can also occur through effects of colonisation on branching (Hooker et al. 1992; Berta et al. 1995; for methods see Hooker et al. 1998). Here we confine methods applied to study internal and external fungal structures. Relatively simple methods can be used to visualize these but quantification is, in both cases, more difficult.

External Mycelium. In the simplest case, Graham et al. (1982) proposed that the weight of a plant-soil root ball would be proportional to the quantity of AMF hyphae that bind the unit together. They thus estimated the relative quantity of

AMF hyphae by weighing the plant roots with and without adhering soil. However, this view has subsequently been considered too simple as many other factors are also involved in the attachment of soil to plant roots (Kough and Linderman 1986). The determination of chitin content has also been used (Bethlenfalvay and Ames 1987). However, because chitin is also present in other fungi and insects, the use of the technique is limited to only very specific controlled situations where they can be eliminated. Schüepp et al. (1987) estimated the extension of AMF hyphae in soil by using a membrane which prevents the passage of roots but allows hyphae to pass. By using receiver plants, the roots of which became colonized by the hyphae, it was possible for them to calculate the distance that hyphae travel through the soil as well as the speed of growth. However, it is likely to be difficult to accurately calculate the latter because of the sensitivity of the technique to time for colonization of the receiver root.

Direct methods involve three steps: extraction, detection and measurement. The first, extracting the hyphae from the soil is arguably the most difficult. A sample of soil is normally first suspended in solution with water. Sieving (250 μm) is then used to separate the hyphae from the soil particles (Abbot et al. 1984). Although the concept is simple and straightforward, it is difficult to apply it in practice and recovery can be poor and uncertain. To increase recovery, Vilarino et al. (1993) immersed a rotating wire in the suspended soil. The hyphae present became attached to the wire and were subsequently washed off with water.

Hyphae of AMF must be distinguished from other hyphae in the extracted sample. The diameter of hyphae has been the most often used criterion for detection (Abbot and Robson 1985; Bethlenfalvay and Ames 1987). The variation in morphology of AMF hyphae makes distinguishing AMF on this basis very difficult and in some cases impossible. Thus, although it may be possible to distinguish the hyphae of AMF from other fungal hyphae in some very specific situations, in most it is not and data obtained using such methods should be interpreted cautiously. Hyphae are often, for example, stained with trypan blue or acid fuschin before identification. Staining also helps in the measurement of hyphae length with the gridline intersect method (Giovannetti and Mosse 1980). The future hope lies in the possibility of using immunological or molecular techniques to routinely distinguish between AMF and non-AMF hyphae in samples extracted from soils.

Internal Structures. Plants vary in the extent to which it is possible to visualize AMF in their roots. In some plants such as cucumber the transparency of the roots makes it possible to observe structures using only a dissection microscope, without preparation or staining. However, in other species, e.g. tree roots, it is often very difficult to observe AMF structures within the root, essentially due to the deposition of tannins and the poor transparency of cell walls. There is thus no prescriptive method for preparing roots in order to observe AMF.

The simplest methods are mechanically teasing roots open (Redhead 1977) or staining by immersing roots in solutions of cotton blue (Gallaud 1905) methylene blue (Daft and Nicolson 1966) or acid fuschin (O'Brien and McNaughton 1928). Improvements were subsequently made to staining by using lactophenol (Mosse and Hepper 1975), extracting phenolic compounds from roots using potassium hydroxide (Bevege 1968) and bleaching in hydrogen peroxide (Philips and Hayman 1970). For many years the most common method applied was that described by Philips and Hayman (1970), often modified by removing phenol from the procedures (Kormanik and McGraw 1982). Nowadays the most commonly used method is a modification of the Philips and Hayman technique by Koske and Gemma (1989). The procedure essentially involves six steps (Box 2.11). It is best carried out on fresh material, although roots can be stored in 70% ethanol for short periods.

BOX 2.11. Preparation of Roots for Observing AMF Structure:

1. Heat roots in 2.5% KOH for 3 min at 121 °C or 10–30 min at 90 °C;
2. Rinse roots in water;
3. Bleach roots, if necessary, for 10 to 30 min in alkaline H_2O_2 and then rinse in water;
4. Soak roots in 1% HCl for 1 to 24 h;
5. Stain roots in a solution of acidic glycerol and trypan blue for 3 min at 121 °C or 10 to 30 min at 90 °C;
6. Destain roots and store in acidic glycerol.

Most laboratories will further modify and optimize this technique for specific situations but it is often useful to begin with this protocol, inspecting roots at each stage using a microscope. When making assessments, even when optimized, it is essential to inspect stained roots very closely. A dissection microscope is frequently used but experience shows that the use of a higher power compound microscope is frequently necessary. This is particularly important because not all parts of the AMF are always stained, even when the method is applied very carefully. Another problem is that it is extremely difficult to visually differentiate between the hyphae of AMF and other fungi. Sometimes differentiation is possible because of the presence of arbuscles and or vesicles. Immunological techniques are available which will permit differentiation (Hahn et al. 1994) but they require much preliminary work and are not appropriate for the screening of large numbers of roots. Quantification of AMF is obtained by estimation of percentage occupancy of roots, usually using a gridline intersect method (Giovannetti and Mosse 1980). Measures of colonization by AMF have usually been made only with all AMF structures, i.e. grouped, with no separation into the different structures. However, there is now a consensus amongst many researchers that a breakdown of colonization into different structures is likely to be informative. More detailed studies of histology are possible following sectioning (Gianinazzi and Gianinazzi-Pearson 1992). Procedures again need careful adaptation to the tissue being studied. Finally, it should be remembered that although the above methods inform on the presence of AMF structures they do not identify function. This may be determined, of course, by bioassay but these data do not provide any information on localization. Techniques are now available to detect functional enzyme activity, e.g. alkaline phosphatase (Tisserant et al. 1993) and the more widespread use of this and similar techniques is likely to provide valuable data.

Spores. The quantification of spores extracted from the soil can be a useful measure. However, it should be emphasized that extrapolations between spore numbers (relatively easy to obtain) and the extent of colonisation by AMF or the external mycelium cannot be made, because they are likely to be unreliable and misleading.

The first step is to obtain a suitable soil sample. This should be taken from the field using a spade or soil corer, sterilized using 70% ethanol and rinsed in water. Soil from a pot(s) should be mixed before sampling. The precise number of samples will depend upon individual circumstances. However, it should be noted that the distribution of spores can be highly heterogeneous and that the design of an appropriate sampling regime should be a priority. The final sample size should be between 50 and 100 g. It is best to extract spores from fresh soils but dried soils can also be used. A procedure for quantification of spores is given in Box 2.12.

2.4.3.2 Ectomycorrhizas

Ectomycorrhizal fungi (EMF) colonise the roots of many important forest trees. Ectomycorrhizas are thus widely distributed in forest ecosystems. Many different morphological features are distinguishable. Undoubtedly, the most distinct structures are the fruiting bodies, which are frequently observable under forest trees. These structures have been widely studied (see Brundrett et al. 1996 for methods) for accurate identification of genera and/or species. Here we limit the description of methods to observation of structures within the root and on its surface.

BOX 2.12. Quantification of Spores:

1. Remove stones and large woody roots from the sample by hand. We do not recommend using a 2-mm sieve, as this will also remove fine roots which are likely to have spores attached.
2. Add the soil (50–100 g) to approximately 9 l distilled water in a suitable container (a bucket is ideal) and stir *by hand*; we recommend the use of plastic gloves. Ensure that all aggregates are broken apart by pressing between fingers. This is an important part of the process and is likely to take at least 10 min.
3. Stir the soil/water solution well and then following agitation pour through a series of 750, 200, 100 and 50 μm sieves.
4. Collect what is retained on each sieve by back washing with distilled water (plastic bottles normally used for distilled water in the laboratory are ideal for this) and pour into a centrifuge tube. Make up to 50 ml volume with distilled water. Prepare one tube for material retained on each sieve i.e. 750, 200, 100 and 50 μm.
5. Carry out steps 6 onwards to each tube,
6. Centrifuge at 2500 rpm (swing out head) for 5 min,
7. Discard the supernatant and any organic material (which should be floating on the surface),
8. Re-suspend the pellet in 40% sucrose and then centrifuge at 2500 rpm for approximately 60 s (the precise time will need to be determined but this is a good guide),
9. Pour the supernatant on to a 50 μm filter and remove the sucrose (this is particularly important if you want to use the spores to inoculate plants) by washing through a large volume (approximately 1 l) distilled water,
10. Back wash as before to remove spores from the sieve and collect onto filter paper (we use nitrocellulose membranes, gridded when we wish to count),
11. Carefully inspect sieves at all stages of the procedure using a stereomicroscope to ensure no spores are lost. If necessary the procedures should be modified.
12. Spores should now be examined under a stereomicroscope and counted, if appropriate. If the aim is to provide identification then extreme care must be taken. Excellent advice can be obtained from the website of La Banque Europèenne des Glomales (BEG) at http://www.ukc.ac.uk/beg/ or the International Culture Collection of Arbuscular and Vesicular Arbuscular Mycorrhizal Fungi at http://invam.caf.wvu.edu/

Ectomycorrhizal fungi frequently cause quite dramatic changes to the morphology of colonised roots. Therefore, one of the most common measurements of EMF is to count the number of mycorrhizal tips or the length of mycorrhizal roots in a sample. This is relatively easy to do because the roots are often extensively modified, highly branched and are thicker than non-colonised roots. However, it is often necessary to spend some time looking closely at the morphology of the root system (examining roots under the microscope) before accurate measurements can be made. Different EMF can result in very different modifications, so some experience is necessary for making accurate measurements. Roots are then measured using the gridline intersect method and the number or length of mycorrhizal tips counted.

Isolated tips (which should be fresh, but can be preserved in FAA or gluteraldehyde) can then be mounted on slides in 0.1% cotton blue (w/v) in 10% lactophenol (w/v), for general observations on the extent of colonisation. Examination of the Hartig Net in particular often provides useful information as to the identity of the fungus involved (for further details see Ingleby et al. 1990).

Sections of ectomycorrhizal root tips can display mantle layers with different kinds of auto-fluorescence. Thickness of sections and required staining depend upon the information needed from study. Median longitudinal sections of ectomycorrhiza are more informative than cross-sections (Agerer 1991) because:

1. They include the ectomycorrhizal mantle from tip,
2. They clearly show the position of Hartig net initiation zone; and where it is not fully developed,
3. The orientation of the cortical cell is oblique, in cross-sections the two neighbouring cells are cut,
4. The shape of the cortical cells enveloped by a Hartig net is more characteristic in longitudinal sections than in cross sections.

Normarski differential interference contrast microscopy also gives good results in anatomical characterization of ectomycorrhiza. It shows definite mantle or rhizomorph layers without interference from hyphal layers beneath or above the focused level.

Sections of ectomycorrhiza embedded in LR white resin are reported to provide good results because of its very low viscosity, providing good infiltration of both symbionts (Massicotte et al. 1987). Because of its hydrophilic nature, it can be used in most histochemical procedures. The results with transmission electron microscopy may not be always satisfactory (Peterson 1991). One of the most useful techniques for demonstrating the Hartig net and mantle hyphae in LR white-embedded ectomycorrhiza is a simple fluorescence technique modified from Culling (1974). Observation with Normarski differential interference contrast optics gives particularly clear images of the Hartig net.

Nuclear magnetic resonance spectroscopy (NMR) can be applied for specialized studies of carbon, nitrogen and phosphorus metabolism by mycorrhizal fungi by using ^{13}C, ^{15}N and ^{31}P respectively. Martin (1991) has summarised the prospects of using NMR in ectomycorrhizal fungi. NMR allows the identification and quantification of multiple metabolites (amino acids, carbohydrates and organic acids) from the same spectrum and because of non-invasiveness, serial measurements of intact mycelium and roots are possible. High operational costs limit its use for routine investigations.

References

Abbot LK, Robson AD (1985) Formation of external hyphae in soil by four species of vesicular-arbuscular mycorrhizal fungi. New Phytol 99: 245–255

Abbot LK, Robson AD, De Boer G (1984) The effect of phosphorus on the formation of hyphae in soil by the vesicular-arbuscular mycorrhizal fungus *Glomus fasciculatum*. New Phytol 97: 437–446

Agerer R (1991) Characterization of ectomycorrhiza. In: Norris JR, Read DJ, Varma AK (eds) Methods in microbiology vol 23. Academic Press, London, pp 25–75

Altemüller HJ, van Vliet-Lanoe B (1990) Soil thin section fluorescence microscopy. In: Douglas LA (ed) Soil micromorphology. Elsevier, Amsterdam

Baluška F, Parker JS, Barlow PW (1992) Specific patterns of cortical and endoplasmatic microtubules associated with cell growth and tissue differentiation in roots of maize (*Zea mays* L.) J Cell Sci 103: 191–200

Berta G, Trotta A, Fusconi A, Hooker JE, Munro M, Atkinson D, Giovannetti M, Morini S, Fortuna P, Tisserant B, Gianinazzi-Pearson V, Gianinazzi S (1995) Arbuscular mycorrhizal induced changes to plant growth and root system morphology in *Prunus cerasifera*. Tree Physiol 15: 281–294

Bethlenfalvay GJ, Ames RN (1987) Comparison of two methods for quantifying extraradial mycelium of vesicular mycorrhizal fungi. Soil Sci Soc Am J 51: 834–837

Bevege DI (1968) A rapid technique for clearing tannins and staining intact roots for detection of mycorrhizas caused by *Endogone* spp. and some records of infection in Austral-asian plants. Trans Br Mycol Soc 51: 808–810

Bhuvaneswari TV, Solheim B (1985) Root hair deformation in the white clover/*Rhizobium trifolii* symbiosis. Physiol Plant 63: 25–34

Blancaflor EB, Hasenstein KH (1993) Organisation of cortical microtubules in graviresponding maize roots. Planta 191: 231–237

Blancaflor EB, Hasenstein KH (1997) The organization of the actin cytoskeleton in vertical and graviresponding primary roots of maize. Plant Physiol 113: 1447–1455

Boyde A, Maconnachie E (1981) Morphological correlations with dimensional change during SEM specimen preparation. Scanning Electron Microsc IV: 27–34

Brundrett M, Bougher N, Dell B, Grove T, Malajczuk N (1996) Working with mycorrhizas in forestry and agriculture. Australian Centre for International Agricultural Research, Canberra. ACIAR Monograph 32, 374 pp

Brundrett MC, Piche Y, Peterson RL (1984) A new method for observing the morphology of vesicular-arbuscular mycorrhizae. Can J Bot 62: 2128–2134

Brundrett MC, Enstone DE, Peterson CA (1988) A berberine-aniline blue staining procedure for suberin, lignin and callose in plant tissue. Protoplasma 146: 133–142

Caradus JR (1979) Selection for root hair length in white clover (*Trifolium repens* L). Euphytica 28: 489–494

Care D (1995) The effect of aluminium concentration on root hairs in white clover (*Trifolium repens* L.). Plant Soil 171: 159–162

Carlson H, Stenram U, Gustafsson M, Jansson HB (1991) Electron microscopy of barley root infection by the fungal pathogen *Bipolaris sorokiniana*. Can J Bot 69 (12): 2724–2731

Clark G (ed) (1981) Staining procedures, 4th edn. Williams and Wilkins, Baltimore, 512 pp

Culling DFA (1974) Modern microscopy: elementary theory and practice. Butterworths, London

Daft MJ, Nicolson TH (1966) Effect of *Endogone* mycorrhiza on plant growth. New Phytol 65: 343–350

Dhingra OD, Sinclair JB (1995) Basic plant pathology methods, 2nd edn. CRC, Boca Raton, 434 pp

Edwards HH, Yeh YY, Tamowski BI, Schonboum GR (1992) Acetonitrile as a substitute for ethanol/propylene oxide in tissue processing for transmission electron microscopy: comparison of fine structure and lipid solubility in mouse liver, kidney and intestine. Microsc Res Technique 21: 39–50

Egerton RF (1986) Electron energy loss spectroscopy in the electron microscope. Plenum Press, New York

Esau K (1965) Plant anatomy, 2nd edn. Wiley, New York

Esau K (1977) Anatomy of seed plants, 2nd edn. Wiley, New York

Fahn A (1990) Plant anatomy, 4th edn. Pergamon Press, Oxford

Fåhraeus G (1957) The infection of clover root hairs by nodule bacteria studied by a simple glass technique. J Genet Microbiol 16: 374–381

Fischer JMC, Peterson CA, Bols NC (1985) A new fluorescence test for cell vitality using Calcofluor white M2R. Stain Technol 60: 69–79

Fitzpatrick EA (1990) Roots in thin sections of soils. In: Douglas LA (ed) Soil micromorphology, vol 19. Elsevier, Amsterdam, pp 9–23

Föhse D, Jungk A (1983) Influence of phosphate and nitrate supply on root hair formation of rape, spinach and tomato plant. Plant Soil 74: 359–368

Gaff DF, Okong'o-ogola O (1971) The use of non-permeating pigments for testing the survival of cells. J Exp Bot 22: 756–758

Gahan PB (1984) Plant histochemistry and cytochemistry. An introduction. Academic Press, London, 301 pp

Gahoonia TS, Nielsen NE (1997) Variation in root hairs of barley cultivars doubled soil phosphorus uptake. Euphytica 98 (3): 177–182

Gahoonia TS, Nielsen NE (1998) Direct evidence on participation of root hairs in phosphorus (^{32}P) uptake from soil. Plant Soil 198: 147–152

Gahoonia TS, Care D, Nielsen NE (1997) Root hairs and acquisition of phosphorus by wheat and barley cultivars. Plant Soil 191: 181–188

Gallaud I (1905) Etudes sur les mycorrhizes endotrophes. Rev Genet Bot XVII: 123–127

Gardner RO (1975) An overview of botanical clearing technique. Stain Technol 50 (2): 99–105

Gerlach D (1984) Botanische Mikrotechnik, 3rd edn. Thieme, Stuttgart, 311 pp

Gerrits PO, Zuideveld R (1983) The influence of dehydration media and catalyst systems upon the enzyme activity of tissues embedded in 2-hydroxyethyl methacrylate. An evaluation of the three dehydration media and two catalyst systems. Mikroskopie 40: 321–328

Gerrits PO, van Leeuwen MBM, Boon ME, Kok LP (1987) Floating on a water bath and mounting glycol methacrylate and hydroxypropyl methacrylate sections influence final dimensions. J Microsc 145 (1): 107–113

Gianinazzi S, Gianinazzi-Pearson V (1992) Cytology, histochemistry and immunocytochemistry as tools for studying structure and function in endomycorrhiza. In: Methods in Microbiology, vol 24. Academic Press, London, pp 109–139

Giovannetti M, Mosse B (1980) An evaluation of techniques for measuring vesicular arbuscular mycorrhizal infection in roots. New Phytol 84: 489–500

Glauert AM (1975) Practical methods in electron microscopy. Fixation, dehydration and embedding of biological specimens. North Holland/American Elsevier, Amsterdam

Grace C, Stribley DP (1991) A safer procedure for routine staining of vesicular-arbuscular mycorrhizal fungi. Mycol Res 95: 1160–1162

Graham JH, Linderman RG, Menge JA (1982) Development of external hyphae by different isolates of mycorrhizal *Glomus* spp. in relation to root colonization and growth of Troyer citrange. New Phytol 91: 183–189

Guttenberg Hv (1968) Der primäre Bau der Angiospermenwurzel. Handbuch der Pflanzenanatomie VIII, 5. Gebrüder Bornträger, Berlin

Hahn A, Gianinazzi-Pearson V, Hock B (1994) Characterisation of arbuscular mycorrhizal fungi by immunochemical methods. In: Gianinazzi S, Schüepp H (eds) Impact of arbuscular mycorrhizas on sustainable agriculture and natural ecosystems. Birkhäuser, Basel, pp 25–39

Hamel C, Fyles H, Smith DL (1990) Measurements of development of endomycorrhizal mycelium using three different vital stains. New Phytol 115: 297–302

Heidstra R, Geurts R, Franssen H, Spaink, van Kammen, Bisseling T (1994) Root hair deformation activity of nodulation factors and their fate on *Vicia sativa*. Plant Physiol 105: 787–797

Herr JM Jr (1971) A new clearing-squash technique for the study of ovule development in angiosperms. Am J Bot 58: 785–790

Hooker JE, Munro M, Atkinson D (1992) Vesicular- arbuscular fungi induced alteration in poplar root system morphology. Plant Soil 145: 207–214

Hooker JE, Berta G, Lingua G, Fusconi A, Sgorbati S (1998) Quantification of AMF induced modifications to root system architecture and longevity. In: Varma A (ed) Mycorrhizal methods. Springier, Berlin Heidelberg New York

Ingleby K, Mason PA, Last FT, Fleming LV (1990) Identification of ectomycorrhizas Institute of Terrestrial Ecology, Publ No 5. HMSO, London

Jenny H, Grossenbacher K (1963) Root-soil boundary zones as seen in the electron microscope. Soil Sci Soc Am Proc 27: 273–277

Jensen WA (1962) Botanical histochemistry. WH Freeman, San Fransisco

Johansen DA (1940) Plant microtechnique. McGraw-Hill, New York

Jones DL, Shaff JE, Kochian L (1995) Role of calcium and other ions in directing root hair tip growth in *Limnobium stoloniferum*. I. Inhibition of tip growth by aluminium. Planta 197 (4): 672–680

Knox RB (1970) Freeze-sectioning of plant tissues. Stain Technol 45: 265–272

Kormanik PP, McGraw AC (1982) Quantification of vesicular-arbuscular mycorrhizae in plant roots. In: Schenck NC (ed) Methods and principles of mycorrhizal research. American Phytopathological Society, St Paul, Minnesota, pp 37–45

Koske RG, Gemma JN (1989) A modified procedure for staining roots to detect VA mycorrhizas. Mycol Res 92: 486–505

Kough JL, Linderman RG (1986) Monitoring extra-matrical hyphae of a vesicular-arbuscular mycorrhizal fungus with an immunofluorescence assay and the soil aggregation technique. Soil Biol Biochem 18: 309–313

Kuck KH, Tiburzy R, Hänssler G, Reisener HJ (1981) Visualization of rust haustoria in wheat leaves by using fluorochromes. Physiol Plant Pathol 19: 439–441

Lamont B (1983) Root hair dimensions and surface/volume/weight ratios of roots with the aid of scanning electron microscopy. Plant Soil 74: 149–152

Lillie RD (1977) H.J. Conn's biological stains, 9th edn. Williams and Wilkins, Baltimore, 692 pp

Lindauer R (1972) Die Technik des Handschnittes. Mikrokosmos 61: 144–151

Lund ZF, Beals HO (1965) A technique for making thin sections of soil with roots in place. Soil Sci Soc Am Proc 29: 633–635

Luxová M, Ciamporová M (1989) Root structure. In: Kolek J, Kozinka V (eds) Physiology of the Plant Root System. Kluwer Academic Publishers, Dordrecht, pp 31–81

Lyon H (ed) (1991) Theory and strategy in histochemistry. Springer, Berlin Heidelberg New York

Lyshede OB (1977) A method for removing starch from plant tissue with bacterial amylase. Mikroskopie (Wien) 33: 241–245

Lyshede OB (1979) Effect of bacterial amylase on the ultrastructure of potato tuber storage cells. Mikroskopie 35: 314–318

MacKay AD, Barber SA (1984) Effect of soil moisture and phosphate level on root hair growth of corn roots. Plant Soil 86: 321–331

Marschner H (1995) Mineral nutrition of higher plants. Academic Press, London

Martin FM (1991) Nuclear magnetic resonance studies in ectomycorrhizal fungi. In: Norris JR, Read DJ, Varma AK (eds) Methods in microbiology, vol 23. Academic Press, London, pp 121–148

Massicotte HB, Melville LH, Peterson RL (1987) Scanning electron microscopy of ectomycorrhizae. Potentials and limitations. Scanning Microsc 1: 1439–1454

Mauseth JD (1988) Plant anatomy. Benjamin Cummings, Menlo Park

McCully M (1995) Water efflux from the surface of field grown grass roots. Observations by cryo-scanning electron microscopy. Physiol Plant 95: 217–224

Mosse B, Hepper C (1975) Vesicular-arbuscular mycorrhizal infections in root organ cultures. Physiol Plant Pathol 5: 215–223

Neergaard E de (1997) Methods in botanical histopathology. Danish Institute of Seed Pathology for Developing Countries, Copenhagen, 216 pp

Norenburg JL, Barrett JM (1987) Steedman's wax embedment and de-embedment for combined light and scanning electron microscopy. J Electron Microsc Technol 6: 35–41

Norris JR, Read DJ, Varma AK (eds) (1991a) Techniques for the study of mycorrhiza. In: Methods in microbiology, vol 23. Academic Press, London

Norris JR, Read DJ, Varma AK (eds) (1991b) Techniques for the study of mycorrhiza. In: Methods in microbiology, vol 24. Academic Press, London

O'Brien DG, McNaughton EJ (1928) The endotrophic mycorrhiza of strawberries and its significance. Research Bulletin 1. The West of Scotland College of Agriculture. Edinburgh, 35 pp

O'Brien TP, McCully M (1981) The study of plant structure. Principles and selected methods. Termarcarphi Pty, Melbourne, Australia

O'Brien TP, von Teichman I (1974) Autoclaving as an aid in the clearing of plant specimens. Stain Technol 49: 175–176

Oprisko MJ, Green RL, Beard JB, Gates CE (1990) Vital staining of root hairs in 12 warm season perennial grasses. Crop Sci 30: 947–950

Peterson LR (1991) Histochemistry of ectomycorrhiza. In: Norris JR, Read DJ, Varma AK (eds) Methods in microbiology, vol 23. Academic Press, London, pp 107–120

Peterson LR, Farquhar ML (1996) Root hairs: Specialized tubular cells extending root surfaces. Bot Rev 62: 1–40

Philips JM, Hayman DS (1970) Improved procedures for cleaning roots and staining parasitic and vesicular-arbuscular mycorrhizal fungi for rapid assessment of infection. Trans Br Mycol Soc 55: 158–161

Purvis MJ, Collier DC, Walls D (1964) Laboratory techniques in botany. Butterworths, London

Redhead JF (1977) Endotrophic mycorrhizas in Nigeria: species of the *Endogonaceae* and their distribution. Trans Br Mycol Soc 69: 275–280

Robards AW, Wilson AJ (eds) (1993) Procedures in electron microscopy. John Wiley, Chichester

Sass JE (1961) Botanical microtechnique. The Iowa State University Press, Ames

Schaffer GF, Peterson RL (1993) Modifications to clearing methods used in combination with vital staining of roots colonized with vesicular-arbuscular mycorrhizal fungi. Mycorrhiza 4: 29–35

Schüepp H, Miller DD, Bodman M (1987) A new technique for monitoring hyphal growth of vesicular-arbuscular mycorrhizal fungi through soil. Trans Br Mycol Soc 89: 429–435

Sieverding E (1991) Vesicular-arbuscular mycorrhizal management in tropical ecosystems Technical Cooperation, Eschborn, Germany

Smith SE, Read DJ (1997) Mycorrhizal symbiosis. Academic Press, San Diego

Spurr AR (1969) A low-viscosity epoxy resin embedding medium for electron microscopy. J Ultrastruct Res 26: 31–43

Strausbaugh CA, Murray TD (1989) Use of epidermal cell responses to evaluate resistance of winter wheat cultivars to *Pseudocer cosporella herpatrichoides*. Phytopathology 79 (10): 1043–1047

Tippkötter R, Ritz K, Darbyshire JF (1986) The preparation of thin sections for biological studies. J Soil Sci 37: 681–690

Tisserant B, Gianinazzi-Pearson V, Gianinazzi S, Gollotte A (1993) In planta histochemical staining of fungal alkaline phosphatase activity for analysis of efficient arbuscular mycorrhizal infections. Mycol Res 97: 245–250

Trolldenier G (1965) Fluoreszenzmikroskopie in der Rhizosphärenforschung. ZEISS-Inf 56: 68–69

Vilarino A, Arines J, Schuepp H (1993) Extraction of vesicular-arbuscular mycorrhizal mycelium from sand samples. Soil Biol Biochem 25: 99–100

Watt M, van der Weele CM, McCully ME, Canny MJ (1996) Effects on local variations in soil moisture on hydrophobic deposits and dye diffusion in corn roots. Bot Acta 109: 492–501

Watteau F, Villemin G, Mansot JL, Ghanbaja J, Touain F (1996) Localization and characterization of brown cellular substances of beech roots by electron energy loss spectroscopy. Soil Biol Biochem 28: 1327–1332

CHAPTER 3

Control and Measurement of the Physical Environment in Root Growth Experiments

W.R. Whalley[1], J. Lipiec[2], W. Stępniewski[2,3], and F. Tardieu[4]

[1] Silsoe Research Institute, Wrest Park, Silsoe, Bedford MK45 4HS, UK
[2] Polish Academy of Sciences, Institute of Agrophysics, Doświadczalna 4, 20-290 Lublin 27, Poland
[3] Technical University of Lublin, 40 Nadbystrzycka Str. 20-618 Lublin, Poland
[4] INRA, Laboratoire d'Ecophysiologie, 34060 Montpellier, France

CONTENTS

A.L. Smit et al. (Eds.), Root Methods

3.1 Introduction

Analyses of the responses of root growth rate to the plant environment in the last 10 years have greatly increased our understanding of the main mechanisms which control root growth and development, and have led to the analysis of root system architecture. However, we currently lack a global approach to plant growth in varying environments, including growth of the root system. This leaves a large, and nearly empty 'ecological niche' for studies carried out at the whole plant level, which analyse root system growth as a function of the environment. This chapter provides guidance on how to manipulate the rooting environment.

Mechanisms which control root elongation at the root tip level have been analysed in detail by several groups. We have methods to analyse cell division rate and expansion spatially within the growing zone of a root (Sharp et al. 1988; Fraser et al. 1990), thereby allowing us to analyse the effect of environmental conditions on these processes. Seminal roots are probably the tissues most frequently studied for responses of cell elongation to temperature, water deficit and salinity. These studies all lead to the conclusion that the mechanical properties of cell walls largely determine the response of root growth to environmental conditions such as temperature (Pritchard et al. 1990) or water deficit (Saab et al. 1992). Proteins involved in the process of elongation are starting to be identified (e.g. Saab et al. 1995; Cosgrove 1993). Turgor of growing cells, which is the driving force for elongation, tends to be buffered against the same environmental conditions (Spollen and Sharp 1991 for water deficit, Pritchard et al. 1990 for temperature) and increases in the case of mechanical impedance (Clark et al. 1996). Overall, these studies suggest that changes in elongation rate are a primary process with their own control and can in many cases be seen as a cause, rather than a consequence, of changes in local concentrations of solutes (Sharp et al. 1990), especially in carbohydrates (in the case of mechanical impedance, see Atwell 1990). These studies, carried out at the single root level, now need to be extended to the whole plant level, thereby integrating processes over longer timescales, and taking into account the branching rate and branch abortion as constitutive mechanisms of root development.

Analyses and models of root system architecture have been developed, and now give a reasonable approximation of root branching, root development rate and 3-dimensional root geometry (e.g. Pellerin and Pages 1994; Jourdan et al. 1995) in cases without appreciable environmental constraints. However, the effect of environmental changes on the parameters of these models has received little attention (e.g. Aguirrezabal and Tardieu 1996). Systematic analyses of the effect of root environment on architecture are crucial for a better prediction of root system growth under variable and fluctuating environments.

This chapter deals with the methods available for these studies at the whole plant level. They essentially deal with the manipulation of water deficit, mechanical stress and oxygen stress under laboratory conditions, and with the measurement of these variables in the field, where they usually cannot be manipulated as easily.

Questions arising on control and manipulation of the environment in root studies:

1. Is the control of root growth local (i.e. depending on the conditions of a few mm around the root) or integrated (i.e. depending on the conditions experienced by the plant as a whole)?
   This question is crucial for determining the relevant measurements of the plant environment in growth studies. A root system does not usually sense uniform conditions over the whole soil profile. Under common water deficits, some roots experience a severe water deficit, while deeper roots still grow in soil at near field capacity. To what extent do unfavourable conditions sensed by part of the root system, or by shoots, affect the growth of roots subjected to favourable conditions? In addition to their role in water and nutrient uptake, roots act as sensors and can send chemical or hydraulic messages to the rest of the plant (see Davies et al. 1994). This is certainly the case for water deficit, during which roots synthesise abscisic acid which is translocated in the plant and affects growth of distant organs. It has been argued (Tardieu 1994) that this is also the case for mechanical impedance: the changes in characteristics of the whole root system cannot be accounted for by local effects alone. It is therefore relevant to consider the effect of local changes in the root environment on the growth of the whole plant, both roots and shoots. Split root techniques, in which part of the root system is subjected to a constraint while the rest of it is not, are a powerful tool for integrated analyses, and are described in this chapter for several environmental conditions. A careful analysis of field experiments, where 'split roots' develop naturally because of uneven soil depletion, can also be very useful, provided that accurate measurements of soil physical conditions are carried out in the whole soil profile.
2. To what extent does root growth depend on shoot development?
   Roots depend on shoots for carbohydrate supply. Several studies suggest that root development can vary appreciably even in constant and favourable soil conditions, when the availability of carbohydrates to roots varies. This is the case when the intercepted radiation varies, between days or between growing seasons (Aguirrezabal et al. 1994), in plants with periodic shoot growth (Thaler and Pages 1996) or if carbon dioxide concentration changes in the air (Idso and Kimball 1992). Changes in leaf area and intercepted light are, therefore, a crucial variable for analysing root growth. This is obviously the

case when air temperature or radiation varies among experiments, but may also be the case when an adverse condition sensed at the root level alters leaf growth (e.g. Beemster et al. 1996 for soil compaction; Ben Haj Salah and Tardieu 1996 for water deficit). In this case, it is essential to consider light interception as an integrating mechanism at the plant level. Adverse conditions sensed by roots can affect leaf growth, which in turn affects assimilate availability to roots and root growth.

We have not presented methods for studying light interception in this chapter as this is a common method in ecophysiology, but we consider it a crucial measurement to be taken into account in root studies, especially in the field.

3. What spatial and temporal resolution of environmental conditions is acceptable in a study of root growth?
Due to the low conductivity of the soil to fluxes of water, heat, oxygen, etc., local conditions are often variable. To what extent is it necessary to characterise this variability, and to what extent is it acceptable to measure mean conditions in a soil volume? Physical processes in the soil are frequently non-linear (e.g. responses of soil water potential and hydraulic conductivity to soil water content), and local conditions sensed by a part of the root system can affect the growth of the whole plant. In this case, measuring mean environmental conditions may not be adequate, depending on the problem under study. It is therefore safe, especially in the field, to combine local measurements (e.g. given by tensiometers or electrodes) with more integrated measurements (such as pre-dawn leaf water potential for water deficit). The location of sensors, in relation to the studied phenomenon, and the choice of integrated vs. local measurements, are frequently causes of success or failure of analyses. We give suggestions in this chapter for taking appropriate decisions.

3.2 Laboratory Control of the Growth Environment

3.2.1 Control of Water Stress

The control of water stress has been reviewed by Krizek (1985), where the various options for the experimental control of water stress are described. They are the regulation of water supply; the control of the matric potential of the soil water; and the use of osmotica.

Before consideration of these approaches in more detail, it is pertinent to consider the conclusions of Krizek (1985). He considered that withholding water was the most natural, but gave results that were difficult to interpret. In contrast, the use of osmotica gave more precise control of water potential, but care should be taken to avoid the possibility of toxic side effects. However, since Krizek (1985) wrote his review, the control of soil matric potential has been developed further. High molecular weight polyethylene glycol, PEG (e.g. 20 000) is relatively common in current literature, in comparison with the widespread use of PEG (6000) as described by Krizek (1985).

3.2.1.1 Control of Water Supply

Controlling water supply by either reducing the frequency of watering, withholding water altogether or by a controlled irrigation is the simplest method of applying water stress to plants growing in pots. However, the results depend on soil type and the size of container, both of which influence the amount of water available to the plant.

Simplistically, a large pot containing clay soil will result in a gradual increase in water stress while a small pot containing sandy soil will result in a rapid onset of water stress. A significant disadvantage of the method is that it is not possible to determine the potential at the root-soil interface because water is continually being extracted.

The choice of soil type or alternative growth medium is an important consideration. Schwaegerle (1983) observed that this water withholding method should only be used with soils with large unsaturated conductivities and with plants with low water requirement, if a uniform water content is to be achieved. It is possible to use other growth media; Sharp et al. (1988) equilibrated vermiculite at potentials between –0.03 and –1.60 MPa. The water release characteristics of the vermiculite used by Sharp et al. (1988) are shown in Table 3.1. From this data it can be seen that a small change in water content can lead to a large change in potential. For example, as the vermiculite dries from 0.15 to 0.08 $g\,g^{-1}$ the water potential decreases by a factor of approximately 2. Therefore the uptake of only a small amount of water can have a large effect on matric potential at the root surface. In the experiment of Sharp et al. (1988) the plants were in a water-saturated environment so that transpiration was negligible.

Table 3.1. Water release characteristics of vermiculite used by Sharp et al. (1988) in root growth experiments

Water content $g\,g^{-1}$	Water potential kPa
3.72	–30
0.6	–200
0.15	–810
0.08	–1600

3.2.1.2 Control of Soil Matric Potential

The requirement for subjecting plants to water stress in a controlled way whilst avoiding possible toxic effects of osmotica, has led to the development of sophisticated methods for controlling soil matric potential. There are two principal methods of controlling water stress, but in each case the soil must have a large hydraulic conductivity in order to be certain that the potential at the soil-root interface is the same as that of the bulk soil. This requirement restricts the use of this technique to relatively coarse soils at high water potentials (usually no lower than –20 kPa), if accurate control of the potential at the soil-root interface is required.

BOX 3.1

An experimental system suitable for controlling matric potential is shown in Fig. 3.1. This represents a refinement of the system described by Snow and Tingey (1985) in which the growth medium was maintained above a fixed water table. In the apparatus of Lipiec et al. (1988) ceramic tubes are connected to a water reservoir in a closed loop. The water reservoir is a sealed vessel and can be subjected to a pressure below atmospheric. The pump circulates water to prevent air collecting in the ceramic tubes, which would seriously limit the performance of the apparatus. The water uptake by roots can be estimated directly from the level in the reservoir, provided the system is initially in equilibrium and there is no change in water storage. Taps can be closed to replenish the water reservoir.

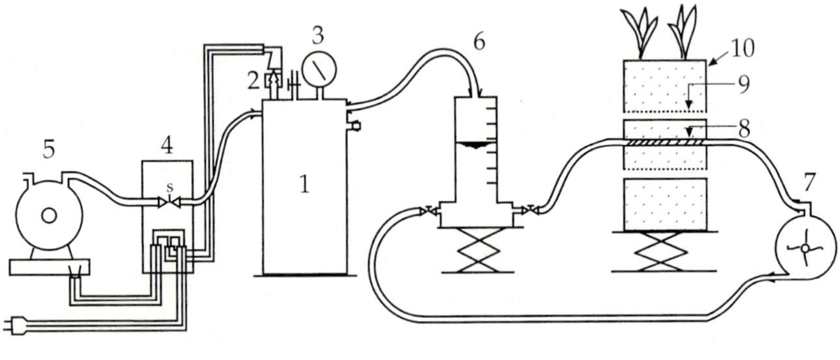

Fig. 3.1. An experimental system for controlling matric potential of soil (*9*). Water in reservoir (*6*) is held under a negative pressure and recirculated using a pump (*7*) through ceramic tubes (*8*) in contact with the soil, which operate in a similar way to tensiometers. The pressure in the reservoir (*6*) is controlled by an automatic pump and pressure control device (s, 1, 2, 3, 4 and 5). (From Lipiec et al. 1988)

Experimental systems as shown in Fig. 3.1 (Box 3.1) have been found to be useful for studying root infection by pathogens. Iwama et al. (1994) used a similar system to study the effect of water potential on the infection of turnips (*Brassica rapa* L.) with *Plasmodiophora brassicae* which causes club root disease. They found that for the soil they used (humic volcanic ash) a potential greater than a critical value of about −11.2 kPa resulted in the initiation of club root disease. The critical potential was thought to be related to pore size which dictates the upper limit of the size of mobile spores.

An analogous experimental system is described by Painter (1966). In this system the soil is packed between two dialysis membranes which are then surrounded by PEG (20 000). With this approach it is theoretically possible to extend the range of potentials to below the −85 kPa limit (i.e. tensiometer limit) with the system described by Lipiec et al. (1988). However, care should be taken to ensure that steep gradients in water potential do not result in more negative matric potentials at the root surface.

Hsieh et al. (1972) described how the surface of a soil connected to a water table could be covered with a fine mesh with openings of 28–30 μm, to enable water uptake by root hairs to be quantified. They grew bean (*Phaseolus vulgaris* L.), maize (*Zea mays* L.) and barley (*Hordeum vulgare* L.) on the surface of the mesh with no apparent ill-effects. They used gamma attenuation to obtain water content profiles with a resolution of 1 mm. Fig. 3.2 shows a time sequence of potential profiles (estimated from water release curves) following the disconnection of the soil water table. This diagram gives a clear illustration of the uncertainties of the potential at the soil-root interface when the water-withholding method is used to apply water stress.

3.2.1.3 The Use of Osmotic Solutions

For convenience, osmotic solutions are widely used to subject plants to water stress. The osmotic agents most widely used are polyethylene glycol (PEG) and mannitol. Caution should always be exercised when osmotic agents are used, because it is necessary to demonstrate that any effect is caused by water stress and not toxicity or an other property of the osmoticum. PEG is available commonly in molecular weights between 200 and 20 000. Lawlor (1970) made a detailed study of the problems associated with PEG and mannitol solutions and, as far as we are aware, it is still the most comprehensive. PEG has been criticised because of the impurities that certain grades sometimes contain. Lawlor (1970) concluded that this is not a serious problem, although Janes (1974) observed that the toxicity of PEG 4000 can differ between different batches.

Figure 3.3 shows the effect of root damage on PEGs entry into the root for PEGs of different molecular weights. When undamaged roots were placed in

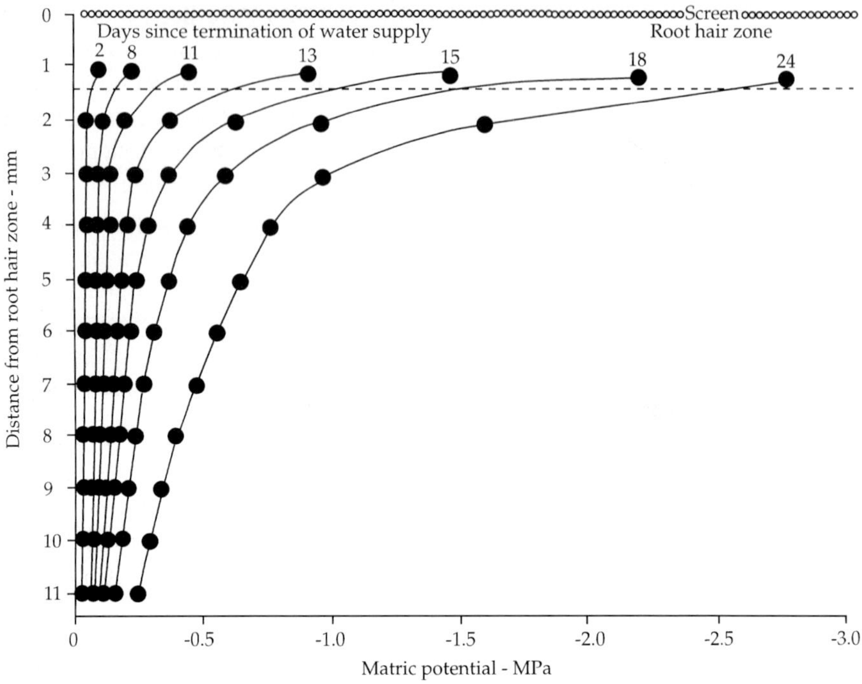

Fig. 3.2. Matric potential as a function of distance from the root hair screen during the growth of maize. Parameter is the time elapsed since the termination of water supply. (From Hsieh et al. 1972)

PEG at −200 kPa, little PEG entry occurred for molecular weights of 1000, 4000 and 20 000, while in damaged roots, PEG 1000 and 4000 entered the root. At a potential of −1.0 MPa, even in damaged roots, the entry of PEG 20 000 was only small. Lawlor (1970) observed that PEG entry was proportional to the amount of damage, which is presumably likely to occur in ageing annual plants subjected to mechanical impedance or in perennial plants due to increasing root permeability.

Both PEG 200 and mannitol can enter undamaged roots, but they have different effects on the leaf compared with higher molecular weight PEG. Zhu and Boyer (1992) demonstrated that for giant algal cells, mannitol disrupts the normal process of cell wall metabolism. The use of mannitol to stress roots has produced some convincing results (e.g. Pritchard et al. 1990). However, it is uncertain whether some other metabolic effect of water stress or the effect of mannitol is being observed.

Mexal et al. (1975) have demonstrated that oxygen solubility is decreased in PEG and the solubility decreases with increasing PEG concentration and increasing molecular weight. They considered that this problem was sufficiently

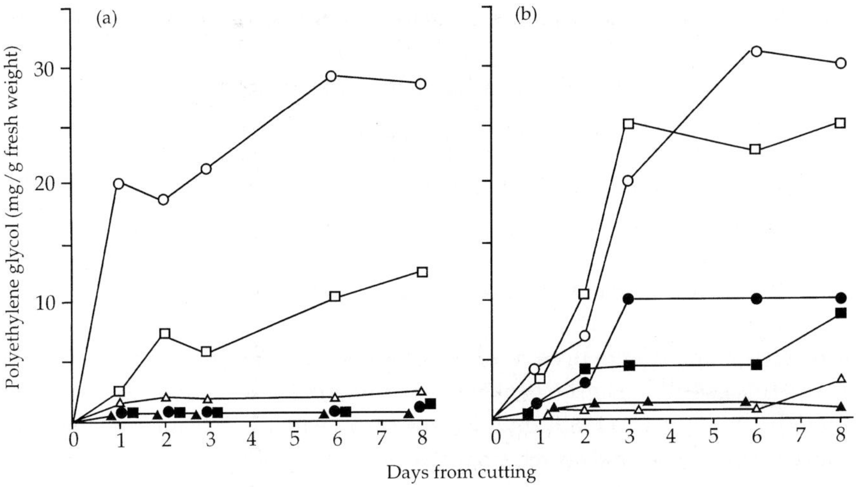

Fig. 3.3. The effect of cutting the root systems of cotton plants on the entry of different molecular weights from **a** −0.2 MPa and **b** −1.0 MPa PEG solutions. Cut roots: *open triangles* 20 000, *open square* 4000, *open circle* 1000. Uncut roots: *closed triangle* 20 000, *closed square* 4000, *closed circle* 1000. (From Lawlor 1970)

serious that plants growing in nutrient solution containing PEG could suffer from oxygen stress. Mexal et al. (1975) suggested that solutions of PEG should be aerated which might allow oxygen to reach the roots directly in a gaseous form. Verslues et al. (1998) have made a detailed study of oxygen concentrations around roots growing in PEG (8000) aerated with gas with an oxygen partial pressure of 20.8 kPa, 30.4 kPa or 45.9 kPa. From measurements of oxygen concentration profiles around the surface of roots, they concluded that PEG should be equilibrated with a gas with a higher than ambient oxygen partial pressure to give root surface oxygen partial pressures equivalent to roots grown in moist vermiculite. Root elongation rate was observed to be higher when PEG was aerated with gases containing supplemental oxygen and this effect was particularly noticeable at water potentials higher than −800 kPa. By making comparisons between the elongation rates of roots growing in vermiculite equilibrated at various potentials and PEG, Verslues et al. (1998) concluded that similar elongation rates could not be taken as evidence against hypoxia. This was because the elongation rate of roots growing in vermiculite was lower than that of those growing in PEG aerated with gas with an ambient oxygen partial pressure at the same water potential. Thus comparisons between the effects of water stress induced by PEG and other methods (e.g. equilibrated vermiculite) must be made with caution. However, Verslues et al. (1998) do conclude that PEG can be used without the complication of oxygen stress if the solution is aerated with a gas containing supplemental oxygen. The work of Verslues et al.

(1998) is an important reference for those intending to induce water stress using PEG.

The water potentials of PEG and other osmotica are commonly measured with a psychrometer. The calibration of concentration against potential is dependent both on the temperature of the solution and on the molecular weight of the particular osmotic agent. There are several published calibrations for PEG (e.g. Malcolm and Rawlinson 1957; Williams and Shaykewich 1969; Steuter et al. 1981; Materechera et al. 1992), and a description of the technique for using a psychrometer is given by Mullins (1991).

Osmotic solutions are commonly used in hydroponic systems. These can be relatively simple as in the case of Gergely et al. (1980) who grew apple (*Malus sylvestris* L.) seedlings in 1 ℓ beakers containing mixtures of PEG and nutrient solutions. West et al. (1980) grew ornamental plants in 500 ml containers of coarse washed sand, which were irrigated with PEG solutions. When plants are transpiring, care must be taken to ensure that the concentration of PEG does not change due to water uptake by roots. Materechera et al. (1992) grew seedlings of several monocotyledons and dicotyledons in PEG 20 000, by mixing sand with PEG and placing the mixture on a sintered glass plate at a matric potential of –2.5 kPa. They used PEG with osmotic potentials of –0.25, –0.50 and –1.00 MPa, in addition to water at –2.5 kPa. The matric potential was applied to control the air-filled porosity of the PEG-sand-air mixture. For all the dicotyledon species the root diameter was thicker under –1.00 MPa of water stress than under –2.5 kPa. In contrast, for monocotyledons, the root diameter was thicker in the sand-water mixture at –2.5 kPa. At –1.00 MPa the elongation rate of the roots was less than 5% of the elongation rate of unstressed roots.

Čiamporová and Luxová (1976) found that at osmotic potentials low enough to stop elongation in maize, radial enlargement occurred. They found that exposure to PEG 4000 over 24 h caused reversible effects, while a 48-h exposure resulted in irreversible changes. PEG is widely used in germination studies. For example, Dahal and Bradford (1994) used PEG (8000) to study the effect of water stress on tomato (*Lycopersicon esculentum* L.) seed germination.

To overcome many of the problems with using PEG, such as possible oxygen stress (Mexal et al. 1975; Verslues et al. 1998) and reduced P uptake (Emmert 1974), Tingey and Stockwell (1977) proposed the use of a semi-permeable membrane to separate the elongating roots from the PEG. Tingey and Stockwell (1977) claimed that the use of PEG (20 000), combined with a membrane with a molecular weight exclusion limit of 6000 to 8000, reduced many problems associated with use of PEG. Roots of beans were grown in vermiculite sandwiched against the semi-permeable membrane. Table 3.2 shows how the

Table 3.2. Influence of various membrane types and thickness on leaf water potential and the PEG concentration in the root-vermiculite mass. (From Tingey and Stockwell 1977)

Membrane combinations[a]	Leaf water potential kPa	PEG 20 000 content mg/g Fresh wt[b]
Single dialysis membrane	–770	3.8
Double dialysis membranes	–840	2
Triple dialysis membranes	–980	0.9
Spectrapor[c]	–790	0.4

[a] Minimum molecular weight exclusion limits for dialysis membrane and Spectrapor are 12–14 000 and 6–8000 respectively.

[b] Expressed as g fresh wt. of the combined root and vermiculite mixture.

[c] The PEG was extracted from the root-vermiculite mass of plants grown in the membrane systems for 7 days at –600 kPa. Each mean is based on seven observations. SE = 0.3 for both leaf water potential and PEG content.

exclusion limit of the membrane affects PEG content of the root-vermiculite mass.

3.2.1.4 Applying Water Stress to Split Root Systems

Split root systems for studying the effect of water stress on root and plant growth can be relatively simple. The work of Coutts (1982) is a good example of a split root experiment. Coutts (1982) described how the roots of a 2-year-old sitka spruce (*Picea sitchensis* L.) seedling were divided between two plastic tubes, 605 mm tall and 109 mm in diameter. The tubes were packed with sandy loam soil equilibrated to given matric potentials. An example of the treatment Coutts used is as follows. Both sides watered to –6 kPa (determined using tensiometers), one side watered to –6 kPa and the second side kept to –60 kPa or both sides kept at –60 kPa.

The vertical variation in soil water potential is usually much greater than the lateral variation. Sommer (1988) describes how it is possible to isolate compartments in a soil monolith, with the water potential in each compartment being controlled by a system similar to that shown in Fig. 3.1 and described by Lipiec et al. (1988). Water movement between adjacent compartments at different water potentials was prevented with a 2–3 mm thick wax layer between compartments. A similar experimental system has been used by Vetterlein and Marschner (1993) to study hydraulic lift in sandy soil planted with pearl millet. Vetterlein and Marschner (1993) used two compartments, separated by a 2 cm perlite layer, to prevent capillary flow of water. The lower layer was wetted by capillary movement of water and the upper layer was not watered.

3.2.2 Control of Mechanical Stress

3.2.2.1 *Controlling Mechanical Impedance at Non-Limiting Stress Levels*

When studying the effect of mechanical stress on soil, the principal experimental objective is to change soil strength independently of soil aeration, water potential and hydraulic conductivity. Voorhees et al. (1975) show that, when soil is used as a growth medium for roots, the range of water contents and bulk densities used must be carefully selected to avoid poor aeration.

Table 3.3 gives the elongation rate of pea seedling seminal roots in relation to bulk density, water potential, air-filled porosity and penetrometer resistance. Any credible experiment with compressible soil which is designed to study the

Table 3.3. Primary root elongation rates, probe measurements, and porosity characteristics of a Sverdrup sandy loam. (From Voorhees et al. 1975)

Soil core bulk density $g\,cm^{-3}$	Primary root elongation rate $cm\,hr^{-1}$	Total point resistance, σ_p MPa		Normal point resistance, σ_p MPa		Volumetric water content	Air filled porosity
		$\alpha = 5°$	$\alpha = 30°$	$\alpha = 5°$	$\alpha = 30°$		
Water potential – 33 kPa							
1.23	0.062	1.62	1.56	0.39	1.09	0.23	0.31
1.24	0.076	1.09	1.05	0.25	0.74	0.24	0.29
1.25	0.079	1.04	1.00	0.24	0.70	0.22	0.31
1.29	0.080	1.00	0.96	0.23	0.67	0.25	0.26
1.38	0.065	1.50	1.44	0.36	1.01	0.27	0.21
1.45	0.050	2.24	2.15	0.56	1.50	0.31	0.15
1.54	0.032	4.48	4.30	1.18	3.01	0.33	0.08
1.61	0.017	6.76	6.44	1.81	4.54	0.30	0.09
Water potential – 100 kPa							
1.23	0.073	1.56	1.50	0.38	1.05	0.20	0.34
1.25	0.084	0.99	0.95	0.22	0.66	0.19	0.34
1.29	0.075	1.10	1.06	0.26	0.74	0.20	0.32
1.38	0.055	1.98	1.90	0.49	1.33	0.21	0.27
1.54	0.029	5.10	4.90	1.40	3.43	0.24	0.18
1.61	0.017	7.55	7.25	2.08	5.08	0.27	0.12

effect of mechanical stress on root elongation should include such data. Voorhees et al. (1975) used two types of penetrometer, both conical, but one with a 5° semi-angle and one with a 30° semi-angle. As discussed by Bengough (1991) and Greacen et al. (1968) the use of a 5° penetrometer allows the pressure between the root and soil to be estimated. This is because a sharp penetrometer fails soil in a cylindrical failure zone which is thought to be similar to that caused by roots. The use of penetrometers to estimate resistance to root elongation is discussed in detail by Bengough and Mullins (1990, 1991) and Richards and Greacen (1986).

Lipiec et al. (1988) described how it is possible to pack clay soil to a range of densities and to use the system for controlling matric potential shown in Fig. 3.1, to ensure adequate aeration and water supply.

Abdalla et al. (1969) described how a triaxial cell can be used to apply mechanical impedance to roots. In this system, the soil is formed into a cylinder and placed inside a thin cylindrical rubber membrane to which an external confining pressure is applied. A similar approach is described by Barley (1962) in which the root was pressed against a porous plate with a pressurized diaphragm. Gordon et al. (1992) developed a specialised cell which allowed monitoring of the impeded root by time-lapse photography. However, in general, triaxial cells are a specialist apparatus used in soil mechanics and are only suitable for small seedling roots. A similar approach is described by Scott-Russell and Goss (1974). In their system, a cell with flat sides was used that contained a layer of ballotini a few centimetres thick confined within a flexible membrane. The ballotini were bathed in aerated nutrient solution and compressed by an external confining pressure. As Scott-Russell and Goss (1974) note, a major advantage of all these pressurized systems is that they allow the effect of mechanical stress per se to be studied. To relate the external confining stress to the pressure a root needs to exert to deform the ballotini, the reader should refer to Richards and Greacen (1986) and Bengough and Mullins (1991).

Materechera et al. (1992) used a static load placed on an incompressible sand to apply a confining stress to the root. Care must be taken to minimise friction at the sand/pot interface, by lining the pot with a low friction PTFE plastic, if uniform stresses and repeatable results are to be obtained. This approach is attractive because it is relatively straightforward to implement, while providing a similar degree of mechanical stress control to the triaxial apparatus.

To separate out the effects of mechanical impedance from those of oxygen stress and water stress, Gardner and Danielson (1964) used wax of various strengths to form layers of known penetration resistance. The degree of aeration and water potential can thus be controlled independently of the strength of the impeding layer. This work provides an excellent example of the care necessary to isolate the effects of mechanical impedance. The wax layer method for generating controlled impedances has been used more recently by Yu et al. (1995) as a screening technique for rice. In their paper they describe in detail how to modify the strength of the wax layers.

Recently, new developments have been described by Bengough and Mackenzie (1994) and Bengough et al. (1994) which allow the root growth force to be controlled whilst the elongation rate is measured simultaneously. The system they describe applied an axial force to the root tip and resulted in slowing of root elongation under surprisingly small loads. Further research is required in order to reconcile the low growth pressures that severely reduce root elongation in the apparatus with the relatively strong soils that roots can deform. However, such apparatus offers the opportunity to increase the understanding of the cellular response to mechanical stress. An excellent review of mechanical impedance and root growth is given by Bengough and Mullins (1990).

3.2.2.2 Effect of Pot Size

When selecting pot size there is an obvious compromise which must be accepted between having a pot which is large enough to provide unrestricted root growth and a pot which is small enough from the perspective of practicalities. A rule of thumb is that an unimpeded main root axis can grow up to 2 or 3 cm per day. Therefore, for a 10-day experiment one should allow a pot depth of 30 cm (3 cm/day $\times$ 10 days), to avoid the main root becoming restricted. In most experiments, soil strength will prevent roots growing at their maximum rate, so this is a useful guide.

Examples of pot sizes which have been used successfully are 15 cm diameter and 27 cm long (Veen and Boone 1990) and 8.5 cm diameter and 19 cm long (Masle 1992). In those experiments maize was grown for up to 9 days (Veen and Boone 1990) and wheat (*Triticum aestivum* L.) and barley for up to 20 days (Masle 1992). For older plants, Arvidsson and Jokela (1995) grew barley for up to 35 days in cores 29.5 cm in diameter and 50 cm deep.

Ruff et al. (1987) give an excellent account of how restricted rooting volume can reduce the shoot growth of tomato plants, when grown for several weeks (up to 10). They grew tomato plants in either a 0.45 ℓ or a 13.5 ℓ rooting container, which were both well watered with nutrient solution. The plants with a restricted rooting volume were dwarfed. The reader should also refer to Marschner (1995, p. 530).

3.2.2.3 Complete Impedance

Measurements of maximum axial growth pressure (σ_{max}), which is the maximum force per unit area the root can exert, (σ_{max}), are reported widely

(Gill and Bolt 1955; Eavis et al. 1969; Pfeffer 1893; Misra et al. 1986; Souty and Stępniewski 1988; Whalley and Dexter 1993; Whalley et al. 1994; Clark et al. 1996). Several different apparatus for measuring σ_{max} have been reported, all of which are "home-made". The principle of the measurement is to allow a root to elongate into a small blind hole in a piece of ceramic which will completely impede the root. If the root is anchored securely, the maximum growth force can be measured using a force transducer attached to the ceramic (Fig. 3.4). The rate of growth force development is dependent on temperature (Bengough et al. 1994; Whalley et al. 1994). At 25°C the roots of pea approach a maximum growth force in about 24h. Although growth force may continue to increase gradually beyond that point, the value of σ_{max} is stable. The value of σ_{max} in pea was not found to be sensitive to temperature

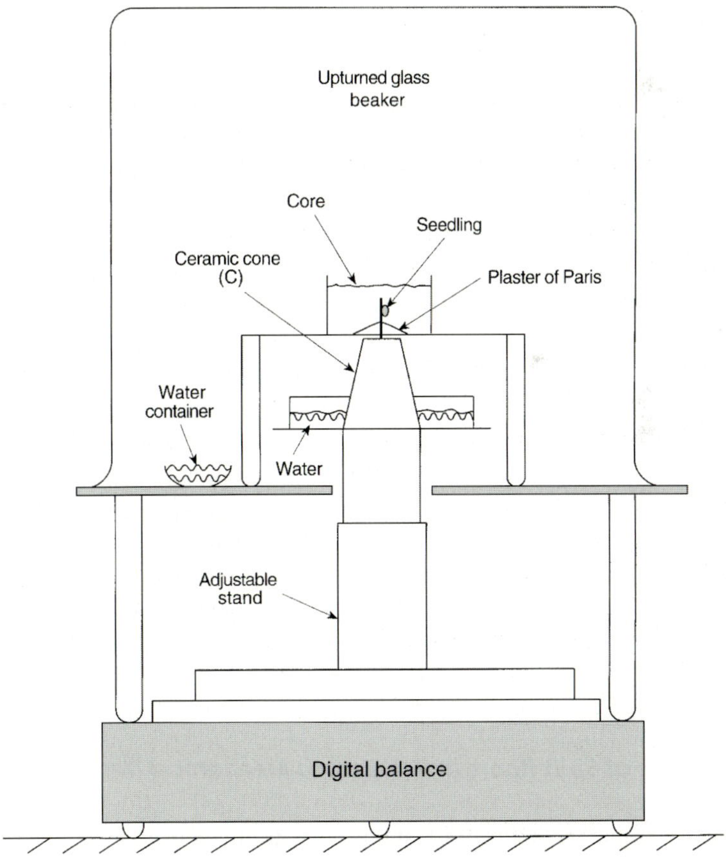

Fig. 3.4. Apparatus for measuring the maximum growth force generated by a seedling root. (From Whalley and Dexter 1993)

over the range 5 to 30 °C (Bengough et al. 1994; Whalley et al. 1994). A major limitation of this approach of measuring root growth pressure is that it can only be easily used with seedlings. An important advance is described by Misra (1997) who measured σ_{max} in both lateral and primary roots of pea and eucalypt (*Eucalyptus nitens*) seedlings. As far as we are aware, measurements of σ_{max} have not been made on the mature roots of seedlings more than a few days old.

One of the main benefits of maximum growth pressure measurements has been to provide data which help explain how individual roots respond to environmental stresses. The approach has the advantage of providing a direct estimate of root growth pressure. The most important data have been provided by experiments in which σ_{max} was measured as a function of oxygen, water and temperature stress. Eavis et al. (1969) have shown that in the case of pea, σ_{max} was not significantly affected when the oxygen partial pressure was reduced from 21 kPa to 3 kPa and was in the range 1.1 to 1.4 MPa. In the case of cotton (*Gossipium hirsutum* L.), σ_{max} was reduced from 1.1 MPa to 0.5 MPa when the oxygen partial pressure was reduced from 21 kPa to 3 kPa. In experiments on young maize, Souty and Stępniewski (1988) varied oxygen partial pressures from 0 to 20 kPa either around the whole seedling or just the radicle. The values of σ_{max} decreased with oxygen partial pressure in both cases, although the decrease was less severe when only the radicle was subjected to oxygen shortage, probably because oxygen could diffuse inside the root. Eavis et al. (1969) measured no significant decrease in σ_{max} for pea and cotton roots when they were subjected to matric potentials of –0.03 or –0.01 MPa. Whalley et al. (1998) extended the range, of water stress for pea down to –0.45 MPa. Over this larger range, the value of σ_{max} almost halved (i.e. from 0.66 MPa to 0.35 MPa).

Clark et al. (1996) combined measurements of maximum growth pressure with micro-pressure probe measurements of cell turgor. The micro-pressure probe technique is described by Pritchard et al. (1990) and it was based on the use of a micro-pipette which is connected to a pressure transducer. By making turgor pressure measurements in impeded pea roots, mounted in a specially adapted device for measuring σ_{max}, Clark et al. (1996) were able to demonstrate that when pea roots become impeded, turgor increased from 0.55 to 0.78 MPa and confining pressure due to the cell wall tension decreased from 0.55 to 0.26 MPa.

3.2.2.4 Use of Split Roots in Relation to Mechanical Impedance

As far as we are aware, split root systems have not been used widely in relation to mechanical impedance per se. Split root systems have been used to study the effect of water relations on plant growth. Gowing et al. (1990) split the roots of young apple plants between two pots 300 mm long and 60 mm in diameter and were able to demonstrate the role of hormones in root-shoot interaction in plants which were growing in soil, when the strength increased as a consequence of soil drying.

Bar-Yosef and Lambert (1981) used two coaxial acrylic cylinders, in which the roots were split between six 2.55 ℓ compartments. A range of soil strengths were obtained by varying either soil bulk density (by packing) or matric potential (by surface irrigation and suction). The experiment was performed with Norfolk sandy loam packed to bulk densities between 1.35 and 1.80 $g\,cm^3$ and matric potentials –3 to –60 kPa. The authors found that at penetrometer resistances greater than 2 MPa, roots continued to grow under a matric potential of –16 kPa and soil density of 1.85 $g\,cm^3$ but not under a matric potential of –60 kPa and soil density of 1.45 $g\,cm^{-3}$. For a given penetrometer resistance, the root elongation rate was always lower when the matric potential was more negative.

3.2.3 Oxygen Stress

3.2.3.1 Indicators of Soil Aeration Status

The oxygen concentration in the soil atmosphere is determined by the balance between gas diffusion and respiration. Methods for measuring gas diffusion through soil cores in laboratory conditions are well developed (e.g. Ball et al. 1981), but they depend upon well-defined boundary conditions which may be difficult to achieve in root growth experiments. However, within the strict context of roots, oxygen flux can be determined by polarographic methods. This is because the platinum electrode can consume oxygen and simulate a sink of a size similar to a root tip (e.g. 3 mm long and 1 mm in diameter). Polarographic methods for estimating oxygen flux are discussed at length by Armstrong and Wright (1976), Blackwell (1983), Gliński and Stępniewski (1985) and Armstrong (1994). According to McIntyre (1970) the process of oxygen reduction at an electrode in the pH range 5 to 12 is:

$$O_2 + 2H_2O + 4e^- \rightarrow 4OH^-,$$

and for pH < 5

$$O_2 + 4H^+ + 4e^- \rightarrow 2H_2O.$$

Thus, for the complete reduction of one oxygen molecule, four electrons are required. The measurement is made with an electrode made from inert platinum which is negatively polarised in relation to a reference electrode. This reference electrode can be a saturated colomel electrode or a silver/silver chloride electrode. The relationship between the potential at the electrode surface and the current generated from the reduction process is complex (Fig. 3.5). As the potential of the electrode is decreased from 0 to –0.4 V the current will increase, until a plateau region is reached. In this region current is proportional to the rate of oxygen reduction, which is in effect the rate at which oxygen molecules collide with the electrode (i.e. the oxygen flux at the surface of the electrode). As the potential is decreased further, the electrode current increases sharply in response to the reduction of hydrogen ions. The plateau region, in which the current is proportional to the oxygen flux, is normally between –0.4 and –0.7 V.

Platinum electrodes (or any other inert electrode) tend to over-estimate oxygen flux at the surface of a root, because the oxygen concentration at the surface of an electrode is zero. In contrast, at the surface of a root, the oxygen concentration is normally greater than zero.

Measurement of oxygen flux with polarographic methods has several practical limitations. In unsaturated soils there may be no plateau region, and in such circumstances the method should not be used. Gliński and Stępniewski (1985) note that the method is only valid when the electrode is completely covered with a water film. This corresponds to a matric potential of greater than about –0.1 MPa. Armstrong (1994) recommended that polarograms should be frequently checked if there is any doubt. He noted that naked electrodes are very prone to plateau shifts, usually at the upper boundary of voltage potential. With time, carbonates and hydroxides can precipitate on the electrode and impair its

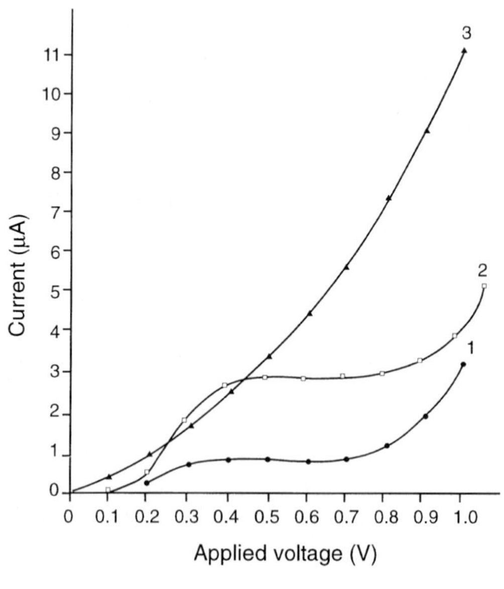

Fig. 3.5. Current versus applied voltage characteristics of a platinum electrode in (1) saturated sandy loam, (2) stirred soil suspension and (3) an unsaturated sandy loam. Note that no plateau is present in the case of the unsaturated sandy loam, and the measurement cannot be used to estimate oxygen flux. From Glinski and Stępniewski (1985)

performance. This is sometimes called electrode poisoning (Gliński and Stępniewski 1985), and for reliable results, the electrodes should be inserted immediately before the measurement is taken.

Redox potential is another useful indication of aeration status in the soil. This is discussed in detail by Gliński and Stępniewski (1985) and it is particularly useful for characterising anoxic conditions, where oxygen flux measurements are of less value. Redox potential can be correlated with various different biological processes. Examples of such processes are the emergence of oats (*Avena sativa* L.) and the elongation rate of roots (Blackwell and Wells 1983; Stępniewski et al. 1991). It is difficult to generalise from such relationships, as the biochemical reactions are specific to the soil type and conditions. Redox potential is measured with a high impedance voltmeter and with platinum electrodes of various sizes and shapes. The potential is measured with respect to a reference electrode of saturated calomel (or silver/silver chloride) and is expressed relative to that of a saturated hydrogen electrode. Electrodes for measuring redox potential can be left in situ for extended periods and still provide good results. The range of water potentials over which the redox potential measurements are reliable is much wider than in the case of oxygen flux measurements. Thus, redox potential measurements appear less problematic than oxygen flux measurements. Devices for monitoring redox electrodes have been reported (Linegarger et al. 1975; Flühler et al. 1976; Callebaut et al. 1980; Blackwell 1983).

3.2.3.2 Controlling Oxygen Stress

Oxygen stress can be controlled directly by flooding the soil air space with different gas mixtures. In this way, the effect of particular components of the soil atmosphere can be studied independently. Examples of this approach are described by Geisler (1967) who considered the effect of the concentration of oxygen and of carbon dioxide and also Drew et al. (1979), Konings and Jackson (1979) and Smith and Robertson (1972) who studied the response of roots to ethylene (ethene). Sarquis et al. (1991) studied the interaction between ethylene and mechanical impedance. They grew maize seedlings in a triaxial cell (see Sect. 3.2.2.1), that was continually flushed with ethylene-free air. Using the same system, Sarquis et al. (1991) studied the effect of applying ethylene exogenously by flushing the triaxial cell with air containing $1\,\mu\ell\,\ell^{-1}$ ethylene at a rate of $1.5\,\ell\,h^{-1}$.

The use of different gases to either flood the pore space of soil or bubble through a soil suspension results in redox potentials varying from +550 to −350 mV (Stępniewski et al. 1991). To obtain a redox potential of 500 mV, air was flushed through a Mississippi sandy loam, flushing the soil with gas containing 10 000 $\mu g\,g^{-1}$ oxygen in nitrogen, resulting in a redox potential of approximately 400 mV, while oxygen concentrations less than 100 $\mu g\,g^{-1}$ in nitrogen all produced redox potentials between 0 and 100 mV. When soil suspensions were used, bubbling with air gave a redox potential of 550 mV, while all other gases containing less than 10 000 $\mu g\,g^{-1}$ oxygen gave redox potentials between –200 and –350 mV. It should be noted that the environment of the soil or soil suspension can take up to 3 days to attain a reasonably stable value of redox potential.

One of the best examples of an experiment designed to estimate the values of oxygen flux that limit root growth is described by Blackwell and Wells (1983). In this experiment oat plants were established in a 1-m-long perspex tube. The tubes were then flooded, by lifting the water table slowly, and tilted so that the elongation of the roots could be observed against the side of the perspex columns. Oxygen stress was induced due to consumption by microbial activity and a low diffusion rate due to water logging. Blackwell and Wells (1983) measured oxygen flux by using both the procedure of Armstrong and Wright (1976) and also by assuming that a single excitation voltage of 0.65 V could be used (i.e. assuming that there is no need to check for a plateau in the voltage-current relationship). Comparison of the results from these two approaches clearly demonstrates the importance of measuring the voltage-current relationship (e.g. Fig. 3.5) to establish that the current flow is limited by the oxygen flux. We recommend Blackwell and Wells (1983) to readers who intend to make laboratory measurements of the value of oxygen flux that limits root growth.

More commonly, indirect approaches are used to modify soil aeration. The water content or the bulk density can be manipulated to change the air-filled porosity (e.g. Stępniewski and Przywara 1992). Typically, air-filled porosities of less than 10% are associated with oxygen stress, but this depends upon soil type. As increased bulk density can result in increased mechanical impedance to root growth, it is more usual to adjust water content to decrease the air-filled porosity.

3.2.3.3 Split Root Techniques in Relation to Oxygen Stress

Van Noordwijk and Brouwer (1993) used a split root system in which the roots were split between two 8 ℓ pots containing nutrient solution. The solution was either permanently aerated in both pots, permanently non-aerated or the aeration was stopped in half the root system at a certain growth stage approximately two weeks before measurements of root porosity. The air-filled porosity of the roots of maize, tomato and gerbera (*Gerbera jamesonii*) was greater in the plants which were not aerated. When half the roots were transferred to a non-aerated nutrient solution, the air-filled porosity of the roots increased in young plants, but not in older plants. Van Noordwijk and Brouwer (1993) noted that aeration of the solution involved complex factors such as agitation due to stirring. They developed a recirculation system in which the oxygen partial pressure could be varied between 1 and 4 kPa. The oxygen partial pressure in the aerated system stayed close to 20 kPa. Again they found that, for most species, older roots did not respond when half the root system was transferred to a decreased oxygen environment.

Air-filled porosity of a root can be measured either with the pycnometer method (Jensen et al. 1969) or by examination of sections of root under a microscope. Both of these methods are compared by Van Noordwijk and Brouwer (1988). For maize, these methods compared well. However, Van Noordwijk and Brouwer (1993) showed that this is not always the case.

Stępniewski and Bennicelli (1992) and Stępniewski et al. (1994) have developed a split root technique based on two rhizotrons filled with soil that can be flooded with various gases (Fig. 3.6). At the end of the experiment, the soil can be washed away, with the roots being held on a pin-board. Stępniewski et al. (1994) observed compensatory growth in the aerated part of the root system. In a more complex experiment, W. Stępniewski and G. Przywara (unpubl.) divided the root system into four parts and modified aeration indirectly by adjusting the water content. Their results are presented in Fig. 3.7. The mass of the root system decreased as the fraction of the root system, that was flooded, increased.

The rooting environment can also be split into independent horizontal strata, with thin layers of wax petroleum separating the layers of soil (e.g. Sect. 3.2.2.1). Soft wax can be used to avoid impeding the root system mechanically.

3.3. Field Measurement of the Root Growth Environment

3.3.1 Field Assessment of Root Conditions

Procedures for inspecting soil in the field are described by Batey (1988). We recommend Batey's handbook (1988) to readers as an excellent practical guide to field inspection of soil conditions. Table 3.4 (Box 3.2) can be used as a rough guide to help decide on the probable causes of poor root growth and then on appropriate measurements which provide data beyond visual observation.

3.3.2 Mechanical Impedance

Mechanical impedance is the most ubiquitous constraint to root growth. Mechanical impedance is usually measured with a penetrometer, although it can be estimated from the bulk density of the soil.

Fig. 3.6. Split-root rhizotron (**a**) exposed and the two (**b**) halves of the root system of the same maize plant grown in aerated (air) and anoxic (nitrogen) conditions. (From W. Stępniewski and R. Bennicelli, unpubl.)

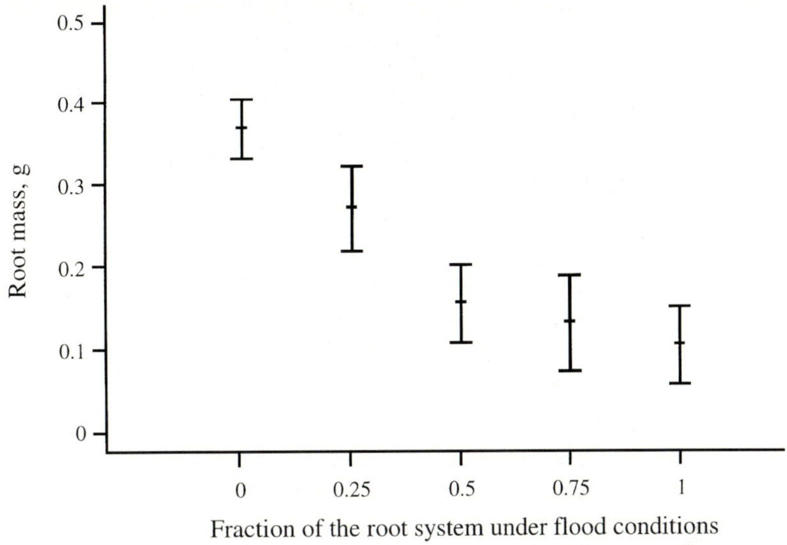

Fig. 3.7. Data from a four part split root system used to study the effect of partial or total flooding on a root system. (From W. Stępniewski and G. Przywara, unpubl.)

The use of penetrometers in relation to root growth is discussed in detail by Bengough (1991). Bennie (1991) has demonstrated that when relative root growth is plotted against penetrometer resistance, there is a remarkable coalescence of data for maize, cotton, wheat and groundnuts. A penetrometer resistance of approximately 3 MPa is generally regarded as the upper limit beyond which root growth is negligible (Boone et al. 1986). This is considerably higher than maximum root growth pressure values reported in Section 3.2.2.3. This difference is thought to be because soil-metal friction is much larger than root-soil friction. In addition, when relatively blunt penetrometers are used (i.e. a cone angle of 30° which is most commonly the case) the failure mode of the soil differs between penetrometers and roots. This is discussed at length by Bengough and Mullins (1990, 1991).

Most field penetrometers have large diameters, with cone semi-angles of 30° and are used either for root studies (Whitely and Dexter 1982; Ehlers et al. 1983) or for engineering purposes (Campbell and O'Sullivan 1991). Large penetrometer cones are relatively insensitive to cracks and heterogenity in the soil at the scale of the root tip. In coarse-textured soils significant mechanical impedance can result from sand particles interlocking to resist displacement (Panayiotopoulos 1989; Gliński and Lipiec 1990). These cannot be detected by penetrometers that are much larger in diameter than sand particles.

Dry bulk density is frequently used as a measure of mechanical impedance. There are several different methods for measuring dry bulk density. The

BOX 3.2 Field Investigation of the Rooting Environment

When inspecting a soil profile, common indicators of a poor rooting environment are sharp structural discontinuities and the presence of grey and brown mottles which can be indicative of periodic saturation and aeration problems. Structural discontinuities are very common in cultivated soils and they can be responsible for limiting deep rooting. A high number of cracks, continuous pores and planes of weakness between peds are important in allowing roots to by-pass regions of high soil strength.

Table 3.4. Estimation of the most likely causes of poor root growth in relation to soil type and water content. This table is intended to give an indication of the likely physical stresses that should be investigated to determine the cause of poor root growth. The likely physical stresses that can lead to poor root growth considered in this table are water stress, mechanical impedance, temperature or aeration. If any of these stresses are likely to be problematic then we have attached a probability of high (H), medium (M) or low (L) to their severity. This is only intended as a rough guide. The effects of poor aeration and excessive mechanical resistance may increase in compacted soil. The effects of poor aeration are often smaller at low soil temperatures. We have not attached any probability to the likely severity of temperature stress because it is too difficult to generalize. This table has been adapted from ISO (1998)

Water status	Clay content		Possible causes of poor root growth	Suggested method of investigation
	>17% clay	<17% clay		
Wilting point	Solid, hard, cannot be moulded, darkens strongly when moistened	Light colour, darkens strongly when moistened	Water stress (H) Mechanical impedance (H)	Water content Small cone penetrometer
Intermediate between field capacity and wilting point	Semi-solid, can be moulded, cracks and crumbles when rolled by hand into 3 mm threads, darkens slightly when wetted	Darkens slightly when wetted	Water stress (L) Mechanical impedance (M)	Water content Small cone penetrometer
Field capacity; free water absent	Plastic, can be rolled into 3 mm threads without cracking, no colour change upon wetting	Fingers get slightly moist, no water escapes soil pores upon compression, no colour change upon wetting	Poor aeration (L) Mechanical impedance (M)	Large cone penetrometer Redox potential
Free water present, saturating part of soil pore space	Soil soft and can be rolled into a thread <3 mm diameter	Fingers moisten rapidly when soil is worked and visible outflow from soil pores upon compression	Poor aeration (M) Mechanical impedance (L) Temperature	Large cone penetrometer Redox potential Temperature
All pores full, free water present	All pores appear water filled and free water present	All pores appear water filled and free water present	Poor aeration (H)	Large cone penetrometer Redox potential
Soil surface covered by water	Soil surface covered by water	Soil surface covered by water	Poor aeration (H)	Large cone penetrometer, redox potential

simplest of these is to take a core of soil of known volume and oven dry it to obtain the mass of dry soil per unit volume. A second group of methods involve excavation of a known mass of soil. The volume of the hole is then obtained by filling it with particles that pack to a known density. If high spatial resolution is required gamma attenuation can be used to measure bulk density. An excellent review of methods for estimating bulk density is given by Campbell and Henshall (1991).

Direct comparisons of bulk density between different soil types are of little value. To overcome this problem, Håkansson (1990) proposed the use of the degree of compactness. This is defined as the ratio of the actual bulk density to the bulk density of the wet soil under a static load of 200 kPa.

An alternative index is defined by Bennie (1991) as:

$$\left[\frac{\rho_{\text{actual}} - \rho_{\min}}{\rho_{\max} - \rho_{\min}}\right],$$

where ρ is soil bulk density and the subscripts refer to actual and reference maximum and minimum densities. The minimum bulk density was estimated from the mass of unsieved dry soil required to fill a $0.2\,\ell$ container and the maximum bulk density was determined with the ASTM Proctor test (Felt 1965). Using the index of Bennie (1991), the values of <0.5, 0.5–0.6, 0.6–0.7 and 0.7 correspond to low, medium, high and very high degrees of compaction respectively. Relative compaction parameters give a useful characterisation of soil compactness in studies of root and crop responses to compaction (van Ouwerkerk and Soane 1994).

3.3.3 Characterising the Effect of Soil Structure on Root Growth

Methods for studying the interaction between soil structure and root growth are perhaps the least well developed. This is because soil structure is a complicated concept which has no simple definition. Dexter (1988) defines soil structure as "the spatial heterogenity of the different components of soil".

To illustrate the care that must be taken when relating soil structure to root growth, it is pertinent to consider two separate classic laboratory experiments. Firstly, Voorhees et al. (1971) used carefully prepared mixtures of aggregates, which differed in density. During the experiments, certain density classes were labelled with ^{32}P. Voorhees et al. (1971) demonstrated that dense aggregates restricted root growth more than porous aggregates. In the case of very dense aggregates, root growth was restricted to the periphery of aggregates, while root growth within the aggregate was affected by the strength and pore size distribution of adjacent aggregates.

Secondly, Scholefield and Hall (1985) demonstrated that roots were able to penetrate pores smaller than their nominal thickness. From the work of Voorhees et al. (1971) and Scholefield and Hall (1985), it is clear that the relationship between root growth and soil structure is far from trivial. A further complication is that the relationship between root structure and soil structure is dynamic. An excellent recent review is presented by Cresswell and Kirkegaard (1995).

Ehlers et al. (1983) studied the growth of oat roots in tilled and untilled soils. They characterised the soil structure in several ways. Penetration resistance profiles were measured weekly with a penetrometer which had a 60° cone angle, 11 mm in diameter, and a relieved shaft. The mean penetration resistance was determined in 5-cm-deep layers down to 40 cm, then in 10-cm-layers down to a depth of 60 cm. Bulk density profiles were also measured in two layers with six samples per layer. Bio-pores (root or earthworm holes) were counted after excavation of the soil to the required depth in an area 50 cm × 50 cm. Ehlers et al. (1983) used a butane flame to heat the excavated surface of the soil and then used suction to lift off the loose dry aggregates, so that the bio-pores could be observed. Ehlers et al. (1983) demonstrated that oat roots exploited bio-pores, and as little as 2% of bio-pores could improve rootability of the soil. A similar study on soybean (*Glycine max* L.) roots was described by Wang et al. (1986).

For more detailed studies, the method of preparing thin polished sections of soil impregnated with resin is particularly useful (McBratney et al. 1992; Van Noordwijk et al. 1992; Krebs et al. 1994). An excellent example of a thin section is shown in Fig. 3.8. Van Noordwijk et al. (1992) note that small thin sections contain insufficient information for quantative analysis, although qualitatively they are extremely useful. In the case of Fig. 3.8, the extension of root hairs into aggregates is clearly visible. Root soil contact is an important input for models of oxygen, water and nutrient uptake (De Willigen and Van Noordwijk 1987).

Methods for studying the inter-relationships between soil structure and root growth need greater development.

3.3.4 Field Measurements of Aeration Status

The polarographic method for measuring oxygen flux described in Section 3.2.3.1 is also suitable for use in the field (Blackwell 1983). Similarly, the measurements of redox potential described in Section 3.2.3.1 can also be used in the field. Armstrong and Wright (1976) and Blackwell (1983) describe devices which enable large numbers of probes to be monitored. Their equipment can be used to measure both redox potential and oxygen flux. Readers should be

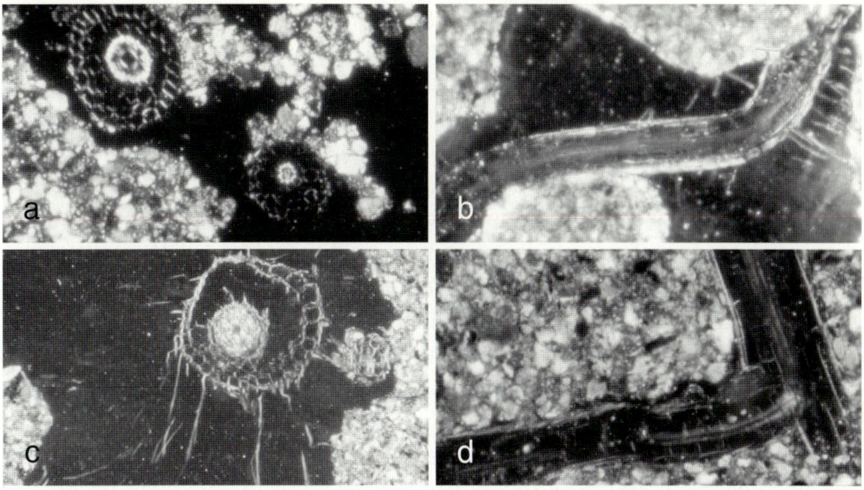

Fig. 3.8. Thin sections of a soil showing different degrees of soil-root contact (*black* void, *light grainy structures* sand particles, *thin lines* root cell walls). **a** Transverse section through two roots growing on the surface of soil peds. **b** Longitudinal section of root in void between soil aggregates, establishing contact with soil via root hairs. **c** Transverse section of root in void with root hairs reaching for the soil. **d** Longitudinal section of root penetrating through and branching in the soil matrix. (Photographs courtesy of Dr. M.J. Kooistra, SC-DLO Wageningen, the Netherlands)

warned that considerable effort is needed to develop a system which can measure oxygen flux and redox potential and then record the data from a number of probes.

Blackwell (1983) compared the use of platinum electrodes to measure oxygen flux and redox potential with methods for gas sampling. The gas sampling methods used either 50 ml (internal volume) sintered bronze samplers or 5 ml polythene and PVC samplers connected to pipes and then to air-tight taps through which gas and soil solution could be extracted. Blackwell (1983) cautions that some plastics are gas permeable and that they should not be used. The samplers were placed at depths of up to 80 cm. Blackwell (1983) used 5 ml gas-tight syringes to sample the chambers and recommended that approximately half of the volume of the gas sampler chamber should be discarded before the sample is analysed for gas concentration in the laboratory. Both gas and liquid phases were analysed to determine their oxygen concentration. The difference in scale between the large chambers for gas sampling and the small platinum electrodes requires soil structure to be considered in interpreting the results. Blackwell (1983) suggests that, because the dimensions of the electrodes are much smaller than those of the soil structural units, the platinum electrodes are much more likely to be embedded in the soil peds. In contrast, chambers for gas sampling will tend to sample the gas between the soil structural units. The gas sampling chamber is a convenient method for obtaining repeated measurements of oxygen concentration in the field and depending on the availability of laboratory methods for gas analysis it can be extended to other gases.

Ball et al. (1994) describe a research apparatus which monitors the diffusion of a radioactive gas from an excavated hole in the soil. The experimental data were used to obtain estimates of relative diffusivity near the soil surface using a numerical simulation based on Fick's equation. They noted that there was poor agreement between field and laboratory tests, which they attributed to variation in air-filled porosity with depth. This technique appears to be promising, but is still being developed. Ball et al. (1994) illustrates clearly how gas diffusion depends on bulk density, water content, and soil type.

Gas diffusion measurements in the field are rarely made because they are labour intensive (Gliński and Stępniewski 1985) and the air-filled porosity is often quoted instead.

3.3.5 Measuring and Modifying Soil Temperature in the Field

Soil temperature can be measured with relatively simple and inexpensive sensors that are widely available. Buchan (1991) discusses temperature measurement in detail.

Soil temperature can be influenced in several ways. Mulching of the soil surface with plant residues, irrigation or tillage of the soil are the most common practices in the field (Voorhees et al. 1981).

Irrespective of other effects, changes in soil temperature are determined largely by soil water content. In dry soil, the small loss of latent heat through evaporation consumes only a small proportion of the incoming radiation, and heat is conducted slowly. This means that the surface of a dry soil warms up quicker than wet soil and provides more favourable conditions for the germination of seeds and initial root growth. Drainage helps the soil warm faster in the early spring, although dry surfaces cool down more rapidly at night and consequently have a higher incidence of night frost.

In the study of Kuchenbuch and Barber (1988), nine years of air temperature data (from Lafayette, Indiana) were analysed in relation to the distribution of root depth in maize. The temperature was expressed in degree days (one DD was a day with an average daily air temperature of 1 °C above a base temperature of 10 °C). At silking, the root length density (RLD) below 30 cm was correlated significantly ($r = 0.82$) with DD for the 2 weeks following planting. The RLD in the 0–15 cm layer was significantly correlated with precipitation but not with DD.

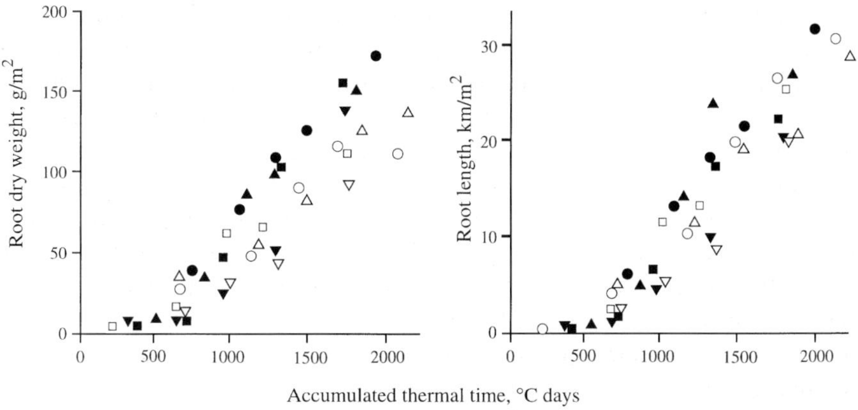

Fig. 3.9. Data for wheat from Barraclough and Leigh (1984), showing the accumulation of total root dry matter or total length as a function of accumulated thermal time. Solid symbols, 1980 Crops; open symbols 1981 Crops; circles, Rothamsted early-sown; triangles, Woburn early-sown; squares, Rothamsted late-sown; inverted triangles, Woburn late-sown

Root growth of winter wheat in England was strongly related to various DDs, accumulated at different rates by varying sowing dates (Barraclough and Leigh 1984). An example of the relation between accumulated thermal time and the root length of winter wheat grown in the UK is shown in Fig. 3.9. The early differences in root length due to sowing date were maintained throughout the season.

The growth trajectory of maize nodal roots is affected by soil temperature until the root is 100 mm long (Tardieu and Pellerin 1991). This root response of young nodal roots to temperature affects the development and proliferation of the root system later in the development of the plant.

Kovar et al. (1992) used two different tillage systems to influence soil temperature and water regimes. Although differences in soil temperature and root length density between the two tillage systems were significant, there was no significant correlation between soil temperature and RLD. The authors suggested that, for prediction of root distribution from the soil temperature and water content of the soil, both parameters should be monitored regularly.

Various aspects of soil temperature in relation to rooting are reviewed by Kaspar and Bland (1992).

3.3.6 Interpreting the Effects of Combinations of Stresses

Root growth in the field is affected by many factors which vary in space and time, and can interact. Establishing which factors are limiting root growth is very difficult. Figure 3.10 shows how both penetrometer resistance and air-filled porosity are related to matric potential and dry bulk density. An air-filled porosity of 10% and a penetrometer resistance of 3 MPa are regarded frequently as critical for root growth. Figure 3.10 illustrates that as bulk density increases, the range of matric potential, in which aeration and mechanical impedance do not limit root growth, becomes narrower. In compacted soils, therefore, relatively small changes in water content (either drying or wetting) can lead to reduced root growth. This has been demonstrated in field experiments by Grath and Håkansson (1994). In areas of compaction (e.g. headlands, wheel tracks), and in depressions, leaf yellowing occurred. Plant species responded differently to combinations of stress (Goss et al. 1989). In wheat, the effect on root elongation of a combination of mechanical impedance and poor aeration could be predicted from measurements of the effects of the stresses acting independently. In contrast, for maize the combined effect was greater than that predicted (see Goss et al. 1989) indicating a significant interaction between stresses.

The effects of soil compaction, temperature, nitrogen and water supply on the growth and distribution of roots of winter wheat were studied by Barraclough et al. (1991). They found that drought-stressed roots grew deeper than roots in irrigated soil, and soil porosity or mechanical impedance appeared to be major factors affecting root distribution.

Dense layers such as ploughpans, claypans or fragipans in soil profiles restrict root growth by large mechanical impedance, insufficient aeration and sometimes reduced water availability to roots (Unger et al. 1981). These layers have large values of bulk density and cone resistance. The thicker the layer and the nearer it is to the soil surface, the greater the restriction of root growth. The negative effect of the dense layers can sometimes be alleviated by compensatory root growth. Barraclough and Weir (1988) showed that when water was not a

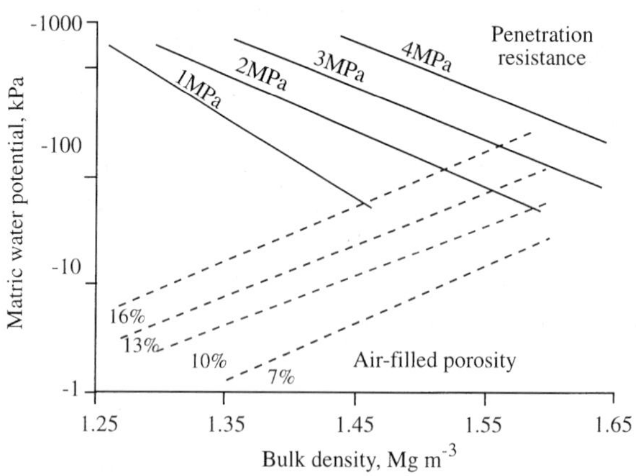

Fig. 3.10. Relationship between penetration resistance of a penetrometer and air filled porosity of a silty loam soil in relation to bulk density and matric potential. (From Lipiec et al. 1991)

limiting factor in the topsoil and N was added, the pan had a negligible effect on the grain yield of winter wheat.

In soils with a water table within the root zone, root development may be affected adversely both by anoxic conditions, and by the lower soil temperature in cold climates. In hot climates, salt in the water table may limit root growth.

3.4 Future Trends

The techniques that we have described for controlling the rooting environment in the laboratory are almost all designed to provide an environment that is constant with time. The exceptions are the withholding of water which provides an increase in both mechanical impedance and water stress with time, but in an uncontrolled way. For a single root axis the approach of Bengough et al. (1994) provides the possibility of changing the mechanical impedance as a function of time. There is a need for more development of systems that allow root growth to be studied in an environment with controlled temporal variation in physical stresses, particularly mechanical impedance and water stress. The data obtained should be related to the root architecture.

The approaches for measuring maximum growth pressure of a single root axis need to be extended to the whole root system. Misra (1997) has made an important contribution to this objective. There is also a requirement for methods which allow the maximum axial growth pressure of roots in mature plants to be measured.

Significant progress has been made in our understanding of the response of roots to water stress by Sharp and co-workers and to mechanical stresses by Bengough and co-workers, but the interaction between these two stresses is less well understood. More work is needed to identify improved methodologies for looking at combinations of these and other stresses.

Laboratory approaches for studying the effect of soil structure on root development are poorly developed. Approaches are needed to study the interaction between root architecture and soil structure which can elucidate the function, both for soil structure and the root system.

There is a need for better integration between field work and laboratory experiments. Both of these approaches are important, but there is a tendency for them to be conducted in isolation.

We hope that the approaches that we have described in this chapter will help to raise the level of awareness of the need to conduct root growth experiments in a carefully controlled environment.

Acknowledgments. W.R. Whalley and J. Lipiec thank the British Council and the KBN in Warsaw who partly funded this work. We thank Dr. A.G. Bengough, Dr.

M.C. Drew, Dr. J. Tisdall and Dr. W.B. Voorhees for the constructive comments made on an early manuscript. W.R. Whalley was supported by Silsoe Research Institute which is grant aided by the Biotechnology and Biological Sciences Research Council.

References

Abdalla AM, Hettiarachi DRP, Reece AR (1969) Mechanics of root growth in granular media. J Agric Eng Res 14: 236–248

Aguirrezabal LAN, Tardieu F (1996) An architectural analysis of the elongation of field-grown sunflower root system. Elements for modelling the effects of temperature and intercepted radiation. J Exp Bot 47: 411–420

Aguirrezabal LAN, Deléens E, Tardieu F (1994) Root elongation rate is accounted for by intercepted PPFD and source-sink relations in field and laboratory-grown sunflower. Plant Cell Environ 17: 443–450

Armstrong W (1994) Polarographic oxygen electrodes and their use in plant aeration studies. Proc R Soc Edinb 102B: 511–527

Armstrong W, Wright EJ (1976) A polarographic assembly for multiple sampling of soil oxygen flux in the field. J Appl Ecol 13: 849–856

Arvidsson J, Jokela WE (1995) A lysimeter study of soil compaction effects on evapotranspiration in a barley crop. Swed J Agric Sci 25: 109–118

Atwell BJ (1990) The effect of soil compaction on wheat during early tillering. II. Concentration of cell constituents. New Phytol 115: 37–41

Ball BC, Harris W, Burford JR (1981) A laboratory method to measure gas diffusion and flow in soil and other porous materials. J Soil Sci 32: 323–333

Ball BC, Galsbey CA, Robertson EAG (1994) Measurement of soil gas diffusivity in situ. Eur J Soil Sci 45: 3–13

Barley KP (1962) The effects of mechanical stress on the growth of roots. J Exp Bot 13: 95–110

Barraclough PB, Leigh RA (1984) The growth and activity of winter wheat roots in the field: the effect of sowing date and soil type on root growth of high-yielding crops. J Agric Sci Cambridge 103: 59–74

Barraclough PB, Weir AH (1988) Effects of compacted subsoil layer on root and shoot growth, water use and nutrient uptake of winter wheat. J Agric Sci Cambridge 110: 207–216

Barraclough PB, Weir AH, Kuhlman H (1991) Factors affecting the growth and distribution of winter wheat roots under UK field conditions. In: McMichal BL, Persson H (eds) Plant roots and their environment. Elsevier, Amsterdam, pp 417–419

Bar-Yosef B, Lambert JR (1981) Corn and cotton root growth response to soil impedance and water potential. Soil Sci Soc Am J 45: 930–935

Batey T (1988) Soil husbandry: a practical guide to the use and management of soils. Aberdeen University Press, Aberdeen

Beemster GTS, Masle J, Williamson RE, Farquhar GD (1996) Effects of soil resistance to root penetration on leaf expansion in wheat: kinematic analysis of leaf elongation. J Exp Bot 47: 1663–1678

Bengough AG (1991) The penetrometer in relation to mechanical resistance to root growth. In: Smith KA, Mullins CE (eds) Soil analysis, physical methods. Marcel Dekker, New York, pp 431–446

Bengough AG, Mackenzie CJ (1994) Simultaneous measurement of root force and elongation for seedling pea roots. J Exp Bot 45: 95–102

Bengough AG, Mullins CE (1990) Mechanical impedance to root growth: a review of experimental techniques and root growth responses. J Soil Sci 41: 341–358

Bengough AG, Mullins CE (1991) Penetrometer resistance, root penetration resistance and root elongation in two sandy loam soils. Plant Soil 131: 59–66

Bengough AG, Mackenzie CJ, Elangiwe HE (1994) Biophysics of the growth response of pea roots to changes in penetration resistance. Plant Soil 167: 135–141

Ben Haj Salah H, Tardieu F (1996) Quantitative analysis of the combined effects of temperature, evaporative demand and light on leaf elongation rate in well-watered field and laboratory-grown maize plants. J Exp Bot 304: 1689–1698

Bennie ATP (1991) Growth and mechanical impedance. In: Waisel Y, Eshel A, Kafkafi U (eds) Plant roots. The hidden half. Marcel Dekker, New York, pp 393–414

Blackwell PS (1983) Measurements of aeration in waterlogged soils: some improvements of techniques and their application to experiments using lysimeters. J Soil Sci 34: 271–285

Blackwell PS, Wells EA (1983) Limiting oxygen flux densities for oat root extension. Plant Soil 73: 129–139

Boone FR, van der Werf HMG, Kroesbergen B, ten Hag BA, Boers A (1986) The effect of compaction of the arable layer in sandy soils on the growth of maize for silage. 1. Critical matric water potentials in relation to soil aeration and mechanical impedance. Neth J Agric Sci 34: 155–171

Buchan GD (1991) Soil temperature regime. In: Smith KA, Mullins CE (eds) Soil analysis, physical methods. Marcel Dekker, New York, pp 551–612

Callebaut F, Balcaen M, Gabriels D, de Boodt M (1980) Data acquisition system for field determination of redox potential, oxygen diffusion rate and soil electrical resistance. Meded Fac Landbouwk Rijksuniv Gent 45: 15

Campbell DJ, Henshall JK (1991) Bulk density. In: Smith KA, Mullins CE (eds) Soil analysis, physical methods. Marcel Dekker, New York, pp 399–429

Campbell DJ, O'Sullivan MF (1991) The cone penetrometer in relation to trafficability, compaction and tillage. In: Smith KA, Mullins CE (eds) Soil analysis, physical methods. Marcel Dekker, New York, pp 399–429

Čiamporová M, Luxová M (1976) The effect of polyethylene glycol-induced water stress on the maize root apex. Biol Plant (Praha) 18: 173–178

Clark LJ, Whalley WR, Dexter AR, Barraclough PB, Leigh RA (1996) Turgor pressure is increased in the apex of pea roots by complete mechanical impedence. Plant Cell Environ 19: 1099–1102

Cosgrove DJ (1993) How do plant cells extend? Plant Physiol 102: 1–6

Coutts MP (1982) Growth of Sitka spruce seedlings with roots divided between soils of unequal matric potential. New Phytol 92: 49–61

Cresswell HP, Kirkegaard SA (1995) Subsoil amelioration by plant roots, the process and the evidence. Aust J Soil Res 33: 221–239

Dahal P, Bradford KJ (1994) Hydrothermal time analysis of tomato seed germination at suboptimal temperature and reduced water potential. Seed Sci Res 4: 71–80

Davies WJ, Tardieu F, Trejo CL (1994) How do chemical signals work in droughted plants? Plant Physiol 104: 309–314

De Willigen P, van Noordwijk M (1987) Roots, plant production and nutrient use efficiency. PhD Thesis, Agricultural University of Wageningen, Wageningen The Netherlands

Dexter AR (1988) Advances in characterization of soil structure. Soil Tillage Res 11: 199–238

Drew MC, Jackson MB, Gifford S (1979) Ethylene-promoted adventitious rooting and cortical air spaces (aerenchyma) in roots may be an adaptive response to flooding in *Zea mays*. L. Planta 147: 83–88

Eavis BE, Ratliff LF, Taylor HM (1969) Use of a dead-load technique to determine axial root growth pressure. Agron J 61: 640–643

Ehlers W, Köpke U, Hesse F, Böhm W (1983) Penetration resistance and root growth of oats in tilled and untilled loess soil. Soil Tillage Res 3: 261–275

Emmert FH (1974) Inhibition of phosphorus and water passage across intact roots by polyethylene glycol and phenylmercuric acetate. Plant Physiol 53: 663–665

Felt EJ (1965) Compactibility. In: Black CA (ed) Methods of soil analysis, Part 1. American Society of Agronomy, Madison, Wisconsin, pp 400–412

Flühler H, Ardakani MS, Szuszkiewicz TE, Stolzy LH (1976) Field measured nitrous oxide concentration, redox potentials, oxygen diffusion rates, and oxygen partial pressures in relation to denitrification. Soil Sci 122: 107–114

Fraser TE, Silk WK, Rost TL (1990) Effects of low water potential on cortical cell length in growing regions of maize roots. Plant Physiol 93: 648–651

Gardner HR, Danielson RE (1964) Penetration of wax layers by cotton roots as affected by some soil physical conditions. Soil Sci Soc Am Proc 28: 457–460

Geisler G. (1967) Interactive effects of CO_2 and O_2 in soil on root and top growth of barley and peas. Plant Physiol 42: 305–307

Gergely I, Korcak RF, Faust M (1980) Polyethylene glycol-induced water stress effects on apple seedlings. 1. Methodology, water consumption, and dry matter production. J Am Soc Horti Sci 105 (6): 854–857

Gill WR, Bolt GH (1955) Pfeffer's studies on the root growth pressures exerted by plants. Agron J 47: 166–168

Gliński J, Lipiec J (1990) Soil physical conditions and plant roots. CRC Press, Boca Raton, Florida

Gliński J, Stępniewski W (1985) Soil aeration and its role for plants. CRC Press, Boca Raton, Florida

Gordon DC, Hettiaratchi DRP, Bengough AG, Young IM (1992) Non-destructive analysis of root growth in porous media. Plant Cell Environ 15: 123–128

Goss MJ, Barraclough PB, Powell BA (1989) The extent to which physical factors in the rooting zone limit crop growth. Aspects Appl Biol 22: 173–181

Gowing DJG, Davis WJ, Jones HG (1990) A positive root sourced signal as an indicator of soil drying in apple *Malus X domestica* Borka. J Exp Bot 41: 1535–1540

Grath T, Håkansson I (1994) A case study on soil compaction and aphanomyces root rot as causes of uneven pea growth. Swed J Agric Res 24: 165–170

Greacen EL, Farrell DA, Cockroft B (1968) Soil resistance to metal probes and plant roots. Trans 9th Congr Int Soc Soil Sci 1: 769–779

Håkansson I (1990) A method for characterizing the state of soil compactness for the plough layer. Soil Tillage Res 16: 105–120

Hsieh JJC, Gardner WH, Campbell GS (1972) Experimental control of soil water in the vicinity of root hairs. Soil Sci Soc Am Proc 36: 418–421

Idso SB, Kimball BA (1992) Seasonal fine-root biomass development of sour orange trees grown in atmospheres of ambient and elevated CO_2 concentration. Plant Cell Environ 15: 337–341

ISO (1998) Soil quality – determination of redox potential – field method. ISO/CD 11271, Helsinki

Iwama H, Osozawa S, Ushiroda T, Kubota T (1994) Analysis of soil water matric potential requirement for infection of turnip with *Plasmodiophora brassicae* using negative pressure water circulation technique. Soil Sci Plant Nutr 40: 293–299

Janes BE (1974) The effect of molecular size, concentration in nutrient solution, and exposure time on the amount and distribution of polyethylene glycol in pepper plants. Plant Physiol 54: 226–230

Jensen CR, Luxmoore RJ, van Grundy SD, Stolzy LH (1969) Root air space measurements by a pycnometer method. Agron J 61: 474–475

Jourdan C, Rey H, Guedon Y (1995) Architectural analysis and modelling of the branching process of young oil-palm root systems. Plant Soil 177: 63–72

Kaspar TC, Bland WL (1992) Soil temperature and root growth. Soil Sci 154: 290–298

Konings H, Jackson MB (1979) A relationship between rates of ethylene production by roots and the promoting or inhibiting effects of exogenous ethylene and water on root elongation. Z Pflanzenphysiol 92: 385–391

Kovar JL, Barber SA, Kladivoko EJ, Griffith DR (1992) Characterization of soil temperature, water content and maize root distribution in two tillage systems. Soil Tillage Res 24: 11–27

Krebs M, Kretzschmar A, Babel V, Chadœuf J, Goulard M (1994) Justifications on distribution patterns in soil: basic and relative distributions of roots, channels and cracks. In: Ringrose AJ, Humphreys S (eds) Soil morphology. Elsevier, Amsterdam

Krizek DT (1985) Methods of inducing water stress in plants. Hortic Sci 20: 1028–1038

Kuchenbuch RO, Barber SA (1988) Significance of temperature and precipitation for maize root distribution in the field. Plant Soil 106: 9–14

Lawlor DW (1970) Absorption of polyethylene glycols by plants and their effects on plant growth. New Phytol 69: 501–513

Linegarger RS, Whisler FD, Lance JC (1975) A new technique for rapid and continuous measurement of redox potentials. Soil Sci Soc Am Proc 39: 375–377

Lipiec J, Kubota T, Iwama H, Hirose J (1988) Measurement of plant water use under controlled soil moisture conditions by the negative pressure circulation technique. Soil Sci Plant Nutr 34: 417–428

Lipiec J, Håkansson I, Tarkiewicz S, Kossowski J (1991) Soil physical properties and growth of spring barley as related to the degree of compactness of two soils. Soil Tillage Res 19: 307–317

Malcolm GN, Rawlinson JS (1957) The thermodynamic properties of aqueoues solutions of polyethylene glycol, polypropylene glycol and droxane. Trans Faraday Soc 153: 921–931

Marschner H (1995) Mineral nutrition of higher plants. Academic Press, Cambridge

Masle J (1992) Genetic variation in the effects of root impedance on growth and transpiration rates of wheat and barley. Aust J Plant Physiol 19: 109–125

Materechera SA, Dexter AR, Alston AM, Kirby JM (1992) Growth of seedling roots in response to external osmotic stress by polyethylene glycol 20 000. Plant Soil 143: 85–91

McBratney AB, Moran CJ, Stewart JB, Cattle SR, Koppi AJ (1992) Modifications to a method of rapid assessment of soil macropore structure by image analysis. Geoderma 53: 255–274

McIntyre DS (1970) The platinum microelectrode for soil aeration measurement. Adv Agron 23: 235–283

Mexal J, Fisher JT, Osteryoung J, Reid CPC (1975) Oxygen availability in polyethylene glycol solutions and its implications in plant water relations. Plant Physiol 55: 20–24

Misra RK (1997) Maximum axial growth pressures of the lateral roots of pea and eucalypt. Plant Soil 188: 161–170

Misra RK, Dexter AR, Alston A (1986) Maximum axial and radial growth pressures of plant roots. Plant Soil 95: 315–326

Mullins CE (1991) Matric potential In: Smith KA, Mullins CE (eds) Soil analysis, physical methods Marcel Dekker, New York, pp 75–109

Painter LI (1966) Method of subjecting growing plants to a continuous soil moisture stress. Agron J 58: 459–460

Panayiotopoulos KP (1989) Packing of sands – a review. Soil Tillage Res 13: 101–121

Pellerin S, Pagès L (1994) Evaluation of parameters describing the root system architecture of field grown maize plants. Plant Soil 164: 155–167

Pfeffer W (1893) Druck und Arbeitsleistung durch wachsende Pflanzen. Abh K Sächs Gese der Wiss 33: 235–474

Pritchard J, Wyn Jones RG, Tomos AD (1990) Measurement of yield threshold and cell wall extensibility of intact wheat roots under different ionic, osmotic and temperature treatments. J Exp Bot 41: 669–675

Richards BG, Greacen EL (1986) Mechanical stresses on expanding cylindrical root analogues in granular media. Aust J Soil Res 24: 393–404

Ruff MS, Krizek DT, Mirecki RM, Inonye DW (1987) Restricted root zone volume: influence on growth and development of tomato. J Am Soc Hortic Sci 112: 763–769

Saab IN, Sharp RE, Pritchard J (1992) Effect of inhibition of abscisic acid accumulation on the spatial distribution of elongation in the primary root and mesocotyl of maize at low water potentials. Plant Physiol 99: 26–33

Saab IN, Ho THD, Sharp RE (1995) Translatable RNA populations associated with maintenance of primary root elongation and inhibition of mesocotyl elongation by abscisic acid in maize seedlings at low water potentials. Plant Physiol 109: 593–601

Sarquis JI, Jordan WR, Morgan PW (1991) Ethylene evolution from maize (*Zea mays L.*) seedling roots and shoots in response to mechanical impedance. Plant Pathol 96: 1171–1177

Scholefield D, Hall DM (1985) Constricted growth of grass roots through rigid pores. Plant Soil 85: 153–162

Schwaegerle KE (1983) A method for maintaining soil moisture availability for potted plants. Soil Sci Soc Am J 47: 608–610

Scott-Russell R, Goss MJ (1974) Physical aspects of soil fertility – the response of roots to mechanical impedance. Neth J Agric Sci 22: 305–318

Sharp RE, Silk WK, Hsiao TC (1988) Growth of the maize primary root at low water potentials. 1. Spatial distribution of expansive growth. Plant Physiol 87: 50–57

Sharp RE, Hsiao TC, Silk WK (1990) Growth of the maize primary root at low water potentials. II. Role of growth and deposition of hexose and potassium in osmotic adjustment. Plant Physiol 93: 1337–1346

Smith K, Robertson PD (1972) Effect of ethylene on root extension of cereals. Nature (Lond) 234: 148–149

Snow MD, Tingey DT (1985) Evaluation of a system for the imposition of plant water stress. Plant Physiol 77: 602–607

Sommer C (1988) Soil compaction and water uptake of plants. Catena Suppl II: 107–111

Souty N, Stępniewski W (1988) The influence of external oxygen concentration on axial root growth force of maize radicles. Agronomie 8: 295–300

Spollen WG, Sharp RE (1991) Spatial distribution of turgor and root growth at low water potentials. Plant Physiol 96: 438–443

Stępniewski W, Bennicelli R (1992) Effect of localized anoxia on maize response. In: Kutschera L, Hubl E, Lichtenegger E, Persson H, Sobotik M (eds) Root ecology and its practical application vol 3. ISRR Symp, Univ Bodenkultur Wien, pp 137–138

Stępniewski W, Przywara G (1992) The influence of soil oxygen availability on yield and nutrient uptake (N, P, K, Ca, Mg, Na) by winter rye (*Secale cereale*). Plant Soil 143: 267–274

Stępniewski W, Pezeshki SR, de Laune RD, Patrick WH Jr (1991) Root studies under variable redox potential in soil using laboratory rhizotrons. In: Kutschera L, Hubl E, Lichtenegger E, Persson H, Sobotik M (eds) Root ecology and its practical application. ISRR Symp, vol 3. Univ Bodenkultur Wien, pp 353–356

Stępniewski W, Bennicelli R, Glinski J, Stepniewska Z (1994) Split rhizotron technique for localized anoxia studies. Trans 15th Congr ISSS, Acapulco, Mexico 2b: 237–238

Steuter AA, Mozafor A, Goodwin JR (1981) Water potential of polyethylene glycol. Plant Physiol 67: 64–67

Tardieu F (1994) Growth and functioning of roots or root systems subjected to soil compaction. Towards a system with multiple signalling? Soil Tillage Res 30: 217–283

Tardieu F, Pellerin S (1991) Influence of soil temperature during root appearance on the trajectory of nodal roots of field grown maize. Plant Soil 131: 207–214

Thaler P, Pages L (1996) Periodicity in the development of the root system of young rubber trees (*Hevea brasiliensis* Muel): relationship with shoot development. Plant Cell Environ 19: 56–64

Tingey DT, Stockwell C (1977) Semipermeable membrane system for subjecting plants to water stress. Plant Pathol 60: 58–60

Unger PW, Eck HV, Musick JT (1981) Alleviating plant water stress. In: Arkin GF, Taylor HM (eds) Modifying the root environment to reduce crop stress. American Society of Agricultural Engineers, St Joseph, MJ, pp 61–66

Van Noordwijk M, Brouwer G (1988) Quantification of air filled root porosity: a comparison of two methods. Plant Soil 111: 255–258

Van Noordwijk M, Brouwer G (1993) Gas-filled root porosity in response to temporary low oxygen supply in different growth stages. Plant Soil 152: 187–199

Van Noordwijk M, Kooistra MJ, Boone FR, Veen BW, Schoonderbeek D (1992) Root-soil contact of maize, as measured by a thin-section techique. 1. Validity of the method. Plant Soil 139: 109–118

Van Ouwerkerk C, Soane BD (1994) Conclusions and recommendations for further research on soil compaction in crop production. In: Soane BD, van Ouwerkerk C (eds) Soil compaction in crop production. Elsevier, Amsterdam, pp 627–642

Veen BW, Boone FR (1990) The influence of mechanical resistance and soil water on the growth of seminal roots of maize. Soil Tillage Res 16: 219–226

Verslues PE, Ober ES, Sharp RE (1998) Root growth and oxygen relations at low water potentials. Impact of oxygen availability in polyethylene glycol solutions. Plant Physiol 116: 1403–1412

Vetterlein D, Marschner H (1993) Use of a microtensiometer technique to study hydraulic lift in a sandy soil planted with pearl millet (*Pennisetum americanum L.* Leeke) Plant Soil 149: 275–282

Voorhees WB, Amemiya M, Allmaras RR, Larson WE (1971) Some effects of aggregate structure heterogeneity on root growth. Soil Sci Soc Am Proc 35: 638–643

Voorhees WB, Farrell DA, Larson WE (1975) Soil strength and aeration effects on root elongation. Soil Sci Soc Am Proc 39: 948–953

Voorhees WB, Allmaras RR, Johnson CJ (1981) Alleviating temperature stress. In: Arkin GF, Taylor HM (eds) Modifying the root environment to reduce crop stress. American Society of Agricultural Engineers, St Joseph, MJ, pp 217–266

Wang J, Hesketh JD, Woolley JT (1986) Pre-existing channels and soybean rooting patterns. Soil Sci 141: 432–437

West DW, Merrigan IF, Taylor JA, Collins GM (1980) Growth of ornamental plants irrigated with nutrient or polyethylene glycol solutions of different osmotic potentials. Plant Soil 86: 99–111

Whalley WR, Dexter AR (1993) The maximum axial growth pressure of roots of spring and autumn cultivars of lupin. Plant Soil 157: 313–318

Whalley WR, Clark LJ, Dexter AR (1994) The temperature dependance of the maximum axial growth pressure of the roots of pea (*Pisum sativum* L.). Plant Soil 163: 211–215

Whalley WR, Bengough AG, Dexter AR (1998) Water stress induced by PEG decreases the maximum growth pressure of the roots of pea seedlings. J Exp Bot 49: 1689–1694

Whiteley GM, Dexter AR (1982) Root development and growth of oilseed, wheat and pea crops on tilled and untilled soil. Soil Tillage Res 36: 370–393

Williams J, Shaykewich CF (1969) An evaluation of polyethylene glycol PEG 6000 and PEG 20 000 in the osmotic control of soil water matric potential. Can J Soil Sci 49: 397–401

Yu L-X, Ray JD, O'Toole JC, Nguyen HT (1995) Use of wax-petrolatum layers for screening rice root penetration. Crop Sci 35: 684–687

Zhu GL, Boyer JS (1992) Enlargement of *Chara* studied with a turgor clamp – growth rate is not determined by turgor. Plant Physiol 100: 2071–2080

CHAPTER 4

Modelling Root System Growth and Architecture

L. Pagès[1], S. Asseng[2], S. Pellerin[3], and A. Diggle[4]

[1] INRA, Unité de Recherches en Ecophysiologie et Horticulture, Site Agroparc, 84914 Avignon cedex 9, France

[2] CSIRO, Division of Plant Industry, Private Bag, PO, Wembley, Western Australia WA 6014, Australia

[3] INRA, Unité d'Agronomie, Domaine de la Grande Ferrade, BP 81, 33883 Villenave d'Ornon cedex, France

[4] Western Australian Department of Agriculture, Baron-Hay Court, South Perth, WA 6151, Australia

CONTENTS

A.L. Smit et al. (Eds.), Root Methods

4.1 Introduction

For any complex system, and particularly for the root system interacting with the rest of the plant and the environment, a model can be a helpful tool for synthesising knowledge to produce more global representations and for testing hypotheses on the interacting mechanisms derived from experimental results. The models presented here are mainly devoted to a description of the kinetics of root system development and the spatial arrangement of roots, with different viewpoints and varying levels of detail according to different objectives. They have been separated into two main categories that correspond to different ways of considering the root system.

In the first category are models of root depth and root length density which aim to describe the root system relative to the soil volume, mapping the presence (or absence) of roots and/or their density (e.g. root length or root mass per soil volume). The root system is considered as an abstract object whose selected characteristic is the spatial distribution of roots, which is described like a concentration throughout the soil. These models are generally one-dimensional, and they allow the simulation of various types of root profiles (especially root length density versus depth). They give a simple representation of the shape of the plant/crop root system in which all roots are equivalent and are not explicitly connected. These models have been developed mainly as part of dynamic hydraulic, plant/crop or ecosystem models. Their ultimate aim is to simulate carbon, water and nutrient cycles.

The second category of models considers the root system as a set of connected axes that grow and branch according to a number of production rules which formalise morphogenetic rules and local (at a root scale) interactions with the environment (mainly the soil). Therefore, they simulate the whole architecture of the system, i.e. not only its shape (spatial arrangement of it components) but also its structure and topology. The structure refers to the specific identification of its components, the roots, and to their relationships (connections, branching patterns, and morphogenetic gradients). Hence, these models are more specifically devoted to the study of root system structure and morphogenesis. The simplified representation which they give allows inference on different aspects of root functioning, which depend on both the shape of the system (e.g. water and nutrient transport from the soil to the roots), and its structure (e.g. uptake and transport within the root system). These root system models are sometimes part of more integrated plant growth models.

4.2 Root Depth and Root Length Density Models

Root depth and root length density models simulate the shape of a root system in terms of root elongation to the depth and root length density distribution over the soil profile. They are based on the assumption that the root system is usually in contact with several parts of the soil profile which differ in water, air and nutrient content, texture, bulk density, organic material and pH, whereas the soil profile is generally considered to consist of layers, and the conditions in each layer are treated separately. These root models commonly abstract from the dynamics of a real root architecture to homogeneously distributed roots in each of the horizontal soil layers. In some cases, only the single process of downward penetration of the deepest roots or the active rooting front (in terms of uptake capacity) is simulated.

Root models vary in their complexity both in simulating a root system and in the response of the roots to their environment and are often part of more complex crop models. Some of them have been reviewed by Klepper and Rickman (1990).

A summarised overview of root models is given in Table 4.1.

4.2.1 Root Depth Models

One of the simplest concepts of root modelling is based upon the assumption that the roots in each centimetre of rooted depth absorb the same amount of water and nutrients, regardless of the quantitative presence of roots and provided that all layers are equally moist. As a result, only the advancing rooting front or deepest root without root distribution is simulated. Some of these models calculate a root system using a predetermined downward penetration rate to a maximum depth (Wessolek 1993; P.D. Jamieson pers. comm. 1996). In others, depth increases as a function of air temperature (T.R. Sinclair pers. comm. 1993) or phenological growth stage (Chapman et al. 1993). Another example in this category of models is the one by Borg and Grimes (1986). They developed a description of the time course of rooting depth as a sine function on a normalised scale without considering any environmental growth impact.

The root models by Stapper (1984), O'Leary et al. (1985), Groot (1987) and Penning de Vries et al. (1989) also simulate a growing root system only as an increasing rooting depth, but they account for some soil constraints. Most of these models include a restriction of the downward penetration due to unfavourable soil moisture. Some additionally consider soil temperature

Table 4.1. Summarised root depth and root length density models according to crop specification, with effects on shoot:root partitioning, root depth and root distribution

Author	Species	Impact on shoot:root	Depth	Distribution
Adiku et al. (1995)	G			θ
Addiscott and Whitmore (1987)	W		T_A	θ, N
Andrew (1987)	W	C_G	R_S, C_{Nu}	
Asseng et al. (1997b)	W	C_P, C_W, C_N	C_W, θ, A, T, R_S	θ, A, N, R_S
Chapman et al. (1993)	Sf		C_P	
Denison and Loomis (1989)	L	C_P	θ, T	θ, T
Groot (1987)	W	C_P	θ, T	
Hoogenboom and Huck (1986)	G	C_W	θ	θ, R_S
Jakobsen and Dexter (1987)	W	C_G	θ, T, R_S	θ, T, R_S
Johnson and Thornley (1985)	Gr	C_P, C_N		
Jones and Kiniry (1986)	M	C_G, C_W		θ, H
Jones et al. (1991)	M, Sb ,W		A, Al, R_S	A, Al, Ca, R_S, H
Keating et al. (1998)	W	C_G, C_W, C_N	C_W, θ, H	θ, H
O'Leary et al. (1985)	W	C_P, T_A	T_A, θ	
Penning de Vries et al. (1989)	G		C_W, A, T, R_S	θ
Ritchie et al. (1985)	W	C_G, C_W		θ, H
Robertson et al. (1993)	Sg	C_G		θ, H
Skiles et al. (1982)	Gr	C_P, C_G		
Spek and Oijen (1988)	M	C_P, C_N		
Stapper (1984)	W	C_P	C_P, C_W, θ, T	
Stol et al. (1993)	W	C_P	θ, T	
Van Keulen and Seligman (1987)	W	C_P, C_W, C_N	C_W, C_N, R_S	
Voit and Sands (1996)	Trees	[a]C_G, [a]C_N		
Williams et al. (1984)	M, Sg, W, B, O, Sf, Sb, L Co, Gn, Gr	T_A	T_A	θ, A, T, Al, R_S

[a] C refers here to trees instead of crops.

[b] Abbreviations used: G general crops, W wheat, B barley, O oat, M maize, Sb soybean, Sg sorghum, Co cotton, L lucerne, Sf sunflower, Gr grasses, Gn groundnut, C_P crop phenology, C_G crop biomass growth, C_W crop water deficit, C_N crop nitrogen deficit, C_{Nu} general nutrient status of crop, T_A air temperature. And for soil properties: θ moisture, A aeration, T temperature, N nitrogen, Al aluminium, Ca calcium, R_s soil strength, H user-defined hospitality.

(Stapper 1984; Groot 1987), soil aeration (Penning de Vries et al. 1989) or soil resistance (Van Keulen and Seligman 1987) at the deepest penetrated soil layer for calculating the advancing root. An acceleration of the downward penetration can result from plant water (Stapper 1984) and plant nitrogen deficit (Van Keulen and Seligman 1987). However, the model by Penning de Vries et al. (1989) reduces downward penetration with increasing plant water deficit.

4.2.2 Root Length Density Models

The simplest empirical descriptions of root systems which result in more than just root depth are models which calculate the distribution of roots with depth under non-limiting growing conditions. Examples are the models by Page and Gerwitz (1974), Gerwitz and Page (1974), Monteith et al. (1989) and Nielsen and Mackenthun (1991). Page and Gerwitz (1974) introduced a mathematical model where roots grow in a manner analogous to the diffusion movement of solutes in a liquid medium, by calculating the radial root distribution starting from a point or line source into a infinite medium. Monteith's model describes declining root length with depth as an inverse square root function of depth, whereas the model by Gerwitz and Page (1974) assumes an exponential decline in root length density with depth, and supplies the percentage of roots between the surface and any soil depth:

$$P_i = 100(1 - e^{-\alpha z}),$$

where P_i is the percentage of roots contained in a soil depth of z cm and α is a parameter, such that $1/\alpha$ is equal to the depth of soil which contains 63% of the total root mass.

In another model by Nielsen and Mackenthun (1991) the root intensity of single trees is calculated (in root mass per soil volume) as a function of the basal stem area and exponentially declines with the distance from the tree.

4.2.3 Root Depth-Root Length Density Models

A more complex class of models simulates a root system by exploring the soil in two separate growth processes: (1) downward penetration of an abstract vertical axis and (2) proliferation of root length in any given soil layer within the penetrated depth. In these models the downward penetration is independently calculated from the proliferation of roots within the soil layers. This approach is employed in a range of crop models, for example the CERES-models by Ritchie et al. (1985) and Jones and Kiniry (1986), the crop models by Hansen (1975), Andrew (1987), Denison and Loomis (1989), Robertson et al. (1993), Asseng et al. (1997b) and Keating et al. (1998), and the stand-alone root models by Hoogenboom and Huck (1986) and Jones et al. (1991). In "root depth-root length density models", the downward penetration serves as the basis for root distribution over the soil profile and therefore limits the root distribution with depth at a given time. Root development over the soil profile is then usually distributed proportional with depth. This characteristic, together with a continuously advancing rooting depth usually results in a negative exponential function of monotonic decrease of root distribution with soil depth.

In root depth-root length density models, calculation of downward penetration is usually similar to the root depth "only" models. Downward penetration can be driven by air temperature without being affected by any soil properties, as for example in the CERES models (Ritchie et al. 1985; Jones and Kiniry 1986) or as in Addiscott and Whitmore (1987), who extended the root distribution algorithm of Gerwitz and Page (1974) by incorporating a downward penetration rate as a function of degree-days. However, the downward penetration rate can also be a constant (Robertson et al. 1993), or additional unfavourable soil properties can be taken into account (Jones et al. 1991). Keating et al. (B.A. Keating, pers. comm.) use a user-defined soil layer specific hospitality factor to account for static soil impacts on the downwards root penetration rate. The carbon supply from the shoot (Denison and Loomis 1989) or the plant water deficit can also be considered for the downward root penetration rate (Asseng et al. 1997b; Keating et al. 1998).

The model by Jones et al. (1991) simulates downward penetration in a similar fashion to proliferation in each layer in that penetration is restricted by the most limiting constraint. The restriction results from soil strength, aluminium toxicity, aeration and coarse fraction at the deepest penetrated soil layer. In the calculation of downward penetration, other models take into account the phenological stage of the plant as well as soil strength (Andrew 1987), soil moisture (Asseng et al. 1997b; Keating et al. 1998) or soil temperature at the deepest rooted layer (O'Leary et al. 1985). The following equation summarises the structure of the rooting depth calculation in most of these models:

$$R_{act}(t) = R_{pot} S_z C(t),$$

where $R_{act}(t)$ is the actual downward penetration rate at time t, R_{pot} is the potential downward penetration rate which might be a function of crop development stage or air temperature, S_z is the soil physical constraint at the deepest rooted depth z and $C(t)$ is the water or nitrogen deficit status of the crop at time t. Note that in the models by Ritchie et al. (1985), Addiscott and Whitmore (1987) and Robertson et al. (1993) soil constraints influence only the root proliferation within soil layers and not the downward root penetration.

4.2.4 Soil Impacts on Root Distribution

Addiscott and Whitmore (1987) have modified the simple root growth model by Gerwitz and Page (1974) to be sensitive to its environment by incorporating a growth factor F_i. This factor considers the effect of varying soil moisture and mineral soil nitrogen contents on root growth distribution, calculated as:

$$F_i(t) = N_i(t)\theta_i(t) / (\overline{N}\ \overline{\theta}),$$

where N_i is the quantity of mineral nitrogen and θ_i the water content in layer *i*. $\bar{N}$ and $\bar{\theta}$ are the means over the rooted soil profile. Note that the model by Addiscott and Whitmore (1987) does not simulate root length density development. However, it defines a fraction of water and solute that is root accessible in a layer *i*. Nevertheless, most root length density models distribute the new growth over the rooted soil profile by using the same or a similar approach. For example Jakobsen and Dexter (1987) partition available carbon between layers according to a potential root growth in each layer, which is driven by a number of soil factors. A slightly different method is used in the model by Asseng et al. (1997b) where, in a first step, all carbon for root growth is offered to the topsoil layer. Only under unfavourable growing conditions in that layer is carbon passed to the next deeper layer and so on, until the layer with the deepest root is reached and all remaining carbon is used in that layer. In case of an aeration deficit in a soil layer reached by the root system, carbon is then given back to the topsoil for root growth.

Impacts of soil moisture on root growth distribution are taken into account in most root distribution models, such as those by Huck and Hillel (1983), Ritchie et al. (1985), Hoogenboom and Huck (1986), Penning de Vries et al. (1989), Asseng et al. (1997b) and Keating et al. (1998). Other effects on root growth considered in these models are soil nitrogen (Asseng et al. 1997b; Keating et al. 1998), soil temperature (Huck and Hillel 1983; Groot 1987; Jakobsen and Dexter 1987), soil resistance (Jakobsen and Dexter 1987; Van Keulen and Seligman 1987; Robertson et al. 1993; Asseng et al. 1997b), soil cracks (Jakobsen and Dexter 1987), soil aeration (Williams et al. 1984; Penning de Vries et al. 1989; Von Pfeil et al. 1992; Asseng et al. 1997a,b), aluminium toxicity (Williams et al. 1984; Hoogenboom and Huck 1986) and calcium deficiency (Hoogenboom and Huck 1986). Simulated root length density distributions affected by different initial soil nitrogen concentrations and its change over time are shown in Fig. 4.1 using the root model by Asseng et al. (1997b).

The root model by Jones et al. (1991), combines all of the above growth impacts, but does not account for soil nutrients and cracks. Additionally, Jones' model considers the effect of coarse structure in the soil on root proliferation. The growth impacts are expressed as stress factors, which range from 0.0 (no growth) to 1.0 (no stress). In cases where more than one property affects a process, the property with the most unfavourable stress is considered limiting in the Jones model. Note also that in the Jones model the soil water content does not directly influence root growth, but indirectly via its effect on soil strength. Some root models employ an extra root soil layer specific hospitality factor, to account for static soil impacts on growth distribution or genotype differences in root morphology. These factors can be user-defined estimates (Ritchie et al. 1985; Keating et al. 1998) or be defined within the model (Jones et al. 1991; Robertson et al. 1993). In the root model by Keating et al. (1998), hospitality

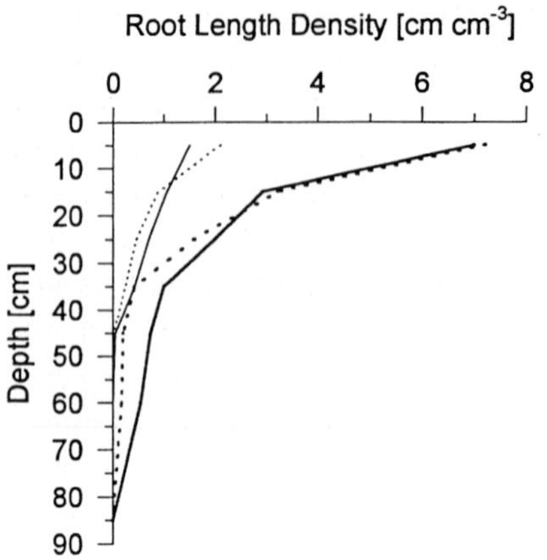

Fig. 4.1. Simulated root distribution under a winter wheat crop in early spring (*thin lines*) and 50 days later (*thick lines*) with an initial high (——) and an initial low nitrogen concentration (.........) in the soil profile. (After Asseng et al. 1997b)

factor for root distribution is defined independently from the hospitality factor for root depth penetration. A few root models consider the previous root distribution in the calculation of new root growth (Hansen 1975; Denison and Loomis 1989; Jones et al. 1991; Adiku et al. 1995; Asseng et al. 1997b).

A quite different approach in estimating the change of root distribution in cereals with time is the one introduced by Wessolek and Gäth (1990). First, root distribution is calculated at anthesis on the basis of field capacity and wilting point of each soil layer. Then root distribution is estimated for pre- and post-anthesis development stages taking into account the soil type and extreme wet and dry seasons. While it is not a dynamic model, calculating the changing root distribution with time in the way that most crop models do, it still allows for estimating root distribution over a growing season.

4.2.5 Shoot-Root Interactions

One style of root modelling involves the partitioning of carbohydrates to dictate below-ground biomass. In some models the partitioning coefficient is assumed to be a constant fraction of the daily carbon produced as for example in the models by Hansen (1975), Jakobsen and Dexter (1987), and Adiku et al. (1995). Robertson et al. (1993) assume a constant conversion factor for root growth from the production of above ground biomass. The models by Stapper (1984) and Van Keulen and Seligman (1987) have a phenologically controlled shoot: root ratio for plant internal carbon distribution. Other models, for example those by Huck and Hillel (1983), Hoogenboom and Huck (1986) and Asseng et al. (1997b) have a phenologically controlled shoot:root ratio component which is additionally modified by plant water deficiency. A further extension used by Johnson and Thornley (1985), Spek and Oijen (1988), Asseng et al. (1997b) and Keating et al. (1998) is the control of the shoot:root ratio by plant nitrogen deficit. The rate at which carbon is utilised by the shoots and roots depends upon their masses and the internal concentration of carbon and N in a model by Thornley (1977). In another model, Reynolds and Thornley (1982) adjust the shoot : root partitioning of carbohydrates according to the ratio of carbon : nitrogen in the plant storage pools. An additional parameter is used which determines the degree of control that the plant exhibits over partitioning. Some of the above approaches to modelling shoot : root carbon partitioning are relevant to trees, if roots are redefined as only the fine roots, and shoots as the foliage (Cannell 1985). In a tree model by Voit and Sands (1996) carbon partitioning to the root system is affected by an internal N level and foliage biomass. A temperature-dependent partitioning is introduced by O'Leary et al. (1985) as temperature either rises or falls from an optimum level.

Under optimal growing conditions, shoot growth is favoured in all of these models, but roots become more favoured as the conditions change. The following equation modified after Adiku et al. (1995) generalises the shoot : root interaction in a range of these models:

$$\Delta W_r(t) = X_r C(t) \Delta W_g(t),$$

where ΔW_r is the availability of carbon for the root system, X_r is a constant fraction of plant-available carbohydrates partitioned to the root which is modified by a phenologically driven shoot : root ratio, water or nitrogen deficit or sub-optimal growth temperature represented by C. ΔW_g is the overall amount of crop-available carbon for growth at time t.

Note that some of the root models in the literature do not take into account any shoot : root interactions like, for example, the models by Jones et al. (1991), Addiscott and Whitmore (1987) and Williams et al. (1984).

Root models, which consider a carbon supply from the shoot, usually transform it into length as a basis for later uptake modelling. This transformation can be done by using a constant parameter for length : mass ratio, hence any environmental impact is not taken into account (Huck and Hillel 1983; Ritchie et al. 1985; Hoogenboom and Huck 1986) or by considering soil strength (Jakobsen and Dexter 1987; Jones et al. 1991; Asseng et al. 1997b) and other constraining soil physical properties such as aluminium toxicity and soil coarse fraction (Jones et al. 1991).

4.2.6 Exudation, Mortality, Respiration and Activity

Some of the root models with carbon coming from the above-ground crop consider a constant (Ritchie et al. 1985; Denison and Loomis 1989) or a plant development dependent (Asseng et al. 1997b) exudation rate which is a percentage of the carbon supplied to the root system for growth.

A mortality rate expressed as a percentage reduction of the current root system can be calculated as a constant (Ritchie et al. 1985; Denison and Loomis 1989; Robertson et al. 1993) or as a fraction which changes with the plant phenological stage (Asseng et al. 1997b) and soil aeration (Asseng et al. 1997a,b) or the internal N level in trees (Voit and Sands 1996). Others consider soil temperature and available carbohydrate reserve levels (Huck and Hillel 1983) or soil aeration and soil moisture (Hoogenboom and Huck 1986; Jones et al. 1991) in calculating the mortality rate. The model by Skiles et al. (1982) describes root mortality and root respiration as a function of soil temperature and plant water deficit only.

In the model by Asseng et al. (1997b) age classes with different uptake capacities are applied to roots grown in a soil layer. As a result, the uptake capacity declines with the advancing age of the roots. Johnson and Thornley (1985) divide the root structure into four compartments (growing, newly expanded, medium-aged and senescent). This compartmentalisation allows separate uptake parameters and respiratory maintenance costs to be inserted for different ages of root.

4.2.7 Conclusions

The model by Gerwitz and Page (1974) is often utilised in different crop models because of its simplicity and low parameter requirement. With a few additional expansions, such as downward penetration and soil impact factors (Addiscott and Whitmore 1987), it can be sufficient for a range of simulation requirements.

Probably the most commonly employed root model is the one incorporated in the CERES-models (Ritchie et al. 1985; Jones and Kiniry 1986). This root model is widely tested as part of its crop model (e.g. Otter-Nacke et al. 1986; Touré et al. 1995) but is also frequently criticised because of its overestimation of downward root penetration (e.g. Savin et al. 1995), which might be due to neglecting soil constraints for the downward penetration. The model by Keating et al. (1998) has derived its root routines from the CERES-models, but incorporates soil water, crop water status and a user-defined hospitality factor for the downward root penetration. In particular, the introduction of an user-defined hospitality factor made this model more applicable for a range of soil types with severe root growth constraints (e.g. Asseng et al. 1998).

Jones et al. (1991) constructed the most comprehensive root length density model in terms of considering soil constraints. It simulates a root system with most of the possible soil constraints which occur on agricultural land and offers a set of required parameters suitable for a range of crops. To be incorporated into a crop model, however, it still needs to be linked via shoot : root interaction routines as introduced in other models.

With more soil impacts incorporated in root models, as well as simulating the shoot : root interactions, it becomes more and more difficult for model users to actually measure the increased number of parameters and to verify the large range of possible scenario combinations. Therefore, the use or further development of root length density models should be closely linked to the target application of the model, the available knowledge of root-soil and root-shoot interactions, and the practical possibilities of verification under field conditions. As a consequence, any further root growth modelling exercise should aim to optimise the balance between complexity and large numbers of parameters on the one hand and simplicity and low parameter requirements on the other.

4.3 Models of the Root System Architecture

These models consider the root system as a set of connected axes, i.e. a tree in the mathematical sense, with nodes (vertices) and links (edges). Thus, they manipulate objects which are more concrete than the root length density models described in the previous section. Their concepts are closer to those used by botanists to describe real root systems, and they allow both geometrical and structural descriptions of the root system.

Root architecture has been described with a more or less dynamic perspective, depending on the objectives and the experimental methods used, thus leading to different types of models. This distinction between static models (aiming to describe root systems as particular branching patterns) and dynamic models (aiming to generate root systems from developmental rules) is very apparent in the literature. Nevertheless, bridges exist between these two approaches, as underlined by Shibusawa (1994) who simulated pseudo-fractal objects with L-Systems (Prusinkiewicz and Lindenmayer 1990).

4.3.1 Describing Branching Patterns

There are several interesting methods for describing the root systems as particular branching patterns.

4.3.1.1 Statistical Descriptions

Henderson et al. (1983*a*) analysed the branching pattern (topology and geometry) of the structural root system of 16-year-old trees. They showed that it could be described by a small number of characteristics (lengths, growth directions, bends, branching angles, and branching frequencies) whose distributions were unexpectedly regular. This regularity contrasted with the global morphology of the root systems, which appears in contrast to be irregular.

The simulation of rooting patterns at the same stage, with a stochastic model based upon these characteristic distributions (Henderson et al. 1983b), confirmed the regularity of the geometric characteristics that were measured. Moreover, the simulations allowed definition of the possible variations of these characteristics for obtaining realistic rooting patterns.

4.3.1.2 Hydraulic Networks

This approach has been developed by Fitter and colleagues, from 1982 (Fitter 1982) onwards (Fitter 1996). The general idea was to characterise the root system architecture, and especially the topology, using the mathematical tools that had been developed in hydrology in order to describe drainage networks.

For that purpose, the root system is defined as a set of links [linear portion between a terminal meristem and a branching point (external link) or between two branching points (internal link)], being characterised by a few topological parameters: its magnitude (number of external links); altitude (the number of links in the longest path between an external link and the base link); topological index (altitude divided by expected altitude in the case of random branching). Two extreme topological types of branching patterns can be distinguished (Fig. 4.2): the "herringbone" pattern, in which branching is localised only on a main axis (high hierarchy), and "dichotomous" pattern, in which branching occurs on every axis, dividing into equivalent axes (low hierarchy). Between these two

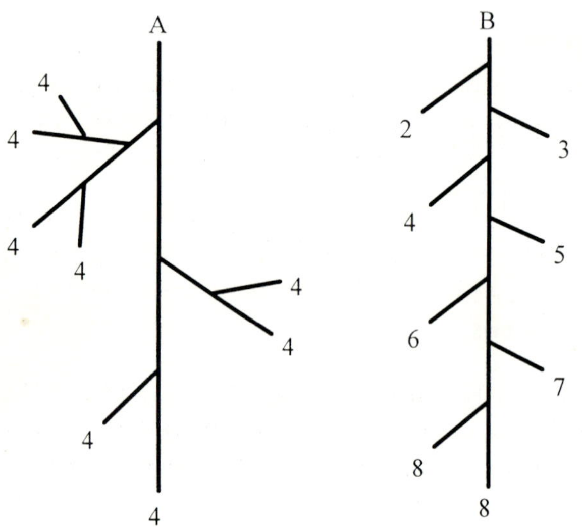

Fig. 4.2. Diagrammatic representations of **A** dichotomous and **B** herringbone topologies. *Numerals* are exterior path lengths, the number of links in the path from each exterior link to the base link. (After Fitter 1991)

extreme types, actual root systems present many intermediate patterns that can be globally quantified by these cited parameters. This method is potentially useful for the description of functional properties of root systems. It has been shown to be particularly suitable for describing excavated root systems on which it is not always possible to distinguish the developmental branching orders.

4.3.1.3 Fractals

Fractals are mathematical objects (Mandelbrot 1983) that were developed to describe natural structures with self-similarities at different observation scales. For root systems, the existence of self-similarities means that the overall branching patterns may be divided into sub-parts that have approximately the same appearance as the overall system, and again these sub-parts may be divided into smaller parts having the same appearance, and so forth. Fractal models are particularly relevant for natural objects if their self-similarities are high, that is when zooming-in does not modify the morphological properties of the object, and when the number of self-similarities levels is high, that is when the number of levels for zooming-in is high. For root systems, the number of self-similarities is determined by the number of branching orders, so it is associated with both genotype and environmental influences. Their "quality" depends on how branches at different levels conform to a common morphological model.

These particular properties of a branching pattern can be evaluated in the plane, for two-dimensional objects, by looking at the relationship between the number (N) of contacts of the object under study with the square cells of a grid, and the side r of these square cells. For fractal objects, the relationship is characterised by the simple equation:

$$N = Kr^{-D},$$

with K being a proportionality coefficient, and D being the fractal dimension. Note that a straight line has a fractal dimension of 1, whereas a plane filling object has a fractal dimension of 2. According to Tatsumi et al. (1989), D is a synthetic parameter to quantify the intricacy of shape.

Tatsumi et al. (1989) showed that root systems from different species could be approximated to fractal objects, with self-similarity within the scale range 0.3–20 mm. Moreover, they estimated fractal dimensions D range from 1.48 to 1.58 depending on the species.

The value of D quantifies an overall space-filling property, and so it may be very useful for quantifying globally colonisation by the root system. The problem is that the relationship between D and biological or geometrical parameters (e.g. length of roots, branching density, number of branching orders) is not straightforward.

Moreover, the application of this method using images, which are two-dimensional objects, for evaluating the fractal dimension of 3-dimensional objects raises a number of questions (Berntson 1994b).

4.3.2 Formalisation of Developmental Rules

Unless previous models devoted to the description of established root systems, these ones aim to generate theoretical growing root systems with a set of production rules (Prusinkiewicz and Lindenmayer 1990) i.e. formal translations of the observed developmental rules.

4.3.2.1 General Rules

These models are essentially single-plant models, simulating each plant separately, even though it is still possible to create theoretical crops by simulating and gathering several plants, providing that the parameters would be estimated in the same field conditions (Pagès et al. 1989). They are more or less stochastic, since they are made up of different sub-models, with some of them being deterministic, and others being stochastic. Consequently, each run gives a different simulated root system.

The existing models (e.g. Pagès and Aries 1988; Fitter et al. 1991; Clausnitzer and Hopmans 1994) work in approximately the same way. At a given time, the virtual root system is coded as a list of linear root segments, which describe and store the characteristics of these segments: geometric coordinates, time of origin, and references to topological position within the root structure (e.g. branching order and connection with other segments; Pagès and Aries 1988; Fitter et al. 1991; Clausnitzer and Hopmans 1994). The development of the root system is simulated in discrete time steps, by applying three formalised developmental processes to the existing root system at each time step. These are root emergence (i.e. generation of new root axes from the shoot system), growth (i.e. elongation of existing axes), and branching (development of new lateral axes). All developing axes of the structure are modified according to the growth and branching they experience at the given time step.

One of the important characteristics of these models, compared with root density models (see previous Sect.), is that they do not assume that growth and branching processes are uniform among root axes. The simulations may take both the topological position and the spatial position into account. This allows the simulation of different developmental behaviours of the axes according to their morphogenetic origin. A preliminary axis typology is necessary to distinguish axis types with analogous behaviours. These types often correspond to the branching orders (Diggle 1988; Pagès and Aries 1988; Fitter et al. 1991; Clausnitzer and Hopmans 1994). They may depend also on inter-node of origin for maize (Pagès et al. 1989), or they may be the result of a more complex typology combining branching order and development process (Pagès et al. 1995). In several recent models (Diggle 1988; Pagès et al. 1989; Clausnitzer and Hopmans 1994), growth and branching are functions of the local conditions as well. These conditions are either fixed at the beginning of the simulation (Diggle 1988; Pagès et al. 1989) or they can be time-dependent (temperature in some models, like Porter et al. 1986, or Clausnitzer and Hopmans 1994), or be the output of another combined model (e.g. water potential in Clausnitzer and Hopmans 1994).

4.3.2.2 Root Emergence

This process generates new root axes, directly originating from the shoot system. They will be called order 1 axes or axile roots in the following.

By the classification of Cannon (1949), two main types of root systems may be distinguished, according to the importance of the root emergence process: primary root systems, and adventitious root systems. Primary root systems originate entirely from a single root – the radicle – which branches and gives rise to lateral roots of order 2, order 3, and so on. In this case, the root emergence process is reduced to emergence of the radicle, some time after germination. Most roots in such systems originate from the branching process. In adventitious root systems, several axile roots are emitted from the shoot system throughout plant development. This type of root system is typical of grasses. In several different species, it has been shown that the root emergence process is highly organised in space and time, and requires a specific modelling approach. In maize, for example, axile roots appear sequentially on the successive basal phytomers from the lowest ones to the highest. The rank of the phytomer on which the emission occurs at a given time was modelled as a linear relationship of growing degree days (Pagès et al. 1989; Pellerin 1993). In a similar way, in wheat, each nodal axis pair is initiated after a given thermal time has elapsed (Diggle 1988). Moreover, a probability of occurrence for each axis of the pair can be used to simulate variations in the number of such axile roots (Diggle 1988).

4.3.2.3 Growth Process

To obtain a three-dimensional representation of a root system in which the trajectories of the roots are not completely straight, the models have to consider two aspects of growth: elongation rate, and growth direction.

Elongation Rate. Most, if not all, of the root architecture models use different growth rates for different root types. Even in a homogeneous environment, between-root variation in growth behaviour is high (Fitter 1991, 1996; Waisel and Eshel 1991), and these models typically allow the representation of this bulk variation.

Moreover, in several models devoted to field simulations (Diggle 1988; Pagès et al. 1989; Clausnitzer and Hopmans 1994), attempts were made to model elongation rate as a root-type-dependent rate modified according to various soil factors.

Elongation rate generally decreases as root branching order increases. Axile roots may have growth rates up to several centimetres per day whereas order 3 branches grow only some millimetres per day (Diggle 1988; Pellerin and Pagès 1996). The basic growth rates are constant (linear growth) in several models (Hackett and Rose 1972; Rose 1983; Porter et al. 1986; Diggle 1988), but more recently, several authors have suggested different refinements for representing either continuous, decreasing, or determinate root growth (Pagès et al. 1992, 1995; Berntson 1994a). Clausnitzer and Hopmans (1994) have suggested representing all types of growth function as series of data points where values between the points are calculated by linear interpolation.

Stochastic growth models have also been developed to describe the large variations in growth rate that can occur within root types (Pagès et al. 1992). Stochastic methods were used because part of this variation in growth rate was considered to have an endogenous origin, which could not be described by a deterministic model at the scale being considered. In these stochastic growth models, some of the parameters of the growth functions are drawn at random for each root from normal or lognormal distributions.

The environmental alteration of the "morphogenetic" component of the growth rate has been simulated in several models (Diggle 1988; Clausnitzer and Hopmans 1994) by multiplying it with an impedance factor (from 0 to 1) calculated from several different soil characteristics. In Diggle's model, the impedance factor (called soil resistance) is specified by the user as a parameter for each soil layer. In Clausnitzer and Hopmans' model, the impedance factor is the result of a submodel calculation that uses soil water content and bulk density as inputs.

Growth Direction. The growth direction of the roots is obviously an important component of root system architecture, because some of the long roots have very specific behaviours (e.g. vertical or horizontal growth) which strongly determine the overall shape of the system, including such important characteristics as overall width and depth (Coutts 1989). At a smaller scale, roots generally exhibit some tortuosity in response to the mechanical constraints which they experience.

The different aspects which determine the trajectories of roots have been simulated in architecture models (Diggle 1988; Pagès and Aries 1988; Pagès et al. 1989; Berntson 1994a; Clausnitzer and Hopmans 1994) by modifying the direction of growth of each root tip at each time step. In these models the influences of the various factors which affect direction are summed to produce a net directional effect (e.g. by adding mathematical vectors, whose modules are parameters that are defined for each root type). In one of the most accomplished descriptions (Clausnitzer and Hopmans 1994), these components include the direction of the root tip at the previous time step, a random deviation representing the space-exploring behaviour of the root tip, geotropism along an angle with the horizontal plane that is user-specified, and the negative soil strength gradient.

This type of direction model can generate quite realistic representations of the major trends of actual trajectories, even though the small scale tortuosity is not very well rendered.

4.3.2.4 Branching Process

The description of the branching process in a root architecture model must specify the locations and times of appearance of branches on the mother root as well as their initial direction. Subsequent changes in direction would be calculated by the growth submodel (see previous subsection).

Branch Location. Most of the models presented here have dealt only with acropetal branching, which is the predominant type of root branching in many species. This acropetal branching takes place on young parts of the root, and typically results in a branching front which follows the apex of the mother root at a more or less definite distance (generally some centimetres). The acropetal branching process has been formalised in slightly different ways by the various authors. Some of them (Hackett and Rose 1972; Rose 1983) described it as a purely temporal process that takes place at given time intervals, and at a given time lag after emergence of the mother root. Conversely, Lungley (1973) or Pagès and Aries (1988) described it by spatial parameters: an inter-branch distance, and a minimal distance to the apex of the mother root. According to these models, branches appear distally from previous branches, at a fixed given inter-branch distance, until the distance from the last branch to the apex is less than a threshold distance (the apical non-branching length).

Other authors (Diggle 1988; Pagès et al. 1992; Clausnitzer and Hopmans 1994) combined both temporal and spatial hypotheses to model the acropetal branching, considering that branches appear at given fixed distances from one to the other (inter-branch distance), and that they can only appear on parts of the mother root having reached a given age (duration of apical non branching).

Fitter et al. (1991) proposed a stochastic model for the location of branches, defining potential nodes for branching which are created during the growth process. These potential nodes are given a branching probability, and branching is tested during a temporal window.

In all these models, the branching parameters are defined for each root type, with branching densities which tend to decrease with developmental branching order. To our knowledge, no models integrate relationships between root branching characteristics and soil properties.

Even though acropetal branching is very important in root system development, other types of branch roots may appear out of the acropetal sequence in some species. On rubber tree for example, Le Roux and Pagès (1994) described "late lateral roots", that appear near the base of the taproot in zones where density is low, and that have an important fate as main and perennial lateral roots. The emergence of such roots has been observed and modelled as a rhythmic process synchronised with the development of shoot flushes (Pagès et al. 1995). This type of branching process can easily be implemented as an extension of the root emergence process that was described previously.

Branch Initial Direction. The initial direction of branches is generally characterised by two angles between branch roots and their mother roots: a radial angle (or azimuth), and an insertion angle (or branch angle). Lateral roots generally appear on a defined number of ranks, which face the internal xylem poles. Nevertheless, the geometrical pattern of root branching is much less clear than the phyllotaxy for the shoot system.

Berntson (1994a) assumed that lateral roots tend to emerge at successive adjacent ranks, resulting in spiral sequences, like phyllotaxy. Other models (Pagès and Aries 1988; Fitter et al. 1991) formalised the emergence as a random sequence on a defined number of ranks. Diggle's branching direction model is still less structured, since branch roots are randomly oriented around the mother root.

The insertion angle is either fixed (Diggle 1988) or drawn at random from a normal distribution (Pagès et Aries 1988).

Root branching distribution, density, and direction, might also be related to tortuosity. It is well known, for example, that where a root is buckled, or even mildly curved, a lateral root will form preferentially on its convex side and not on the concave side.

4.3.2.5 Mortality and Abscission

Root mortality and turnover are important aspects of the development and function of root systems, especially in perennial plants like trees, but to date this process has received little attention in root architecture models (Pagès et al. 1995). This is probably because of the difficulties involved in formalising this mortality process, and also because many of the models have been devoted to the description of young plants, up to some weeks, in which mortality is not yet a dominant phenomenon. The difficulties with formalisation are due firstly to the determinism of mortality, which is complex, and secondly because morphological markers are not reliable and are linked to functional aspects.

In an attempt to include root decay as an aspect of root architecture dynamics in their model on *Hevea brasiliensis* trees, Pagès et al. (1995) considered root abscission to be the final, and most reliable marker of root decay. Root abscission is very clear at the base of old perennial axes that become gradually bare. Decay was modelled using the concept of life expectation. A life expectation value was attributed to each root according to a normal distribution with parameters for each root type. Fine roots are fragile, and have a short life expectation, whereas some of the big roots are perennial. According to this model, a root is partially or completely eliminated after a given duration (life expectation), calculated from the moment when it stops growing. It is completely eliminated when it does not support any living branch roots. Otherwise, only its most distal end, beyond the last living branch, is eliminated.

4.3.3 Parameter Estimation

The problem of parameter estimation is a subset of the more general problem of root observation methods (see other chapters of this book). One major interest of the modelling approach is to drive the observer directly to a selected set of morphological criteria that are worth observing, because they allow estimation of model parameters. Furthermore, the repetition of identical substructures within the root system, which are clearly identified by the model, allows interesting reasoning for sampling design. In some species, some parameters were shown to be very stable over a wide range of environmental conditions, or easily predicted (e.g. the rate of appearance of adventitious roots on maize), so that the parameter estimation can focus on parameters which are more dependant on environmental conditions. The parameters to consider obviously depend on the model under use, but it is possible to distinguish between the parameters that need a dynamic observation of the modelled process (e.g. elongation rate), from those that can be estimated from purely static observations (e.g. trajectories).

In hydraulic networks or fractal models (Fitter 1982, 1986, 1987, 1991, 1996; Tatsumi et al. 1989; Fitter and Stickland 1992; Berntson 1994b), which are devoted to a more or less static description of the root system, most parameters can be obtained from unique observations of root systems. Moreover, the measured variables are purely morphological; they do not require expertise in the underlying developmental processes. They can be measured on detached parts of root systems.

In estimating parameters for developmental models, expertise is always required in order to distinguish between root types, i.e. branching orders, originating phytomers, and so on. These distinctions are relatively easy on young root systems, but they may be difficult on older ones because radial growth and root decay can tend to hide some of the morphological markers of development. Observations of the root system at closely spaced stages of development can obviously help to abstract the continuity of developmental processes.

It is worth noting that many of the parameters of these developmental models can be obtained from purely static observations. Branching densities, branching angles, or trajectory parameters (Tardieu and Pellerin 1990; Berntson 1994a; Pagès and Pellerin 1994; Pellerin and Pagès 1996) are examples of such parameters. Consequently, these parameters are quite easy to obtain in field conditions, even from scattered observations. However, one must take care to allow for the effects of root decay, which can reduce root density on old parts of the roots, especially on perennial plants. Parameters which require dynamic observation are the rate of appearance of roots and their elongation rate. In grasses, the rate of appearance of adventitious roots on successive internodes can be easily recorded from regular sampling of plants. Stable relationships were found with foliar stages on several species, so that this parameter need not be evaluated in each situation (Klepper et al. 1984; Picard et al. 1985). The appearance of laterals was found to occur after a constant lag-time of non-branching behind the apex, whose duration can be estimated in artificial conditions (Pellerin and Tabourel 1995). Estimation of the parameters for the elongation process, which is particularly important because it drives the kinetics of root system extension, is more difficult. Elongation rates of individual roots are highly dependant on environmental conditions, and very few satisfactory methods exist for estimating this parameter under field conditions. The growth rate parameters can be derived from continuous monitoring of root systems, either in the field with mini-rhizotrons or in root observation boxes (see Chap. 8 of this book; Berntson 1992, 1994a; Pagès 1992). They can also be estimated from plant or root populations that are excavated at different stages of their development. Pellerin and Pagès (1994) presented such a study in which they approximated the growth curve of nodal roots in maize. Pagès et al. (1992, 1993) and Tsegaye et al. (1995) have presented general methods for obtaining the growth characteristics of different types of roots using periodic excavation of plants.

Global estimation of parameters, by optimising one or more criteria of the global output of a model (e.g. total numbers or total length of roots, root maps, or root profiles), is possible in some situations. Sometimes, however, this is very difficult, because some aspects of the output may be relatively insensitive to some parameters, or correlations between parameters may be important. Also, the accuracy of globally estimated parameters will be very sensitive to bad conditioning of the model. Since the parameters involved tend to have concrete biological significance, separate estimation seems to be attractive in principle. Furthermore, individual calibration of parameters allows subsequent validation using the global outputs, as they have not been used directly for calibrating the model.

4.3.4 Evaluation of Models of Root System Architecture

An essential aspect of modelling, especially concerning dynamic models, is production of quantitative outputs that can be compared with actual data, in order to test and to modify the underlying hypotheses of the models, and thereby improve them.

Since models of this type are in fact aggregations of several more or less independent sub-models, which formalise different aspects of development, and since the objectives of modellers may differ, evaluation of the models can be done at different levels, using a number of different criteria (e.g. geometry, topology, developmental kinetics). Moreover, the problem is generally not to accept or to refute the model, which contains phenomenological parts which are not really refutable, but to quantify the discrepancies between actual data and model predictions for criteria of interest (Pagès and Pellerin 1996). Given the potential multiplicity of such criteria, and the multiplicity of the modellers' objectives, the evaluation approach is not easy to formalise, and is still nearly virgin territory.

Different authors have suggested global and visual validation of their models, showing pictures of simulated architectures (Fig. 4.3; Diggle 1988; Pagès and Aries 1988; Fitter et al. 1991; Berntson 1994a; Clausnitzer and Hopmans 1994), compared with root system drawings from Kutschera and Lichtenegger (1992), or with some of their specific observations (Berntson 1994a). This procedure makes the structural nature of the simulation more tangible, and gives a global and qualitative feeling about the quality of the model, but it is not very useful for evaluating or discriminating between hypotheses. The criteria under evaluation in this sort of comparison are not explicit, and the stochastic nature of the models raises problems, as each run gives a different picture. Graphical

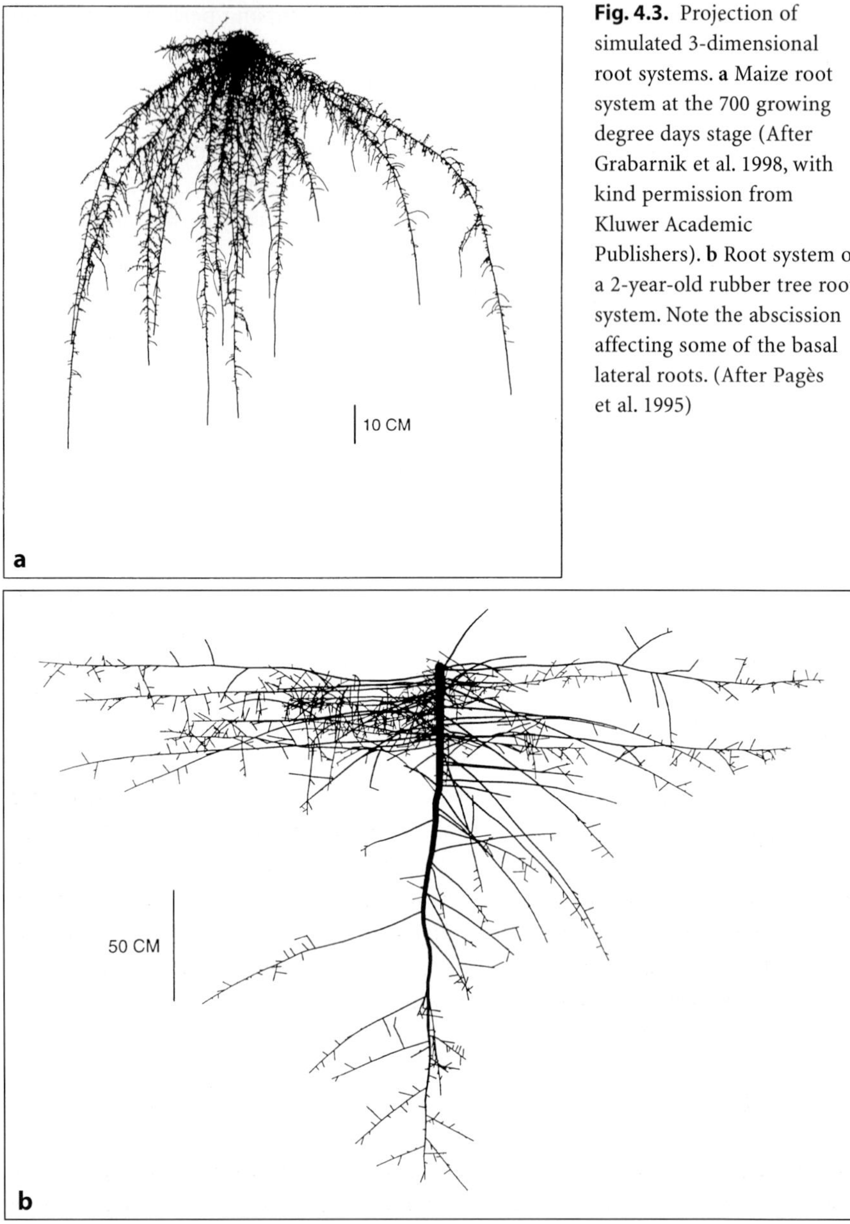

Fig. 4.3. Projection of simulated 3-dimensional root systems. **a** Maize root system at the 700 growing degree days stage (After Grabarnik et al. 1998, with kind permission from Kluwer Academic Publishers). **b** Root system of a 2-year-old rubber tree root system. Note the abscission affecting some of the basal lateral roots. (After Pagès et al. 1995)

aspects of the image, like resolution, colour, or texture rendering are important criteria for making a picture more or less realistic, and they may hide aspects of the simulated geometry that the model should represent.

To evaluate developmental aspects of the models more specifically, Pagès et al. (1992) suggested comparing the observed and simulated kinetics of the

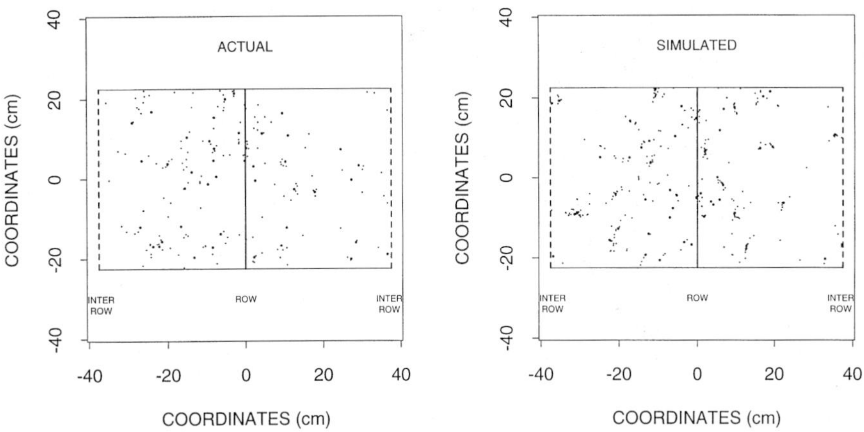

Fig. 4.4. Actual and simulated examples of horizontal root maps. *Big dots* are primary roots, and *small dots* are other roots. Note the clustered pattern. (After Pellerin and Pagès 1996)

length and number of roots of different branching orders. The stochastic nature of the model allowed the simulation of mean relationships as well as confidence intervals. These could be used to test the hypothesis that observed individuals could be accepted as members of the set of all possible runs of the model for a particular set of parameters.

Geometrical aspects of these models have been tested indirectly using one- or two-dimensional summaries of global outputs from these models. Porter et al. (1986) qualitatively compared observed and simulated root length density profiles, and discussed the possible explanations for the discrepancies. More recently, Pellerin and Pagès (1996) compared observed and simulated maize root densities and root aggregation on 2-dimensional horizontal maps, and showed the quality and limits of the geometrical representation produced by a model of root system architecture (Fig. 4.4).

4.3.5 Utilisation

Root architectural models have only been developed in recent years, so their utilisation by biologists is still extremely limited. Several applications are, however, expected in the near future, and there are already some pioneer works.

The most immediate applications of root architectural models are those directly stemming from the modelling approach: synthesis of scattered pieces of information about the root systems of several species (Diggle 1988 on wheat; Pagès et al. 1989 on maize), bringing to the fore apparent lack of knowledge and suggestion of new hypotheses for further experimentation. Sensitivity analyses will help in studying the relative importance of root parameters on the final root distribution (what is the consequence of an alteration of one root parameter on the final soil colonisation?). Summarising root architecture in a limited set of parameters provides a basis for comparison of genotypes (Colin-Belgrand et al. 1989; Lynch and van Beem 1993). At the extreme, fractal branching models provide relationships between one or two easily measured parameters (the proximal root diameter for instance, which is close to the plant stem) and total root length or surface area (Spek and van Noordwijk 1994; Van Noordwijk et al. 1994; Van Noordwijk and Purnomosidhi 1995).

Important applications of three-dimensional architectural root models lie in their combination with classical methods for root observation. Spatial and temporal information about root distribution given by the model may help in selecting an appropriate sampling strategy. This point is discussed in Chapter 5. Additionally, simulations may help in investigating both practical and theoretical problems associated with these classical methods. Using a model, Pagès and Pellerin (1996) have investigated practical problems associated with the trench profile wall method, namely the effect of the roughness of the observation face on the rooting profile observed. Bengough et al. (1992) have used an architectural root model to study the relationships between the number of root intersecting horizontal and vertical planes and the root length density.

Another area of application for root architecture models is in studying the response of the root system to its environment. Root architectural models are well-adapted for simulating interactions of roots with dynamic and spatially variable soil processes. A pioneer work in this area was proposed by Clausnitzer and Hopmans (1994) who have developed a model for simultaneous, dynamic simulation of soil water movement and plant root growth, with the effect of current soil water content on soil mechanical properties and subsequent root growth being considered. A similar work was also proposed by Tsegaye et al. (1995). Important soil features for root growth, such as soil cracks or earthworm channels, are unevenly distributed within the soil volume, so that the geometrical information contained in architectural models is relevant for calculating the probability that roots encounter such features. Architectural models are also a good support for associating responses of individual roots to their local environment, and the whole root system response to the percepted environment, which involves chemical signalling and assimilate distribution between shoots and roots. Aguirrezabal et al. (1994) have shown that the response of roots to carbon nutrition depends on their position in the root system architecture, thus emphasising the idea that architectural models are a necessary tool for understanding and predicting the root system response to carbohydrate availability. An attempt at combining these two categories of processes (local, and general, through carbon distribution) was proposed by

Clausnitzer and Hopmans (1994). An approach of the carbon cost of root systems, based on an architectural model, was also proposed by Nielsen et al. (1994) by associating carbon costs for biomass deposition, respiration and exudation to elementary root growth processes. Root architectural models contain information (for instance the number of growing apices at each time step) which help in understanding the sink strength of the root system for carbon. This is of particular interest in the context of studying plant response to atmospheric CO_2 increase.

A fourth important area of application for architectural models is in the context of water and nutrient uptake and transport. Nearly all existing models for resource capture are based on the assumption that roots are vertically and evenly distributed within the soil volume, and that they all have the same absorption properties. The information contained in architectural models about the root system geometry is an appropriate basis for more precise estimates of the volume of soil explored, allowing for overlaps of depletion zones between competing roots. Competition between adjacent plants for soil resources may be studied too. Fitter (1987), Fitter et al. (1991) and Berntson (1994a) have used architectural models for studying the relative effect of size-dependent, and architectural, size-independent, properties of root systems on their potential for resource acquisition. By combining this approach with a carbon cost approach, Fitter et al. (1991), Nielsen et al. (1994) and Berntson (1994a) have developed a cost/benefit approach of root systems (volume of soil explored per unit of volume of tissue required to construct the root system). Recent progress has also been made in modelling water and nutrient uptake with the actual, uneven distribution of roots being taken into account (Bruckler et al. 1991; Lafolie et al. 1991; Tardieu et al. 1992). Until now, these investigations were performed assuming that all roots have the same uptake properties, but the information contained in a root architectural model about root age and branching order make it possible to relate uptake properties to these variables. Finally, architectural models of the root system provide a basis for considering transport resistance within the root system, as proposed by Fitter (1987).

4.4 Present and Future Trends

4.4.1 Root Depth and Root Length Density Models

Root depth and root length density models have reached an advanced state of development. These models can be used to quantify plant growth effects via the root system of any soil factor and shoot : root interaction in an empirical way. They can make direct predictions of readily measurable characteristics of root systems under field conditions. For this reason, the accuracy of their predictions can be verified from straightforward field trials.

Future development of these models is likely to occur in two areas. The first is in selecting of the particular factors which are most important for any particular situation being examined, and custom tailoring a root model to deal with those factors. This effort will be driven by the limitations which are inherent in increasing the complexity of models. Beyond a certain point, the errors associated with estimating additional parameters outweigh any possible improvements in precision due to refinement of the structure, as discussed by Passioura (1996). Because of these limitations, obtaining accurate predictions from models will depend on skilful selection of the optimal parameter set.

The second area of development of these models is likely to be the continuing quantification of effects of site and season-specific conditions on proliferation of roots. Root depth and density models are highly abstract representations of root systems and, as a consequence, estimating these effects can be complicated. The simulation of increased heterogeneity of root distribution deeper in the soil profile represented with one-dimensional models will be a continuous challenge, in particular when they are the basis for water and nutrient uptake under supply-limited conditions. It may be that architectural root models, which concentrate more on mechanism than on measurability, will prove useful as part of this process.

4.4.2 Models of Root System Architecture

In the case of models of root system architecture, there are a number of areas where further development or new applications are likely to prove fruitful. This potential arises from the richness of the information which models of this sort can store about roots.

Architectural models have complete information about the locations of all parts of the root system. For this reason architectural models of root systems will be increasingly used to simulate interaction of roots with dynamic soil processes. Examples have been presented of the use of these models in concert with models of soil water and soil strength (Clausnitzer and Hopmans 1994). They can, and no doubt will, also be used with models of the distribution patterns and dynamics of mobile and immobile nutrients, pH, oxygen, and soil temperature. The three-dimensional nature of the spatial information in these models will make them particularly relevant to the study of inter-plant competition for resources.

Furthermore, architectural models store complete information about the geometry of roots. This information can and, to some extent, already has been used to examine and add value to various measurement techniques. Many types of root measurement involve counting intersections between roots and planes. Simplifying assumptions are then made about root geometry so that root length density can be inferred. Architectural root models can replace the less sophisticated geometric assumptions currently used in this process. Another area which could benefit from application of geometric information from these models is the design of statistically adequate sampling regimes for any type of measurement. Measurement methods which have been related to model predictions to date include length measurement by the core break technique (Bengough et al. 1992), intersection of roots with pit faces (Pellerin and Pagès 1996), and time of arrival of roots at transparent walls of containers (Tsegaye et al. 1995). Other candidate techniques could be core sampling, mini-rhizotrons and pinboards.

Geometric information can also be used to refine models of movement of material to and from roots by processes such as diffusion. Such processes depend on the relationship between distance from roots and the volume of soil which is within that space. Models which calculate these processes must make some assumption about this relationship. Baldwin et al. (1973) for example discuss models of uptake of nutrients assuming either a regular array of roots or randomly dispersed roots. In fact, of course, roots are not distributed in either of these ways. The relationship between soil volume and distance from roots can be estimated for simulated root systems by repeatedly evaluating the distance between random points in the simulation volume and the nearest part of the root system. The resulting information can then be represented mathematically in calculations of diffusion by using an appropriate geometric entity such as a fractal.

A further class of information which is embodied in architectural models relates to the status of the plant and of all of its parts, and to their interconnection. In plants, the growth and branching behaviour of individual roots depends not only on the local environmental conditions, but also on the conditions experienced by all other parts of the plant at that time and on the collective state of the plant. This information propagates in plants by various means. It impacts on the patterns of allocation of assimilate through time and results in characteristic responses to such stimuli as disease, injury, and uneven distributions of nutrients and water. Architectural models are uniquely suited to examining this sort of relationship and will be applied to studies in this area.

A localised disruption in a root system, as might be caused by some types of insect or fungal infection, can immediately affect parts which are quite remote but directly connected. These models can readily be used to address this type of occurrence.

Architectural root models can also make use of geometric information relating to other objects in the soil. This capability has been used by Brown and

Kulasiri (1994) to detect encounters between growing roots and pieces of root systems infected by fungi. It could also be used to model effects of cracks, old root channels, walls of containers, mini-rhizotrons, or other such structures.

There are also several examples where architectural models of roots are combined with other emerging technologies. Considerable advances have been made in recent times in the architectural modelling of plant tops (i.e. Prusinkiewicz and Lindenmayer 1990), and it is possible to combine these models with root models to produce architectural models of whole plants. By using such a model a unified approach could be taken to calculating growth and development of plant tops and roots, and partitioning of assimilate within and between them. A case where the pattern of interconnection between roots and tops may be important is in the relationship between tillers in wheat and the growth of their associated nodal roots.

It is also possible to produce architectural root models in what is referred to as "voxel space" (Greene 1991). Voxel space consists of an array of rectangular volume elements which are the three-dimensional equivalent of the pixels which make up computer displays. The elements, or voxels, are very small (typically less than 1 mm) and each can have any of several states, such as root tip, root, obstacle, or air. Such a model operates by calculating the state of each voxel in each time step based on its previous state and the state of its neighbours. The simulated roots are made up of a collection of adjoining voxels rather than a collection of line segments, as is the case in other architectural root models. The roots can be made to grow in any direction from any point at any time; so elongation, branching and thickening of roots all occur similarly. Voxel technology uses large amounts of computer memory, but it could be applied to study interactions between roots and very small-scale features of soil. Voxel models are particularly well suited to visualising complex structures because any voxels or types of voxel can be defined to be transparent for display purposes, making elaborate cut-away views possible.

Another class of emerging technology is used for the non-destructive imaging of volumes (e.g. nuclear magnetic resonance and microtomography). As discussed in Chapter 11 the resolution achievable with these techniques is steadily improving and first in-situ images of root systems have been presented. Such technology will very much facilitate the testing of architectural models.

Acknowledgments. We thank Drs P. Barlow, A. Fitter, and M. Robertson, for comments on the manuscript and the Grains Research and Development Corporation for financial support of SA.

References

Addiscott TM, Whitmore AP (1987) Computer simulation of changes in soil mineral nitrogen and crop nitrogen during autumn, winter and spring. J Agric Sci 109: 141–157

Adiku SGK, Braddock RD, Rose CW (1995) Simulating root growth dynamics of cowpea under varying soil conditions. Proc MODSIM95, Int Congr on Modelling and Simulation, The University of Newcastle, 27–30 November 1995, Newcastle, Australia, pp 57–61

Aguirrezabal A, Deleens E, Tardieu F (1994) Root elongation rate is accounted for by intercepted PPFD and source-sink relations in field and laboratory-grown sunflower. Plant Cell Environ 17: 443–450

Andrew SPS (1987) A mathematical model of root exploration and of grain fill with particular reference to winter wheat. Fertil Res 11: 267–281

Asseng S, Keating BA, Huth NI, Eastham J (1997a) Simulation of perched water tables in a duplex soil. Proc MODSIM 97, Int Congr on Modelling and Simulation, 8–11 December 1997, Hobart, Tasmania, Australia, pp 538–543

Asseng S, Richter C, Wessolek G (1997b) Modelling root growth of wheat as the linkage between crop and soil. Plant Soil 190: 267–277

Asseng S, Keating BA, Fillery IPR, Gregory PJ, Bowden JW, Turner NC, Palta JA, Abrecht DG (1998) Performance of the APSIM wheat model in Western Australia. Field Crops Res 57: 163–179

Baldwin JP, Nye PH, Tinker PB (1973) Uptake of solutes by multiple root systems from soil. III. A model for calculating the solute uptake by a randomly dispersed root system developing in a finite volume of soil. Plant Soil 38: 621–635

Bengough AG, Mackenzie CJ, Diggle AJ (1992) Relations between root length densities and root intersections with horizontal and vertical planes using root growth modelling in 3-dimensions. Plant Soil 145: 245–252

Berntson GM (1992) A computer program for characterizing root system branching patterns. Plant Soil 140: 145–149

Berntson GM (1994a) Modelling root architecture: are there trade-offs between efficiency and potential or resource acquisition? New Phytol 127: 483–493

Berntson GM (1994b) Root systems and fractals: how reliable are calculations of fractal dimensions? Ann Bot 73: 281–284

Borg H, Grimes DW (1986) Depth development of roots with time: an empirical description. Am Soc Agric Eng 29: 194–197

Brown TN, Kulasiri D (1994) Simulation of Pinus radiata root system structure for ecosytem management applications. Simulation 62: 286–294

Bruckler L, Lafolie F, Tardieu F (1991) Modeling root water potential and soil-root water transport: II. Field comparisons. Soil Sci Soc Am J 55: 1213–1220

Cannell MGR (1985) Dry matter partitioning in tree crops. In: Cannell MGR, Jackson JE (eds) Trees as crop plants. Titus Wilson, Abbots Ripton, Huntington, 160–193

Cannon WA (1949) A tentative classification of root systems. Ecology 30: 542–548

Chapman SC, Hammer GL, Meinke H (1993) A sunflower simulation model: I. Model development. Agron J 85: 725–735

Clausnitzer V, Hopmans JW (1994) Simultaneous modeling of transient three-dimensional root growth and soil water flow. Plant Soil 164: 299–314

Colin-Belgrand M, Pagès L, Dreyer E, Joannes H (1989) Analysis and simulation of the architecture of a growing root system: application to a comparative study of several tree seedlings. Ann Sci For 46 (Suppl): 288–293

Coutts MP (1989) Factors affecting the direction of growth of tree roots. Ann Sci For 46 (Suppl): 277–287

Denison RF, Loomis RS (1989) An integrative physiological model of alfalfa growth and development. Div of Agricultural and Natural Resources, Oaklands, California, 73pp

Diggle AJ (1988) ROOTMAP – a model in three-dimensional coordinates of the growth and structure of fibrous root systems. Plant Soil 105: 169–178

Fitter AH (1982) Morphometric analysis of root systems: application of the technique and influence of soil fertility on root system development in two herbaceous species. Plant Cell Environ 5: 313–322

Fitter AH (1986) The topology and geometry of plant root systems: influence of watering rate on root system topology in *Trifolium pratense*. Ann Bot 58: 91–101

Fitter AH (1987) An architectural approach to the comparative ecology of plant root systems. New Phytol 106 (Suppl): 61–77

Fitter AH (1991) Characteristics and functions of root systems. In: Waisel Y, Eshel A, Kafkafi U (eds) Plant roots, the hidden half. Dekker, New York, pp 3–25

Fitter AH (1996) Characteristics and functions of root systems. In: Waisel Y, Eshel A, Kafkafi U (eds) Plant roots, the hidden half. 2nd edn. Dekker, New York, pp 1–20

Fitter AH, Stickland TR (1992) Fractal characterization of root system architecture. Funct Ecol 6: 632–635

Fitter AH, Stickland TR, Harvey ML, Wilson GW (1991) Architectural analysis of plant root systems. 1. Architectural correlates of exploitation efficiency. New Phytol 118: 375–382

Gerwitz A, Page ER (1974) An empirical mathematical model to describe plant root systems. J Appl Ecol 11: 773–781

Grabarnik P, Pagès L, Bengough AG (1998) Geometrical properties of simulated root systems: consequences for length density and intersection density. Plant Soil 200, 157–167

Greene N (1991) Detailing tree skeletons with voxel automata. SIGGRAPH '91 Course Notes on Photorealistic Volume Modeling and Rendering Techniques, Association for Competing Machinery, New York

Groot JJR (1987) Simulation of nitrogen balance in a system of winter wheat and soil. Simulation Report CABO-TT, 13. Centre for Agrobiological Research, Wageningen

Hackett C, Rose DA (1972) A model of the extension and branching of a seminal root of barley, and its use in studying relations between root dimensions. I. The model. Aust J Biol Sci 25: 669–679

Hansen GK (1975) A dynamic continuous simulation model of water state and transportation in the soil-plant-atmosphere system. I. The model and its sensitivity. Acta Agric Scand 25: 129–149

Henderson R, Ford ED, Renshaw E, Deans JD (1983a) Morphology of the structural root system of sitka spruce. 1. Analysis and quantitative description. Forestry 56: 121–135

Henderson R, Ford ED, Renshaw E (1983b) Morphology of the structural root system of sitka spruce. 2. Computer simulation of rooting patterns. Forestry 56: 136–153

Hoogenboom G, Huck MG (1986) ROOTSIMU V4.0. A dynamic simulation of root growth, water uptake, and biomass partitioning in a soil-plant-atmosphere continuum: update and documentation. Agronomy and Soils Departmental Series 109. Auburn University, Alabama, 83pp

Huck MG, Hillel D (1983) A model of root growth and water uptake accounting for photosynthesis, respiration, transpiration, and soil hydraulics. Adv Irrig 2: 273–333

Jakobsen BF, Dexter AR (1987) Effect of soil structure on wheat root growth, water uptake and grain yield. A computer simulation model. Soil Tillage Res 10: 331–345

Johnson IR, Thornley JHM (1985) Dynamic model of the response of a vegetative grass crop to light, temperature and nitrogen. Plant Cell Environ 8: 485–499

Jones CA, Kiniry JR (1986) CERES-maize: a simulation model of maize growth and development. Texas A&M University Press, College Station, Texas

Jones CA, Bland WL, Ritchie JT, Williams JR (1991) Simulation of root growth. In: Hanks J, Ritchie JT (eds) Modeling plant and soil systems. Agronomy Monograph 31, American Society of Agronomy, Inc., Madison, Wisconsin, pp 92–123

Klepper B, Belford RK, Rickman RW (1984) Root and shoot development in winter wheat. Agron J 76: 117–122

Kutschera L, Lichtenegger E (1992) Wurzelatlas mitteleuropäischer Grünlandpflanzen. Gustav Fischer, Stuttgart

Lafolie F, Bruckler L, Tardieu F (1991) Modeling root water potential and soil-root water transport: I. Model presentation. Soil Sci Soc Am J 55: 1203–1212

Le Roux Y, Pagès L (1994) Développement et polymorphisme racinaires chez de jeunes semis d'hévéa (*Hevea brasiliensis*). Can J Bot 72: 924–932

Lungley DR (1973) The growth of root systems: a numerical computer simulation model. Plant Soil 38: 145–159

Lynch J, Van Beem JJ (1993) Growth and architecture of seedling roots of common bean genotypes. Crop Sci 33: 1253–1257

Mandelbrot B (1983) The fractal geometry of nature. Freeman, New York

Monteith JL, Huda AKS, Midya D (1989) RESCAP: a resource capture model for sorghum and pearl millet. In: Virmani SM, Tandon HLS, Alagarswamy G (eds) Modeling the growth and development of sorghum and pearl millet. ICRISAT Res Bull 12: 30–34

Nielsen CHN, Mackenthun G (1991) Die horizontale Variation der Feinwurzelintensität in Waldböden in Abhängigkeit von der Bestockungsdichte. Allg Forst- Jagd Ztg 162: 112–119

Nielsen KL, Lynch JP, Jablokow AG, Curtis PS (1994) Carbon cost of root systems: an architectural approach. Plant Soil 165: 161–169

O'Leary GJ, Connor DJ, White DH (1985) A simulation model of the development, growth and yield of the wheat crop. Agric Syst 17: 1–26

Otter-Nacke S, Godwin DC, Ritchie JT (1986) Testing and validating the CERES-wheat model in diverse environments. AGRISTARS Yield Model Development. USDA-ARS Temple, Texas and IFDC Muscle Shoals, Alabama, 147pp

Page ER, Gerwitz A (1974) Mathematical models, based on diffusion equations, to describe root systems of isolated plants, row crops, and swards. Plant Soil 41: 243–254

Pagès L (1992) Mini-rhizotrons transparents pour l'étude du système racinaire de jeunes plantes. Application à la caractérisation du développement racinaire de jeunes chênes (*Quercus robur*). Can J Bot 70: 1840–1847

Pagès L, Aries F (1988) SARAH: modèle de simulation de la croissance, du développement et de l'architecture des systèmes racinaires. Agronomie 8: 889–896

Pagès L, Pellerin S (1994) Evaluation of parameters describing the root system architecture of field grown maize plants (*Zea mays* L.). II. Density, length, and branching of first-order lateral roots. Plant Soil 164: 169–176

Pagès L, Pellerin S (1996) Study of differences between vertical root maps observed in a maize crop and simulated maps obtained using a model for the three-dimensional architecture of the root system. Plant Soil 182: 329–337

Pagès L, Jordan MO, Picard D (1989) A simulation model of the three-dimensional architecture of the maize root system. Plant Soil 119: 147–154

Pagès L, Chadoeuf J, Kervella J (1992) Modélisation stochastique de la croissance et du développement du système racinaire de jeunes pêchers. I. Estimation et validation du modèle. Agronomie 12: 447–458

Pagès L, Kervella J, Chadoeuf J (1993) Development of the root system of young peach trees (*Prunus persica* L. Batsch): a morphometrical analysis. Ann Bot 71: 369–375

Pagès L, Le Roux Y, Thaler P (1995) Modélisation de l'architecture racinaire. Plant Rech Dév 2: 19–34

Passioura JB (1996) Simulation models: science, snake oil, education or engineering? Agron J 88: 690–694

Pellerin S (1993) Rate of differentiation and emergence of nodal maize roots. Plant Soil 148: 155–161

Pellerin S, Pagès L (1994) Evaluation of parameters describing the root system architecture of field-grown maize plants (*Zea mays* L.). I. Elongation of seminal and nodal roots and extension of their branched zone. Plant Soil 164: 155–167

Pellerin S, Pagès L (1996) Evaluation in field conditions of a three-dimensional architectural model of the maize root system: comparison of simulated and observed horizontal root maps. Plant Soil 178: 101–112

Pellerin S, Tabourel F (1995) Length of the apical unbranched zone of maize axile roots: its relationship to root elongation rate. Environ Exp Bot 35: 193–200

Penning de Vries FWT, Jansen DM, Ten Berge HFM, Bakema A (1989) Simulation of ecophysiological processes of growth in several annual crops. Simulation Monographs. Pudoc, Wageningen

Picard D, Jordan MO, Trendel R (1985) Rythme d'apparition des racines primaires du maïs (*Zea mays* L.). I. Etude détaillée pour une variété en un lieu donné. Agronomie 5: 667–676

Porter JR, Klepper B, Belford RK (1986) A model (WHTROOT) which synchronizes root growth and development with shoot development for winter wheat. Plant Soil 92: 133–145

Prusinkiewicz P, Lindenmayer A (1990) The algorithmic beauty of plants. Springer, Berlin Heidelberg New York

Reynolds JF, Thornley JHM (1982) A shoot: root partitioning model. Ann Bot 49: 585–597

Ritchie JT, Godwin DC, Otter S (1985) CERES-wheat: a user oriented wheat yield model. Preliminary documentation. AGRISTARS Publ No YM-U3-04442-JSC-18892. Michigan State University, MI, 252pp

Robertson MJ, Fukai S, Hammer GL, Ludlow MM (1993) Modelling root growth of grain sorghum using the CERES approach. Field Crops Res 33: 113–130

Rose DA (1983) The description of the growth of root systems. Plant Soil 75: 405–415

Savin R, Satorre EH, Hall AJ, Slafer GA (1995) Assessing strategies for wheat cropping in the monsoonal climate of the Pampas using the CERES-wheat simulation model. Field Crops Res 42: 81–91

Shibusawa S (1994) Modelling the branching growth fractal pattern of the maize root system. Plant Soil 165: 339–347

Skiles JW, Hanson JD, Parton WJ (1982) Simulation of above- and below-ground carbon and nitrogen dynamics of *Bouteloua gracilis* and *Agrophyron smithii*. In: Lavenroth WK,

Skogerboe GV, Flug M (eds) Analysis of ecological systems: state of the art in ecological modelling. Elsevier, Amsterdam pp 467–473

Spek LY, Van Noordwijk M (1994) Proximal root diameter as predictor of total root size for fractal branching models. II. Numerical model. Plant Soil 164: 119–127

Spek L, Van Oijen M (1988) A simulation model of root and shoot growth at different levels of nitrogen availability. Plant Soil 111: 191–197

Stapper M (1984) SIMTAG: a simulation model of wheat genotypes. Model documentation. University of New England, Armidale, 108pp

Stol W, van Keulen H, van Kraalingen DWG (1993) The FORTRAN version of the Van Keulen – Seligman CSMP-spring wheat model. Simulation Report CABO-TT 30, 112, Wageningen Agricultural University, Wageningen

Tardieu F, Pellerin S (1990) Trajectory of the nodal roots of maize in fields with low mechanical constraints. Plant Soil 124: 39–45

Tardieu F, Bruckler F, Lafolie F (1992) Root clumping may affect the root water potential and the resistance to soil-root water transport. Plant Soil 140: 291–301

Tatsumi J, Yamauchi A, Kono Y (1989) Fractal analysis of plant root systems. Ann Bot 64: 499–503

Thornley JHM (1977) Root:shoot interactions. In: Jennings DM (ed) Integration of activity in the higher plants. Cambridge University Press, Cambridge, pp 367–389

Touré A, Major DJ, Lindwall CW (1995) Comparison of five wheat simulation models in southern Alberta. Can J Plant Sci 75: 61–68

Tsegaye T, Mullins CE, Diggle AJ (1995) An experimental procedure for obtaining input parameters for the "ROOTMAP" root simulation program for peas (Pisum sativum L.). Plant Soil 172: 1–16

Van Keulen H, Seligman NG (1987) Simulation of water use, nitrogen nutrition and growth of a spring wheat crop. Simulation Monographs. Pudoc, Wageningen

Van Noordwijk M, Purnomosidhi P (1995) Root architecture in relation to tree-soil-crop interactions and shoot pruning in agroforestry. Agrofor Syst 30: 161–173

Van Noordwijk M, Spek LY, De Willigen P (1994) Proximal root diameter as predictor of total root size for fractal branching models. I. Theory. Plant Soil 164: 107–117

Voit EO, Sands PJ (1996) Modeling forest growth I. Canonical approach. Ecol Model 86: 51–71

Von Pfeil E, Hundertmark W, Thies FD, Widmoser P (1992) Calibration of the simulation model "ceres wheat" under conditions of soils with shallow water table and temperate climate. Part 1: Limitations in the applicability of the original model and necessary modifications. Z Pflanzenernähr Bodenkd 155: 323–326

Waisel Y, Eshel A (1991) Multiform behavior of various constituents of one root system. In: Waisel Y, Eshel A, Kafkafi U (eds) Plant roots, the hidden half. Dekker, New York, 39–52

Wessolek G (1993) Einfluss von Klimaänderungen auf den Bodenwasserhaushalt (regionale Fallstudien). Mitt Dtsch Bodenkdl Ges 69: 289–292

Wessolek G, Gäth S (1990) Estimation of root density in modelling nutrient requirements. In: Proc 22th Colloq Int Potash Institute, Soligorsk, USSR

Williams JR, Jones CA, Dyke PT (1984) The EPIC model and its application. Proc Int Symp on Minimum Data Sets for Agrotechnology Transfer, Patancheru, India, 21–26 March 1983, pp 111–121

Further Reading

Jones CA, Bland WL, Ritchie JT, Williams JR (1991) Simulation of root growth. In: Hanks J, Ritchie JT (eds) Modeling plant and soil systems. Agron Monogr 31: 92–124

Klepper B, Rickman RW (1990) Modeling crop root growth and function. Adv Agron 44: 113–132

Lynch J (1995) Root architecture and plant productivity. Plant Physiol 109: 7–13

CHAPTER 5

Sampling Strategies, Scaling, and Statistics

A.G. Bengough[1], A. Castrignano[2], L. Pagès[3], and M. van Noordwijk[4]

[1] Scottish Crop Research Institute (SCRI), Invergrowie, Dundee, DD2 5DA, UK
[2] Instituto Sperimentale Agronomico, Via C. Ulsiani 5,70125 Bari, Italy
[3] INRA, Unité de Recherches en Ecophysiologie et Horticulture, Site Agroparc, 84914 Avignon Cédex 9, France
[4] ICRAF, JI CIFOR, PO Box 161, Situ Gede, Sindang Barang Bogor 16680, 16001 Bogor, Indonesia

CONTENTS

5.1 Introduction

Many experiments aim to understand how some aspect of the aerial and soil environment of a plant influences its growth. Considerable biomass is often allocated to the root system, and it is the roots that absorb most nutrients and water

A.L. Smit et al. (Eds.), Root Methods

(Russell 1977). The property of the root system which is most appro-priate to measure depends on the objective of the experiment. Root length should be measured to calculate the inflow rate of water and nutrients (see Chaps. 6, 13, 14). Root dry mass indicates the carbon allocation to the root system. Branching patterns and the number and lengths of each class of root, together with the distribution of root diameters gives a more complete picture of root architecture, but requires a large investment of labour. Detailed information on root architecture may be of interest in constructing mathematical models of root growth (see Chap. 4), in comparing the structure and function of root systems of different species or genotypes (e.g. Fitter and Stickland 1991), and in evaluating root responses to environmental conditions (e.g. Robinson 1994).

It is not easy to design an appropriate scheme to adequately sample the root system, because of the complex and variably branched structure of roots (see Fig. 5.1), the variability of root distribution in space and, most importantly, the opaque growing environment, which means the experimenter works continually "in the dark". Root researchers face the peculiar problem of only being able to see or sample a small fraction of the root system at any time, using sampling procedures that are often destructive (e.g. see Chaps. 6 and 7). The design of sampling schemes for roots that are adequate and appropriate for different situations has received little study. This is despite the large effort that has been directed, and sometimes misdirected, towards sampling roots in the field. Many studies have considered different techniques for measuring roots, but the problem of where to sample, and how big a sample to collect, is rarely considered (van Noordwijk et al. 1985).

The "problem" of large variation between replicate samples of roots is widely known. This has often been seen as a nuisance, making it necessary to sample many times to obtain an accurate estimate of the mean. However, it is becoming recognized that the variation itself is worthy of study. This is also being seen in the soil sciences, where interest is increasing in the study of soil heterogeneity, and the resources that the root system is trying to exploit (e.g.

Fig. 5.1. Root systems vary widely in architecture and in spatial distribution. Examples are shown of: **a** A 54-year-old Austrian pine (*Pinus nigra*) in sandy soil in the Netherlands – roots weighed a total of 30 kg and extended about 5 m from the base of the tree. **b** An excavated horizontal lateral root of a 3-year-old rubber tree (*Hevea brasiliensis*) in sandy soil in the Ivory Coast. **c** The thick vertical taproot and horizontal lateral roots near the base of the trunk of a 3-year-old rubber tree. In deeper horizons the lateral roots are finer and less horizontal. **d** An excavated nodal axis of maize (*Zea mays* L.) at silking, in a loamy clay soil at Colmar. **e** Maize grown for 690 Kd (above base temperature 6 °C; seven visible leaves present) in aeroponics. **f** and **g** Root systems of bean (*Phaseolus vulgaris*) and rye-grass plants grown in root boxes for 2 weeks under glasshouse conditions. Photos kindly supplied by Drs. F. Egas (**a**), Yannick Le Roux (**b**,**c**), S. Pellerin (**d**,**e**), and F. Meijboom (**f**,**g**)

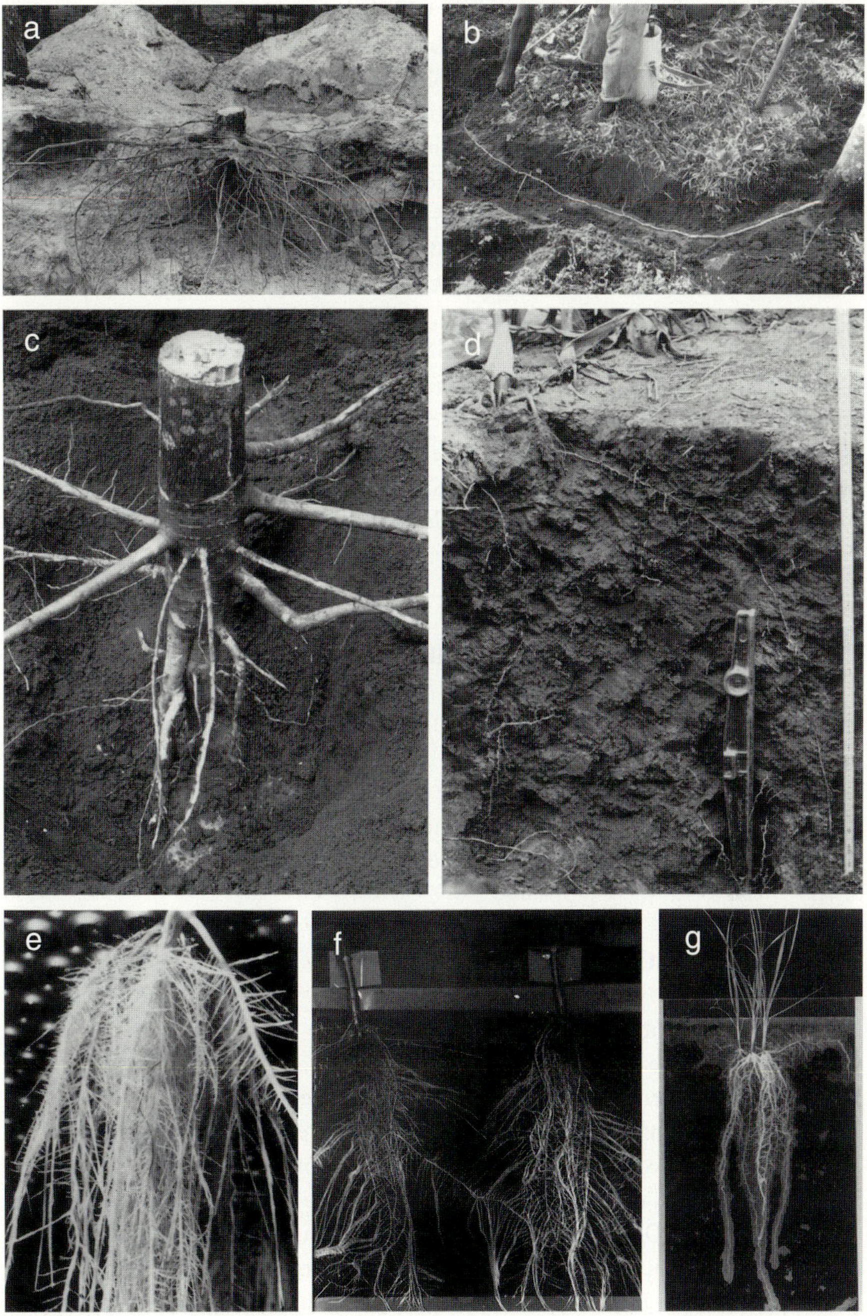

Fig. 5.1a–g

Jackson and Caldwell 1993a,b). A plant with a root system that explores an area of soil with a 50 cm radius may experience a variability in nutrient distribution equivalent to that across a 100 m^2 field plot (Jackson and Caldwell 1993b). A root system that is distributed non-uniformly can extract water and nutrients much faster from areas where the root length density is large. In contrast, resources may be largely inaccessible in regions of the soil that are rooted sparsely. The superposition of soil and root data in space promises to give new insights into how roots explore the soil environment.

The aim of this chapter is to guide the reader towards designing effective schemes for sampling roots in the field. Specific statistical analyses appropriate to particular methods are discussed in the appropriate technique chapter (e.g. cohort analysis in Chap. 9). We examine the spatial properties of root systems, and the problems associated with the techniques for measuring them. The coefficients of variation of root systems of different crops are detailed to help determine the likely number of samples required. The potential uses of root growth models in experimental design are described, together with how geostatistics can be used to study spatial variability.

5.2 Root System Structure and Spatial Distribution

In the following sections, we identify three main properties associated with the spatial distribution of root systems: (1) systematic trends in the root length density of root systems – i.e. large scale (several decimetres) gradual variations; (2) root clustering – variations at a small scale (several centimetres); and (3) root anisotropy – the non-uniform directional distribution of roots.

5.2.1 Systematic Trends

The form of a root system depends both on the plant genotype and on its interaction with the environment. Species may be grouped into classes that have broadly similar forms of root system. Monocotyledons generally form dense fibrous root systems. The seminal and nodal axes of cereals and other grasses produce several successive orders of fine lateral roots. The total root length of a single 16-week-old winter rye plant (*Secale cereale*) was measured to be more than 500 km long (Dittmer 1937). Dicotyledons form root systems often initially centred around a tap root from which many lateral roots emerge (e.g. sugarbeet, *Beta vulgaris*). Trees may also have a taproot system; a more surface "plateroot" system consisting of horizontal main root axes; or a "heartroot" system which consists of more evenly distributed main axes that split into smaller roots (Wilde 1958). Certain species of tree (e.g. *Populus tremuloides*) grow new shoots from buds on roots in the surface layers of soil: Such groups of interconnected trees form very extensive root systems, exploring huge volumes of soil.

The depths and distributions of root systems vary enormously according to species, climatic zone, and soil type. In a comprehensive review of rooting depths around the globe, maximum rooting depths varied between <0.3 m in some tundra species, to 68 m for *Boscia albitrunca* in the Kalahari desert (Canadell et al. 1996). Maximum rooting depths averaged about 9 m in the desert, with only about half of that root biomass being in the top 0.3 m (Jackson et al. 1996). In contrast, tundra vegetation averaged 0.5 m maximum rooting depth, with 80–90% of root biomass being in the top 0.3 m. When plants are grouped across biomes by functional group, trees, shrubs and herbaceous plants had global average rooting depths of 7 m, 5.1 m, and 2.6 m respectively.

The length of root per unit ground surface area has been reviewed briefly by Newman (1969). Perennial grasses had the densest root systems, with 360–3400 cm/cm^2 of root in the top 10 cm layer of soil. This corresponds to 36 to 340 cm/cm^3 volume of soil. Cereals and non-*gramineae* herbs had smaller root lengths per unit area than grasses, but generally greater root lengths per unit area than woody shrubs and trees.

The spatial distribution of roots depends on the spacing and arrangement of individual plants in an ecosystem (van Noordwijk et al. 1985). In grassland, the plants are spaced closely in all directions, so that the root length density varies mainly in one dimension with depth. In cereal crops planted densely in rows, the pattern of root length density is essentially two-dimensional. As distance between plants increases, the pattern becomes three-dimensional. In agroforestry systems, trees, often arranged in rows, are interplanted with grass or an arable crop, leading to two interacting root system structures. Large variations in the height of the soil surface, as occurs across the ridges and furrows of the potato crop, create even more complex three-dimensional patterns (e.g. Parker et al. 1991). Natural ecosystems in which plants are randomly spaced or clumped on an undulating landscape may mean that the plants have to be treated as isolated individuals or small groups, without any of the symmetry afforded by agricultural practice.

Root systems originate from seeds, rhizomes or shoots that are near to the surface of the soil. The root system extends and branches gradually, extending away from this origin. This creates a pattern of root length density that is oriented centrifugally. Root length density decreases with increasing distance from the stem base, both in the horizontal and in the vertical directions (Gajri et al. 1994). Gradients in soil characteristics, such as soil-water content, soil strength, and temperature, interact with the development of the root system (e.g. Tardieu and Pellerin 1990). Such gradients often restrict the spread of the root system with depth but, in practice, it is very difficult to separate the effects of genotype and environment on root distribution.

The gradual decrease of root length density with depth is a feature that is general to most root systems. A negative exponential function has been used to model the relationship between root distribution (measured as length density or root mass) and depth for vegetable crops, cereals, and grasses (Gerwitz and Page 1974). The exponential model accounted for at least two thirds of the variation in 70 of 101 cases.

The variation in root length density across a crop has been measured by several authors (see Fig. 5.2; van Noordwijk et al. 1985; Gajri et al. 1994; Pellerin and Pagès 1996). The variation was greatest near the surface for crops grown in widely spaced rows. The variation was small in established forest stands (Persson 1978). The horizontal variation in root length density tends to decrease with increasing depth, because the "fan" shape of the root systems tends to smooth out fluctuations between neighbouring plants.

5.2.2 Clustering

Large variations in root length density occur across distances of several centimetres because of root clustering or clumping. Roots cluster in space because of their branched connections with daughter roots necessarily concentrated close to their mother root (Logsdon and Allmaras 1991). The constraint caused by the connection of roots, and the consequent aggregation of roots in space is exacerbated in root systems that have many short branches, with short link lengths between branches (e.g. Varney et al. 1991). A large percentage of the total root length is segmented into roots shorter than 10 cm, even in root systems that explore a total volume of several cubic metres (Lyr and Hoffman 1967). Such short roots are generally much thinner than the main axis, and can represent an economical way for the plant to increase the surface area of the root-soil interface, and perhaps exploit a nutrient source before its competitors.

Clustering of roots can also be induced by changes in the local environment of the root. Roots grow preferentially in cracks or biopores in compacted soil, such that the soil space is colonised irregularly (Wang et al. 1986; Tardieu 1988; Logsdon and Linden 1992). Patches of nutrients or residues from previous crops may also cause local increases in root length density, and this localised branching has become the subject of many studies (see review by Robinson 1994).

5.2.3 Anisotropy

Anisotropy is the non-uniform directional distribution of roots. It is the result of both morphogenetic and environmental factors, which interact during

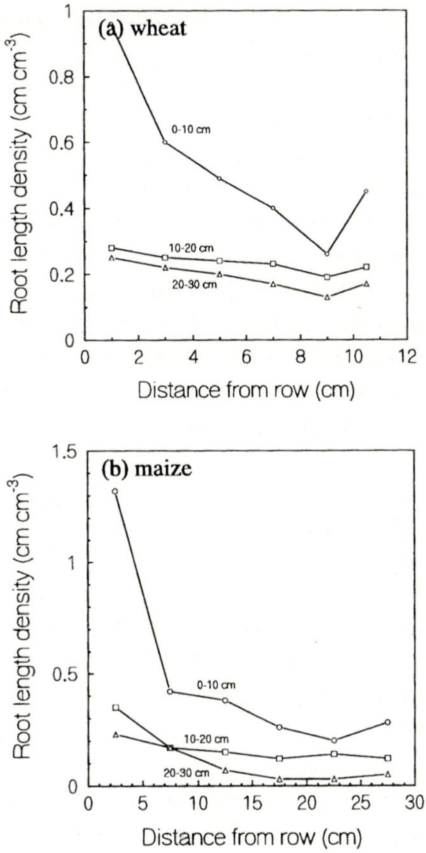

Fig. 5.2. Root length density at different distances from rows of crops of **a** wheat (row spacing 22 cm), and **b** maize (row spacing 60 cm). (Data from Gajri et al. 1994)

the development of the root system. Tropisms, and especially gravitropism, orient root growth in defined directions, according to the root type: the orientation is vertical for orthogravitropic roots, inclined at an angle for plagravitropic roots, and is horizontal for diagravitropic roots (Coutts 1989). The pattern of root branching also makes some root directions more likely. In some diarch surface roots of trees, for example, branching takes place only in a horizontal plane, resulting in a typically anisotropic pattern (Wilson 1964). Many soil properties (e.g. temperature, water content, strength) and morphological features, are structured in space, and contribute to root anisotropy (Logsdon and Linden 1992). Compact layers of soil, such as a tillage pan, can deflect roots horizontally for considerable distances, leading to anisotropy. This anisotropy is particularly marked when the main tap root of a plant encounters an impenetrable layer, as illustrated in Bennie (1996), for the tap root of cotton.

5.3 Choosing a Measurement Technique

The main steps in deciding on an appropriate sampling scheme are discussed in the following sections (see Fig. 5.3). Details of the various root system properties and measurement techniques are given in Table 5.1, with reference to appropriate chapters where the techniques are described.

It is first essential to focus on a clear research question, so that resources are used to maximum effect. Root sampling and measurement is always labour intensive, and the workload may be decreased by concentrating measurements on particular layers of the soil, or on particular samples. In the time that it takes to measure accurately the root length density in one sample containing 5 cm cm^{-3} of root, several samples containing 0.5 cm cm^{-3} may be measured. To determine, for example, the volume of soil from which a root system can extract water

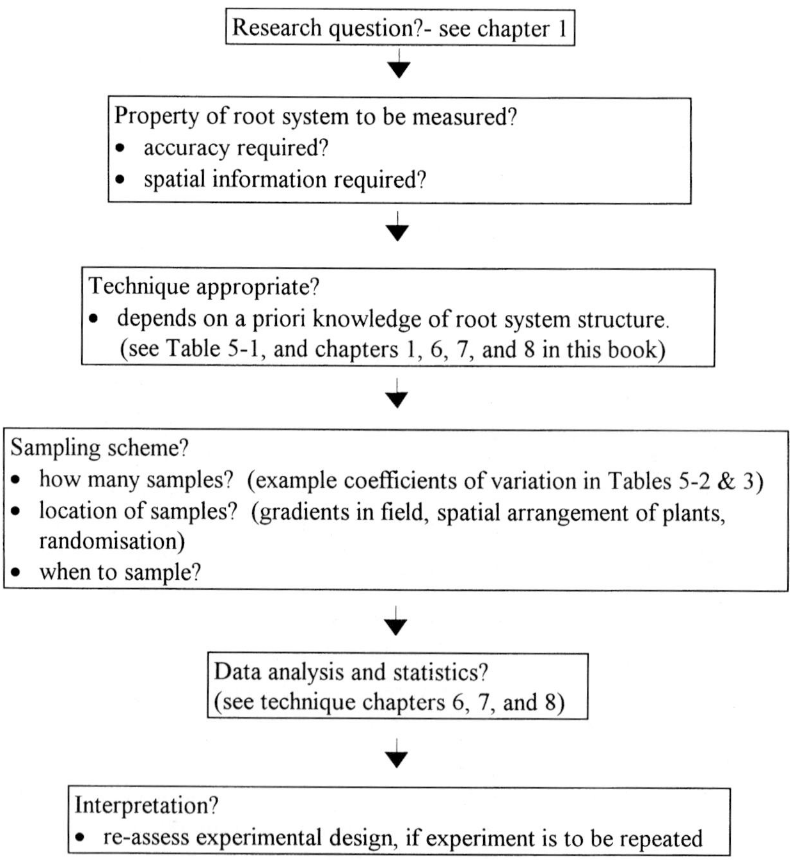

Fig. 5.3. Flow diagram of steps in deciding on experimental technique and sampling procedure

Table 5.1. Root system properties, measurement techniques, and their associated advantages and problems

Property	Technique and description	Advantages	Problems
Root length density in cores	Auger (Chap. 6). Soil cores sampled. Roots washed from cores in the lab	Most accurate method for root length density	Labour intensive. Many (e.g. 30%) roots lost during washing. Sample size determined by auger
Root end distribution in excavated planes	Trench wall and root mapping techniques (Chap. 7). Vertical or horizontal planes are excavated in the soil to expose root ends. Root end locations are plotted on transparent sheet, and number of ends per unit area calculated	Relatively fast. Good spatial resolution	Many fine roots go undetected. Problems converting measurements to root length density
Root end distribution in excavated cores	Core-break technique (Chap. 7). Root ends are counted on the face of a cut soil core	Relatively fast	Many fine roots go undetected. Problems converting measurements to the root length density. Sample size determined by auger
Root end or length distribution on cylindrical surface	Minirhizotron (Chap. 8): roots are observed with an endoscope around a transparent cylinder buried in the soil	Non-destructive, in situ. Can monitor temporal variation	Equipment expensive. Root grow preferentially along tube walls. Poor correlation with root length density

Table 5.1. (*cont.*)

Property	Technique and description	Advantages	Problems
Root diameter and branching patterns	Minirhizotron (above)	(as above)	(as above)
	Rhizotron (Chap. 8): Roots observed at a soil-glass interface	(as for minirhizotron, above)	Expensive facility. Roots growth preferentially along interface.
	Pinboard (Chap. 6): a board containing nails arranged in a regular grid is pushed into an excavated plane. The block of pinned soil is then then washed to expose the roots	Spatial distribution of roots in soil. Interactions between species	Roots disturbed and lost during washing
Whole root system architecture	Excavation and washing (Chap. 6) – with woody root systems, much of the soil may be washed or or blown from the root system to expose the main roots	Three-dimensional root distribution	Much fine root material lost. Roots moved during excavation
Root turnover	Rhizotron/minirhizotron (Chap. 9)	Non-destructive	Preferential root growth along interface. Equipment expensive. Difficult to determine live and dead roots

effectively, it may be more efficient to concentrate on measuring many samples that are rooted sparsely, than on measuring a few densely rooted samples accurately. Under some circumstances it may be appropriate to avoid measuring root growth and, instead, measure some function associated with roots: Gregory et al. (1978) monitored the advancing rooting front in wheat by measuring the movement of the drying front using a neutron probe.

Measurement of washed samples is very rapid using image analysis techniques (discussed in Chap. 10), but much time is required to spread root samples before analysis. Simple visual comparison of washed samples with standards of known length can provide faster but more approximate measurements of root length, with less preparation of samples than is required normally for image analysis (see Chap. 6 for visual comparison techniques).

It is impossible to set rigid rules for balancing all aspects of experimental design. The best approach depends crucially on the particular research question, the priorities and resources of the investigator, and a sound knowledge of the current literature.

5.4 Sampling Design

Two approaches based on different concepts are possible: firstly, we may excavate a given root system, individual root, or part of a root. We shall call this "architecture-guided" sampling. Secondly, we may determine the density distribution of the root system in some volume of soil. We shall consider these two approaches separately. In both cases, good prior knowledge is essential of the way that root systems are structured and distributed.

5.4.1 Architecture-Guided Sampling

It is possible to measure the length and branching density of individual roots, by gradually following and excavating them, starting from the base of the shoot (e.g. Tardieu and Pellerin 1990; Pagès and Pellerin 1994; Fig. 5.1b,d). This type of sampling requires an appreciation of root type, morphology, and behaviour (e.g. Waisel and Eshel 1991). Various criteria can be used to classify roots during this sampling procedure (e.g. Le Roux and Pagès 1994). Roots are often classified by some developmental criteria, such as the origin of the roots (seminal, nodal, branch), or branching order (primary axis; first second or third order lateral). Roots in some developmental categories continue to appear throughout the life of the plant. Age is an important additional criterion for classifying the sub-population, because it affects both root morphology and root physiology. For example, the proportion of primary lateral roots of maize that can be classified as determinate or normal, depends on the age of the parent root to which they are attached (Varney and McCully 1991). This information can be was obtained by careful excavation and sampling of individual nodal axes, followed by histological examination.

Spatial and temporal variation in the soil is superimposed on the endogenous variability between roots. In the field it is often difficult to assess the variation in soil conditions that a root has experienced during its existence. By studying the spatial distribution of a root system, it should also be possible to identify regions of soil of particular importance. For example, regions of root proliferation, or places roots have failed to penetrate could be selected for further study.

For trees and other perennials root architecture can also be approached by a study of fractal branching patterns. The basic assumption is that secondary thickening of roots is based on the demands for transport and thus at any point in the branched system transport capacity is proportional to the amount of fine roots distal to it. The proximal root diameter of a tree root (at the stem base) may contain enough information to reconstruct the total branched structure – provided that a number of simplifying assumptions hold, and that a number of parameters (which probably depend on tree species, but can otherwise be treated as constants) are known. The basic assumption of a "pipe-stem" model is that cross-sectional area (or the sum of diameter squares) is conserved during a branching event. Van Noordwijk et al. (1994) and Spek and van Noordwijk (1994) relaxed this assumption by introducing a proportionality factor alpha, but assumed that alpha is independent of current root diameter. A second parameter, q, describes the share of the largest "daughter" root in the total sum of squares of "daughters". To test the assumptions, one has to trace individual roots and take measurements of diameters before and after branching points (Van Noordwijk and Purnomosidhi 1995), although this is obviously a labour intensive process. If regression of alpha and q against diameter does not reveal any dependency, fractal (self-repetitive) models can be used. To re-construct three-dimensional structures of root systems, additional information on link ("inter node") length and branching angles is needed. Spek (1997) has since developed a visualisation routine by using a model originally developed for complex molecules. The overall prospects for this approach are relatively "quick and not-too-dirty" statements about tree root distribution, which can, if necessary, be backed up by detailed destructive sampling, but which can at least be used to specify sampling schemes in situations where tree roots are important.

5.4.2 Measuring the Distribution of Root Length Density

The root length density can be measured by separating the roots from a soil sample of known volume. Another method of assessing density relies on counting the number of intersections that roots make with a surface of known area in the soil. The properties measured, with a brief description of the techniques, are listed in Table 5.1, together with some advantages and problems associated with the techniques. Fuller information about each technique is given in the appropriate chapters of this book as indicated in Table 5.1.

None of the techniques are entirely satisfactory: the auger technique is the standard method for measuring root length density although, using this technique, up to a third of the fine root length may be lost during washing. The root-mapping technique is better for measuring the spatial distribution of individual roots. It is possible theoretically to calculate the root length density from the density of root intersections with a plane, although correlations obtained from field data suggest that a large proportion of the fine roots are often overlooked using the root-mapping techniques, leading to underestimates in the root length density – this is discussed in Bengough et al. (1992).

5.4.3 Sample Location

When designing any field experiment, it is important to assess the major sources of variation in data. Standard texts on experimental design discuss these sources of variation (e.g. Pearce 1983), which can be divided into patterned and non-patterned sources. Factors, such as slope, soil fertility, soil depth, shading, and variation in soil texture, are sources of patterned variability, and their relative importance will vary with weather conditions between seasons. Much more information is now becoming available on the spatial variation of yield in arable crops, due to the advent of precision agricultural equipment, which makes use of global satellite position references (Robert et al. 1996). The yield data from such systems gives an indication of plant performance, but there is no fixed relation between yield and root system development. Sources of non-patterned variability include errors such as loss of root material during washing and measurement.

When sampling real root systems, it is not possible to recover the whole root system of individual plants. Special consideration must be given to deciding which part of the root system should be sampled. Root systems of neighbouring plants are often intermingled, although this depends on the species concerned: Nelson and Allmaras (1969) found that maize roots intermingled with neighbouring soybean roots to a much greater extent than with neighbouring roots of the same species. It is possible to define a "unit soil area" for crop plants that are spaced regularly (Fig. 5.4, after van Noordwijk et al. 1985). The soil below each unit area is expected to contain a total root length equal to the mean root length per plant. Many of the roots within the unit area may belong to neighbouring plants but, similarly, an equal number of roots from the plant may have extended outside the area.

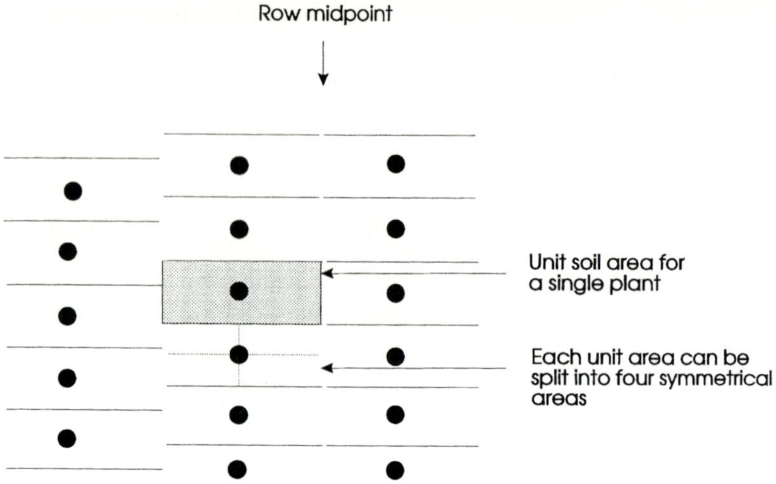

Fig. 5.4. Plan view showing unit area for single plants, represented by *black circles*, in a row crop. By symmetry, each unit area consists of four smaller representative areas. (After van Noordwijk et al. 1985)

The unit area can be divided into four equal parts which are equivalent, because of symmetry. The quarter area represents the fundamental unit in which the distribution of root length density should be studied. The most appropriate sampling schemes have been modelled for a range of crops (Fig. 5.5; after van Noordwijk et al. 1985). The assumptions of root distribution that underlie the choice of these sampling schemes are discussed in detail in van Noordwijk et al. (1985). Systematic trends in the root length density can bias the estimates of mean root length density, if the way that the quarter area is sampled over-represents either the dense or sparsely rooted areas (Fig. 5.5). Simply averaging the root dry weight in the row and the inter-row of cereals can overestimate the total root dry weight by as much as 30%.

5.4.4 How Big a Sample, and How Many Samples?

Clustering of roots, and the associated variability in root length density is an important characteristic of the root system (Tardieu 1988; Logsdon and Allmaras 1991). The size of the sample taken using an auger determines the minimum scale on which the variation of the root length density will be detected. Clustering increases the variance between samples and so, to achieve a given precision, more samples are required.

The average root length density measured for two populations can be compared using a T-test, provided that the root length density data are normally

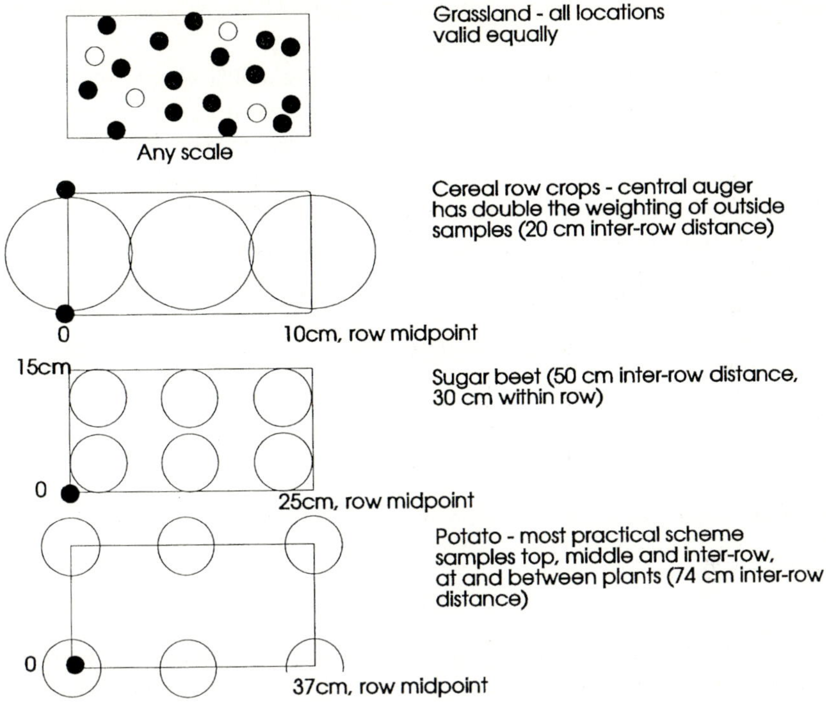

Fig. 5.5. Plan view of locations where samples should be taken in grassland and in crops of cereals, sugar beet and potato. *Black circles* represent the plants, *open circles* represent sampling locations using a cylindrical auger. (After van Noordwijk et al. 1985)

distributed, and that the variance for the two data sets is similar (see Box 5.1). The number of replicates required to give a good chance of finding a significant difference between treatments is discussed in Box 5.1 for a T-test. Approaches to estimating sample number for more complex experimental designs, for example involving blocking, are discussed in Mace (1964, especially Chap. 3) and Cochrane and Cox (1957, especially Chap. 2).

If the assumptions associated with parametric statistics are invalid (e.g. the quantity is not normally distributed), it may be possible to transform the data so that the transformed values can be tested. Failing this, or for rank or root count data, it is necessary to use non-parametric statistical tests, such as the Chi squared test. These tests can be relatively simple to perform, although they use less information than the parametric tests, and so are weaker.

In planning an experiment it is necessary to have some prior estimate of the degree of variation that is expected. In Tables 5.2 and 5.3 the coefficients of variation are listed for measurements made using the auger technique from crops grown under a variety of conditions. In the crops studied the coefficient

BOX 5.1. Number of Samples Required for T-test

The number of replicates that is required to give a 50% chance of detecting the difference between two means is shown in Fig. 5.6. More replicates are required to detect small differences between treatments, or for populations with a large coefficient of variation. For example, at least 25 replicates are needed to detect a difference of 22% between two means, if the coefficient of variation is 40%. To detect a difference of just 10% between two means requires more than 120 replicates, if the coefficient of variation is 40%. More details of T-tests are given in standard statistical texts (e.g. Sanders 1990)

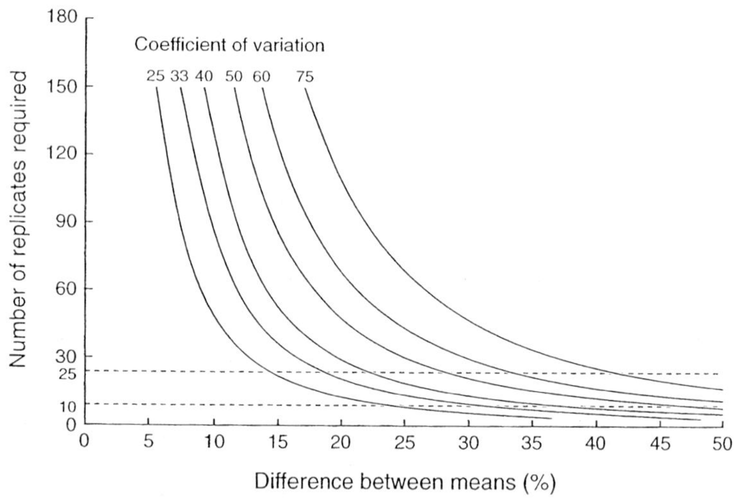

Fig. 5.6. Plot of the number of replicates required to give a 50% chance of distinguishing between two means in a two-sided t-test at 95% significance level. (After van Noordwijk et al. 1985)

of variation increased with depth of sampling (Tables 5.2 and 5.3). The reason for this may be partly associated with the sparse rooting of these deeper layers, with relatively few main axes present, surrounded by associated clusters of lateral branch roots (Grabarnik et al. 1998). Another contributing factor is changes in the soil structure with depth – in compacted subsoils, roots may be confined to continuous cracks and biopores in the soil to a much greater extent (e.g. Ehlers et al. 1983).

The coefficients of variation for root dry weight in auger samples from grassland are typically between 30 and 50% (Table 5.3). There is considerable variation between the coefficient of variation of root dry weights measured in

Table 5.2. a Coefficients of variation for root length density from Kucke et al. (1995). Samples were taken from fields of sugar beet, wheat and rye, using an auger (6.5 cm diameter by 15 cm long) from soils of different texture. **b** Coefficients of variation for the root length density of maize. Samples taken using an auger (5.1 cm diameter by 15 cm deep)

a

Depth (cm)	Coefficients of variation (%)		
	Sand[a]	Loam[a]	Clay[b]
0–15	12	27	21
15–30	28	8	54
30–45	76	12	57
45–60	129	60	41
60–75	74	71	69
75–90	113	35	60

[a] Sugar beet followed by winter wheat.
[b] Sugar beet followed by rye.

b

Depth (cm)	Coefficients of variation (%)	
	45 Days after emergence	59 Days after emergence
15	42	30
30	42	22
45	33	49
60	71	61
75	54	45
90	69	53

different studies – they average 38% for studies j to l in Table 5.3, but 59% for m1 and n1. Data from samples taken in the row should not be pooled with data from samples taken between the rows, as the root samples represent distinct populations. This is shown by the increased variance when the data is pooled (see columns headed m3 and n3 in Table 5.3). The root weight is large in the surface layer of soil immediately around the stem base. The ratio of the mean root dry weight in the row to that between rows is generally bigger than for the ratio of root length densities. This is because the root mass per unit length is normally greatest in the topsoil at the base of the stem. The coefficient of variation for root length density of cereals is often between 30 and 70% (Tables 5.2a,b), depending on the particular study.

Table 5.3. Coefficients of variation of root dry weight in auger samples of grassland and cereals.[a] (After van Noordwijk et al. 1985)

Depth (cm)	Grassland										Cereals								
	Ref	a	b	c	d	e	f	g	h	i	j	k	l	m1	m2	m3	n1	n2	n3
	Samples	100	100	20	20	20	20	20	50	50	25	25	20	4	4	8	47	32	103
	Diameter	7	4	7	7	7	7	7	4	4	7	7	7	7	7	7	105	36	70
0–5				34	30	51	45	41	29	43	30	43		41	89	100	36	31	33
5–10		33	38	34	36	33	48		29	37				83	56	73	42	44	43
10–20				30	45	44	30	55	36	50	34	27	47	40	32	37			63
20–30				36	41	40	38	75	43	56	31	28			64	65			59
30–40				35	41	55	38	53	35	38	36	37	43	45		69			46
40–50		35	41	52	31	48	47	59	46	31	35	29				78			45
50–60				44	44	54	54	76	51	53	39	35	50	54		46			62
60–70								100	56	39	56	47				48			106
70–80								85	75	46	53	53				43			125
80–90								76			61	63				72			
90–100								76			44	50				54			

[a] The key to column headings is as follows: a, b homogeneous grassland (1949); c, d young grassland (at Gilze, 1966); e, f same fields as c and d, 4 years later (June 1970); g established grassland (1976); h, i established grassland: root dry weight and root counts (estimates from Schuurman and Knot 1957); j, k oats: root dry weight and root counts (estimates); l winter wheat on cracking clay soil (Biddinghuizen, May and June 1977); m, n spring wheat on clay loam and sandy loam, respectively (Ulrum 1957); m1, n1 samples in row; m2, n2 samples between rows; m3, n3 equal number of row and between row samples combined.

5.5 New Developments

5.5.1 Using Root Growth Models to Improve Sampling Strategies

Schemes for sampling root systems must be designed carefully to obtain good estimates of the total dry weight or root length density in a layer of soil below a crop. Schemes used traditionally for cereal crops, for example, can result in a bias of up to 30% in the total root dry weight (van Noordwijk et al. 1985). Experiments that compare different sampling schemes are labour intensive and subject to experimental errors that are difficult to quantify and are associated with the excavation, washing and measurement of roots. An alternative approach, which is complementary to experimental comparisons, is to use a model of root distribution based on existing knowledge. Examples are becoming more common in the literature of using models to evaluate the sampling procedure for root systems, and this is a promising area in experimental design.

The types of model used range from simple exponential functions describing root distribution (van Noordwijk et al. 1985) to simulations of biomass fluctuation with time (Singh et al. 1984), and sophisticated simulations of root architecture in three-dimensional space (Nielsen et al. 1997; Pagès and Bengough 1997; Grabarnik et al. 1998). The basic approach using these models is the same, although the sophistication and applications of the models may vary: the model is used to generate a theoretical root distribution in space or time. This root distribution is then sampled or measured in a particular way, to simulate some sampling procedure. The simulation can be repeated many times very rapidly, as it is performed numerically on a computer. The sensitivity of sampling schemes to changes in the root distribution can be quantified by systematically varying model-input parameters.

The effect of sample timing and frequency on estimates of root biomass production was investigated using a model of root biomass for prairie grasses (Singh et al. 1984). The model was used to simulate variations in biomass with time, and the effects of sample timing, variability, and replication were studied. The estimate of biomass production was compared using two methods of calculation: firstly, from the difference between maximum and minimum biomass values and, secondly, from the summation of increments in biomass across the season. Estimates of biomass production were between 1.8 and 7 times greater than the actual production, suggesting that existing sampling schemes were inadequate. The use of a model allowed the effects of sample timing and replication to be investigated in more detail than is practical in most field experiments, although the details of the study were controversial (Lauenroth et al. 1986; Vogt et al. 1986).

Relations between root length density and root intersections with planes have been investigated with models that simulate root architecture in three-dimensions (Bengough et al. 1992). The use of minirhizotron tubes to measure rooting depth in a maize crop has been simulated using a three-dimensional model of root architecture of a small plot of 51 maize plants (Pagès and Bengough 1997). It was shown that the maximum rooting depth measured using minirhizotrons was very variable and could strongly underestimate the true rooting depth. Underestimation was a particular problem if the tube radius was smaller than 3 cm, and when the tube was close to the vertical. In this simulation study, however, the interactions between the growing root and the tube were not modelled. Some aspects of root growth and orientation alongside the minirhizotron tube may be modified by the presence of the tube itself, but there was insufficient detailed information on root-tube interaction to enable a realistic model of the interaction to be developed.

The fractal dimensions of two-dimensional projections of root systems have been compared with the three-dimensional fractal dimension of simulated bean root architecture (Nielsen et al. 1997). The model showed that the three-dimensional fractal dimension differed from that in two-dimensions, suggesting that the washing and flattening of the root system is not an acceptable way of measuring the fractal dimension of root systems in situ. The use of root intersection data with horizontal and vertical planes was found to give accurate estimates of the fractal dimension in three-dimensions, suggesting that this is a much better way of characterising root distribution in soil.

5.5.2 Applications of Geostatistics to Study Spatial Variability

Geostatistics is a relatively new technique that can be used to study the spatial variation of roots in a particular depth layer across a field site. It can also be used to study spatial variation in root systems with depth, although any systematic trends in vertical root distribution must first be subtracted. A major disadvantage of geostatistical techniques is that large numbers of samples (typically >100, but sometimes 300 or more; Jackson and Caldwell 1993b) are required, each at a known location, and at a range of separations.

The traditional statistical tests used in biology assume that the data in the populations being tested have the same distribution, and that each datum is independent of all other data. Roots form interconnected branched structures that may be correlated spatially and temporally: the presence of a root in a particular volume is often more likely if a root is present in a neighbouring volume, or was present when the volume was sampled previously. The scale and timing at which the sampling is performed determine whether such spatial and temporal correlations are present.

Geostatistics is a branch of applied statistics that can be used to detect, model and estimate spatial patterns. It was developed originally for mining and geology, but has been used more recently in the plant and soil sciences (e.g. Castrignano and Lopez 1988; Castrignano et al. 1994; Bourgault et al. 1997). In this chapter we give a very brief introduction to some geostatistical techniques. Concise introductory works that use examples from soil science and agronomy are those by Vieira et al. (1983), Burgess and Webster (1980) and Webster (1985) while Rossie et al. (1992) is a comprehensive review of geostatistical tools for ecology. Only a few studies have applied geostatistical techniques to root systems (Aiken 1992; Jackson and Caldwell 1993a,b). The techniques of variography and kriging are potentially very relevant: Variography models spatial dependence of variables, whilst kriging interpolates between the measured locations.

5.5.2.1 Semivariograms

Geostatistical techniques quantify spatial correlations and can be used to detect clustering of roots. Semivariance is a measure of the variability between pairs of observations and, in semivariograms, is plotted against the separation between those observations (Box 5.2).

Semivariograms have been used to study the spatial distribution of soil properties and roots (Jackson and Caldwell 1993a,b). A positive increase in semivariance with distance (Fig. 5.8) indicates clustering. The separation distance at which semivariance approaches a constant value is called the range, and is the length of spatial correlation, corresponding roughly to the dimension of clustering. Roots of an individual plant may experience very different conditions due to the heterogeneity of soil conditions. The distribution of nutrients (nitrate, ammonium, phosphate and potassium), has been studied on a scale of 0.1–10 m around perennial *Artemisia* and *Pseudoroegneria* plants (Jackson and Caldwell 1993b): nitrate and ammonium contents varied by between two and three orders of magnitude within the 120 m^2 plot, with phosphate and potassium showing smaller variation. Clustering of these nutrients occurred on a scale of <0.5 m (corresponding to the range), with no further increase in variance at scales greater than 1 m.

The semivariance at a separation equal to the range is known as the sill, and is an estimate of the population variance for observations separated by distances greater than the range. The semivariance extrapolated to zero separation is called the nugget, and represents all unaccounted for spatial variability at distances smaller than the smallest sampling distance.

The nugget includes both variation due to measurement errors and spatial variation that occurs over distances shorter than the sample spacing. The

BOX 5.2. Semivariance and Semivariograms

A semivariogram is constructed by plotting the semivariance as a function of the distance separating pairs of observations (Fig. 5.7). Semivariance is defined as

$$\gamma(h) = \frac{1}{2N(h)} \sum_{i=1}^{N(h)} [R(i) - R(i+h)]^2, \tag{5.1}$$

where $\gamma(h)$ is the semivariance at separation distance (h), N(h) is the number of paired observations separated by distance h, R(i) is some root variable (e.g. root number, length density or dry weight) at location i, and R($i + h$) is the same variable separated from i by a distance h (Vieira et al. 1983)

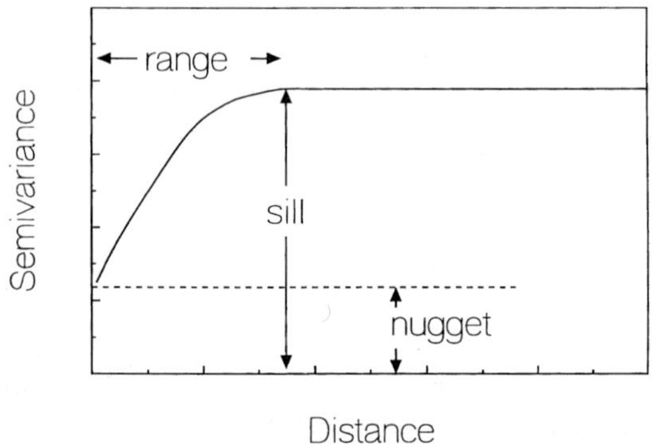

Fig. 5.7. Semivariogram showing idealised behaviour

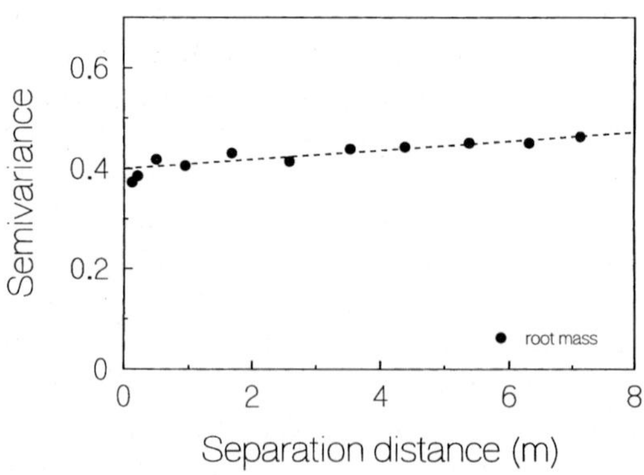

Fig. 5.8. Semivariogram from root mass data measured by Jackson and Caldwell (1993a)

difference between the semivariance at the sill and that at the nugget represents the variance that can be modelled as a spatial dependence, based on the available sampling grid. A small nugget/sill variance ratio indicates spatial correlation.

When values of semivariance are being compiled into a semivariogram, it is assumed that there are no systematic trends in the quantity measured (Aiken et al. 1991). If such trends exist, functions characterising these trends must first be subtracted from the data. For example, vertical trends must be removed before generating a variogram of root intersection density with minirhizotrons or washed root length density with depth. Semivariograms constructed from data containing trends show increasing semivariance with increasing distance between observations.

5.5.2.2 Kriging and Cokriging

Kriging is an interpolation procedure that uses the semivariogram to estimate values at locations where measurements were not made. The values estimated $[Z^*(x_o)]$ using kriging are linear functions of the known values $[Z(x_i)]$ measured at N other locations:

$$Z^*(x_0) = \sum_{i-1}^{N} [\lambda_i Z(x_i)], \tag{5.2}$$

where the weight factors (λ_i) are chosen such that the mean value of the difference $[Z^*(x_o) - Z(x_o)]$ is zero, and the variance of the difference is minimised. The semivariogram determines the values of λ_i and the relative weight of observations decreases with distance from the interpolation points.

Cokriging can be used where two or more variables are correlated in space. Measurements of one variable may be used to produce predictions of how a second variable is distributed. For example, very hard regions of the soil may be rooted sparsely, and so the root length density would be correlated negatively with penetration resistance. Measurement of one variable may be used to produce predictions of how a second variable is distributed.

5.5.2.3 Conditional Simulations

Kriging is a procedure for estimating optimal values at intermediate points on a map. However, no interpolation procedure can produce information that has not been surveyed. The real spatial variation in the quantity studied includes an extra component of variation that is not accounted for in the kriged values.

Conditional simulations show realistic images of the variability of the quantity studied. By running many of these simulations, the variance and mean values can be mapped and this map can provide a reliable basis for calculating root biomass, deciding on root sampling strategies, or for simulating water or nutrient uptake. Further details of the theory of conditional simulation are given in Matheron (1973).

5.6 Conclusions

Root systems are complex branched structures that vary in space and time. The structure of the root system must be appreciated before it can be studied in the field. It is important to consider the aims of any experiment carefully when choosing an appropriate measurement technique from the many available, as root sampling is very labour intensive. Auger sampling gives the most reliable estimate of root length density, but losses of about 30% can occur during washing procedures (see Chap. 6). Abias of about 30% for cereals can result in estimates of root length density, per plant or per layer of soil, made using traditional sampling schemes. The scientific basis of where to sample roots is still understood relatively poorly, and there is a need for new experimental and theoretical studies of root distribution to address this problem. Modelling is starting to become a useful aid in investigating the consequences of different temporal and spatial sampling schemes. Geostatistics is also providing new ways of studying spatial heterogeneity in roots, although large numbers of samples (e.g. 100 plus) are often required to provide sufficient data for the analyses.

Acknowledgements. We thank Drs Allmaras and Berntson for their valuable comments on an earlier draft of this manuscript. AGB and LP thank NATO for a collaborative research grant. The Scottish Crop Research Institute receives grant-in-aid from the Scottish Executive Rural Affairs Department.

References

Aiken RM (1992) Functional relations of root distributions with the flux and uptake of water and nitrate. PhD Dissertation. Michigan State University, East Lansing

Aiken RM, Jawson MD, Grahammer K, Polymenopoulos AD (1991) Positional, spatially correlated and random components of variability in carbon dioxide efflux. J Environ Qual 20: 301–308

Bengough AG, MacKenzie CJ, Diggle AJ (1992) Relations between root length densities and root intersections with horizontal and vertical planes using root growth modelling in 3-dimensions. Plant Soil 145: 245–252

Bennie ATP (1996) Growth and mechanical impedance. In: Waisel Y, Eshel A, Kafkafi U (eds) Plant roots, the hidden half. Marcel Dekker, New York pp 453–470

Bourgault G, Journel AG, Rhoades JD, Corwin DL, Lesch SM (1997) Geostatistical analysis of a soil salinity data set. Adv Agron 58: 241–292

Burgess TM, Webster R (1980) Optimal interpolation and isarithmic mapping soil properties. I. The semivariogram and punctual kriging. J Soil Sci 31: 315–331

Canadell J, Jackson RB, Ehleringer JR, Mooney HA, Sala OE, Schulze ED (1996) Maximum rooting depth of vegetation types at the global-scale. Oecologia 108: 583–595

Castrignano A, Lopez G (1988) La variebilita delle proprieta del suolo e la sua applicazione nelle ricerche agronomiche. Proc Conf Italian Soc of Soil Science, Verona, 1988 pp 714–728

Castrignano A, De Giongio D, Stelluti M, Rizzo V (1994) A geostatistical approach to characterize spatial variability of yield in a durum wheat submitted to four tillage treatments. Proc 13th Int Conf Int Soil Tillage Research Organisation, Aalborg, 1994 pp 997–1004

Cochran WG, Cox GM (1957) Experimental designs 2nd edn. John Wiley New York

Coutts MP (1989) Factors affecting growth direction of tree roots. Ann Sci For 46: 277–287

Dittmer HJ (1937) A quantitative study of the roots and root hairs of a winter rye plant (*Secale cereale*). Am J Bot 24: 417–420

Ehlers W, Kopke U, Hesse F, Bohm W (1983) Penetration resistance and root-growth of oats in tilled and untilled loess soil. Soil Tillage Res 3: 261–275

Fitter AH, Stickland TR (1991) Architectural analysis of plant root systems 2. Influence of nutrient supply on architecture in contrasting plant species. New Phytol 118: 383–389

Gajri PR, Arora VK, Kumar K (1994) A procedure for determining average root length density in row crops by single-site augering. Plant Soil 160: 41–47

Gerwitz A, Page ER (1974) An empirical mathematical model to describe plant root systems. J Appl Ecol 11: 773–781

Grabarnik P, Pagès L, Bengough AG (1998) Geometric properties of simulated maize root systems: consequences for length density and intersection density. Plant Soil 200: 157–167

Gregory PJ, McGowan M, Biscoe PV (1978) Water relations of winter wheat. 2. Soil water relations. J Agric Sci 91: 103–116

Jackson RB, Caldwell MM (1993a) Geostatistical patterns of soil heterogeneity around individual perennial plants. J Ecol 81: 683–692

Jackson RB, Caldwell MM (1993b) The scale of nutrient heterogeneity around individual plants and its quantification with geostatistics. Ecology 74: 612–614

Jackson RB, Canadell J, Ehleringer JR, Mooney HA, Sala OE, Schulze ED (1996) A global analysis of root distributions for terrestrial biomes. Oecologia 108: 389–411

Kucke M, Schmid H, Spiess A (1995) A comparison of four methods for measuring roots of field crops in three contrasting soils. Plant Soil 172: 63–71

Lauenroth WK, Hunt HW, Swift DM, Singh JS (1986) Overestimation of net root production – a real or imaginary problem – reply. Ecology 67: 580–582

Le Roux Y, Pagès L (1994) Développement et polymorphisme racinaire chez de jeunes semis d'hévéa (*Hevea brasiliensis*). Can J Bot 72: 924–932

Logsdon SD, Allmaras RR (1991) Maize and soybean clustering as indicated by root mapping. Plant Soil 131: 169–176

Logsdon SD, Linden DR (1992) Interactions of earthworms with soil physical conditions influencing plant growth. Soil Sci 154: 330–336

Lynch JP, Nielsen KL, Davis RD, Jablokow AG (1997) Simroot: modelling and visualization of root systems. Plant Soil 188: 139–151

Lyr H, Hoffmann G (1967) Growth rates and growth periodicity of tree roots. Int Rev For Res 2: 181–236

Mace AE (1964) Sample-size determination. Reinhold Publishing, New York

Matheron G (1973) The intrinsic random functions and their applications. Adv Appl Probabil 5: 438–468

Nelson WW, Allmaras RR (1969) An improved monolith method for excavating and describing roots. Agron J 61: 751–754

Newman EI (1969) Resistance to water flow in soil and plant. J Appl Ecol 6: 1–12

Nielsen KL, Lynch JP, Weiss HN (1997) Fractal geometry of bean root systems: correlations between spatial and fractal dimension. Am J Bot 84: 26–33

Pagès L, Bengough AG (1997) Modelling minirhizotron observations to test experimental procedures. Plant Soil 189: 81–89

Pagès L, Pellerin S (1994) Evaluation of parameters describing the root system architecture of field-grown maize plants (*Zea mays* L.). II. Density, length, and branching of first-order lateral roots. Plant Soil 164: 169–176

Parker CJ, Carr MKV, Jarvis NJ, Puplampu BO, Lee VH (1991) An evaluation of the minirhizotron technique for estimating root distribution in potatoes. J Agric Sci 116: 341–350

Pearce SC (1983) The agricultural field experiment: a statistical examination of theory and practice. John Wiley, New York

Pellerin S, Pagès L (1994) Evaluation of parameters describing the root system architecture of field-grown maize plants (*Zea mays* L.). I. Elongation of seminal and nodal roots and extension of their branched zone. Plant Soil 164: 155–167

Pellerin S, Pagès L (1996) Evaluation of in-field conditions of a 3-dimensional architectural model of the maize root-system – comparison of simulated and observed horizontal root maps. Plant Soil 178: 101–112

Persson H (1978) Root dynamics in a young Scots pine stand in central Sweden. Oikos 30: 508–519

Robert PC, Rust RH, Larson WE (1996) Precision agriculture. Proc 3rd Int Conf on precision agriculture, Minneapolis, 1996

Robinson D (1994) The responses of plants to non-uniform supplies of nutrients. New Phytol 127: 635–674

Rossi RE, Mulla DJ, Journel AG, Franz EH (1992) Geostatistical tools for modeling and interpreting ecological spatial dependence. Ecol Monogr 62: 277–314

Russell RS (1977) Plant root systems. Their function and interaction with the soil. McGraw Hill, Maidenhead

Sanders DH (1990) Statistics: a fresh approach (4th edn). McGraw-Hill, New York

Schuurman JJ, Knot L (1957) Het schatten van hoeveelheden wortels in voor wortelonderzoek genomen monsters. Versl Landbouwk Onderz 63: 31

Singh JS, Lauenroth WK, Hunt HW, Swift DM (1984) Bias and random errors in estimators of net root production – a simulation approach. Ecology 65: 1760–1764

Spek LY (1997) Generation and visualization of root-like structures in a three-dimensional space. Plant Soil 197: 9–18

Spek LY, Van Noordwijk M (1994) Proximal root diameters as predictors of total root system size for fractal branching models. II. Numerical model. Plant Soil 164: 119–128

Tardieu F (1988) Analysis of the spatial variability of maize root density II. Distances between roots. Plant Soil 107: 267–272

Tardieu F, Pellerin S (1990) Trajectory of the nodal roots of maize in fields with low mechanical constraints. Plant Soil 124: 39–45

Van Noordwijk M, Purnomosidhi P (1995) Root architecture in relation to tree-soil-crop interactions and shoot pruning in agroforestry. Agrofor Syst 30: 161–173

Van Noordwijk M, Floris J, de Jager A (1985) Sampling schemes for estimating root density distribution in cropped fields. Neth J Agric Sci 33: 241–262

Van Noordwijk M, Spek LY, De Willigen P (1994) Proximal root diameters as predictors of total root system size for fractal branching models. I. Theory. Plant Soil 164: 107–118

Varney GT, McCully ME (1991) The branch roots of *Zea*. II. Developmental loss of the apical meristem in field-grown roots. New Phytol 118: 535–546

Varney GT, Canny MJ, Wang XL, McCully ME (1991) The branch roots of *Zea*. I. First order branches, their number, sizes and division into classes. Ann Bot 67: 357–364

Vieira SR, Hatfield JL, Nielsen DR, Biggar JW (1983) Geostatistical theory and application to variability of some agronomical properties. Hilgardia 51: (3)

Vogt KA, Grier CC, Gower ST, Sprugel DG, Vogt DJ (1986) Overestimation of net root production – a real or imaginary problem? Ecology 67: 577–579

Waisel Y, Eshel A (1991) Multiform behavior of various constituents of one root system. In: Waisel Y, Eshel A, Kafkafi U (eds) Plant roots, the hidden half. Marcel Dekker, New York pp 3–24

Wang J, Hesketh JS, Wooley JT (1986) Preexisting channels and soybean rooting patterns. Soil Sci 141: 432–437

Webster R (1985) Quantitative spatial analysis of soil in the field. Adv Soil Sci 3: 1–70

Wilde SA (1958) Forest soils: their protection and relation to silviculture. Roland Press, New York

Wilson BF (1964) Structure and growth of woody roots of *Acer rubrum* L. Harv For Pap 11: 14

CHAPTER 6

Auger Sampling, Ingrowth Cores and Pinboard Methods

Maria do Rosário G. Oliveira[1], M. van Noordwijk[2], S.R. Gaze[3], G. Brouwer[4], S. Bona[5], G. Mosca[5], and K. Hairiah[6]

[1] University of Évora, Apartado 94, 7002-554 Évora, Portugal
[2] ICRAF, JI CIFOR, PO Box 161, Situ Gede, Sindang Barang Bogor 16680, 16001Bogor, Indonesia
[3] University of Reading, PO Box 233, Reading, RG6 6DW, United Kingdom
[4] Plant Research International, Postbus 16, 6700 AA Wageningen, The Netherlands
[5] Università di Padova Instituto di Agronomica, Dipt. di Agronomia Ambientale e Produzioni Vegetali, Agripolis – Via Romea, 16, 35020 Legnaro (Padova), Italy
[6] University Brawijaya, Faculty of Agriculture, Department of Soil Science, Jl. Veteran, 65145 Malang, Indonesia

CONTENTS

A.L. Smit et al. (Eds.), Root Methods

6.1 Introduction

This chapter outlines those methods for assessing root systems structure and function in the field which are based on washing roots free from the soil in which they grew. Some of these methods are included in previous reviews (Kolesnikov 1971; Böhm 1979). The methods are either disruptive or totally destructive to the root system being studied and to the immediate environment (Taylor et al. 1991).

The spatial variability of root systems requires that a large number of replicates be sampled to obtain reasonably accurate estimates of root parameters. Intensive, destructive sampling disrupts the remaining roots, which may in turn affect results in the future. Consequently, due to the large variability of results, it is not possible, in practice, to estimate root turnover from frequent sampling schemes (Van Noordwijk 1993). Nevertheless, soil cores from auger sampling or pinboards often yield the best quantitative information on root system biomass and root length per volume of soil (Caldwell and Virginia 1991). These methods are therefore the basis for calibrating other techniques, including those for estimating root turnover.

Monolith samples can be taken using a pinboard, which enables the complete root system to be observed after washing the soil away. In contrast, augered samples are taken at discrete locations in relation to the horizontal and vertical distance in the soil profile from the plant (stratified sampling). Auger sampling provides a soil-root sample of a limited volume from the root zone.

The ingrowth core technique (Steen 1984; Fabião et al. 1985) is based on mesh bags filled with root-free soil, placed at different depths and removed at prescribed intervals. Differences between soil conditions inside the bags and in the surrounding undisturbed soil, such as soil strength and mineral nutrition status, can change the pattern and quantity of roots. However this method is particularly suitable for recording root response to localised fertiliser application or other heterogeneity in the soil (Cuevas and Medina 1988; Hairiah et al. 1991).

The destructive sampling methods described in this chapter are the basis for quantifying root length, diameter, surface area, volume, biomass, disease, mycorrhizal association and ion content. Depending on the goal of the further analysis, however, one may need correction factors to account for the changes in root properties during sampling, handling and storage.

Table 6.1 indicates the information that can be obtained from different methods of root measurement. Auger sampling, ingrowth cores and pinboard methods are described in this chapter. Core break and root mapping methods are described in Chapter 7. Minirhizotron and related methods are described in Chapter 8.

6.2 Methods of Root Sampling

6.2.1 Auger Sampling

The best sampling technique for obtaining volumetric soil-samples is the auger method. With this method soil samples are taken from the field using hand-operated or mechanical samplers and washed to separate roots from soil.

Table 6.1. Aim of different methods for root measurement[a]

Parameter	Auger sampling	Ingrowth cores	Pinboard	Core break method	Root mapping	Minirhizotron
Root length $cm\,cm^{-3}$	+++	+++	++(+)	+	+	+
Root weight $g\,cm^{-3}$	+++	+++	++(+)	–	–	–
Distribution pattern	+	–	+	–	++	–
Root growth and decay	–	++	–	–	–	+++
Root diameter	++(+)	++(+)	++(+)	–	–	+
Branching pattern	–	+	+	–	–	+
Time consuming d (8 hours)	7–20	5–15	15–30	2	4	4–10
Number of samples	120	60	2 Boards $0.5\,m^2$	120 = 12 Boreholes	6 Maps $0.5\,m^2$	1 Minirhiz./year[b]

[a] Symbols are as follows: – no information on parameter, + qualitative interpretation may be drawn, ++ semi-quantitative interpretation through a ranking, +++ fully quantitative information obtained.

[b] length of minirhizotron is 1–2.1 m.

6.2.1.1 Hand Sampling

The simplest method for taking soil samples is to use a hand-driven corer. One of the most commonly used corers is that described by Schuurman and Goedewaagen (1971; Fig. 6.1). It comprises a cylindrical tube 15 cm long with an inside diameter of 7 cm. A T-handle at the top of the auger shaft facilitates rotation of the auger to aid penetration into and removal from the soil. Many authors have modified this simple hand auger, mainly by varying the tube diameter, or by using mass impact to drive the auger into the soil.

Albrecht (1951) and Albrecht et al. (1953) describe an auger consisting of two halves held together by a metal ring. These halves can be separated to allow for recovery of intact soil cores. A new manual coring system has been proposed by Prior and Rogers (1992, 1994) which includes a manual driver of adjustable weight, a manual core extractor and steel core tubes with clear plastic liners which encase the soil core for retrieval and transport.

Various materials have been used for the augers but the most frequently utilised are metal (stainless steel) or Plexiglas, bevelled at one end to minimise soil disruption during sampling.

6.2.1.2 Mechanised Techniques

When core samples must be recovered from depth or from difficult soils, mechanised equipment is used. One of the simplest ways is to use an auger driven into the soil by a hand-held motorised drop-hammer and removed by a puller or a screw-jack (Böhm 1979). Soil core samplers can also be mounted on tractors, typically on the drawbar, which has hydraulic power for the movement (upward and downward). It is also possible to utilise special machines constructed for agricultural engineering purposes, such as for drainage studies and soil survey mapping. Hydraulic devices transported by large tractors have reduced the time and the labour required for taking soil cores but their use is limited to situations where damage of a large portion of the plot is not a problem.

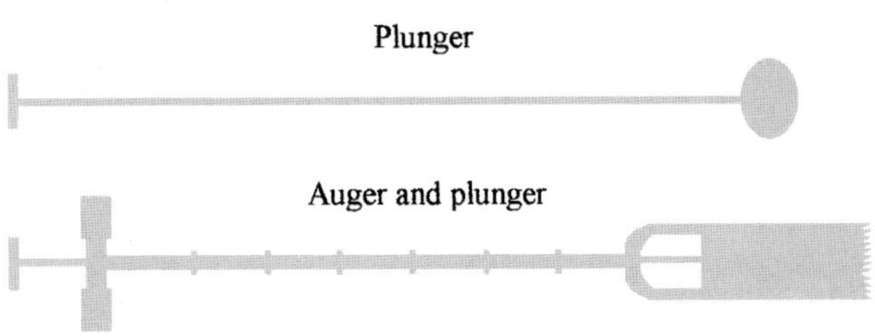

Fig. 6.1. Design of a simple auger based on Schuurman and Goedewaagen (1971)

The method proposed by Baarstad et al. (1993), where a hydraulic probe is attached to a knuckle boom mounted on a truck, would reduce soil compaction, although its cost could be a problem.

6.2.1.3 Core Diameter

One of the most important factors in field sampling is the core diameter (see Böhm 1979). The core must be large enough to obtain a reasonable sample volume, yet small enough to enable the cores to be obtained by the sampling method to be used. With decreasing core diameter an increasing number of replicates must be taken to maintain sampling accuracy (see Chap. 5). Small diameter cores can be a particular problem where there are low rooting densities. Furthermore, Schuurmann and Goedewaagen (1971) reported average root length densities were lower using a 4 cm diameter core than with a 7 cm diameter core. The difference could not be accounted for by the actual soil volume sampled and was attributed to the increased frictional resistance between the soil core and the bore of the tube in the smaller diameter tube. They recommended a core diameter of 7 cm, and the most commonly used core diameters range from 5 to 8 cm (Van Noordwijk 1993).

In dry sandy soils, a smaller auger diameter may be required to avoid soil loss during core extraction. In very wet soils special augers may be needed which allow air entry into the mud (Schuurman and Goedewaagen 1971).

6.2.1.4 Sampling Strategy in the Field

Position of Sampling. Sample position depends on the purpose of the measurements. If the aim is to obtain an average value for the field, a form of stratified sampling may be desirable, as rooting density will vary with the spatial arrangement of plants. In grassland, or other systems where plants are not grown in rows, a completely randomised design may be followed, unless there is reason to expect specific spatial patterns in the field (e.g. due to distance to the nearest drain, or irrigation water patterns), when such factors must be considered in the sampling design. For row crops, sampling must be stratified within and between rows. Crozier and King (1993) suggest sampling in a transect perpendicular to the row, within the rooting zone, while Van Noordwijk et al. (1985) propose special sampling schemes, depending on row spacing. In more complex plant arrangements, for example in agroforestry systems, additional levels of stratification are required and in the case of very complex systems or in natural vegetation, samples are no longer stratified but rather collected along a larger number of randomly assigned transects. Within each stratum, core position must be assigned at random. With any stratified sampling scheme care must be taken when estimating a field average value from the data. Serious biases may be introduced if the average value for all samples is assumed to be the average for the field.

For treatment or variety comparisons, one sampling position per treatment may be sufficient, though this would need to be replicated throughout the experiment. For more detailed process or model calibration studies it may be necessary to quantify the total root expansion or the root depth or the soil layers in which the root length density is above a certain threshold (Passioura 1980).

Depth of Sampling. Ideally the core should be driven to the limit of rooting depth but the deepest layers are difficult to reach, and within those horizons the variability is often high. The maximum sampling depth can be estimated in some cases by extrapolating from the root length densities in the upper soil using a negative exponential equation (Van Noordwijk 1993). In any case, all soils must be sampled to a minimum depth of 30 cm or the bottom of the plough layer because most of the roots will probably be concentrated in this layer.

If clearly differentiated soil horizons are present in the sampled soil, the first approximation is to separate the different soil layers and then to subdivide the cores in standard length intervals. The most widely used core length is 10 cm. However, in grasslands and no till systems, it may be worth separating the 0–10 cm layer into 0–5 and 5–10 cm. Shorter core lengths will increase sample variability and it is difficult to cut portions of the core shorter than the diameter of the core itself.

Number of Replications. Particular attention must be given to the number of replicates (see Chap. 5). It is clear that by increasing the number of replicates the variability of data decreases, but the standard error of 25 samples is only five times smaller than that of a single sample (van Noordwijk 1993).

The number of replications adopted by different authors varies from three to ten per experimental unit, similarly to what is done in aboveground measurements. Van Noordwijk et al. (1985) estimated a coefficient of variation in grassland root weight for individual auger samples of 385 cm^3 (10 cm height and 7 cm diameter) of at least 40%. Values up to 100% are, however, not uncommon. The number of soil cores required for each plot can be determined at the beginning of the sampling program by extracting the roots from a number of soil cores coming from the same plot (Vogt and Persson 1991). In this way it is possible to have an idea of the variability of the samples and optimise the sampling process.

In order to reduce the number of samples to be processed, Schroth and Kolbe (1994) propose a method consisting of the combination and homogenisation of several cores from a plot, with consequent subsampling for root extraction. This method, similar to the composite sampling schemes usually followed for soil chemical analyses, allows an increase in the number of soil samples and/or a reduction of processing time. However, the method introduces a new source of error dependent on the homogeneity of the total sample before subsampling and it is not appropriate when the root turnover is studied.

6.2.1.5 Special Techniques for Tree Roots

Sampling tree roots requires special techniques because of the large range of root diameters within a tree root system and the large volume of soil occupied by the roots. Knowledge of the species' root system structure is necessary if it is desired to sample a particular portion of the root system (see Chap. 1). Veller (1971) proposed a sampling scheme consisting of concentric circles, centred on the stem of the tree, the diameter of each circle being chosen according to the nature and age of the tree under study.

Generally, for taking soil cores in tree root systems it is necessary to utilise mechanised sampling techniques because of the hardness of the large roots.

For some purposes it is relevant to investigate the finer parts of the woody roots (e.g. spatial distribution, biomass turnover). Methods defined by Kalela (1950) or Nielsen (1990, 1994) are suitable. The "Nielsen sieve" is 45 × 45 cm with 25-cm-high walls, and ribs, 0.5 cm thick, placed in two perpendicular layers, defining holes of 3 × 3 cm squares. Depending on the soil and the hypothesis to be tested, the soil material is extracted separately from a number of soil horizons to the sieve. Stones and soil material are sieved in the field, and the collected root material is washed, sorted, dried and weighed in the laboratory. Roots are sorted into biomass and necromass, the living roots being sorted to the following diameter classes (root material below 0.1 cm diameter is removed after drying): fine roots (<0.2 cm), thin roots I (0.2–0.5 cm), thin roots II (0.5–1 cm), thin roots III (1–2 cm) and coarse roots (2–5 cm). Coarse roots can not be adequately sampled with any auger sampling scheme and one may have to resort to methods predicting ratios of roots in different diameter classes based on fractal branching analysis (Spek and van Noordwijk 1994; Van Noordwijk et al. 1994; Van Noordwijk and Purnomosidhi 1995).

6.2.1.6 Drawbacks

Auger sampling requires a large number of samples since the volume of auger samples is small compared with the total soil volume. In very small plots problems may arise if coring damages a relatively large portion of the plots.

Sampling depth using hand-operated apparatus is seldom greater than 100 cm. On stony soils or soils with large amounts of woody roots, inserting the auger may be difficult. Also, on dry clay soils the penetration resistance presents a problem. Sometimes it is helpful to rewet the surface soil before coring, if this does not interfere with the purposes of sampling.

The use of heavy mechanised equipment for core sampling may be a risk since it can compact the soil cores, shortening the core itself. This must be taken into account when cutting the core and interpreting the results.

The total time needed for a single core depends on soil conditions and depth of sampling, but generally the time needed for taking samples in the field is much less than the subsequent processing in the laboratory.

6.2.2 Ingrowth Cores

6.2.2.1 Concept

The basic concept of the root ingrowth core technique is to present root-free soil to the growing root system for a fixed length of time, so that root growth during that period can be quantified (Steen 1984, 1991; Hansson et al. 1991). The method reduces uncertainty about the time interval during which the roots in a particular soil core developed, but may introduce biases, as it is not possible to create root-free soil which is identical to the undisturbed soil conditions. It is therefore best to use the method mainly as a *relative* indication of the main periods of root growth, by comparing mesh bags placed at various times or left in the soil for various time intervals.

Practical instructions for using ingrowth cores are given in Box 6.1.

6.2.2.2 Drawbacks

The main difficulty in interpreting data from soil ingrowth cores arises from the unknown degree of disturbance upon introducing the mesh bags. Some of the roots growing into the core may be branch roots of roots severed by the coring, and this may cause an overestimate of the root growth which would otherwise have occurred in this volume of soil. More importantly, however, the process of sieving and re-packing the soil modifies soil conditions. The alteration of sandy soil properties is small but on other soil types, the disturbance to the aggregate structure accelerates mineralization of soil organic matter and affects soil aeration. When results are compared for mesh bags placed at different time periods, one has to be cautious of the fact that the degree of disturbance may vary with soil moisture content, even if exactly the same

BOX 6.1. Procedure for Installing the Mesh Bags for the Root Ingrowth Cores Technique

The procedure is summarised in Fig. 6.2. Remove long cores of soil and sieve the soil to remove any roots. Then, place a tubular nylon or polypropylene mesh bag (0.5–0.7 cm mesh) in the hole, loosely attached to a plastic pipe which is slightly smaller in diameter than the soil core. Fill the hole with the sieved soil via this tube (using a funnel), re-packing the soil to the original bulk density. The filling procedure is critical and results depend on soil texture, structure and moisture content. The best results are obtained by measuring the amount of soil to be filled per 1 cm layer (or less) and pressing the soil layer by layer with a pestle which fits into the tube. After pressing a layer, scrape the surface to create roughness which allows good contact with the next layer of soil.

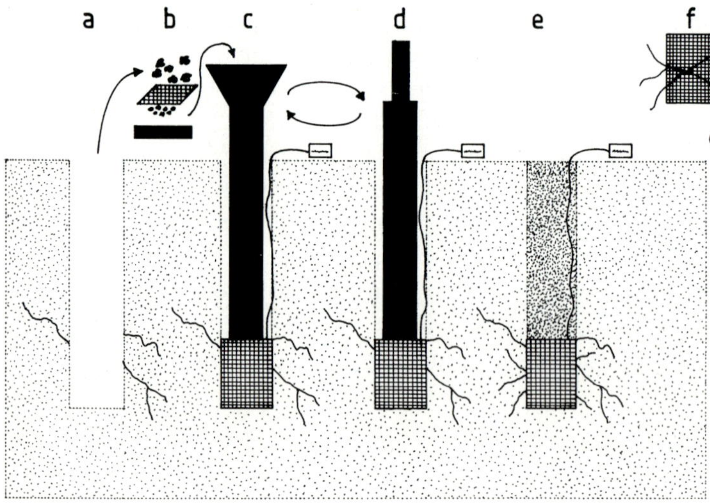

Fig. 6.2. Procedure for installing the mesh bags for the root ingrowth cores technique

It is convenient to attach a rope to the bottom of the mesh bag which can be used to pull out the mesh bag at harvest time. The other end of the rope is labelled and remains on the soil surface.

When the bags are recovered, it is often necessary to cut woody roots entering the bag with a knife (or auger of diameter slightly larger than the one originally used). Non-woody roots can be trimmed after mesh bag recovery. Wash the mesh bag contents as per a normal auger sample.

procedure is used for placing the mesh bags. If large variations in soil moisture contents are to be expected, it may be better to compare different intervals for root ingrowth into bags placed at the same time, rather than constant time intervals for bags placed at different times.

6.2.2.3 Results and Perspectives

With sufficient precautions, the method produces a reasonable estimate of the *relative* seasonal pattern of fine root growth. The results obtained from this method can usefully complement data from destructive sampling, which should be conducted at least once during the period of study. It may be worth collecting the roots sieved from the soil before placing it back into the mesh bag to correlate new root growth and previous root mass in the core, but sieving individual soil cores in the field is time consuming and often soil obtained from the same soil layer elsewhere in the experimental plot is used. Using the same, homogenised soil for all replicate ingrowth cores may reduce the variability of the results, but it depends on the purpose of the measurement whether or not this is desirable.

An alternative use of the ingrowth core technique is to compare different soil types, soil with different organic or inorganic amendments and/or packed to different bulk densities. For these applications one accepts that the soil inside the ingrowth core will be different from the surrounding bulk soil. Results may indicate the "local response" of root development to soil heterogeneity. Hairiah et al. (1991) used this method to investigate whether shallow root development of *Mucuna pruriens* on an acid soil was due to an inherent characteristic of the subsoil, or to the position of the subsoil. In other situations addition of specific soil organisms (symbionts, rhizosphere organisms, and rhizovores) may be compared. In this way "split-root" experiments can be conducted under field conditions with the advantage of keeping other conditions equal and focusing on local effects on root development. Most previous split-root studies have been restricted to pot experiments (e.g. De Jager 1982; Hairiah et al. 1993). The ingrowth core technique may be especially relevant for trees and other perennials where other experimental techniques are difficult to implement.

If dead roots are included in the ingrowth core, root decomposition can be studied under field conditions, but results may depend on whether the root material was naturally senesced or killed for the purposes of measurement. Root growth and senescence may be measured during root ingrowth studies when alive and dead roots are easily distinguished and a series of measurement are made over time.

6.2.3 Pinboard Method

6.2.3.1 Concept

Monolith samples can be obtained with pinboards ("fakir beds"), where the pins hold the root system in an approximately correct position during washing. The equipment needed to take a pinboard sample is shown in Box 6.2 and Fig. 6.3.

BOX 6.2. Pinboard Construction and Other Requirements

A pinboard ('fakir bed') can be made by inserting U-shaped stainless steel pins through plywood, to form a regular grid of steel pins (Fig. 6.3; Schuurman and Goedewaagen 1971; Böhm 1979). These pins can be made from bicycle/motorbike spokes bent into a U-shape (*A*) with a 5-cm-base and upright length equal to the sum of the required length plus the thickness of the board (a convenient total upright length is 12 cm). Before inserting the pinboard in to the soil, slide a plastic coarse-mesh screen over the pins. Other requirements (Fig. 6.3) are trowels (*B*), metal blades to smooth the profile wall (*D*), a rubber hammer, a car jack (*C*) adjusted to support the board, metal string or a steel cable with a diameter of 2 mm (e.g. motorcycle brake cable) with two handles (*E*) for cutting the soil.

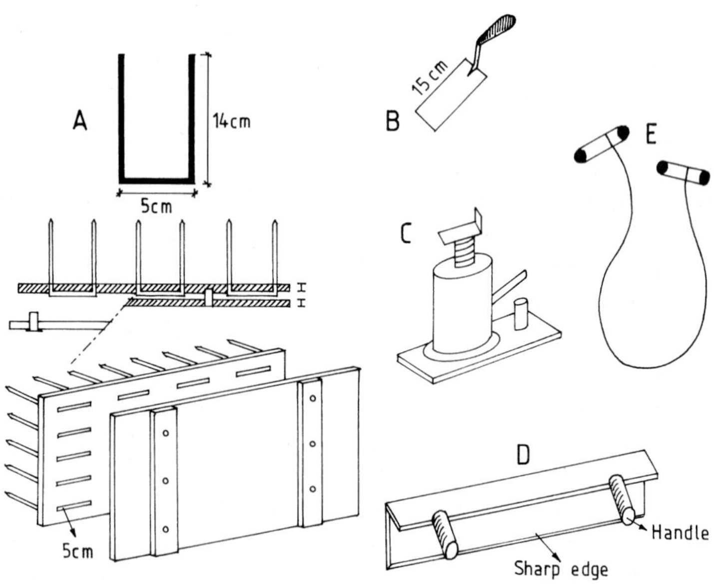

Fig. 6.3. Construction of a pinboard and some of the other equipment needed for taking samples. *A* pins, *B* trowel, *C* car jack, *D* blade, *E* string

The method has recently been improved by putting a coarse mesh screen (e.g. with openings of 0.5 cm) on the pins before the board is pushed into the soil, and gradually lifting this screen while washing away the soil. This screen prevents the movement of roots by "surface runoff" during the washing process.

Pinboard samples can be taken of individual plants, of row crops (usually perpendicular to the crop row), or of grassland vegetation. The size of the pinboard is determined by the vegetation rooting habit and practical considerations (samples of 100 × 60 × 10 cm of soil weigh about 100 kg).

6.2.3.2 Procedure

Practical guidelines for taking a sample are detailed in Box 6.3 and Fig. 6.4.

A set of pictures showing different steps of the pinboard method for sampling crop root systems is presented in Fig. 6.5.

An alternative sampling device for monolith samples, which avoids the need for a soil pit, is described by Floris and Van Noordwijk (1984). A metal

BOX 6.3. Procedure for Taking Pinboard Samples

Select and mark an appropriate sample location either at random or according to specified criteria (e.g. restricting samples to plants of a certain size).

Dig a soil pit just outside this sample location; the size of the soil pit depends on the dimensions of the pinboard. For a 60-cm-wide pinboard, a pit 100 cm wide allows for sufficient manoeuvrability to set and recover the pinboard. When plants are grown in rows, the pit is normally dug perpendicular to the row.

Smooth off the profile wall at about 5 cm from the plant (for 12 cm pins) so that the plant is in the middle of the pins. Harvest the aboveground parts of the plant and describe the soil profile.

Place the pinboard vertically with the pins against the profile face, and adjust so that the top row of pins is at ground level, and then push the pinboard into the soil by hammering it in. Alternatively, a jack or similar hydraulic device can be used to push it in smoothly.

Support the pinboard with a car jack (Fig. 6.4) and remove about 15 cm (a few centimeters beyond the tips of the pins) of soil underneath the pinboard with a knife. Cut away soil on both sides of the board, also a few centimeters further than the tips of the pins and put the steel cable in place (Fig. 6.3). Tie the plastic mesh screen around the sample to prevent soil loss during the next step. In a sawing movement, cut the last surface connecting the monolith to the surrounding soil mass, leaving the pinboard supported by the jack. Normally one person

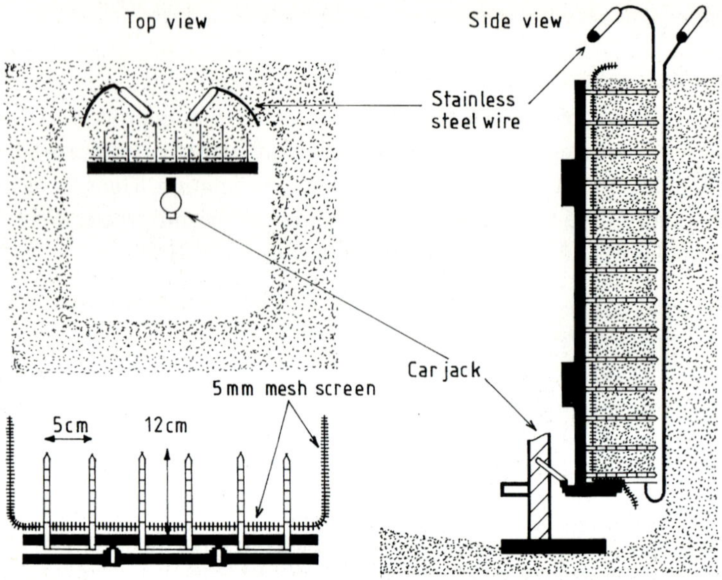

Fig. 6.4. Procedure for taking pinboard samples: after pushing the pinboard into a smoothed profile wall, the board is supported on a car jack; soil below and on both sides of the board is cut away by sawing movements with a stainless steel wire

stands in the soil pit to control the process while two people pull the cable. When loose, pull the board backwards and carefully lift it from the pit. Cut away the edge of the soil monolith to the level of the tips of pins.

Label the sample and cover it with a plastic bag for transport to the laboratory.

box is driven into the soil by a hydraulic cylinder and supported by a strong metal frame, anchored into the soil. When the desired depth is reached, the bottom is cut by pulling a cable (as in the pinboard sample) and a pair of claws is inserted into the soil by tightening screws on top of the box. A small metal tube provides air and thus prevents suction on the soil while the box is pulled to the soil surface. In constructing the box a delicate balance must be maintained between keeping the walls thin and smooth enough to allow easy insertion into the soil, and providing enough grip on the soil that the sample will not be lost from the box while it is pulled up. In practice, the equipment described by Floris and Van Noordwijk (1984) is no more time efficient than the pinboard method, but it involves less destruction of the surrounding soil (except for the soil anchors) and may thus be desirable in long-term experi-

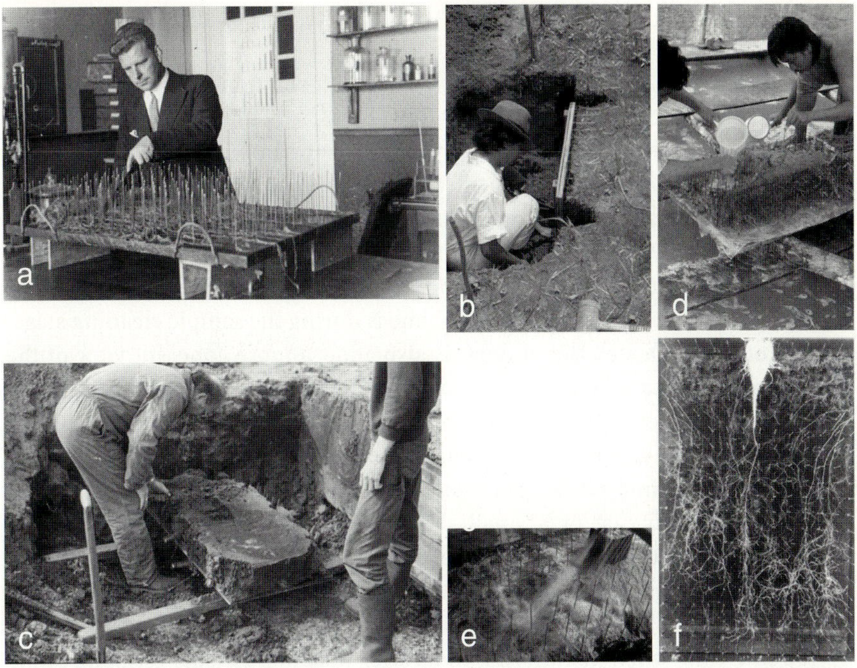

Fig. 6.5. Pinboard method for sampling crop root systems. **a** M.A.J. Goedewaagen (author of a classical description of root research methods) in the 1930s with a washed out sugarbeet (*Beta vulgaris*) root system. **b** Pinboard just pushed into a soil profile. **c** Cutting away redundant soil after the monolith has been cut free from the profile wall. **d** Washing a pinboard sample directly in the field (a small stream in the forest). **e** Washing a pinboard sample in the laboratory with a rotating sprinkler. **f** Sugarbeet root system washed free. (Photographs **a** and **f** courtesy of archive AB-DLO, Haren, the Netherlands, **b**–**e** Meine van Noordwijk)

ments with a restricted plot size. If a similar box is mounted on a tractor's hydraulic power system, the weight of the tractor is generally sufficient to push the box into the soil. This reduces the destruction of the experimental field to that caused by a tractor.

6.2.3.3 Drawbacks

Taking the pinboard samples requires some skill and takes more labour than sampling with augers. The major drawbacks for many situations are probably the destruction caused by the soil pit and the limits to sample size by the weight of the soil. Compared with auger samples, larger losses of fine roots during the washing process can occur, especially for roots entering the sample volume from the outside, instead of from the plant in the centre of the board.

6.2.3.4 Results and Perspectives

Overall, the pinboard method provides useful information per unit effort. The distinction between live and dead roots is easier with pinboard sampling than with methods where the root system is not sampled in its entirety. In comparison with other sampling methods, based on blocks of soil or auger samples, pinboard samples contain more coherent parts of a root system and can thus be washed on a coarse mesh screen rather than on a fine mesh sieve. This facilitates the washing process and leads to root samples without much organic debris. The main gain in time is during the sample cleaning stage. Apart from being faster, the pinboards also allow a more coherent view of the branching pattern and allow the tracing of individual roots to their origin. For an analysis of intercropping systems or in situations with a prominence of "weed" roots, this can be a clear advantage. The washed system can be photographed to obtain a visual impression or can be conserved as such. If combined with a resin cast of a small slice of the soil profile (Schuurman and Goedewaagen 1971) a visual link can be made between soil conditions and root response. Examples of root systems washed free on pinboards are presented in Fig. 6.6.

To save time when quantifying roots after washing, a visual estimate of the amount of roots per block of soil ($5 \times 5 \times 12\,cm^3$) can be made, with a representative selection of samples recovered to calibrate the visual ranking. Results then depend on the consistency of the person doing the visual assesment and

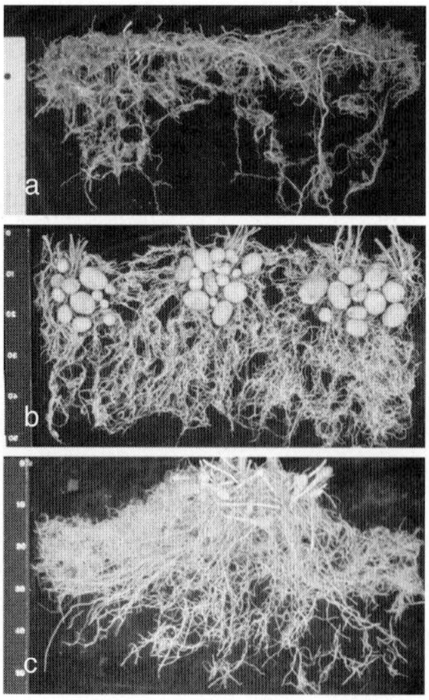

Fig. 6.6. Examples of root systems washed free on pinboards. **a** *Calopogonium mucunoides*, a legume cover crop on acid soils in Nigeria. **b** Potato (*Solanum tuberosum*) in a pinboard sample taken along the ridge. **c** *Helianthus tuberosus* in a pinboard sample perpendicular to the ridge. (Photographs courtesy of Meine van Noordwijk and archive AB-DLO, Haren, the Netherlands)

on the characteristics of the roots. Where large differences in root diameter occur it is more difficult to obtain accurate estimates. If root data are collected for a well-specified purpose, the analysis may concentrate on root densities in a certain range, and a visual estimation procedure can prove helpful. For example, models of nitrate uptake suggest that root length densities above 0.5 cm cm^{-3} are redundant for uptake of available N from a soil layer. If nitrate uptake is the main interest, analysis of the pinboard samples can focus on the layers or zones around and below this critical point. Generally, cleaning and measuring the upper soil layers with large amounts of roots takes a large share of the total time, and for certain questions it may be enough to rank root densities as few, intermediate or many.

6.3 Procedures for Root Washing

Once sampling in the field is finished the roots must be separated from the soil. This is done in different ways but usually soil is washed away from the root samples. Ideally, the best approach is to wash roots from the samples immediately upon arrival from the field but core samples may be stored in sealed polyethylene bags under refrigeration, or frozen until processing.

Roots may be separated from soil by dry-sieving but this method is only appropriate for very sandy soils, and its use is limited to tree root studies, where roots with a diameter greater than 2 mm are to be investigated (Böhm 1979).

6.3.1 Hand Washing

Soil Cores. Roots must be washed gently to minimise loss or damage. The water flow of the sprinkler must be very low because the roots are easily damaged by high water flows. Large roots may be removed by hand.

If roots and soil have not separated with sprinkling, it may be necessary to soak and gently agitate the sample. The soil particles should then settle while the roots and other organic matter float to the top and can be poured into another bucket or sieved for further cleaning. The procedure must be repeated as many times as necessary until a clean root sample is obtained. The sieve size used is critical to minimising root loss during washing (see Sect. 6.3.4.1), though the sieve size used will depend on the type of root material and the objective of the research. If there is a lot of organic matter, a coarse sieve (e.g. 0.2 cm mesh) can be used to split the sample into a first fraction, with roots and large organic particles, and a second fraction, with fine short roots and debris.

When the main purpose of the research is chemical analysis of roots, it is important to reduce the washing time and a coarse sieve only can be used to save time.

Washing Pinboard Samples. Washing the soil from pinboard monoliths requires a different procedure than that described for auger or ingrowth samples.

Washing pinboard samples is facilitated by soaking overnight in water, deep freezing (for clay soils), soaking in oxalic acid (for soils with free calcium carbonate) or soaking in hexametaphosphate, preferably under vacuum. Whatever the pretreatment used, gentle washing must follow; a rotating or oscillating sprinkler is ideal, but pinboard samples can also be washed under more primitive conditions with a hose attached to pressurised water or a locally constructed "water tower" or by pouring small buckets of water over the sample. The pinboard is put at a slight angle and washing starts at the lower part. During the procedure some caution is need to avoid the formation of sharp ridges in the profile, which can cause the breaking of roots. Even when the roots do not break, there is a change in their positions, moving downwards which affects root distribution. To avoid this, the coarse mesh screen is lifted, and supported with wooden strips when the soil layer becomes very thin. The roots then become fixed on the mesh screen and are not pushed downwards by the water. Root losses from a pinboard can be collected by placing a large sieve under the waste-pipe. This indicates how carefully washing is being conducted and how well the root system is secured. Some losses are acceptable if the main purpose is the rooting pattern, but for quantitative studies care must be taken to minimise root displacement.

During the washing process large-sized organic debris can be carefully removed from the board using forceps. Soil layers or other feautures which become visible during the washing may be marked with string using the pins as points of reference.

After washing away the soil, the root system is lifted by the mesh screen, photographed (on a black cloth as background) and/or cut according to soil layers, depth zones and/or distance to the plant to obtain root biomass and/or root length. Total plant root biomass can be estimated by integrating root weight densities measured per zone and depth over the relevant soil volume.

6.3.2 Automatic Washing

Various root wash devices have been developed (Böhm 1979). The first to be commercially available (the Gillison root washer), was a hydropneumatic elutriation system (Fig. 6.7) developed by Smucker et al. (1982).

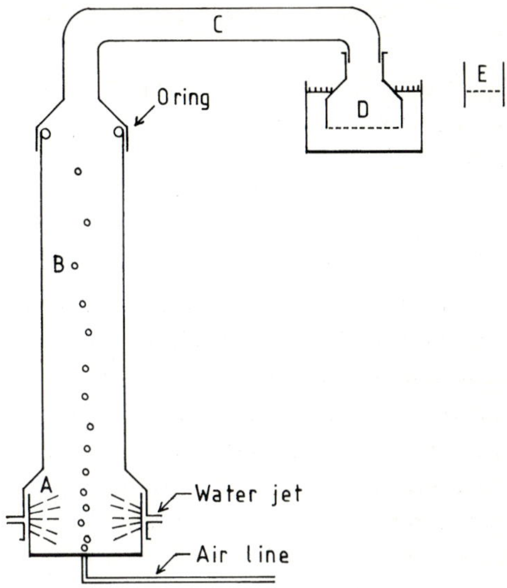

Fig. 6.7. Schematic representation of Smucker's system. *A* High kinetic-energy washing chamber, *B* elutriation chamber, *C* transfer tube, *D* low kinetic-energy primary sieve, *E* secondary sieve

This system consists of a high-kinetic energy water vortex, generated in a chamber in which roots and organic plant residues are separated from soil. Floating roots and associated materials pass onto low-kinetic energy separation screens (multiple screens may be stacked) submerged in water. The mechanical separation of roots is achieved by a closed system using water pressure and compressed air to isolate and deposit roots on a submerged sieve (Byrd et al. 1976; Smucker et al. 1982; Mackie-Dawson and Atkinson 1991).

According to Smucker et al. (1982) and Smucker (1992) this method retains almost all roots >0.05 mm diameter, with attached root hairs and nodules (if present), as well as some fruiting bodies of fungal pathogens. Nevertheless, roots can sometimes remain somewhere in the system and a periodic check of washing water and residues should be made to see if fine roots are being lost. Such losses should be kept less than 5% of root weight (Van Noordwijk 1993).

More recently another device, based on a design from the CSIRO Cunningham Laboratory, in Australia, has become commercially available (Delta-T root washer). The method proposed by Pallant et al. (1993) is also suitable for collecting very fine roots, with a minimum diameter of 0.025 mm, and is an inexpensive method.

The main advantage of automatic washing is the standardisation of the process, even when different operators do this work. Labour time is also saved by this washing system, but after elutriation roots must still be sorted by hand, and this may take several hours (Van Noordwijk 1993). Another advantage is that automatic washing can be used in research with radioisotopes, since the system is totally closed.

6.3.3 Chemical Dispersing Agents

For soils with a high clay content, chemical dispersing agents can be used to help disperse the soil particles and facilitate root washing. A list of some of these chemicals is presented in Table 6.2. Some form of sodium metaphosphate is the most commonly used chemical. Weak acids can be used on soils containing calcium carbonates as binding agents.

Chemical dispersing agents can damage the root system, creating problems for further observations. For example, Van Noordwijk (1993) mentions that soaking the samples in a 5% sodium hexametaphosphate solution speeds up the process of washing roots from clay soil, but the roots become discoloured, particularly in soil with high organic matter content, making the subsequent identification of live roots more difficult. So, the decision of when to use these chemicals must depend on the objectives of a given study.

6.3.4 Errors

The errors that occur during the washing and cleaning of samples are loss of fine roots, loss of dry weight; changes in nutrients content of the roots; incomplete separation of root, soil and debris; poor distinction between live and dead roots; and personal screening differences. Precautions to solve the problems can lead to conflicting situations. To quantify errors, samples can be split or a few samples analysed as a control, checking the losses.

6.3.4.1 Loss of Fine Roots

It is almost inevitable that not all the roots in the soil-root sample are recovered during the washing and cleaning stages. Most of the losses during this stage

Table 6.2. Listing of some dispersing chemicals used for root studies by different authors

Chemicals	Reference	Concentration
$Na_4P_2O_7$	Schuurman and Goedewaagen (1971)	0.27%
Na_3PO_4	Cassel et al. (1995)	0.2 M
$(NaPO_3)_6$	Van Noordwijk (1993)	5%
NaCl	Tardieu and Manichon (1986)	Sat. solution
HCl	Böhm (1979)	3–5%
$(COOH)_2$	Heringa et al. (1980)	10 g/l
CH_3COOH	Mitchell et al. (1993)	0.25–2.5 M
$NaHCO_3 + CH_3COOH$	Pallant et al. (1993)	Not given
Detergents	Kücke et al. (1995)	Not given

are fine roots, which can comprise a major loss in terms of root length but are not so critical in terms of root weight.

The mesh size of screens used to collect roots plays an important role in determining the amount and size of the recovered roots. Amato and Pardo (1994) compared 2 and 0.2 mm mesh sieves (4 and 0.04 mm^2), and found that recovered root length for wheat and field beans tended to be an order of magnitude higher when using the smaller mesh sieve. Roots collected on the 2 mm sieve represented on average 55% of the weight but only 10% of the total length collected using a 0.2 mm sieve. With a 1 mm sieve the recovered root weight was 75%, but the recovered root length was still only 34% of the 0.2 mm sieve. Caldwell and Fernandez (1975) recovered more than twice as much root biomass using a mesh of 0.03 mm^2 instead of 0.2 mm^2. The use of coarse screens is therefore more acceptable for root weight than for root length determinations and reduces the time needed for cleaning the samples from organic debris.

6.3.4.2 Loss of Dry Weight

Loss of root dry weight during washing and storage of root samples has been observed for several crops grown in nutrient solution (Van Noordwijk and Floris 1979; Floris and De Jager 1981; Grzebisz et al. 1989). Losses were typically 20–40% of the original root dry weight. Some combinations of washing and storage treatments resulted in up to 50% loss of the original root dry weight. Five to 10% losses in dry weight was reported over the 24 h immediately after harvest (stored at 20 °C in a moist environment), prior to any washing or longer term storage treatments being imposed. This initial loss was attributed to root respiration (Floris and De Jager 1981). Using sodium pyrophosphate as a dispersing agent tended to increase losses at the washing stage by 10–15% compared with washing without a dispersing agent.

Grzebisz et al. (1989) found that, for most of their washing and storage treatments, loss of cell wall material was limited. They therefore concluded that most of the dry matter loss comprised cell contents, not root tissues such as the epidermis or cortex. This agrees with the lack of change in root diameter during washing and storage reported by Floris and De Jager (1981).

It has been suggested that a correction factor in the range 1.4 to 2.0 could be used to account for dry matter losses of 30–50% due to handling of root samples (Grzebisz et al. 1989). As the authors point out, these correction values were obtained from roots grown in nutrient solution, which may result in a different root structure and chemical composition from field-grown roots. However, Grzebisz et al. (1989) suggest relative root dry weight losses from field-grown crops could be even higher, particularly when grown in conditions of limited water or nutrient availability.

Variation in dry weight losses can be reduced by treating all samples uniformly. When working with a large number of samples it is recommend that the samples be frozen after washing. This treatment gives high dry weight losses but extra handling of the roots after thawing does not involve new dry weight losses.

6.3.4.3 Change in Chemical Composition

There are conflicting reports concerning the importance of nutrient loss from roots during washing. Böhm (1979) cites work showing nutrient loss during washing of agricultural root systems was not significant. In contrast, G. Brouwer (unpubl. data) has observed that the nutrient content (N, P, Ca, K), as a percentage of dry matter remaining, can change considerably in a short time. For calcium, manganese and iron contents, Evdokimova and Grishina (1968) report a decrease of 10–15% in samples that were washed for 2 h compared with samples which were washed quickly. Grzebisz et al. (1989), comparing different washing and storage treatments, found that N losses were smaller than dry matter losses. Consequently, the root N content (as a percentage of dry matter remaining) in some cases increased by up to 10% of the original value. This can lead to overestimates of N content.

Böhm (1979) suggested that nutrient loss during root washing depends mainly on how old the roots are. Brown, decaying roots lose more nutrients than white, young roots.

Lignin content typically increases as a percentage of dry matter during decomposition of organic residues (Parr and Papendick 1978; Jawson and Elliot 1986).

When the root biomass is used for chemical analyses, quick washing on a coarse sieve, with some control samples is preferred. Samples should not be frozen before the effect of freezing on the analytical results is known. Nutrient contents are generally only reliable if sample handling is completed within 1 day (Van Noordwijk 1993).

6.3.4.4 Presence of Debris and Soil Particles

The amount of debris and soil particles present in washed roots is variable, depending on the conditions under which roots are grown. In the last stage of cleaning the distinction between small roots and debris can be difficult and for this reason it is better to try to standardise the cleaning procedure as much as possible.

When washing by floatation (as opposed to sprinkling) it is possible to separate the majority of debris from roots by repeating the decanting process.

Debris which are lighter than living roots float more readily than roots and can be picked up. Some care during this process is needed because some fine roots can come together with the debris.

The type of soil can modify the amount of roots separated from the soil itself. In general, the coarser the soil the easier is the recovery of roots. Small particles can adhere more closely to the roots, especially when the roots have an abundance of root hairs. This can break the root hairs during washing.

The adherence of fine soil particles to the fine root fraction in particular, despite careful washing, can be a problem and results in an overestimate of root dry weight. Böhm (1979) reported that soil particles could account for up to 50% of the sample weight. This error can be avoided, or at least checked, by determining the weight of the ash-free organic root matter (see Sect. 6.5).

In organic soils, where the microbial load is usually high, the roots present in the soil cores can be degraded rapidly.

Separation of soil particles and organic debris from roots still remains a big problem which has led some investigators to concentrate on developing techniques to determine accurately the total length of roots in samples without removing the debris. Dowdy et al. (1995) developed a technique for partitioning live root images from debris using imaging techniques, based on the assumption that the length : width ratio for root segments differs greatly from that of soil and organic debris.

6.3.4.5 Presence of Dead Roots

It can be difficult to distinguish between dead and live roots. Probably the best method is to check root elasticity during the washing procedure. Annual roots, which break when stretched a little, are dead and those with some elasticity are alive. With the auger method it is useful to have a fallow treatment to test if roots are still present from the previous crop, especially if sampling early in the growth of the current crop.

Many staining techniques are available (see Chap. 10), not only for increasing the discrimination between live and dead roots but also for increasing the contrast for automatic measurement of roots. The most appropriate technique for any particular experiment should be determined by testing a few samples taken before the main sampling date.

6.3.4.6 Operator Differences

Differences between operators in sample handling, mainly during the stages of decantation and cleaning up the root samples, can cause errors. The resolution of the human eye, time pressure, operator fatigue, different criteria for distinction between dead and live roots, the decision to stop cleaning, etc. are reasons for these errors.

It is preferable if the same person does the work, since this gives more comparable data. When several people are involved in the same work care should be taken to avoid bias in the measurement results. It is advisable to record the operator for each sample, to allow calculation of a correction if necessary.

6.3.5 Time Investment

The time invested in cleaning roots, often recognised as a problem in root systems studies, is highly variable. It depends on: the procedure used; the crop itself; the stage of development of plants; and on the content of organic debris and other materials in the sample. Whereas for the upper layers of a grass stand it can take 1 day to clean up one or two samples, in the same period of time it is possible to wash and clean 40 samples of leek (G. Brouwer, unpubl. data).

Böhm et al. (1977) found that 9.5 person h were required to collect, wash out and measure soybean roots from twelve 15 cm increments of a 180 cm core from a silt loam soil.

T.C. Kaspar (pers. comm.) and Kaspar et al. (1995), working with core samples 10 cm in diameter and 15 cm long, for maize roots grown in a silty clay loam soil, has estimated that it required 0.16 person h/sample to take a core with a hydraulic sampler and an eight person crew, 0.30 person h/sample to wash out the soil using a hydropneumatic elutriator based on Smucker's design and a four person crew, and 0.30 person h/sample to remove organic debris.

6.4 Methods of Storage

Ideally, root measurements should be taken immediately after sampling. However, the logistics of processing the large number of samples that are required from an experiment mean that it is often necessary to store roots before measurement, sometimes for long periods.

6.4.1 Storage Before Washing

As a rule, samples should be washed immediately after being taken from the field and then stored in a suitable manner. This minimises the period for root respiration and microbial degradation of the roots after sampling. If it is not possible to wash samples immediately, they can either be cold-stored or dried.

Cold-storage should only be considered for temporary storage (<2 to 3 days) because the treatment does not stop root decay but only slows it down. Drying at 60–75 °C (Schuurman and Goedewaagen 1971) offers a longer-term storage option. An alternative long-term storage option is to freeze the samples (e.g. Kücke et al. 1995), but the sheer bulk of material that would usually need to be stored often precludes this in practice. Whichever option is used, because of their bulk it will probably take at least several hours before the samples attain the desired temperature, during which time some root degradation will occur.

6.4.2 Storage After Washing

Cold Storage. Washed fresh roots can be stored for up to 48 h in a refrigerator (Gregory et al. 1992). During storage, the roots should be kept moist to avoid desiccation. This can be achieved either by wrapping samples in a damp paper towel (Gregory et al. 1992), or by storing the samples in suspension with water. If the storage period is longer than 2 days, a chemical preservative should be added to the suspension (Table 6.3).

Freezing. Freezing at –20 °C (Schuurman and Goedewaagen 1971) provides a long-term storage option. Samples can be stored in small plastic bags or boxes, with a little water. The addition of water helps to avoid root damage during freezing.

Drying. Drying the root samples is another long-term storage option. Samples should be dried at 60–75 °C to avoid the pulverisation that can occur when roots are dried at 100 °C (Schuurman and Goedewaagen 1971).

Table 6.3. Listing of some chemicals used for root storage by different authors

Chemicals	Reference	Concentration
Formalin and thymol	Schuurman and Goedewagen (1971)	3–5% and 3%
Ethanol, formalin, acetic acid and water (FAA)	Tennant (1976)	50%; 6.5%; and 2.5%
Ethanol	Böhm (1979)	15–20%
Methanol	Henry and Deacon (1981)	70%
Formalin	Tanaka et al. (1993)	10%
Thymol	Marcum et al. (1995)	3%
Isopropanol	Kücke et al. (1995)	15%

Chemicals. A range of chemical agents have been reported in the literature as satisfactory for long-term root storage (Table 6.3). Formalin and alcohol are the most commonly used preserving agents. A 5% formalin solution or a 15–20% alcohol solution can preserve roots satisfactorily for several months at 10 °C (Böhm 1979). The alcohol concentration required is proportional to the storage temperature. The higher the storage temperature, the greater the alcohol concentration required. Combinations of low temperature with chemical preserving agents seem to prove most satisfactory.

6.4.3 Effect of Storage Methods on Root Properties

Storage methods can affect the condition of the roots. Ideally, a storage method should be chosen that does not influence the root property that is to be measured. All samples must be stored in the same manner.

For chemical analyses, roots need to be dried as quickly as possible before storage. When this is impossible, it is recommended that a zero control be taken to quantify the changes prior to drying. Samples must not be stored in a freezer because soluble nutrients can be lost from the roots when subsequently thawing the sample for analysis. Nutrient content, as a percentage of dry matter, begins to change after a few days in cold storage at 4 °C.

For physical analyses, a wide range of storage methods is acceptable, though the best method is to store the fresh samples in a freezer. Alternatively, samples can be dried at 60–75 °C (Schuurman and Goedewaagen 1971) and then stored. On re-wetting, the roots more or less completely recover their original root length and diameter. Any effect of drying and re-wetting on the physical parameters should be similar for all the samples. Williams and Baker (1957) report that storing roots in formalin solution had no significant effect on herbaceous root dry weight.

6.5 Root Quantification

A number of parameters may be estimated from roots obtained with the methods described in this chapter, the most frequent being root weight, length, diameter, surface area and branching pattern. The parameters measured depend on the purpose of the experiment and how the researcher intends to use the root data. For example, when studying the absorptive capacity of the root system, root length and surface area are important parameters. Net ion influx into roots is influenced by root diameter. In other studies, it might be pertinent to know the proportion of below-ground dry matter relative to the above-ground dry matter of the plant, and hence the weight of the root samples would be recorded. Whatever parameter is measured, it is usually related to the volume of the soil sample, which can be the volume of the entire monolith obtained with the pinboard, or a smaller volume, such as that obtained with auger sampling.

6.5.1 Root Weight

Root weight is estimated either from fresh roots or, more frequently, from oven-dried roots. Fresh weight is often used in plant pathology studies, for example investigating nematodes and fungi in roots (Böhm 1979). For determination of root fresh weight, it is recommended that a standardised procedure be applied to remove all the water adhering to the roots, before weighing. This can be achieved by submitting the roots to low-speed centrifugation, typically for 30 s (van Noordwijk and Floris 1979; Grzebisz et al. 1989).

Root dry-matter is a useful measure of plant investment in the root system and also of the contribution of roots to the soil humus, particularly when related to shoot biomass. However, it is not valid to assume that root weight is correlated with root activity (Böhm 1979). Root length, particularly of fine roots, is generally considered a better indication of root activity.

Root dry weights are generally recorded for root samples which have been oven-dried at 65–75 °C until there is no further change in weight (typically 24 h). Adherence of fine soil particles to the roots is sometimes a problem. In these situations the weight of the ash-free organic root matter is determined. The procedure involves placing the weighed oven-dry roots in a muffle furnace at 650 °C for 5 h to burn off the organic matter. The weight of the residue is then recorded. The difference between root dry weight and the weight of ash residues represents the ash-free organic dry weight.

6.5.2 Root Length

Various techniques, both manual and automated, are available for determining root length. This section considers primarily manual methods. Automated methods, in particular the use of image analysis software, are considered separately in Chapter 10.

6.5.2.1 Direct Measurement

Direct measurements of root length can only practically be made on large diameter roots, e.g. tree roots. Direct measurements of herbaceous plant root lengths are rarely made, and then only primary roots (e.g. Sivakumar and Salaam 1994). It is simply not feasible to undertake the direct measurement of the total root length (including fine roots) of a plant, or even of a root sample.

6.5.2.2 Line Intersect Method

The root length of a sample can be estimated using intercept counting techniques (Newman 1966; Tennant 1975). Newman's technique is based on the relationship between root length and the number of intercepts between roots, spread over a surface of defined area, and randomly arranged lines of known length in that area. Marsh (1971) simplified Newman's method, and Tennant (1975) tested and popularised this modified method.

The modified line intersect method has become the standard manual technique for estimation of root length (See Box 6.4). The line intersect method gives an estimation of the root length of the sample, and is not a direct measurement. Sources of error arise from the random arrangement of the roots on the grid, root visibility, the definition of an intersection, and operator bias and fatigue. With care, it is possible for a single individual to obtain coefficients of

BOX 6.4. The Modified Newman Line-intersect Method

Spread the roots, with the minimum overlap possible, over a regular grid of indeterminate dimensions, but of known regular grid units (e.g. 1 cm grid squares). Where necessary, tease the root material apart or cut it into smaller pieces, to avoid overlap. Placing the samples on the grid in a shallow layer of water in a dish (or on a glass plate) aids separation of the roots.

Count the total number of intersections between the roots and the horizontal (H) and vertical (V) grid lines. Assign a count of 1 to a root crossing a line, a root ending touching a line and a curved root portion touching a line. Curved root portions which lie on or along a grid line are assigned a count of 2 (Tennant 1975).

The relationship between the root length and the number of root intersections with the grid lines is given by:

$$\text{root length (cm)} = \pi/4 \times \text{no. of intersects} \times \text{grid unit (cm)}$$

To improve the contrast of roots it can be helpful to stain them. Alternatively, place a coloured background under the roots. To optimise working efficiency at an acceptable random error level, choose the grid size to obtain about 400 intercepts (200 H + 200 V) per sample (Van Noordwijk 1993).

variation of the root length estimate of 5% or less (Tennant 1975). However, this is the minimum likely error. Coefficients of variation for the line intersect method are typically 10 to 15% (Böhm 1979; Bland and Mesarch 1990; Farrell et al. 1993). Bland and Mesarch (1990) reported a bias in root length estimates by observers, with some observers yielding consistently low or high estimates relative to the mean.

Steps can be taken to minimise the error associated with the technique. These include: training new observers and comparing their results against experienced observers; ensuring adequate lighting; using a magnifying glass to aid root visibility; improving the colour contrast between the roots and the background; and ensuring the observers have regular rest breaks to avoid fatigue.

The line intersect technique can be mechanised, using photoelectric counters such as those of Rowse and Phillips (1974), Richards et al. (1979) and Wilhelm et al. (1982). The automation eliminates operator bias from the estimation and should therefore reduce error. However, root samples do need to be cleaned more carefully since it is difficult for automated systems to distinguish between roots and debris. Details of automated methods for root analysis are given in Chapter 10.

6.5.2.3 *Visual Estimation Method*

The visual estimation of the root length is a rapid low-cost method for estimating root length. At its simplest the sample root length can be described by a subjective rating where, for example, 1 indicates low root length and 9 indicates high root length. This method has been used to screen for cucumber root length as part of a cultivar evaluation programme (Walters and Wehner 1994). Visual estimation may also be used to obtain a quantitative root length estimate of samples recovered using the pinboard method or from auger samples (Naab 1994; Gaze 1996). This essentially involves visual comparison of root samples of unknown length with standard root samples of known length, enabling a reasonable estimate of the unknown root length to be made (see Box 6.5).

BOX 6.5. Visual Estimation of Root Length

After washing, store the roots in a chemical preserving solution (Table 6.2) until required. Comparison of samples for root length estimation is easier when the roots are suspended in solution.

Select a sub-set of the washed samples which covers the range of root lengths in all the samples, based on a visual subjective rating. Determine the root length of the samples in this sub-set as accurately as possible using the line intersect method (see Box 6.4). These samples comprise the standard samples of known root length.

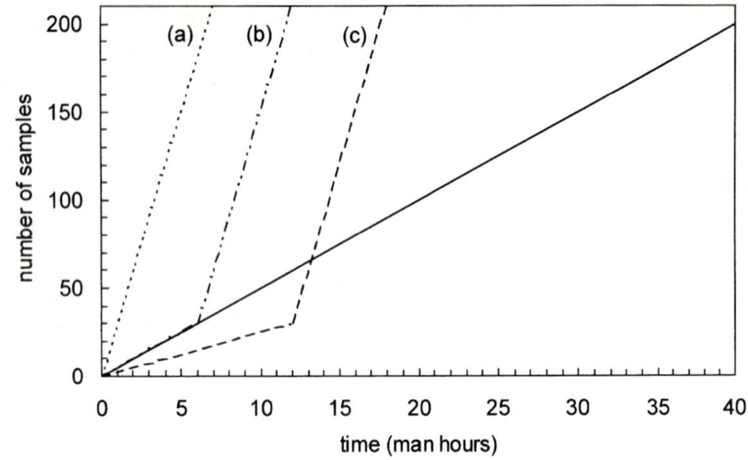

Fig. 6.8. Time required to determine root length using the line intersect method (*solid line*) and visual estimation (*broken lines*). The comparisons assume that 5 samples per h can be counted using the line intersect method, and that 30 samples per h can be estimated using the visual estimation method. For the visual estimation method: line *a* assumes standard samples are already available; line *b* assumes 30 standard samples are first counted once using the line intersect method; line *c* assumes that 30 standard samples are counted twice before visual estimation of the remaining samples

Estimate the root length of each of the remaining samples by comparing them against the standard samples. If necessary, estimate the root length by interpolation between the two standard samples which appear most similar to the root length of the sample being estimated. If the roots are stored correctly to prevent degradation, it is possible to use the same standard samples from one sampling date to the next. Standards should be for the same species (and possibly cultivar), have roots of similar condition and size, and should be stored in an identical solution volume and sample container to those samples against which they will be compared.

The major advantage of the visual estimation of root length is the time saved when compared with the line intersect method. Gaze (1996) reported that a single person could score 30–40 samples per h, compared with counting about five samples per h using the line intersect method. Figure 6.8 illustrates the major time-saving that can be achieved using the visual estimation method.

There is inevitably some loss in the accuracy of the estimation of root length when using a visual estimation method as opposed to the line intersect method. However, other measurements of root length, such as the line intersect method and image analysis techniques, are themselves in fact only root length estimates. For visually estimated root lengths of millet [*Pennisetum glaucum* (L.) R.Br.], the coefficient of variation was reported as just over twice that of the line intersect method (Gaze 1996). No systematic bias was found between observers using the visual estimation method. There can, however, be a tendency to underestimate root length using the visual method at high root lengths (Walters and Wehner 1994; Gaze 1996). This upper limit is probably a function of root length density in the sample solution volume, rather than absolute root length.

6.5.3 Root Diameter and Surface Area

Root diameter can be measured on samples spread out on a grid, using a binocular microscope with an ocular micrometer. According to Van Noordwijk (1993) at least 20 readings per sample are required.

Total root surface area has been estimated from other root parameters, like root length and diameter, by solution adsorption measurements, or by using photoelectric devices (Böhm 1979).

Recently, the availability of image analysing computers has made the determination of root length, diameter or surface area possible with more accuracy and with less time investment (see Chap. 10).

6.5.4 Presentation of Root Data

Root length and root weight are often related to soil volume and are referred to as root length density – L_v ($cm_{root}\,cm^{-3}{}_{soil}$) or root weight density ($g_{root}\,cm^{-3}{}_{soil}$). These are particularly useful for describing the spatial distribution of the root system, and are also useful data for soil-vegetation-atmosphere transfer models.

Root length and weight can also be related to soil surface area – L_a ($cm_{root}\,cm^{-2}{}_{soil}$) and ($g_{root}\,cm^{-2}{}_{soil}$). This facilitates comparison of root data with measurements of above ground plant production, which are typically made per unit soil surface area (e.g. yield, leaf area index).

The specific root length ($cm_{root}\,g^{-1}{}_{root\ dry\ weight}$) is used to give an indication of root thickness.

Comprehensive summaries of published rooting data have been made by Van Noordwijk and Brouwer (1991), Canadell et al. (1996) and Jackson et al. (1996).

6.6 Conclusions and Perspectives

Rapid developments in computing hardware and software hold much promise for assisting in the measurement of root length, root diameter and root branching patterns and architecture. However, to use these systems, the researcher must already have taken the samples from the field and washed the soil away from the roots. For the foreseeable future, this aspect of root sampling is likely to remain relatively time consuming and tedious when compared with taking measurements of the aerial parts of plants. Nevertheless, quantifying the location, amount and role of plant roots is essential to understanding plant growth on an individual plant basis, as well as understanding the competition and complementarity between plants and between flora and fauna in the wider ecosystem. The methods described in this chapter (together with those in Chapter 7) are likely to remain the basis of quantitative root research for many years to come.

References

Albrecht D (1951) Verbesserung der Spatendiagnose. Dtsch Landwirtsch 2: 41–43

Albrecht D, Fritzsche KH, Winkler S (1953) Weitere Entwicklung des Strukturbohrers. Dtsch Landwirtsch 4: 206–208

Amato M, Pardo A (1994) Root length and biomass losses during sample preparation with different screen mesh sizes. Plant Soil 161: 299–303

Baarstad LL, Rickman RW, Wilkins D, Morita S (1993) A hydraulic soil sampler providing minimum field plot disruption. Agron J 85: 178–181

Bland WL, Mesarch MA (1990) Counting error in the line-intercept method of measuring root length. Plant Soil 125: 155–157

Böhm W (1979) Methods of studying root systems. Ecological Studies: Analysis and Synthesis, vol 33. Springer, Berlin Heidelberg New York

Böhm W, Maduakor H, Taylor HM (1977) Comparison of five methods for characterizing soybean rooting density and development. Agron J 69: 415–419

Byrd DW Jr, Barker KR, Ferris H, Nusbaum CJ, Griffin WE, Small RH, Stone CA (1976) Two semi-automatic elutriators for extracting nematodes and certain fungi from soil. J Nematol 8: 206–212

Caldwell MM, Fernandez OA (1975) Dynamics of Great Basin shrub root systems. In: Hadley NF (eds) Environmental physiology of desert organisms. Dowden, Hutchinson and Ross, Stroudsburg, Pennsylvania, pp 38–51

Caldwell MM, Virginia RA (1991) Root systems. In: Pearcy RW, Ehleringer J, Mooney HA, Rundel RW (eds) Plant physiological ecology: field methods and instrumentation. Chapman and Hall, London, pp 367–398

Canadell J, Jackson RB, Ehleringer JR, Mooney HA, Sala OE, Schulze ED (1996) Maximum rooting depth of vegetation types at the global scale. Oecologia 108: 583–595

Cassel DK, Raczkowski CW, Denton HP (1995) Tillage effects on corn production and soil physical conditions. Soil Sci Soc Am J 59: 1436–1443

Crozier CR, King LD (1993) Corn root dry matter and nitrogen distribution as determined by sampling multiple soil cores around individual plants. Commun Soil Sci Plant Anal 24 (11&12): 1127–1138

Cuevas E, Medina E (1988) Nutrient dynamics within Amazonian forests: Part II. Fine root growth, nutrient availability and litter decomposition. Oecologia 76: 222–235

De Jager A (1982) Effects of localized supply of H_2PO_4, NO_3, SO_4, Ca and K on the production and distribution of dry matter in young maize plants. Neth J Agric Sci 30: 193–203

Dowdy RH, Smucker AJM, Dolan MS, Ferguson JC (1995) Root length measurements of segments washed from soil cores without removal of extraneous debris. Proc 14th Long Ashton Int Symp: Plant roots – from cell to systems. Sept 1995. Long Ashton, Bristol, England

Evdokimova TI, Grishina LA (1968) Productivity of root systems of herbaceous vegetation on flood plain meadows and methods for its study. In: Methods of productivity studies in root systems and rhizosphere organisms. Int Symp USSR 1968. USSR Academy of Sciences, Nauka, Leningrad, pp 24–27

Fabião A, Persson HA, Steen E (1985) Growth dynamics of superficial roots in Portuguese plantations of *Eucalyptus globulus* Labill. studied with a mesh bag technique. Plant Soil 83: 233–242

Farrell RE, Walley FL, Lukey AP, Germida JJ (1993) Manual and digital line-intercept methods of measuring root length: a comparison. Agron J 85: 1233–1237

Floris J, De Jager A (1981) Een schatting van het verlies aan drogestof en van de verandering in diameter van wortels van engels raaigras (*Lolium perenne*) dor bemonsteren, bewaren en spoelen. Inst Bodemvruchtbaarheid Haren-Gr Rapp: 1–81

Floris J, Van Noordwijk M (1984) Improved methods for the extraction of soil samples for root research. Plant Soil 77: 369–372

Gaze SR (1996) Water balance of farmer-managed millet and fallow-savannah on sandy soils in south west Niger. PhD Thesis, Department of Soil Science, University of Reading, Reading

Gregory PJ, Tennant D, Belford RK (1992) Root and shoot growth, and water and light use efficiency of barley and wheat crops grown on a shallow duplex soil in a Mediterranean-type environment. Aust J Agric Res 43: 555–573

Grzebisz W, Floris J, Van Noordwijk M (1989) Loss of dry matter and cell contents from fibrous roots of sugar beet due to sampling, storage and washing. Plant Soil 113: 53–57

Hairiah K, Van Noordwijk M, Setijono S (1991) Tolerance to acid soil conditions of the velvet beans *Mucuna pruriens* var. *utilis* and *M. deeringiana*. I. Root development. Plant Soil 134: 95–105

Hairiah K, Van Noordwijk M, Stulen I, Meijboom FW, Kuiper PJC (1993) P nutrition effects on aluminium avoidance of *Mucuna pruriens* var. *utilis*. Environ Exp Bot 33: 75–83

Hansson A, Andrén O, Steen E (1991) Root production of four arable crops in Sweden and its effect on abundance of soil organisms. In: Atkinson D (ed) Plant root growth: an ecological perspective. Blackwell, Oxford, pp 247–266

Henry CM, Deacon JW (1981) Natural (non pathogenic) death of the cortex of wheat and barley seminal roots, as evidenced by nuclear staining with acridine orange. Plant Soil 60: 255–274

Heringa JW, Groenwold J, Schoonderbeek D (1980) An improved method for the isolation and the quantitative measurement of the crop roots. Neth J Agric Sci 28: 127–134

Jackson RB, Canadell J, Ehleringer JR, Mooney HA, Sala OE, Schulze ED (1996) A global analysis of root distributions for terrestrial biomes. Oecologia 108: 389–411

Jawson MD, Elliot LE (1986) Carbon and nitrogen transformations during wheat straw and root decomposition. Soil Biology and Biochemistry, 18 (1): 15–22

Kalela EK (1950) On the horizontal roots in Pine and Spruce stands. Acta For Fenn 57 (2): 62–68

Kaspar TC, Logsdon SD, Prieksat MA (1995) Traffic pattern and tillage system effects on corn root and shoot growth. Agron J 87: 1046–1051

Kolesnikov VA (1971) The root system of fruit plants. Mir Publishers, Moscow

Kücke M, Schmid H, Spiess A (1995) A comparison of four methods for measuring roots of field crops in three contrasting soils. Plant Soil 172: 63–71

Mackie-Dawson LA, Atkinson D (1991) Methodology for the study of roots in field experiments and the interpretation of results. In: Atkinson D (ed) Plant root growth, an ecological perspective. Blackwell, London, 25–47

Marcum KB, Engelke MC, Morton SJ (1995) Rooting characteristics of buffalograsses grown in flexible plastic tubes. HortScience 30 (7): 1390–1392

Marsh B.a'B (1971) Measurement of length in random arrangements of lines. J Appl Ecol 8: 265–267

Mitchell AR, Shouse PJ, Rechel EA (1993) Using acetic acid to wash roots from the soil. Soil Sci Plant Anal 24: 15–16

Naab JB (1994) Interaction of canopy and root system in water use of potato. PhD Thesis, University of Reading, Reading

Newman EI (1966) A method for estimating the total length of root in a sample. J Appl Ecol 3: 139–145

Nielsen CCN (1990) Methodische und ökologische Untersuchungen zur Sturmfestigkeit der Fichte. In: Einflüsse von Pflanzenabstand und Stammzahlhaltung auf Wurzelform, Wurzelbiomasse, Verankerung sowie auf die Biomassenverteilung im Hinblick auf die Sturmfestigkeit der Fichte, Schriften aus der Forstl Fak Uni Göttingen ud Niedersächs Forst Versuchsanst. JD Sauerländers, Frankfurt

Nielsen CCN (1994) Aspects of sustainability by afforestation of agricultural set-aside areas: Development of roots and root/shoot-ratios (AIR3-ct93-1269), 1st Progress Report, Royal Vet Agric University, Kirkegaardsvej 3A, DK-2970 Hoersholm

Pallant E, Holmgren RA, Schuler GE, Mccracken KL, Drbal B (1993) Using a fine root extraction device to quantify small diameter corn roots (≤0.025 mm) in field soils. Plant Soil 153: 273–279

Parr JF, Papendick JR (1978) Factors affecting the decomposition of crop residues by mircooorganisms. In: Iscwald WR (ed) Crop residue management systems. American Society of Aoronomy, Madison, pp 101–129

Passioura JB (1980) The transport of water from soil to shoot in wheat seedlings. J Exp Bot 31: 335–345

Prior SA, Rogers HH (1992) Portable soil coring system that minimizes plot disturbance. Agron J 84: 1073–1077

Prior SA, Rogers HH (1994) A manual soil coring system for soil-root studies. Commun Soil Sci Plant Anal 25: 517–522

Richards D, Goubran FH, Garwoli WN, Daly MW (1979) A machine for determining root length. Plant Soil 32: 69–76

Rowse HR, Phillips DA (1974) An instrument for estimating the total length of root in a sample. J Appl Ecol 11: 309–314

Schroth G, Kolbe D (1994) A method of processing soil core samples for root studies by subsampling. Biol Fertil Soils 18: 60–62

Schuurman JJ, Goedewaagen MAJ (1971) Methods for the examination of root systems and roots, 2nd edn. Pudoc, Wageningen

Sivakumar MVK, Salaam SA (1994) A wet excavation method for root shoot studies of pearl-millet on the sandy soils of the Sahel. Exp Agric 30: 329–336

Smucker AJM (1992) Contemporary analytical methods for quantifying plant root dynamics. In: Hübl E, Lichtenegger E, Persson H, Sobotik M (eds) ISSR Symp. Root ecology and its practical application. Klagenfurt, Austria, pp 721–726

Smucker AJM, McBurney SL, Srivastava AK (1982) Quantitative separation of roots from compacted soil profiles by the hydropneumatic elutriation system. Agron J 74: 500–503

Spek LY, Van Noordwijk M (1994) Proximal root diameters as predictors of total root system size for fractal branching models. II. Numerical model. Plant Soil 164: 119–128

Steen E (1984) Variation of root growth in a grass ley studied with a mesh bag technique. Swed J Agric Res 14: 93–97

Steen E (1991) Usefulness of the mesh bag method in quantitative root studies. In: Atkinson D (ed) Plant root growth: an ecological perspective. Blackwell, Oxford, pp 75–86

Tanaka S, Yamauchi A, Kono Y (1993) Cultivar difference in response of root system to nitrogen application in rice plant. Jpn J Crop Sci 62: 447–455

Tardieu F, Manichon H (1986) Caractérisation en tant que capteur d'eau de l'énracinement du maïs en parcelles cultivées. I. Discussion des critères d'étude. Agronomie 6: 345–354

Taylor HM, Upchurch DR, Brown JM, Rogers HH (1991) Some methods of root investigations. In: Mc Michael BL, Persson H (eds) Plant roots and their environment. Elsevier, London, pp 553–562

Tennant D (1975) A test of a modified line intersect method of estimating root length. J Ecol 63: 995–1001

Tennant D (1976) Root growth of wheat. I. Early patterns of multiplication and extension of wheat roots including effects of levels of nitrogen, phosphorus and potassium. Aust J Agric Res 27: 183–196

Van Noordwijk M (1993) Roots: length, biomass, production and mortality. Methods for root research. In: Anderson JM, Ingram JSI (eds) Tropical soil biology and fertility, a Handbook of Methods. CAB International, Wallingford, pp 132–144

Van Noordwijk M, Brouwer G (1991) Quantitative root length data in agriculture. In: McMichael BL, Persson H (eds) Plant roots and their environment. Proc ISRR-Symp, Aug 21–26, 1988, Uppsala, Sweden. Elsevier, Amsterdam, pp 515–525

Van Noordwijk M, Floris J (1979) Loss of dry weight during washing and storage of root samples. Plant Soil 53: 239–243

Van Noordwijk M, Purnomosidhi P (1995) Root architecture in relation to tree-soil-crop interactions and shoot pruning in agroforestry. Agrofor Syst 30: 161–173

Van Noordwijk M, Floris J, De Jager A (1985) Sampling schemes for estimating root density in cropped fields. Neth J Agric Sci 33: 241–262

Van Noordwijk M, Spek LY, De Willigen P (1994) Proximal root diameters as predictors of total root system size for fractal branching models. I. Theory. Plant Soil 164: 107–118

Veller F (1971) A method for studying the distribution of absorbing roots of fruit trees. Exp Agric 7: 351–361

Vogt KA, Persson H (1991) Measuring growth and development of roots. In: Lassoie JP, Hinckley (eds) Techniques and approaches in forest tree ecophysiology. CRC Press, Boca Raton, pp 477–501

Walters SA, Wehner TC (1994) Evaluation of the US cucumber germplasm collection for root size using a subjective rating technique. Euphytica 79: 39–43

Wilhelm WW, Norman JM, Newell RL (1982) Semiautomated x-y-plotter-based method of measuring root length. Agron J 74: 149–152

Williams TE, Baker HK (1957) Studies of root development of herbage plants. I. Techniques of herbage root investigations. J Br Grassl Soc 12: 49–55

CHAPTER 7

Trench Profile Techniques and Core Break Methods

M. van Noordwijk[1], G. Brouwer[2], F. Meijboom[2], M. do Rosário G. Oliveira[3], and A.G. Bengough[4]

[1] ICRAF · JI CIFOR, PO Box 161, Situ Gede, Sindang Barang Bogor 16680, 16001Bogor, Indonesia

[2] Plant Research International, Postbus 14, 6700 AA Wageningen, The Netherlands

[3] University of Évora, Apartado 94, 7002-554 Évora, Portugal

[4] Scottish Crop Research Institute (SCRI), Invergrowie, Dundee, DD2 5DA, UK

CONTENTS

A.L. Smit et al. (Eds.), Root Methods

7.1 Introduction

This chapter describes methods for root observations based on mapping or counting root intersections with planes of observation in the soil. Normally these planes of observation are either vertical or horizontal. Compared with the methods based on washed root samples discussed in Chapter 6, these "profile wall" methods have advantages as well as disadvantages. A major disadvantage of the profile wall methods is that only a small part of a root is visible on such an intersection and it is not easy to distinguish between roots of different species, or between live or dead roots. Even the question of whether a whitish thread-like object sticking out of a plane is a root and not an enchytraeid (pot worm) or other soil organism may take some experience to answer (potworms move when touched). Creating access to planes of observation via trenches can be a rather destructive activity which is not welcome on small experimental plots, especially those intended for long-term experiments. On the positive side, however, profile wall methods can give a quick estimate of overall root distribution and can give detailed information on spatial patterns of roots in their interaction with physical, chemical and biological characteristics of the soil profile. If maps are made of root occurrence as well as any other readily observable feature, the toolbox of geographical information systems and quantitative map analysis can be used to analyze patterns, be it in only two dimensions.

In this chapter we will describe practical aspects of preparing profile walls for observations and various methods for recording data (Sect. 7.2). In Section 7.3 we will focus on the analysis of data obtained, especially for root maps, and discuss the inferences which can be made about three-dimensional reality on the basis of two-dimensional observations.

7.2 Practical Aspects

7.2.1 Preparing Profile Walls for Root Observations

Field methods for root mapping include the following steps: a plane of observation must be chosen, and an access trench dug (Fig. 7.1A, B). The number of profile walls observed per treatment is usually one but further observation planes can be obtained from the same trench, a few centimeters (depending on plant space in the row) beyond the previous profile wall. The surface of the plane is then prepared and root locations recorded on plastic overlays or as tabulated data (Fig. 7.1C, D). We will consider these steps in turn, as there is a range of options for each of these.

Choosing an Observation Plane. Usually maps are made on vertical planes of observation, perpendicular to crop rows and near the base of a plant, where

Fig. 7.1. Direct root observations on trenches and profile walls. **a** Direct observations in a deep trench perpendicular to hedgerows, **b** direct observations in a plough layer of an arable system, with minimum disturbance to the soil profile, **c** brushing away soil from a sugarbeet root system to observe its interactions with different layers in the soil profile, **d** mapping root intersections on a polythene sheet, with a grid system as a guide. (Photographs: courtesy of Meine van Noordwijk)

applicable. Horizontal planes give additional information on root orientation, but this makes removing the last layers of soil difficult.

Preparing Access. For "plough layer" observations a trench 0.3 m wide and 0.3 m deep may be enough, as the observations can be made without the observer having to get into the trench. Otherwise, a trench about 1 m across is needed for access by the observer, with a depth and width depending on the size of the maps to be made. To reduce damage to long-term experimental plots, sampling locations are normally chosen close to the border of plots; this may compromise on the representativeness of the sample. The soil from the soil pit or trench should be put on the outer side, so as not to influence the observations. If topsoil and subsoil are separated by canvas sheets during excavation, disturbance to the soil profile can be reduced, but the sampling sites should be marked and avoided in future.

Smoothing the Observation Plane. A long knife blade is normally used to obtain a flat and smooth plane of observation. No further preparations are necessary if the soil and root have very good contrast in colour, provided that the observer examines the profile closely.

Removing Last Layers of Soil. On most soils it is advisable to wash a few millimetres of soil from the prepared surface to facilitate observations. This can be done with a knapsack sprayer, working from the top of the profile downwards; a disadvantage of spraying is that in a wet soil 'smearing' of surfaces can easily cover fine roots. On heavy clay soils washing can be facilitated by covering the profile with a cloth soaked in sodium pyrophosphate (see Chap. 6) to disperse the clay. As an alternative one can carefully blow away fine soil with a "vacuum cleaner" (used in reverse) – live roots are remarkably resilient to this and will generally remain visible.

Recording Roots and Profile Features. The number of root intersections can be counted using a sampling grid, or the positions of the roots can be mapped on polythene (plastic) overlays. This latter method gives most flexibility in subsequent analysis of the data. Appropriate light conditions should be created for observation. Avoid direct sun-light, as it dries out roots quickly and gives too sharp a contrast and will lead to water condensation if plastic overlays are used for mapping. An umbrella or screen can be used to shade the plane of observation and give diffuse light.

7.2.2 Root Counts on a Grid

The simplest way of recording root occurrences is by overlaying the plane of observation with a grid and counting the number of roots crossing the plane. The size of the grid then depends on the purpose of the study and the level of detail needed. Fine grids require more work, but allow data to be converted to those of coarser grids (multiples of the fine grid), while the reverse process is not possible. French researchers in particular (Tardieu and Manichon 1986; Tardieu 1988) score the presence/absence of roots on a fine grid, rather than counts of root sections seen on the plane. A frame of strings can be used as a grid, either prepared in advance, or created on-site with pins, string, and pieces of lead as weight. It is usually worth distinguishing between root diameter classes in the counts.

Two types of criterion have been used for including roots in the counts. In the "Reymerink method" all roots exposed after removing a specified layer of soil are counted (Schuurman and Goedewaagen 1971; Böhm 1976). More recently, roots are only counted at the point where they emerge from the soil in the plane of observation. The latter can be more easily related to root length density per unit volume of soil (see below), but may not be as visually rewarding as drawings of roots hanging on the profile wall recorded with the first method.

7.2.3 Root Maps on a Plastic Overlay

Maps on plastic overlays can be made with permanent marker pen; roots are represented as 'dots', with different colours used for diameter classes or roots of different species where these are recognizable. Other features can be included such as soil horizons, cracks or finer soil structural features, or local accumulations of organic residues (Fig. 7.2). By spraying suitable dyes onto the excavated soil surface, spatial patterns in other features of the profile wall such as soil pH or redox status may be recorded. Water infiltration patterns via macropores can be visualized with dyes such as methylene blue or fluorescein (the latter requires UV light for observations and may be poorly visisble when adsorbed to the soil). Macropores can also be made visible by infiltrating an iodine solution and covering the soil surface with a fine layer of starch powder; a blue colour will appear where the iodine has reacted. Root maps should be wrapped in paper on a roll, rather than being folded to avoid a "printing" process during storage.

As a special case of profile wall observations, resin-impregnated blocks can be prepared and used for more detailed observations. This is similar to the "thin sections" used in soil micromorphology (Kooistra and Van Noordwijk 1996; Ringrose-Voase 1996).

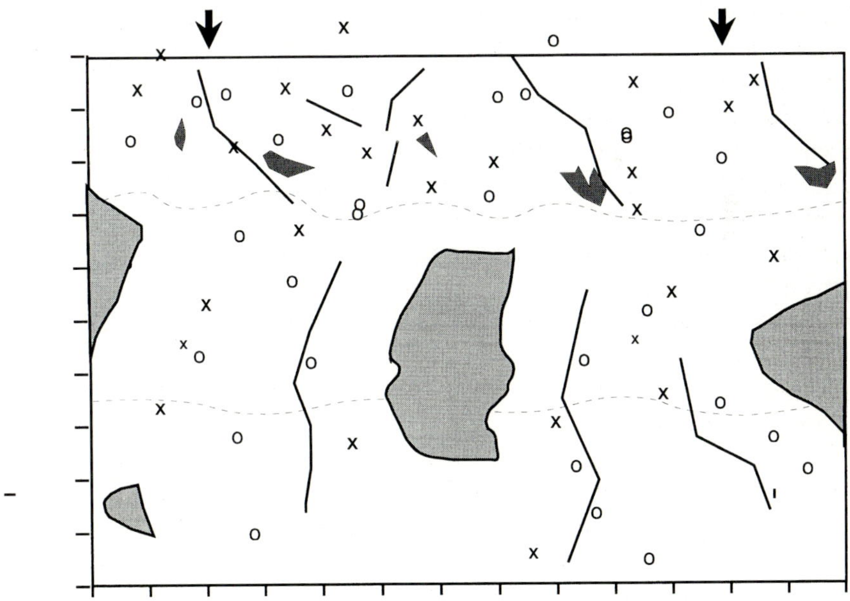

Fig. 7.2. Example of a root map with a number of features: *x* and *o* indicate two types of roots (different species or diameter), the *continuous lines* indicate major structural features of the soil, the *broken lines* profile layers, the *hatched areas* recognizable plant residues ploughed into the topsoil, and the *shaded areas* reflect anaerobic areas (bluish-grey)

7.2.4 Calibration Samples

The number of root intersections per unit area of observation plane can be calibrated against the root length density per unit volume in the soil directly behind the plane of observation by taking shallow sample boxes from well-marked places in the profile (Fig. 7.3). The box size should correspond with (a multiple of) the grids used for direct counts; where root maps are made, the size of the sample box is arbitrary, but it should be marked on the root map. Shallow sample boxes (about 2 cm deep) are better where large gradients of root length density are expected. In very shallow samples, however, the root segments obtained may be so small that washing losses increase; variability in actual sample size may also increase when the boxes are made shallow. The calibration samples can be washed and treated as described for other washed samples in Chapter 6. Oliveira-Carvalheiro and Nepstad (1996) made profile wall observations and empirically calibrated these against small block samples up to a depth of 9 m.

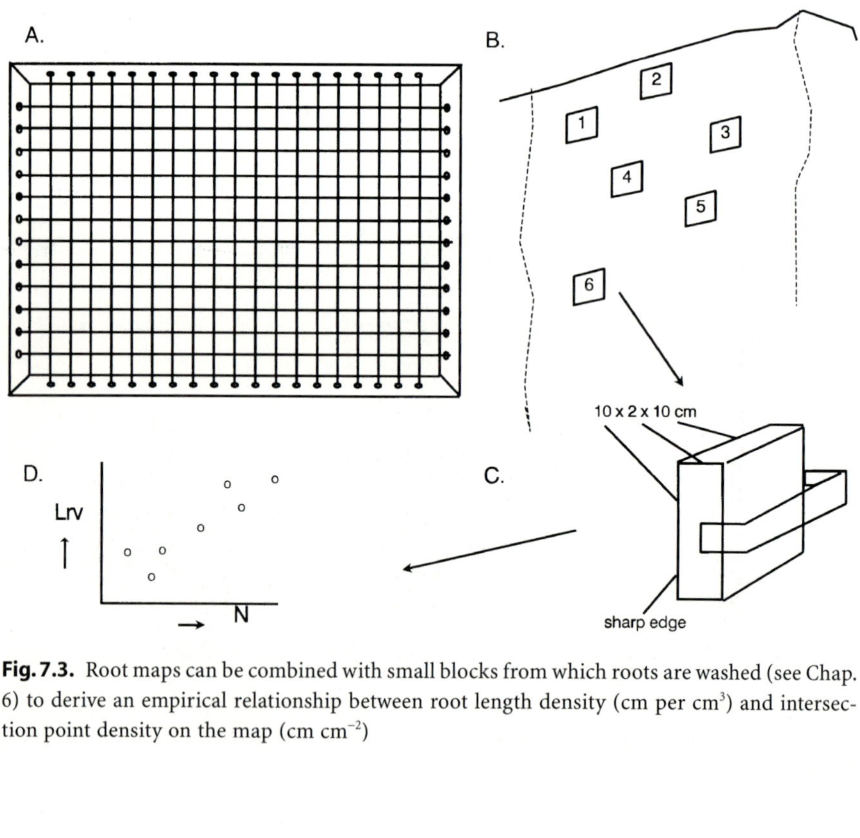

Fig. 7.3. Root maps can be combined with small blocks from which roots are washed (see Chap. 6) to derive an empirical relationship between root length density (cm per cm^3) and intersection point density on the map (cm cm^{-2})

7.2.5 Core-Break Method

The core-break method (Schuurman and Knot 1957; Schuurman and Goedewaagen 1971; Böhm 1979; Drew and Saker 1980) is a rapid and simple method of observing and recording the presence of roots as a function of depth; the method is treated here rather than in Chapter 6, as the basic observation of number of root intersections per specified area of observation is the same as in the profile wall methods. Cylindrical soil cores (see Chapter 6 for auger design) are broken in half and the number of roots sticking out of both soil surfaces is recorded (Fig. 7.4). The two results should be added together, as the same root cannot be exposed on both sides after breakage; usually roots will break some distance from the plane of observation and only show on one side. If root branches are observed these should be ignored, as the count refers to a 'plane of observation' and not to a volume.

Advantages of the core-break method are that samples are small, allowing many replicates to be taken. Direct counts are faster than for washing the roots from a whole sample, and a higher sample variability (the coefficient of variation for root counts may be 50–100% higher than for root length in washed auger samples) can be compensated for by larger sample numbers; the time needed per sample is considerably less. A selection of the samples can be

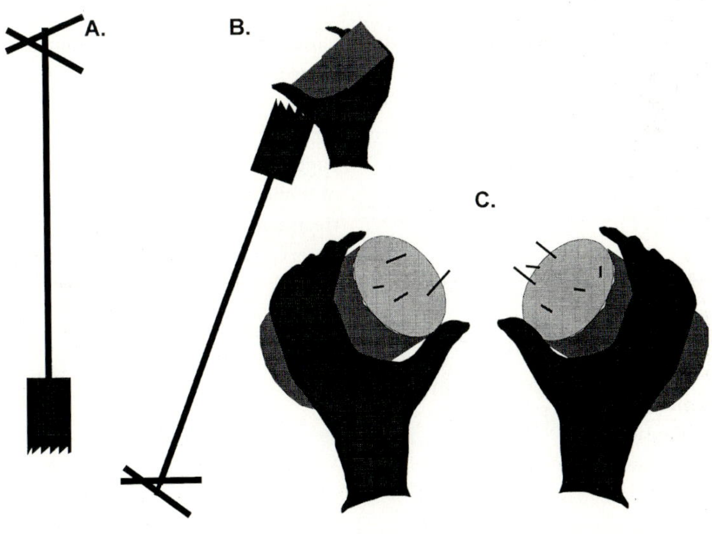

Fig. 7.4. Core-break method for estimating density of roots intersecting with a horizontal plane of observation

washed to obtain a direct calibration of root length versus counts. Schuurman and Goedewaagen (1971) gave a series of standardized dot maps with logarithmic increase in density which can be used for rapid assessment ("Knot method"), especially in grasslands where point densities will be high in the upper layers.

Disadvantages of the core-break method are that, similar to washed auger samples, no correlation can be made between local root density and other features of the soil, as recorded in the profile wall method. Bland (1989, 1991) discussed a number of sources of error: effects of preferential root orientation (see below), systematic gradients of root length density within a core making the break in the middle non-representative (this can be ignored unless there are very steep gradients in root length density), random variation of number of roots intersecting the plane of observation, and counting errors. Between different people scoring core-break results, however, systematic differences can be reduced or avoided by mutual cross-checking during a learning period.

7.2.6 Time and Labour Requirements of the Methods

A rough estimate of the time involved is given in Table 7.1. The number of replicates to be used, especially for the calibration of samples depends on

Table 7.1. Estimated time involved in profile wall and core-break method; estimates are based on an experienced crew and soils which are easy to work; they serve as a first approximation only

	Person-days per sample	Suggested number of replicates per treatment/observation period	Total time per treatment in person-days
Profile wall			
Fieldwork to make maps of $1 \times 1\,m^2$	0.25	4	7
Digitizing and analysis	1.5		
Washed calibration samples: processing	0.15	16	2.5
Core break			
Fieldwork	0.08	25	2
Calibration of samples: processing	0.25	10	2.5

the number of treatments or situations to be compared, as data may be pooled for analysis. The estimates given refer to three to five treatment comparisons. Time for processing data in the lab exceeds (as usual) the amount of time spent in the field. With efficient organization of digitizing and standard software for data processing, the total time required can be reduced.

7.3 Data Analysis

Data from trenchwall observations can be analyzed in a number of ways to derive information on spatial distribution of roots in the plane of observation, as well as to make inferences on the three-dimensional reality from which samples were obtained. The first step in any data analysis is digitizing the root maps at an appropriate level of precision for further analysis. The theoretical relation between point density on a map and the three-dimensional density and distribution of cylinders has been well studied, but empirical correlations make clear that often considerable "errors" are involved in the observation process. After describing these, we will discuss a range of techniques for describing root distribution with depth and horizontal distance from the plant, and spatial correlations between roots and other mapped features.

7.3.1 Digitizing Root Maps and Overlays

The purpose of digitizing is to derive a list of x,y coordinates of roots intersecting the plane of observation; this corresponds with grid counts with an infinitely small grid size. A range of technical options exist, from digitizing tablets where every point has to be touched with a pen to record its position, to "scanners" which read a section of a map and can be linked with an image analysis routine which returns the "centres of gravity" to retain a single-pixel representation of root occurrence. Overlays can be scanned separately to digitize other mapped features as lines or areas. The list of x,y coordinates can be analyzed in various ways.

The points can be grouped into any grid size and shape for the types of analysis discussed in section 7.3.2. Dividing the x,y coordinates by a grid size and taking integer values classifies the roots into a grid cell; a subsequent tabulation can yield point densities in a grid. Similarly, classification in circular sections can be obtained by transforming the rectangular coordinates to circle coordinates around plant rows or single trees as the point of origin.

7.3.2 Relation Between Intersection Point Density and Root Length Density

For a set of randomly oriented line pieces, the basic relationship between length density per unit volume L_v and the point density of intersections with any plane of observation N, was derived as $L_v = 2N$ by Lang and Melhuish (1970) and Melhuish and Lang (1968). The derivation has its origin in the 18th Century and is known as Buffon's problem (Marriot 1972). The derivation considers the probability that a randomly positioned and randomly oriented line section (of unit length) with its centre within a specified volume of space will intersect with a given plane of observation (any plane or a plane bordering the volume), and then integrates over all possible positions and orientations, to obtain an average probability of being observed per unit root length. For a population of line pieces, the expected number of intersections per unit area of observation plane can then be related to the total length of line. Taking the inverse of this equation, the length can be derived from the number of intersections. The fact that root systems in a unit volume of soil are not independent little sections of line, but tend to be interconnected may cause some doubt on the applicability of the $L_v = 2N$ result, but the connections will increase the variance rather than affecting the expected value and mean.

Random orientation can, however, not be taken for granted in root systems and a modified form of the equation is $L_v = 2XN$, where the factor X depends on root orientation (see Box 7.1). If the N in the equation represents the average for three mutually perpendicular planes of observation ($N_m = (N_{x1} + N_{x2} + N_y)/3$), the deviations of X from 1.0 are relatively small (Lang and Melhuish 1970; Baldwin et al. 1971, 1972). In practice, however, one wants to estimate L_v from observations in one (or maximum two) plane(s) of observation.

Information on root orientation, or at least on the N_x/N_z ratio (e) is obviously needed to convert intersection point density measurements on profile walls or core-break planes into estimates of L_v. Counting root intersections in both vertical and horizontal planes may give the most direct approach. Available estimates show that this ratio will at least vary in the 0.5–4 range, and may thus cause substantial deviations from the $L_v = 2N$ equation.

Theoretically, one can reconstruct root angles from observations of the elliptical shapes of root intersections, e.g. on resin-imbedded soil thin sections. In practice, however, slight deviations of root shapes from a cylinder shape and errors in measurement make this approach unreliable (Van Noordwijk et al. 1992).

A number of authors (Bengough et al. 1992; Pagès and Pellerin 1996; Pellerin and Pagès 1996; Jourdan and Rey 1997; Pagès and Bengough 1997; Lynch et al. 1997; Grabarnik et al. 1998) have used three-dimensional root distribution models (cf. Chap. 4) to predict root intersections with any plane of

BOX 7.1. Effects of Preferential Root Orientation (Anisotropy)

In the following we denote the point density of intersections of roots with three mutually perpendicular planes of observations as N_x, N_y and N_z, respectively, with N_m as their mean. If we can assume the two vertical planes of observation to yield equal results, we may write $N_x = N_y = \lambda\, N_z$, and hence $N_m = N_z(2\lambda + 1)/3$. Van Noordwijk (1987) normalized the anisotropy factor defined by Melhuish and Lang (1968), to obtain:

$$a_n = \sqrt{\frac{(1-N_x/N_m)^2 + (1-N_y/N_m)^2 + (1-N_z/N_m)^2}{6}}$$
$$= \frac{|1-\lambda|}{2\lambda+1}, \qquad (7.1.1)$$

(this means that $a_n = 0$ for $N_x = N_y = N_z = N_m$ ($\lambda = 1$), $a_n = 1$ if two of the three observation planes have no roots at all, $\lambda = 0$) and a_n approaches 0.5 if one of the planes has virtually no roots ($\lambda \to \omega$).

Marriot (1972) derived estimates of the factor X (in $L_v = 2\,X\,N_m$) as function of a_n. The functions for X can be adequately approximated (Van Noordwijk 1987) by

$$X = 0.5a_n^2 + 1, \quad \text{and} \quad X = 0.8a_n^2 + 1, \qquad (7.1.2)$$

where the first equation applies to a 'linear' and the second to a 'planar' situation with (0,0,1) and (1,1,0) roots in the extreme cases, and $\lambda < 1$ and $\lambda\, \Omega\, 1$, respectively. Root anisotropy will have a relatively mild effect on the calculated L_v values based on this X, provided that the average point density N_m is used. For the core-break method we are, however, mainly interested in the L_v/N_z relationship and for the profile wall observations in L_v/N_x, or L_v/N_y. In these ratios we find much stronger effects of anisotropy, as λ also influences N_m/N_z or N_m/N_x. Van Noordwijk (1987) derived from the above equations that for preferentially vertical root orientations with $0 \le \lambda \le 1$:

$$\frac{L_v}{N_z} = \frac{3\lambda^2 + 2\lambda + 1}{2\lambda + 1} \quad \text{and} \quad \frac{L_v}{N_x} = \frac{3\lambda^2 + 2\lambda + 1}{2\lambda^2 + \lambda}, \qquad (7.1.3)$$

and for roots with a preferentially horizontal orientation and $\lambda > 1$:

$$\frac{L_v}{N_z} = \frac{16\lambda^2 + 8\lambda + 6}{10\lambda + 5} \quad \text{and} \quad \frac{L_v}{N_x} = \frac{16\lambda^2 + 8\lambda + 6}{10\lambda^2 + 5\lambda}. \qquad (7.1.4)$$

One can see that for $\lambda = 1$ the equations return to the $L_v = 2\,N$ form; for $\lambda = 0$ we obtain $L_v = N_z$ (and L_v/N_x is infinite), which may represent the parallel, vertically oriented cylinders of many root uptake models.

The equations for L_v/N_x (profile wall) show a sharp decrease in the trajectory up to $\lambda = 0.5$, stabilize around 2.0 for $0.5 < \lambda < 2$ and show a mild decrease for $\lambda > 2$ (Fig. 7.5). The equations for L_v/N_z (core-break) are monotonously increasing functions of λ starting at 1.0 for $\lambda = 0$.

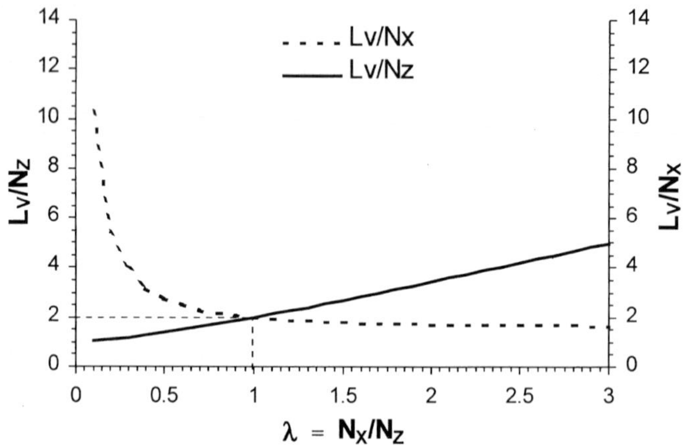

Fig. 7.5. Theoretical relationship between root length density per unit volume (L_v) (cm cm^{-3}) and intersection point density N on the map (cm cm^{-2}) of horizontal and vertical planes of observation for cases where the densities in the two possible mutually perpendicular vertical planes of observation are equal

observation. This has provided a valuable tool for exploring the effect of the type of deviations from random root orientation to be expected for real-world root systems.

7.3.3 Empirical Correlation for Core-Break and Profile Wall Observations

Real world calibrations of the method contain observation errors (in both profile wall and washed samples) and may deviate from the theoretical values which primarily depend on root orientation. Table 7.2 presents a selection of published data on empirical calibration factors. Schuurman and Goedewaagen (1971) gave calibration results for "root coverage" (related to N) versus root dry weight for grassland roots.

A point of warning may be that not all these studies have corrected for trends of root length density and comparisons may be based on intersection density on a plane external to the volume for which L_v data are obtained rather than for a plane halfway through this volume (Bengough et al. 1992).

Table 7.2. Examples of published values of calibration factor X in $L_v = 2\,X\,N$ relation between root intersection section N and root length density L_v. (Extended from Bengough et al. 1992)

Method	Crop	X for Nv	X for Nh	Comments	Reference
Profile wall	Maize/ soybean	8			Logsdon and Allmaras (1991)
	Maize	1.9			Vepraskas and Hoyt (1988)
	Tobacco	4.1			Vepraskas and Hoyt (1988)
	Mixed vegetation	0.41		Profile walls up to depth of 9 m	Oliveira-Carvalheiro and Nepstad (1996)
	Potato	1.0–4.5			Parker et al. (1991)
	Maize	3.0			Van Noordwijk et al. (1995)
Core-break	Oats		2.25		Bragg et al. (1983)
	Wheat		3.3–4.4		Drew and Saker (1980)
	Cotton/ sorghum		0.7–1.2		Bennie et al. (1987)
	Maize	0.8–1.1		X increases with depth	Van Noordwijk et al. (1995)
Impregnated soil blocks	Cotton	1.0			Melhuish and Lang (1968)
	Maize	1.0			Van Noordwijk et al. (1992)

7.3.4 Root Counts as Function of Depth and Horizontal Distance

Grid counts can be made from root maps in the laboratory by using a plastic sheet overlay. Grid size can be adjusted, and more flexibility is retained than with field grid counts. Non-rectangular grids, e.g. following soil horizons, can be used, as long as the surface area of each section is measured so that results can be expressed as number of points per area mapped. Point densities of root intersections can be converted to estimates of root length density L_v on the basis of empirical correlations or estimates of root anisotropy (Table 7.2).

Recording the x and y coordinates of each root is equivalent to a very fine grid, where the data get a presence-absence character rather than a number of points per grid. When x,y coordinates are recorded one can easily derive a grid classification (by taking the integer part of the coordinates divided by grid size). It usually is best to use grid sizes in which inter-row distances are multiples.

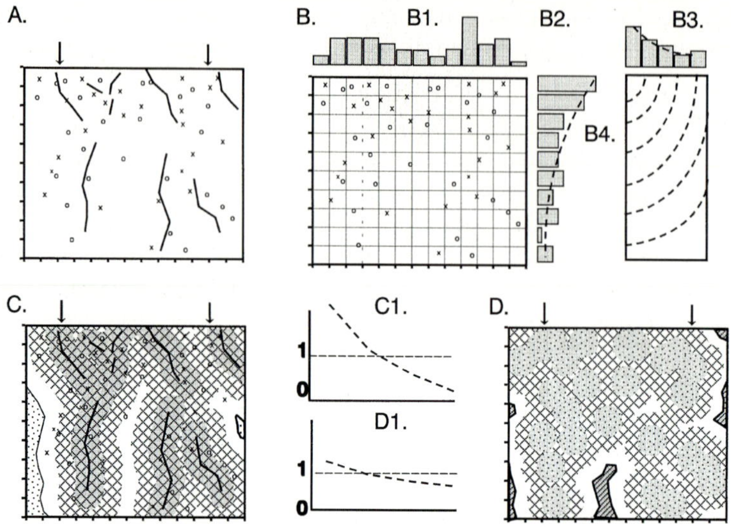

Fig. 7.6. Methods for analyzing root maps obtained in a vertical plane (**A**). **B** By overlaying the map with a grid, we can obtain histograms of intersection point density (number of intersections per unit area) as a function of horizontal distance (**B1**), vertical depth (**B2**) and distance to the nearest plant (**B3**). **C** By classifying the map by distance to cracks (or any other mapped feature) one can test the homogeneity of the relative point density for the strata so obtained (**C1**). **D** If two types of roots were mapped (*x* and *o* in this example) the point density of *x* can be sampled by strata differing in distance to the nearest o to test their spatial correlation (**D1**)

When a specified layer of soil is washed away from the profile wall, the number of points per grid or soil depth can also be converted into root length per unit of soil surface area (Böhm 1976). This can be achieved for a known length of profile wall by assuming that each root has the same length in the removed layer. Calibration of this technique with washed samples (blocks or cores) is very desirable.

Various root distribution functions can be fitted to the data (see Chap. 4), once grid counts are made. The grids are classified by their depth z below the soil surface and radial distance r to the nearest plant and root intersection. Densities are converted to root length densities on the basis of the conversion factor $X(z,r)$, which may depend on depth z and distance r. In soils without major restrictions to root growth an exponential decrease of root density with increasing distance from the plant may be expected. Van Noordwijk et al. (1996) expanded on the one-dimensional form of Gerwitz and Page (1974) and obtained:

$$L_v(z,r) = L_a|_{r=0}\,\alpha e^{-\alpha\sqrt{z^2+\beta r^2}}\,. \tag{7.1}$$

This function has three parameters: $L_a|_{r=0}$ is the total root length per unit area ($cm\,cm^{-2}$) at a distance $r = 0$ from the tree stem, α is the parameter (m^{-1}) governing the decrease with depth of root length density (for $r = 0$, at a depth of $0.699/\alpha$ the root length density has half of its value at the soil surface), and β is the dimensionless parameter governing the shape of the root system; values less than 1 indicate shallow-but-wide root systems, values of 1 give a circular symmetry, and values >1 indicate deep-but-narrow root systems.

If β does not differ significantly from 1 and we ignore possible variation in $X(z,r)$ with distance r, the equation may be simplified to a one-dimensional distribution of root length density with depth:

$$L_v(z) = L_a \alpha e^{-az}.$$

7.3.5 Root Distribution Pattern Within a Zone: Nearest Neighbour Distances

Root maps in the horizontal and vertical plane (trench method) not only yield information on average root length density per soil layer, but also allow quantification of the distribution pattern within each layer (Barley 1970; Diggle 1983). The 'null model' of root distribution, against which one would usually want to test, is of independent roots with a homogeneous probability of occurrence within (a section of) the plane of observation. A Poisson distribution in the observation grid thus gives the appropriate comparison. A number of tests have been proposed, especially in plant ecological literature (Pielou 1969). Several of these use the fact that in Poisson distributions the mean of the number of points per cell equals the variance; the ration of mean and variance in a sample can thus be compared with the confidence intervals for finite samples. A variance/mean ratio above 1 indicates clustering, a ratio below 1 regularity (Fig. 7.7).

A different category of test is based on a comparison of "point-root" and "root-root" distances, where "points" are chosen randomly in the plane of observation. If roots behave as mutually independent points, the two distributions should be essentially identical. Where roots tend to cluster, root-root distances tend to decrease while point-root distances increase. Where root patterns tend towards regularity, the point-root and root-root curves will change in an opposite direction. The difference between these two distributions thus gives a sensitive test (Pielou 1969).

To derive the frequency distribution of nearest neighbour distances, two basic approaches are:

1. Start with a "source" point of measurement, calculate the distance to all neighbours and select the shortest distance (nearest neighbour);

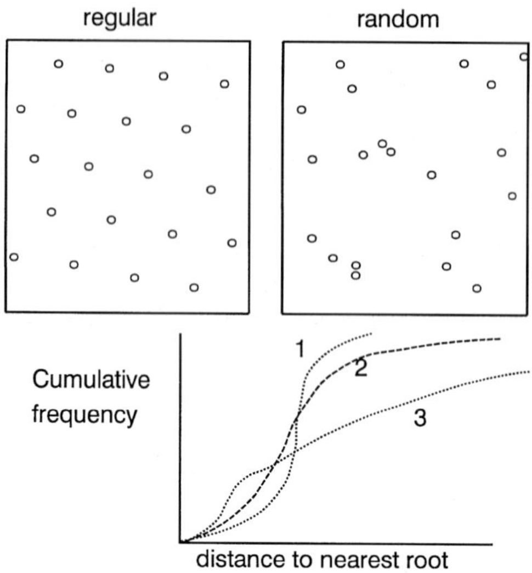

Fig. 7.7. Basic terminology for spatial point patterns and a nearest neighbour distance frequency diagram which can be used to test randomness of the pattern; if the lines for random points to the nearest root are not essentially different from those for roots-to-roots, the root pattern may be called random itself

2. Start with all target points and classify the area surrounding them by distance – this is the equivalent of drawing circles of increasing diameter around all "target points".

In the past, method 1 was most used, as it can be defined as a straightforward algorithm combining logical steps and a distance equation. Within this approach there is the option of calculating the distance from each source to all targets and select the shortest one, or use some prior sorting of target points which directly helps to select a subset of target points to be considered for a given source point. The choice between these approaches depends on the size of the data set (Diggle 1983). With the progress made in image analysis techniques, the approach based on drawing circles around all target points, lines or polygons can now be easily implemented. Image analysis or GIS software may have a built-in "distance transform" which can be used, or a distance-like operation can be based on a sequence of 'filters' using four or eight-neighbour expansion steps (Van Noordwijk et al. 1993a,c). With the latter one can evaluate the effects of 'barriers' on distance measurements. With some GIS software one can, similarly, incorporate information on 'resistance' in distance measurements (similar to the use of roads of different qualities).

If statistical tests indicate that observed patterns differ from "random", the question is what alternative distribution should be used. Diggle (1983) discusses the use of 'stratified Poisson' processes, which may be useful in root research.

Essentially, these distributions assume a random set of "mother" points, with an increased probability of finding 'daughters' in a local neighbourhood. Two parameters of such a process are then the average number of daughters per mother, and the distance around the mother point in which daughters are to be found. By modifying these two parameters the spectrum from highly clustered to truly random patterns can be generated.

Two interpretations of these parameters are: (1) that the area was a priori heterogeneous in its attractiveness to root development, and (2) that branch roots develop necessarily close to the axes from which they originate. The pattern as such does not allow us to distinguish between these two reasons for clustering.

7.3.6 Spatial Correlation of Mapped Features

Spatial correlation can be tested by implementing a stratified sampling of roots with strata chosen with increasing distance to mapped features such as cracks (Fig. 7.6C; Van Noordwijk et al. 1993b,c).

The following steps can be taken in the analysis:

1. Digitize root (X,Y) coordinates and map other features of interest (here indicated as focal phenomena),
2. Determine the point density of roots coinciding with the focal phenomena, by counting the number of roots and measuring the area,
3. "Expand" the image of the focal phenomena by including all the area within one unit distance of the previous map of focal phenomena,
4. Determine the point density of roots for the previous distance increment,
5. Repeat steps 3 and 4 until the whole map is covered,
6. Transfer the data to a statistics package and test for spatial correlation; the null hypothesis of independent random events, is equivalent to the absence of a significant trend in local point density with distance to other features. A model of point density as a function of distance can be fitted and used to test the null hypothesis. When fitting a distance function, one should acknowledge that the data will have a Poisson rather than normal distribution (an example, using the Genstat statistical package is provided by Van Noordwijk et al. 1993c)

7.3.7 Root Position Effectivity Ratio, R_{per}

The frequency distribution of nearest-neighbour distances of random points to the nearest root on such a map indicates the possibilities for transport by diffusion of all soil resources (Van Noordwijk 1987; Rappoldt, 1990).

BOX 7.2. Steps in the Derivation of the Root Position Effectivity Ratio, R_{per} (Van Noordwijk et al. 1993c):

1. Digitize a map of root distribution in a horizontal or vertical plane to derive a list of X,Y coordinates and estimate the corresponding root length density L_v,
2. If desired, introduce 'barriers' in the root map, e.g. representing incomplete root-soil contact or cracks in the soil,
3. Derive a cumulative frequency distribution (Fig. 7.8A, B) of distances from points in the soil to the nearest neighbour root, taking into account the existence of barriers (this can be done by 'expanding' the area taking roots as starting points and measuring surface area after each distance increment),
4. Derive an 'annulus fraction' representation (Fig. 7.8C) of these nearest neighbour distances by dividing the fraction of soil in each distance increment by that for an equivalent annulus,
5. Transfer the annulus fractions into fractions f_i of complete circle models with radius R_i ('cutting the pie'; Fig. 7.8C).
6. Calculate the total transport capacity to a sum of cylinders with radius R_i for a given uptake model; since the zero-sink uptake of water as well as nutrients is proportional to a G-function (De Willigen and Van Noordwijk 1987a; cf. Chap. 15):

$$Gsum = \frac{\sum_i f_i G(R_i) R_i^2}{\sum_i f_i R_i^2}, \qquad (7.2.1)$$

with:

$$G(Ri) = 0.5\left[\frac{1-3\rho^2}{4} + \frac{\rho^4 \ln \rho^2}{\rho^2 - 1}\right] \quad \text{and} \quad \rho = \frac{2R_i}{D_r} = \frac{2}{D_r\sqrt{\pi L_v}}, \qquad (7.2.2)$$

where D_r is the assumed root diameter,

7. Find the R^* for which $G(R^*) = Gsum$, and the corresponding L_v^*,
8. $R_{per} = L_v^*/L_v$.

The R_{per} factor (root position effectivity ratio) was introduced (Van Noordwijk et al. 1993a; Van Noordwijk and Brouwer 1997) to derive a correction factor, such that when L_v is multiplied with this factor an equivalent root length density L_v^* is obtained for a simple model geometry (regularly spaced, parallel cylinders), with the same opportunities for transport towards the root surface. The method (Box 7.2; Fig. 7.8) can potentially be applied to any root distribution pattern, but it assumes that the soil resources themselves are homogeneously distributed (initially). For homogeneous resource distributions, a regular root spacing maximizes uptake, so R_{per} will be less than 1.0 for non-regular root distribution. If resources are non-homogeneous, non-regular root distributions may be superior if roots and recources tend to coincide.

$$L_v^* = R_{per} L_v. \tag{7.3}$$

Van Noordwijk and Brouwer (1997) reported values of R_{per} in the range of 0.3 for winter wheat and sugar beet in the plough layer of arable land. Haberle et al. (1994) explored how R_{per} varies with the plane of observation based on three-dimensional root branching models (See Chap. 4).

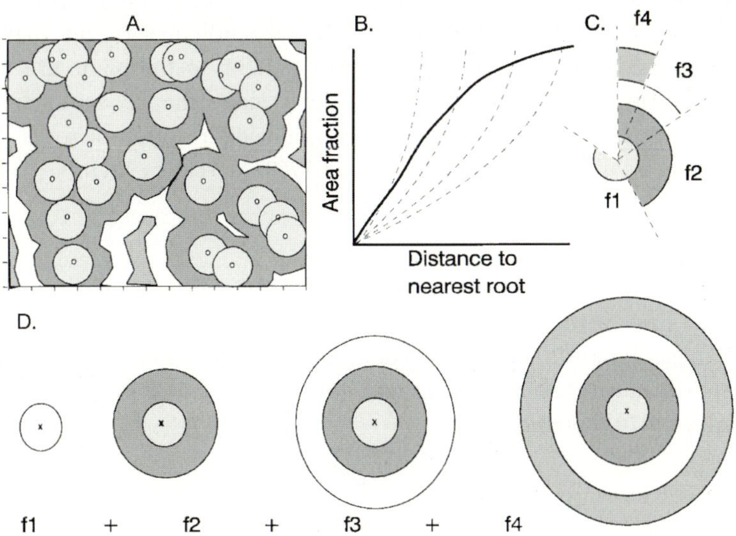

Fig. 7.8. Analysis of root distribution pattern. By applying a 'distance transform' to the map on the basis of all roots mapped; measurement of the nearest neighbour distances (**B**) can be used to derive the area in each distance class for an 'average' root (**C**); by 'cutting the pie' we can derive the relative frequency of cylindrical models with 1 . . . x layers which together may represent the possibilities for uptake by the root pattern (**D**); for a given root uptake function, we can finally derive the equivalent homogeneous root density which would allow the same uptake to occur and express the ratio of this hypothetical and the measured point density as the root position effectivity ratio, R_{per} (Van Noordwijk et al. 1993a,c)

A major assumption in this approach is that a classification of soil according to the nearest root (i.e. a Dirichlet tessellation) is acceptable. If all roots start to function at the same time, this assumption leads to only small errors (De Willigen and Van Noordwijk 1987b). In the model representation, the cumulative frequency distribution of transport distances should be conserved. For any frequency distribution of distances to a spatial point pattern, a frequency distribution of cylinders can be derived which fulfills this criterion (Van Noordwijk 1987; Rappoldt 1990; Van Noordwijk et al. 1993a).

7.3.8 Transition from 2-D to 3-D Distance

A partly unresolved problem in interpreting root maps is the fact that the real three-dimensional distances are smaller (except for roots growing perpendicular to the mapped plane) than the two-dimensional distances measured on the map. For randomly orientated roots, the frequency distribution of three-dimensional distances coincides with that for 0.71 times the two-dimensional distances (Van Noordwijk 1987). For non-random root orientation the problem is unresolved; a serious complication is the fact that a plane represents a biased sample of roots, as roots growing almost perpendicular to the plane are over-represented and roots growing almost parallel to the plane are under-represented. For a 2-dimensional map of the Z-plane, the frequency distribution of point-root distances in case of a random distribution of roots can be derived from a Poisson distribution (Pielou 1969; Marriott 1972):

$$P[d < D_2] = 1 - e^{-\pi N_z D_2^2}, \tag{7.4}$$

where D_2 is the two-dimensional distance and d is the distance of a point on the map to the nearest root.

For three-dimensional distances of points to randomly oriented and spaced cylinders (lines with density L_v per unit volume and radius R_0) Ogston (1958) and Barley (1970) derived that:

$$P[d < D_3] = 1 - \pi L_v R_0^2 - e^{-\pi L_v \left(D_3^2 + 4/3 \Lambda D_3^3\right)}, \tag{7.5}$$

where D_3 is three-dimensional distance, and Λ is the number of root tips per unit root length.

The term $L_v\ R_0^2$ refers to the volume of the cylinders themselves and is normally negligible. The last term in the equation (4/3 $\Lambda\ D_3^3$) refers to a half sphere around the root tip for 'end-contacts' and has a considerable modifying effect on the curve, which is mainly determined by the tangential contacts of the middle term. For $\Lambda = 0$ the 3-D equation becomes comparable to the 2-D one if:

$$D_3 = \frac{D_2}{\sqrt{2}} = 0.71 D_2 . \tag{7.6}$$

This result strictly depends on random orientation of the roots with regard to the plane in which two-dimensional distances were measured. If D_2 is measured in a plane perpendicular to a parallel set of cylinders, D_3 will equal D_2.

7.4 Conclusions and Perspectives

The profile wall methods discussed in this chapter for observing roots span a wide range of applications: from "quick and not-too-dirty" impressions of overall spatial patterns, which can be much faster than methods based on washed samples, to methods aimed at extracting specific spatial information which cannot be obtained from washed samples at all.

The geometrical relations between three-dimensional root systems and observations on any two-dimensional plane of observation are far from trivial and further progress is to be expected from links between root architectural models, and predicted and observed point patterns on a range of planes of observation. For several root ecological questions analysis of "spatial correlation" of roots and other phenomena (roots and cracks and/or anaerobic spots, roots of species A and those of species B, roots and concentrations of organic inputs, roots and decaying root channels formed by a previous vegetation) is crucial, but refined observation techniques may be needed and the standard "root map" may be too coarse.

References

Baldwin JB, Tinker PB, Marriott FHC (1971) The measurement of length and distribution of onion roots in the field and the laboratory. J Appl Ecol 8: 543–554

Baldwin JP, Tinker PB, Nye PH (1972) Uptake of solutes by multiple root systems from soil. II. The theoretical effects of rooting density and pattern on uptake of nutrients form soil. Plant Soil 36: 693–708

Barley KP (1970) The configuration of the root system in relation to nutrient uptake. Adv Agron 22: 159–201

Bengough AG, Mackenzie CJ, Diggle AJ (1992) Relations between root length densities and root intersections with horizontal and vertical planes using root growth modelling in 3-dimensions. Plant Soil 145: 245–252

Bennie ATP, Taylor HM, Georgen PG (1987) An assessment of the core-break method for estimating rooting density of different crops in the field. Soil Tillage Res 9: 347–353

Bland WL (1989) Estimating root length density by the core-break method. Soil Sci Soc Am J 53: 1595–1597

Bland WL (1991) Root length density from core-break observations: sources of error. In: McMichael BL, Persson H (eds) Plant roots and their environment. Elsevier, Amsterdam, pp 565–569

Böhm W (1976) In situ estimation of root length at natural soil profiles. J Agric Sci 87: 365–368

Böhm W (1979) Methods of studying root systems. Berlin Heidelberg New York

Bragg PL, Govi G, Cannell RQ (1983) A comparison of methods, including angled and vertical minirhizotrons, for studying root growth and distribution in a spring oat crop. Plant Soil 73: 435–440

De Willigen P, Van Noordwijk M (1987a) Roots for plant production and nutrient use efficiency. Doct Thesis, Agricultural University, Wageningen, 282 pp

De Willigen P, Van Noordwijk M (1987b) Uptake potential of non-regularly distributed roots. J Plant Nutr 10: 1273–1280

Diggle PJ (1983) Statistical analysis of spatial point patterns. Academic Press, London

Drew MC, Saker LR (1980) Assessment of a rapid method, using soil cores, for estimating the amount and distribution of crop roots in the field. Plant Soil 55: 297–305

Gerwitz A, Page ER (1974) An empirical mathematical model to describe plant root systems. J Appl. Ecol 11: 75–96

Grabarnik P, Pagès L, Bengough AG (1998) Geometric properties of simulated maize root systems: consequences for length density and intersection density. Plant Soil 200: 157–167

Haberle J, Van Noordwijk M, Brouwer G (1994) Comparison of root position effectivity ratio of field-grown and modelled maize root systems. Poster abstract. In: Proc 3rd Congr Eur Soc for Agronomy, 1994, Abano-Padova

Jourdan C, Rey H (1997) Modelling and simulation of the architecture and development of the oil palm (*Elaeis guineensis* Jacq.) root system. II. Estimation of root parameters using the RACINES postprocessor. Plant Soil 190: 235–246

Kooistra MJ, Van Noordwijk M (1996) Soil architecture and distribution of organic carbon. In: Carter MR, Stewart BA (eds) Structure and organic carbon storage in agricultural soils. Advances in Soil Science. CRC Lewis, Boca Raton pp 15–57

Lang ARG, Melhuish FM (1970) Lengths and diameters of plant roots in non-random populations by analysis of plane surfaces. Biometrics 26: 421–431 (corrections in Biometrics 28: 626–627)

Logsdon SD, Allmaras RR (1991) Maize and soybean root clustering as indicated by root mapping. Plant Soil 131: 169–176

Lynch JP, Nielsen KN, Davis RD, Jablokow AG (1997) SIMROOT: modelling and visualization of root systems. Plant Soil 188: 139–151

Marriot FHC (1972) Buffon's problem for non-random distribution. Biometrics 28: 621–624

Melhuish FM. Lang ARG (1968) Quantitative studies of roots in soil. I. Length and diameters of cotton roots in a clay-loam soil by analysis of surface-ground blocks of resin-impregnated soil. Soil Sci 106: 16–22

Ogston AG (1958) The spaces in a uniform random suspension of fibres. Trans Faraday Soc 54: 1754–1757

Oliveira Carvalheiro K de, Nepstad DC (1996) Deep soil heterogeneity and fine root distribution in forests and pastures of eastern Amazonia. Plant Soil 182: 279–285

Pagès L, Bengough AG (1997) Modelling mini-rhizotron observations to test experimental procedures. Plant Soil 189: 81–89

Pagès L, Pellerin S (1996) Study of differences between vertical root maps observed in a maize crop and simulated maps obtained a model for the three-dimensional architecture of the root system. Plant Soil 182: 329–337

Parker CJ, Carr MKV, Jarvis NJ, Puplampu BO, Lee VH (1991) An evaluation of the minirhizotron technique for estimating root distribution in potatoes. J Agric Sci 116: 341–350

Pellerin S, Pagès L (1996) Evaluation in field conditions of a three-dimensional architectural model of the maize root system: comparison of simulated and observed horizontal root maps. Plant Soil 178: 101–112

Pielou EC (1969) An introduction to mathematical ecology. Wiley-Interscience, New York

Rappoldt C (1990) The application of diffusion models to aggregated soil. Soil Sci 150: 645–661

Ringrose-Voase AJ (1996) Measurement of soil macropore geometry by image analysis of sections through impregnated soil. Plant Soil 183: 27–47

Schuurman JJ, Goedewaagen MAJ (1971) Methods for the examination of root systems and roots, 2nd edn. Pudoc, Wageningen

Schuurman JJ, Knot L (1957) Het schatten van hoeveelheden wortels in voor wortelonderzoek genomen monsters. [Estimating root quantities in auger samples; in Dutch] Versl Landbouwkd Onderz 63: 31

Tardieu F (1988) Analysis of the spatial variability of maize root density. I. Effect of wheel compaction on the spatial arrangement of roots. II. Distances between roots. Plant Soil 107: 259–266 and 267–272

Tardieu F, Manichon H (1986) Caracterisation en tant que capteur d'eau de l'enracinemnent du mais en parcelle cultivee. I. Discussion des criteres d'etude. II. Une methode d'etude de la repartition verticale et horizontale des racines. Agronomie 6: 345–354 and 415–425

Van Noordwijk M (1987) Methods for quantification of root distribution pattern and root dynamics in the field. 20th Colloq Int Potash Institute, Bern, pp 263–281

Van Noordwijk M, Brouwer G (1997) Roots as sinks and sources of carbon and nutrients in agricultural systems. In: Brussaard L, Ferrera-Cerrato R (eds) Soil ecology in sustainable agricultural systems. CRC Lewis, Boca Raton, pp 71–89

Van Noordwijk M, Kooistra MJ, Boone FR, Veen BW, Schoonderbeek D (1992) Root-soil contact of maize, as measured by thin-section technique. I. Validity of the method. Plant Soil 139: 109–118

Van Noordwijk M, Brouwer G, Harmanny K (1993a) Concepts and methods for studying interactions of roots and soil structure. Geoderma 56: 351–375

Van Noordwijk M, De Ruiter PC, Zwart KB, Bloem J, Moore JC, Van Faassen HG, Burgers S (1993b) Synlocation of biological activity, roots, cracks and recent organic inputs in a sugar beet field. Geoderma 56: 265–276

Van Noordwijk M, Brouwer G, Zandt P, Meijboom FW, Burgers S (1993c) Root patterns in space and time: procedures and programs for quantification. IB-Nota 268, IB-DLO Haren (the Netherlands)

Van Noordwijk M, Driel W, Brouwer G, Schuurman W (1995) Heavy metal uptake by crops from harbour sludge covered by non-contaminated topsoil. II. Cd uptake by maize in relation to root development and distribution of metals. Plant Soil 175: 105–113

Van Noordwijk M, Lawson G, Groot JJR, Hairiah K (1996) Root distribution in relation to nutrients and competition. In: Ong CK, Huxley PA, eds. Tree-crop interactions – a physiological approach. CAB International, Wallingford. pp 319–364

Vepraskas MJ, Hoyt GD (1988) Comparison of the trench profile and core methods for evaluating root distributions in tillage studies. Agron J 80: 166–172

CHAPTER 8

Root Observations and Measurements at (Transparent) Interfaces with Soil

A.L. Smit[1], E. George[2], and J. Groenwold[1]

[1] Plant Research International (Wageningen UR), Postbus 16, 6700AA Wageningen, The Netherlands

[2] Hohenheim University, Institute for Plant Nutrition, Fruwirthstr. 20, 70593 Stuttgart, Germany

CONTENTS

A.L. Smit et al. (Eds.), Root Methods

8.1 Introduction

Glass windows are often installed in rhizotron facilities to observe root growth. The same principle has been applied under field conditions, using "root windows" and "minirhizotrons" (transparent glass or plastic tubes pushed into the soil: Bates 1937), through which roots are observed non-destructively in situ at the interface between the glass and the soil. The possible observations include recording root growth and decay – processes which are difficult to measure otherwise – by still photography or videos, thereby providing time series that reveal the dynamics of root growth. Root researchers also use transparent interfaces when estimating the rooting characteristics of the surrounding bulk soil, such as root length intensity (L_v) and root mass per unit volume of soil (W_v), although this often necessitates a (cumbersome) conversion from 2-D to 3-D. All these methods and tools used to view roots are reviewed in this chapter.

8.2 Minirhizotrons

8.2.1 Introduction

Inserting clear tubes into the soil through which roots can be observed in situ with various items of equipment (see Fig. 8.1), has several advantages over other methods of root quantification.

Fig. 8.1. A common minirhizotron set-up with observation equipment. (Photograph courtesy J. Bartz)

One such advantage is that studying root dynamics at the interface between a transparent viewing medium and the soil is usually less labour-intensive than *core sampling*, *in-core growth* or *profile wall methods*. Another is that individual roots (and their properties) can be followed in time. Root characteristics such as the growth rate, branching, longevity and the presence of root hairs can best be assessed through an in situ viewing medium. Moreover, this technique makes it possible to quantify infection with mycorrhizae, fungi or nematode species and to assess the diameter distribution of the root system in situ (Majdi and Nylund 1996).

Minirhizotrons allow the same area/volume of soil and the same population of roots to be monitored continuously, but questions have been raised about how representative these observations are compared with bulk soil methods.

8.2.2 Types of Minirhizotrons

8.2.2.1 Solid Tubes

Tube Material. Tubes made of glass (borosilicate, trade name e.g. Pyrex), acrylic (Perspex, Plexiglas), polybutyrate and polycarbonate (Lexan) have been used as minirhizotrons. Glass tubes are cheaper than commercially available polybutyrate tubes, but may break if large roots and rocks are encountered in the profile, or if the soil is subject to severe freeze-thaw and shrink-swell movements.

McMichael et al. (1992) mention that polybutyrate is sometimes preferable to acrylic tubes because of its flexibility and resistance to excessive scratching during installation. Kloeppel and Gower (1995) described the construction of tubes from inexpensive acrylic tubing (according to them this is 42% cheaper than glass), marking them with an easily constructed etcher and installing the tubes with a tree seedling auger.

Although Taylor and Böhm (1976) suggested that soil adhered more readily to glass windows than to acrylic plastic windows, they did not specifically compare the two materials. They also suggested that if the soil touching the plastic dries, it will shrink at higher soil water contents than if it is touching glass. They concluded that the resulting cracks and voids encouraged higher root densities at the plastic face compared with the bulk soil.

It has also been suggested, although not quantified (Voorhees 1976), that using acrylic material might allow electric charges to develop, which would interfere with root growth. Although many researchers have voiced concern about the material used to create an interface to observe root growth, the materials have never been explicitly compared. This suggests that the material itself is not a crucial factor for the minirhizotron method. In many cases it is the soil and environment that determine the choice of material.

Dimensions. Upchurch (1987) argued that larger tube diameters increase the sample size, which improves precision and reduces variability. Mackie-Dawson and Atkinson (1991) mentioned that a 5.6 cm o.d. (outer diameter) tube would "sample" a volume of 352 cm^3, assuming a view 2-mm deep. If a strip of 2 cm only is observed at the upper side of the minirhizotron, only about 40 cm^3 would be "sampled", a small volume compared with the volume of soil sampled by soil coring. Coring to a depth of 1 m yields samples over 2000 cm^3 of soil, depending on the diameter of the corer. But note that when Itoh (1985), cited by Brown and Upchurch (1987), evaluated tube sizes from 1–3 cm, he found that accuracy increased with decreasing diameter.

The preferred diameter for round tubes is about 5 cm, mainly because this will accommodate commercially available video-camera systems. Square tubes have also been used (Vos and Groenwold 1987; van Noordwijk et al. 1994) but are not a standard stock item and do not allow the viewing equipment to be rotated. A further disadvantage is that they require a custom-made soil probe for creating the minirhizotron access hole.

It has been mentioned that it is more difficult to uniformly illuminate in round tubes than square tubes (Brown and Upchurch 1987) but modern video equipment seems to have overcome this problem.

Marking System/Depth Registration. The observation system must be positioned accurately along the tube; this is particularly imporant if individual roots have to be monitored over time. Ferguson and Smucker (1989) described an indexing handle for controlling and identifying the camera position when making repeated video recordings from minirhizotrons in field experiments. One problem with an indexing handle is that it is probably not accurate enough to enable the images to be semi-automatically processed by image analysis, as horizontal or vertical displacement can still occur (pers. observ.). This can be overcome by putting a registration mark on the tube, so that the frames can be properly aligned with image analysis software. The spring-loaded etching trough, described by Kloeppel and Gower (1995) which etches accurate, equidistant circles along acrylic tubes in minutes, would be useful for this.

8.2.2.2 Flexible or Pressurised Tubes

One factor crucial to the success of the minirhizotron method is the contact with the surrounding soil. Imperfect contact prevents an accurate estimate of the bulk soil root density. If there are voids, the roots will grow into them or will track along the surface of the tube. One of the signs of poor contact between tube and soil is condensation on the tube surface and a tendency for roots to produce more root hairs when growing in the space between the tube and soil. In soils with layers which hinder root penetration, a tunnelling effect can be created, with these gaps promoting deeper rooting along the tube than in undisturbed soils (Vos and Groenwold 1987).

An alternative minirhizotron design has been developed, with improved contact with the surrounding soil. The pressurised wall minirhizotron developed by Merrill et al. (1987) and Merrill (1992) consists of (see Fig. 8.2):

1. A central tube of clear, extruded acrylic plastic,
2. A 0.05-cm-thick cylindrical piece of clear polyvinyl sheeting fitted over this central tube and clamped to both its ends, and
3. A PVC collar at the upper end of the tube, through which air pressure tubing is introduced.

Fig. 8.2. A pressurised wall minirhizotron system (after Merrill 1992 (*left*)) and a solid minirhizotron (*right*). Courtesy S. Asseng, CSRIO

These minirhizotrons of 7.5 cm o.d. were installed in holes of 8.9 cm excavated with a rotary auger and finished with a bucket auger to achieve a final diameter of 9.6 cm. A system of low-pressure air pumps ensures that there is constant air pressure (10–20 kPa) between the sheeting (outer wall) and the tubing (inner wall). An advantage of the pressurised wall system is the relative ease with which the uninflated minirhizotron tubes can be inserted into their access holes.

Gijsman et al. (1991) proposed another method for establishing close contact between the observation interface and the surrounding soil. Based on the same idea, they used an inflatable flexible wall, made from a modified motorcycle tyre. The method involved inserting a square aluminum or stainless steel frame consisting of ribs 1 cm wide into a hole in the ground. An inflatable rubber tube, made from a cut motorcycle tyre and stretched by a metal rod, was introduced into the frame and pressurised. The tyre bulges between the ribs and presses against the soil. Before the interface is observed the tyre is deflated and carefully removed from the frame, in order not to change the positions of the roots and to ensure that roots continue to adhere to the soil.

This design has two additional advantages over Merrill's system:

1. Improved visibility of the roots because it is not limited by the opaqueness of the minirhizotron wall, nor do the tubes generate disturbing reflections, and
2. Roots can be sampled/collected in situ with appropriate devices, e.g. to determine N content, presence of VA mycorrhizae, and physiological functioning (dead/living).

These aspects underline the potential usefulness of the system in research.

Gijsman et al. (1991) checked for the presence of voids by examining stereoscopic pairs of photographs. They concluded that no condensation or smears from alga growth obscured the visibility of the roots and that all roots were growing in one plane of the soil/tube interface. They advised taking special care at the points where the frame ribs meet the soil, because voids will occur here if the frame is too thick or the pressure too low. The transition can be made more gradual by bevelling the ribs of the frame to prevent gaps from forming.

Volkmar (1993) compared an adjusted pressurised tyre system with rigid minirhizotrons (transparent polybutyrate plastic tubes (5.4 cm o.d.). From his observations of root growth at horizontal minirhizotrons at bulk densities increasing from 1.5 to 1.7 g cm^{-3} he concluded that significant differences in root counts between the two systems became apparent with increasing density. When he examined the relationship between root counts (minirhizotron) and root density of the bulk soil, he found that with rigid tubes, root counts were higher at the higher bulk densities. With the flexible tubes, the relationship between root counts and root density remained more or less stable with different bulk densities. This implies that rigid tubes probably overestimated rooting, especially at higher bulk densities. Volkmar (1993), however, also noted that inflatable tubes have a drawback compared with rigid tubes: the pressure is difficult to maintain and even under controlled environment conditions the tubes gradually lose pressure or deflate suddenly if the rubber wall develops a weakness. For this reason, minirhizotrons have to be checked daily after installation. In the pressurised system described by Merrill (1992) the pressure is maintained by a solar-driven air pressure system. Both these sophisticated modifications increase the cost of the apparatus and its maintenance, often beyond research budgets.

8.2.3 Installation Technique

8.2.3.1 Installation Equipment

The technique for installing tubes (glass and plastic) in an undisturbed soil profile in the field is critical. The main concerns are to establish good contact between the tube and the surrounding soil, to avoid compaction of the soil during the installation of the tube in the soil and to minimise the creation of voids at the tube/soil interface.

Hand augers can be used in conjunction with an angled support (see Fig. 8.3A), to make a hole slightly smaller than the outside diameter of the minirhizotrons. Box et al. (1989) used hydraulic soil coring machines for this purpose (see Fig. 8.3B).

They took soil cores of 5.5 cm o.d. prior to installing a tube of 5.7 cm o.d. The cutting points of the soil sampling tubes were machined, so that soil displaced by the sharpened tip was compressed toward the interior of the sampling tube, thus preventing compaction at the interface. The angle of the soil

Fig. 8.3. a Installation of minirhizotrons with hand auger and an angled support. (Courtesy S. Asseng, CSRIO). b Mechanical hydraulic coring machine in use at Michigan State University. (Courtesy A. Smucker)

probe was maintained by using a large wooden guide constructed at a specific angle. Afterwards the hole was brushed with a 5.5 cm fine circular steel brush. The tubes were then forced into the soil manually. Similar procedures have been used by others (Gregory 1979; Vos and Groenwold 1987; Majdi et al. 1992).

Hummel et al. (1989) studied the effect of the installation procedure on the nature of the contact zone and investigated the importance to the success of the procedure of stabilising the coring unit, guiding the augering and coring devices and of coring bit size. Following field installation in a silty loam they found no voids, but coring and tube installation did cause soil compression which was uniform around the tube at a given depth and varied between depths. The compression was minimised when diameter of the coring bit was 0.16 cm less than the minirhizotron diameter.

In general, data from horizontal minirhizotrons in containers with uniformly repacked soil (which allows minirhizotrons to be installed without voids) best agreed with data from auger sampling (Vos and Groenwold 1983, 1987; Smit et al. 1994; Meyer and Barrs 1985). This underlines the importance of avoiding voids next to the tubes.

8.2.3.2 Installation

Angle of Installation. A wide range of angles to the vertical has been used when inserting the tubes. Originally, researchers inserted the tubes vertically, but later various angles were used. Minirhizotrons have even been positioned horizontally (Smit et al. 1994; Dubach and Russelle 1995). Most studies, however, have used a range of angles between 30° and 45° to the vertical. Bragg et al. (1983) claimed that installing the tubes vertically would induce preferential tracking of the roots alongside the tube and that this could be eliminated with an angled insertion. They reached this conclusion after comparing 45° and vertical tubes with soil coring. When the data from the top 30 cm of soil were omitted, angled tubes showed a better relation with coring than vertical tubes. However, it has been pointed out (De Ruijter et al. 1996), that this conclusion is not supported by the data shown in the original publication. Overall, considering all depths, vertical tubes showed even better correlation with coring than angled tubes.

In a pot experiment with Sitka spruce and sycamore, root tracking was found to occur to the same extent on vertical tubes and on tubes angled at 45° (Mackie-Dawson et al. 1989). The root counts were higher at vertical tubes, especially in the first 7 cm, but this was attributed not to tracking but to a more horizontal root growth in this layer. Evidence was also found with vertical tubes that there was greater variance associated with the measurement of root length (Mackie-Dawson and Atkinson 1991).

Many investigators have installed their tubes at an angle to the vertical, usually 45°, although vertical positioning is still used (Campbell et al. 1994; De Ruijter et al. 1996), particularly when conditions are unfavourable for angled tubes. Horgan et al. (1993) acknowledge the importance of the installation angle, especially if root growth orientation deviates from isotropic (isotropic = all orientations are equally likely) and if the observations of the 2-D-surface are to be converted to a 3-D environment. If isotropy cannot be assumed, the "point process" will depend on the orientation of the surface on which the observations are made. Horgan et al. (1993) derived an optimal angle of 54° theoretically. At this angle the influence of a non-random orientation of root growth would be minimal (bias should not exceed 3% whatever the distribution of root angles). They also suggested two further strategies to minimise the influence of root anisotropy: one was to give the horizontal count per unit root length twice the weight of the equivalent vertical count if horizontal *and* vertical tubes can be used, as then bias will be close to zero. Another strategy would be to have groups of three tubes of equal length; two positioned perpendicular to each other in horizontal planes and the third positioned vertically. Both strategies allow examination of variation in root density by depth.

Spatial Distribution in Relation to Rows. It is suggested that in *row crops,* tubes should be inserted parallel to the row (Upchurch and Ritchie 1983, 1984; McMichael and Taylor 1987) to avoid a bias with depth. To ascertain the lateral distribution of roots, more tubes should be installed at various distances between the rows (Brown 1984 cited by McMichael and Taylor 1987). It is very difficult to assess the average rooting characteristics of individual plants (including forest trees): in such cases, minirhizotrons may be inserted at a constant distance from the plant.

8.2.3.3 Light Leaks and Temperature Effects

Light entering a minirhizotron through a gap in the soil/tube interface can travel down the minirhizotron tube to substantial depths and influence roots approaching the tube wall. The standard treatment to exclude light in minirhizotron studies is to wrap black tape around the above-ground section of the tube and cap the end with a painted aluminium can which slides snugly over the tube. Levan et al. (1987) investigated the efficacy of various additional measures to prevent light leakage: gaskets made of rubber matting (0.16 cm thick rubber, 26 × 16 cm) pulled over the minirhizotron flush with the soil surface and extending about 4 cm laterally around the minirhizotron. The outer 2 cm of the rubber shield was punctured with pinholes to improve water infiltration. Finally, a piece of 0.15 cm black plastic was wrapped over the top of the tube, covering the junction between the tube and the cap and just reaching the soil line. The standard treatment (described above) resulted in serious light leaks in 25% of the minirhizotrons. The minirhizotrons with the extra light exclusion treatment were found to have fewer roots in the upper 10 cm compared with the standard treatment. The difference decreased with depth in the profile.

Levan et al. studied the effect of light on root growth in controlled environment rooms and reported that at unshielded windows the root numbers were 54% of those at shielded windows. The inhibitory effect of light on root growth is obvious and is therefore also given as one of the reasons why minirhizotrons usually exhibit fewer roots in the upper soil layers compared with soil coring. Böhm (1979, Sect. 7.3.4) gives more references on the negative effects of light on root growth but concludes that the short time that roots are exposed to the light during recording has no strong influence on the results.

To exclude light *and* solar warming, Majdi et al. (1992) painted the 15 cm of tube which remained above the soil black, and then white. Others (Smit et al. 1994) have put insulation foam inside the tube, to equalise the temperature at the minirhizotron interface.

8.2.4 Observation Equipment

8.2.4.1 Scopes

The scopes developed to inspect interior parts of instruments and human bodies include endoscopes, borescopes and duodenoscopes. They consist of an objective lens, a light source, a pathway (flexible or rigid) from the lens to the eyepiece (e.g. optic fibre) and an eyepiece. The scopes are available in various lengths, and the objective lens can have various magnifications and fields of view (Brown and Upchurch 1987).

8.2.4.2 Video Cameras

In contrast to scopes, modern video camera systems are specifically designed (e.g. by Bartz Technology Corporation, Santa Barbara CA, USA) to be used in minirhizotrons in order to view root development and the activity of organisms and soil conditions non-destructively. Many improvements have been made in resolution, lighting, the indexing handle and the depth registration since the early developments described by e.g. Upchurch and Ritchie (1984).

The latest commercially available developments include:

1. A camera which can change the field of view from 18 mm horizontal to 3.5 mm horizontal, while observing a position on the minirhizotron,
2. A camera system in which a computer moves the camera up and down the tube. The computer can be programmed to repeat runs at any interval. This is useful for the "zoom camera" because the length of the interval between index locations varies, depending on whether the top of the picture must match the bottom edge of the previous picture, and
3. A digital camera enabling the video pictures to be stored in digital format. This makes it easier to arrange individual observations (frames) in a time series (consecutive images of one position on a minirhizotron) and dispenses with the time spent editing and taping (cf. Dubach and Russelle 1995).

The Use of Ultra Violet Light in Root Observation. It has been reported that the roots of many plant species fluoresce (emit light in the blue wavelengths) under UV light, and that there is a relationship between root functionality or age and the intensity of UV fluorescence (Dyer and Brown 1983). Although some modern minirhizotron camera systems are equipped optionally with a UV lighting system, few results obtained with this observation method have been reported. UV-transmission through the minirhizotron material used should be checked.

Wang et al. (1995) investigated whether UV lighting was better than visible light in estimating the proportion of living roots. Their results implied that visible light should be chosen in grass-dominated communities and UV light in communities dominated by broad-leaved herbs or woody plants. Under UV light some dead roots apparently still fluoresced and were easily misidentified as living.

Smit and Zuin (1996) studied two plant species: Brussels sprouts (fast growing, many roots) and leeks (slow growing, low rooting intensity). Leek roots fluoresced more than Brussels sprouts roots. With increasing age, the UV fluorescence decreased in Brussels sprouts roots but increased over time in leek roots. The authors therefore concluded that UV fluorescence could not be used as a universal indicator of root age or root functionality. However, they suggested that the technique might be used to identify the roots from a species in mixed cropping experiments.

8.2.4.3 Telescopes, Mirrors and Still Cameras

An Inexpensive System. Recently, a system for photographing within 8 × 8 cm square minirhizotrons was described by Poelman et al. (1996). The system is based on a telescopic lens instead of an endoscope. The main advantage of this system is that its components are cheap, are usually in stock and can be repaired easily. Additional advantages are that it is portable, shockproof and can be operated by a single person. The equipment consists of a long tubular frame with a mirror angled at 45° at one end and a camera with a telescopic lens at the other end (see Fig. 8.4). Lighting within the system is achieved with an electronic flash for taking photographs and a halogen light for focusing. The strength of this system is its simplicity, as no high-tech parts are involved. The length of observation tubes is limited to approximately 140 cm when using a 600 mm telelens.

Using a camera as described above, Van Noordwijk et al. (1985) obtained a 3-D view of the roots at the minirhizotron surface, using stereoscopic pairs of photographs of a 6-cm-diameter circle on a square 8 × 8 cm Lexan minirhizotron wall, taken every 2.5 cm. They concluded that in most of the photographs the gap between the tube and the soil was considerable (several times the diameter of a root). Most of the roots grow directly along the Lexan tube, forming abundant root hairs. A stereoscopic view of the roots can add important information about the stage of decay of roots, especially when information about root growth dynamics is wanted, and makes it easier to assess the functionality of a root.

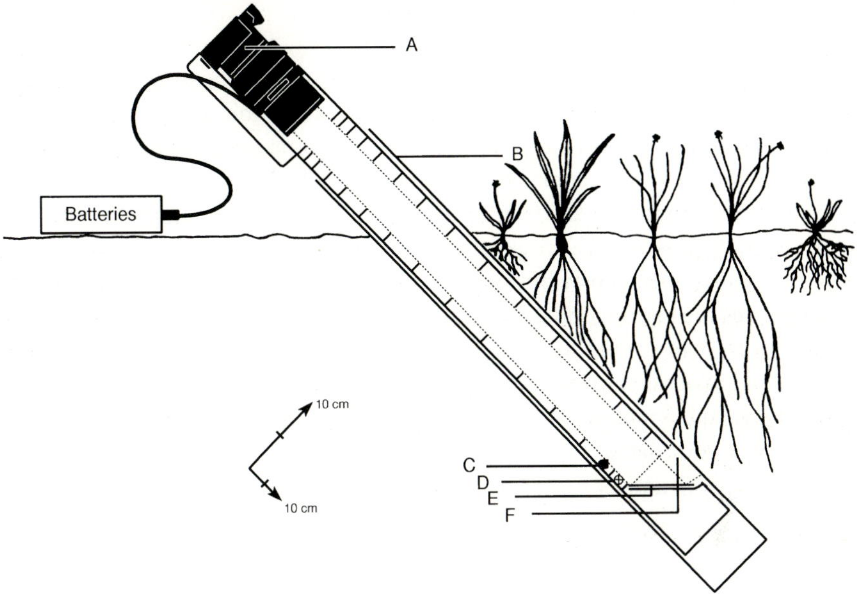

Fig. 8.4. Basic set-up of inexpensive observation equipment for minirhizotrons. *A* camera with telelens, *B* minirhizotron wall, *C* flash, *D* halogen light, *E* mirror, *F* viewing aperture. See Poelman et al. (1996) for a detailed description. (Fig. courtesy Dr. Johan van de Koppel)

8.2.5 Observations with Minirhizotrons

8.2.5.1 *Introduction*

Researchers are interested in responses of roots in the bulk soil. However, only the roots in contact with the minirhizotron surface can be observed with minirhizotrons. Therefore, the root parameters measured at this interface should be *directly* related to the desired bulk soil root parameter, and the influence of the interface on the measured parameter should be minimised (Upchurch 1987). So, the minirhizotron surface must *represent* a transect through the soil rather than a root collection or inhibition zone (Mackie-Dawson and Atkinson 1991). For this reason it is important to be clear about what you want to use the minirhizotron method for:

1. To estimate a bulk soil parameter like volumetric root length density, root distribution etc., or
2. To observe a root characteristic or property such as longevity, diameter, branching, which is visible at the minirhizotron surface.

In both cases, but particularly in the first case, where a translation from 2-D to 3-D is necessary, the effect of the interface is important.

8.2.5.2 Estimation of Bulk Soil Parameters

Root Distribution in the Profile. Minirhizotrons have been widely used to estimate the rooting distribution in the surrounding soil. However, there have been numerous reports of discrepancies when comparing minirhizotron data with soil coring (Bragg et al. 1983; Hansson and Andren 1987; McMichael and Taylor 1987; Hansson et al. 1992; Andren et al. 1993; Heeraman and Juma 1993; Wiesler and Horst 1994). In general, fewer roots are found in the upper soil layers when the root distribution in the soil profile is assessed with minirhizotrons, but the rooting intensity in deeper layers is overestimated. Researchers have resorted to excluding data from the uppermost soil layers in order to achieve a positive correlation between minirhizotron data and soil coring (Heeraman and Juma 1993). Various explanations have been given for this discrepancy in root data:

1. Light leakage effects (Levan et al. 1987),
2. Voids, which allow roots to track along minirhizotrons and encourage deeper rooting (Upchurch and Ritchie 1983; Bragg et al. 1984; Heeraman and Juma 1993),
3. Disrupted hydrology in the soil above an angled minirhizotron (loss of contact with the subsoil; Vos and Groenwold 1987),
4. Loss of structure of the soil caused by the installation of minirhizotrons (Hansson et al. 1992), and
5. Temperature differences (McMichael and Taylor 1987).

The interference of voids with rooting observed with minirhizotrons seems to be the most likely explanation for the discrepancy with coring at deeper layers. A possible explanation for the discrepancy in shallow layers might be the more horizontal direction of root growth at those depths, where roots would be less likely to encounter a minirhizotron. If roots do grow predominantly horizontally, it seems likely that fewer roots will be seen when observing the upper side of a minirhizotron, especially if the minirhizotron has been installed horizontally (Smit et al. 1994). Beyrouty et al. (1987) suggest root growth orientation as a possible explanation for observing less root growth in the upper 9 cm with rice, but they used vertical tubes, which would in fact be preferable for observing horizontal root growth.

Rooting Depth. Several definitions of "rooting depth" exist in the literature:

1. The first root to appear at a certain depth (Bland 1993). However, in the field this criterion is subject to large variation,
2. When a critical level of root number/length per cm^3 or per cm^2 (minirhizotrons) has been exceeded, or
3. The depth at which 95% of the roots (length or mass) are present.

Smit et al. (1996) used data from horizontal minirhizotrons and applied an iterative statistical regression procedure to calculate rooting depth, assuming that rooting increases linearly with (thermal) time over a certain period.

Andren et al. (1993) used the following formula to calculate the *mean rooting depth*, although a better description of this calculation might be *the average depth of a root*:

$$z_m = \sum (N_i \cdot z_i) / \sum N_i, \tag{8.1}$$

in which z_m is depth in cm (average depth of a root), z_i is depth in cm, and N_i is the root number per cm^2 at depth z_i.

Root Length Density and Numbers of Roots. The following observations can be made at the tube interface. However, to estimate values of the surrounding soil conversion procedures are required:

1. Length of roots ($cm\,cm^{-2}$ minirhizotron surface, L_a). The length of the roots on the minirhizotron surface can be calculated by counting the number of intersections on a grid for each observation. The grid may be inscribed on the tube or projected on a recorded image. Using the modified Newman-line-intersect method (Tennant 1976), the number of roots intersecting the grid is then converted to length and can be expressed per cm^2 surface.

   Some researchers (e.g. Heeraman and Juma 1993) have subsequently assumed a certain "depth of view" of 0.2 or 0.3 cm and calculated the volumetric root length density (L_v) directly. Although simple, the assumption about the depth of view seems rather arbitrary, as good contact between the soil and tube would imply a depth of view of 0 cm!

   When assessing this parameter (length) it should be considered whether the length of roots on the minirhizotron surface is influenced by the minirhizotron itself, as roots may grow slower or faster than in the bulk soil. If this is so, root length is not the most appropriate parameter to assess. It has been suggested that the *number of roots arriving* at the surface (Upchurch 1987) or *first intersections* as described by Mackie-Dawson and Atkinson (1991) would show a better relationship with the rooting intensity in the bulk soil.
2. Number of roots on the minirhizotron surface (N_{ra}). This parameter is probably less influenced by the conditions at the interface than length and is independent of any property expressed by the root after that root has touched the minirhizotron. The effect of the interface on results obtained is therefore probably minimal. Upchurch and Ritchie (1983) proposed the following procedure. Given a certain minirhizotron area, all roots that intercept the tube are counted. Each individual root is assigned a count of 1; branched roots

receive a score of 1 for the primary root and an additional score for each branch. The hypothesis is that the density of the roots in the surrounding soil determines the number of arrivals of roots at the interface. This is in contrast to the length of the root at the interface, which is a function of both the arrival of the root and growth of the root after it has touched the minirhizotron.

The minirhizotron root counts can be converted to L_v using either an empirical calibration obtained from the regression of root numbers on length of roots washed out of soil samples, or a theoretical model. Several theoretical models have been developed to predict the relationship between the number of roots on the minirhizotron surface and the root length density in the surrounding soil,

$$L_v = c \cdot N_{ra}. \tag{8.2}$$

The theories, with different assumptions, have resulted in a wide range of values for c (Table 8.1).

Bland and Dugas (1988) compared these c-values in a study with cotton and sorghum and concluded that c = 2 provided the best estimate of L_v for cotton and the data were sufficient to reject the c = 1 model (empirically, c was found to be 2.8). However, no simple relationship between minirhizotron data and measured root length was found for sorghum.

Using horizontal minirhizotrons in the Wageningen Rhizolab, Smit et al. (1994) found that most crops largely followed the c = 2 relationship, when minirhizotron counts had been calibrated with core sampling.

Studies with wheat have yielded c-values of 2.8 (Bragg et al. 1983), 3.5 (Meyer and Barrs 1985) and as large as 20 (Belford and Henderson 1985). Values of c larger than 15 were also found by Murphy et al. (1994) for creeping bentgrass and annual bluegrass turfs; they concluded that a single value of c probably does not apply throughout the growing season. Parker et al. (1991) found likewise for potatoes, as did Smit et al. (1994) when working in the Wageningen Rhizolab with horizontal minirhizotrons where no voids were present. However, the Wageningen researchers found a stable relationship in time for other crops (leeks, wheat, Brussels sprouts). There are reports of the reverse effect, however: Samson and Sinclair

Table 8.1. Range of values for c with relevant authors

Value of c	Reference
1	Upchurch and Ritchie (1983)
2	Melhuish and Lang (1968)
3.3	Upchurch (1985)
3.0–3.8	Merrill and Upchurch (1994)

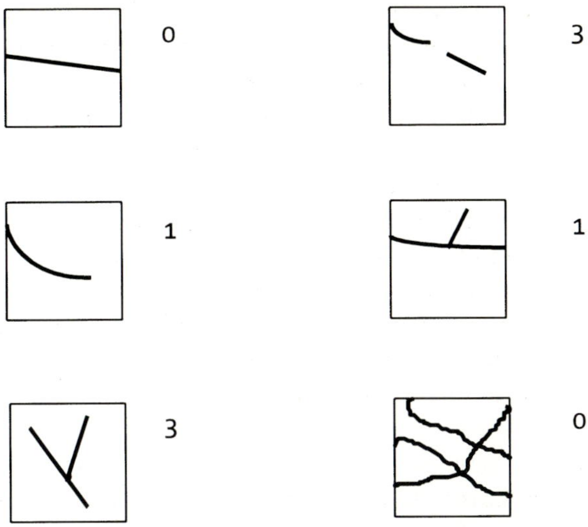

Fig. 8.5. Points of contact on minirhizotron frames, according to Buckland et al. (1993)

(1994) and Wiesler and Horst (1994) found a shift over time for maize and potatoes.

3. *Points of contact.* Instead of counting all root segments (with or without branches), Buckland et al. (1993) proposed counting only the *first and last points of contact* with the minirhizotron (see Fig. 8.5). A root that intercepts a tube, grows away from it, then intercepts it again is considered to have two points of contact, so that it is not necessary to determine whether one root intercepts the tube more than once in this method. A tracking root "passing by" is not counted at all.

   Conversion to root length density is done under the assumptions that: (1) root growth orientation is random, and (2) tubes do not affect the growth of the root except by diverting.

   Volumetric root length density is estimated by:

$$L_v = N/(p\,\pi\, T_r\, T_l) = n/(p\,\pi\, T_r), \tag{8.3}$$

   in which T_l is the total length of tubes in the sample, T_r is the radius of each tube, N is the total count of points of contact of roots with tubes in the sample (comprising first and last points of contact and root tips), and p is the proportion of the tube surface that is counted. $n = N/T_l$ gives the average number of counts per unit length of tube.

   According to Buckland et al. the point of contact method agreed well with measured root length densities in cores for wild cherry roots, but not for roots of pasture species.

Conclusions. Although minirhizotrons are perhaps not the most appropriate tools for estimating root length density in the bulk soil, several researchers have tried to evaluate the success of minirhizotrons by comparing the results with soil coring methods. In general, soil coring is considered as more representative for measuring the distribution and intensity of rooting (Hansson et al. 1992). Heeraman and Juma (1993) concluded that destructive sampling is still the best method to quantify root growth in the top 10 cm of soil. Underestimation of rooting in shallow layers and overestimation at deeper layers is a particular problem when studying the effects of bulk density using minirhizotron methods. De Ruijter et al. (1996) concluded that soil compaction had a major effect on root length in the bulk soil, which was not reflected in the root number observed using minirhizotrons. The most appropriate parameter seems to be to assess the number of root segments or points of contact.

At the moment there is no universal theory to predict the relationship between L_v and N_{ra}. Different models point to different numbers for the conversion coefficient c when converting root number to the root length density of the bulk soil. However, questions remain about the stability of the relationship with time. This means that the method has to be recalibrated for each crop, and for different times of the year, which negates the benefits of using minirhizotrons instead of soil coring.

8.2.5.3 Estimation of Root Morphology and Root Dynamics

Root Properties. The resolution of a typical modern video camera system (VCR) is probably too low to obtain a precise figure for *root diameter* and its variation. After transfer from VCR to PC, a root of 0.01 cm diameter would comprise only 4 pixels, which considerably reduces the accuracy of measuring the diameter with image analysis programs. Better results can be expected using the newest camera systems with zooming facilities (field of view reduced from 18 to 3.5 mm). According to the manufacturer (Bartz Technology Company, USA) the resolution of this camera would allow detailed examination of objects of 3 μm, e.g. infection with mycorrhizal fungi, nematodes and root hairs. Note that when observing these root properties the same prerequisites exist as for root length: conditions at the interface should not differ from the conditions in the bulk soil.

Although *branching* characteristics can easily be assessed with minirhizotrons, we know of no published results. However, because roots cannot be traced into the soil, their order of branching cannot be ascertained.

Meisner (1991) observed and quantified root hair characteristics for groundnut under water stress.

Assessment of Root Directionality. Upchurch et al. (1988) developed a mathematical model, based on probabilistic assumptions about root growth directions, to characterise the dominant root system orientation. By determining the ratio of the number of roots that intercept the top of an angled minirhizotron to the number that intercept the bottom, the model predicts whether the root system is oriented more horizontally or more vertically. Later, Merrill et al. (1994) extended the theory to a situation where upward root growth occurs. After summing the root observations on the upper and lower sides of the minirhizotron per treatment, the ratio between the two sums then gives the root directionality ratios (RD). Merrill et al. gave the theoretical RD ratio for four hypothetical probability distributions of root orientation at minirhizotron installation angles of 5° to 60° from the vertical (See Table 8.2).

Using *vertical* minirhizotrons and an endoscope De Ruijter et al. (1996) counted the number of roots intersecting a horizontal line on the tubes every 2 cm. The roots intersecting a vertical line on the minirhizotrons were counted at eight compass points. The average orientation of the roots, in degrees from the vertical was calculated from the ratio between horizontal and vertical counts. If horizontal and vertical lines are not making a square grid, the horizontal or vertical counts have to be corrected for unequal increments (the authors used a 6 cm o.d. minirhizotron; in this case the horizontal counts were divided by 0.375π).

After correction, the average root angle (α) is then:

$$\alpha = \arctan\left(\frac{\text{vertical counts}}{\text{horizontal counts}}\right). \qquad (8.4)$$

This method gives the exact average orientation of roots when growing directions are distributed symmetrically around one angle, otherwise a small error is made.

Table 8.2. Root directionality ratios (upper side/ lower side) for four hypothetical probability distributions of root orientation at minirhizotron installation angles from 5° to 60°

Minirhizotron angle	Patterns of root directionality distribution			
	Random	→ Increasing downward	→	More vertical
5°	1.04	1.07	1.12	1.22
15	1.13	1.23	1.40	1.79
25	1.22	1.40	1.77	2.66
35	1.31	1.65	2.60	4.07
45	1.39	1.73	3.00	6.51
55	1.46	1.88	4.14	11.34
60	1.49	1.94	5.00	16.76

Rooting Dynamics. Root observations at transparent interfaces enable root dynamics to be studied, because individual roots can be followed through time. However, the main problem is that there are no objective criteria for deciding whether a root is dead or alive. This decision has to be made visually, using morphological characteristics of the root; very often this will be the *colour* of the root (a change from white to grey or black roots). Unfortunately, the change in colour varies with crop and soil conditions, so comparisons are difficult to make. Smit and Zuin (1996) observed that the physical decay of a root is often preceded by a *permanent shrinking* of the root and so used this as an extra indication that the root had ceased functioning.

Keeping track of individual roots is laborious, but so far automatic processing of images with image analysis software has not been entirely satisfactory because roots cannot yet be fully distinguished from the background automatically or the software cannot recognise the physiological condition of the root. Not surprising, even experienced staff have difficulty in processing observations. However, image analysis software can be used to make it easier to assess root dynamics such as mortality and longevity by eye (See also Chap. 10). One such aid is the MSU ROOTS software developed by Enslin et al. (1994). This is a video-digitising program with advanced capabilities for capturing, displaying, and analysing minirhizotron images of plant root systems and creating a database for the study of plant root dynamics. After capturing an image, an operator can digitise the outline of individual roots or entities. The program can be used to identify and measure the length of individual roots within cohorts at particular sites, while allowing for comparisons between same site identifications and measurements over time.

Visual comparisons can be made between data digitised earlier and existing root images. Furthermore, a root identification number, physiological condition code and branch order number can be assigned. The *MSU ROOTS* program allows the user to perform additional analytical operations using database management systems like FoxPro and DBaseIV.

Smit and Zuin (1996) used the *NIH-Image* program on a Macintosh computer to assess root dynamics in Brussels sprouts and leeks. This is a public domain image analysis program written by Wayne Rasband at the US National Institute of Health and available from the Internet. It has also been used to capture the images from a VCR, to make time series, to align images and to measure length, grey value etc. from root segments in time. Also available for Macintosh computers and specifically developed for processing minirhizotron images is *RooTracker* (Duke University, USA).

Two approaches that can be used to quantify root dynamics in terms of root longevity are:

BOX 8.1. Calculations on Root Dynamics

If the fate of individual roots is followed it is possible to graph total roots produced in time (expressed in length, numbers or mass) and roots which have decayed versus time. Smit and Zuin (1996) regressed the root data on time with an ordinary logistic model of the form:

$$y_{n,t} \text{ and } y_{d,t} = a + \frac{c}{1 + e^{(-b(t-m))}},$$

where y denotes the total amount of new roots ($y_{n,t}$) and dead roots ($y_{d,t}$) at time t. The sum $a + c$ is the total amount to be reached ultimately, b is a shape factor, and the value of m is the date at which 50% of the roots are produced.

The *standing root intensity*, the amount of roots available at time t can be calculated as:

$$y_{s,t} = y_{n,t} - y_{d,t}.$$

This approach is applicable for annual crops, but for perennial crops and for forestry where some roots are already present at the start of the observation period, a variation has been proposed (Van Noordwijk 1987), i.e.:

The total amount of roots ($y_{t,t}$) present at time t, is:

$$y_{t,t} = y_{n,t} - y_{I,t},$$

in which $y_{I,t}$ is the amount of roots present at the beginning of the observation period.

In this case the *standing living root intensity* is:

$$y_{s,t} = y_{t,t} - y_{d,t}.$$

At the end of the season at time = e, data are obtained for $y_{n,e}$, $y_{t,e}$, $y_{d,e}$, and $y_{s,e}$. From these data two *ratios* can be defined which may summarise the root dynamics throughout the growing season:

$$T = y_{d,e} / y_{t,e},$$

where T is the turnover of root length/number during a growing season or a set time period (–)

and

$$R = y_{n,e} / y_{s,e},$$

where R is the root replacement ratio during a growing season or a set time period (–).

1. To follow cohorts of roots in time, as explained in detail in Chapter 11, see also Hendrick and Pregitzer (1992a,b, 1993) and Cheng et al. (1990, 1991), and
2. To use the cumulative "production" curves of new and dead roots (length, number), e.g. Rygiewicz et al. (1997), see Box 8.1. See Box 8.2 for some considerations of subsampling minirhizotrons.

BOX 8.2. How or when to Subsample with Minirhizotrons
Minirhizotrons are particularly promising tools for assessing root turnover, but the large number of individual observations suggest that much labour is required. This raises the question of how or when to subsample. Dubach and Russelle (1995) described that with their system (a VCR and horizontal minirhizotrons) each replicate of depth (a total of 80 images in one horizontal tube) required 29 h of labour. But a large part of the processing time (greater than two thirds) is needed for editing tapes (making a time-series from the individual tapes from different observation dates). In the near future the time needed for this editing will probably be considerably reduced, thanks to the availability of robust, read-write optical disks that can record images digitally, together with an identification code as the file name. Dubach and Russelle (1995) also concluded further that accuracy and precision of results were not decreased by using only every second image along both sides of the tube (40 images), but that precision declined when either only one side of the tube or only 20 consecutive images on both sides were used. Recording less frequently reduced labour requirements even more: recording intervals could be lengthened to 2 weeks for shallow depths and even longer for greater depths without sacrificing accuracy or precision. These changes led to a 70% saving in the labour required for estimating root turnover.

8.3 Root Windows

Root windows are transparent viewing planes installed in soils, allowing the non-destructive monitoring of root growth and detailed studies of individual roots and of rhizosphere soil. In contrast to minirhizotrons (see Sect. 8.2), the viewing area is usually much larger and not curved. In addition to viewing roots, roots and soil can be assessed, so that small-scale manipulations and samplings are possible in addition to continuous measurements. Root windows are much cheaper to install than rhizotron facilities, and allow roots and soils to be observed in situ. They have to be installed with great care, to avoid water or light penetration into the window-soil interface.

8.3.1 Introduction

The idea of viewing roots growing in field or forest soil behind a transparent panel is quite old. In most early experiments, glass plates were installed against vertical soil profiles. Horizontal glass plates were also used to observe root growth in forest topsoil (McDougall 1916). Root windows allowed the study of in situ root growth periodicity in perennial plant species, for example in fruit trees (Head 1966), forest trees (McDougall 1916), shrubs and bushes (Fordham 1972; Fernandez and Caldwell 1975), in grassland plant communities (Speidel and Weiss 1974; Hayes and Seastedt 1987) or in crop plants (Pearson and Lund 1968). Root windows allow a continuous and long-term observation of roots and soil with minimum disturbance. Non-destructive techniques can be combined with coring techniques to increase their validity by helping to assess the best time to core. The seasonal pattern of root growth and the periods of minimal and maximal root growth can be determined in root windows (Vogt et al. 1986).

Non-destructive techniques are also important when dynamic changes of roots in response to the environment are to be studied (McMichael et al. 1992). Because their field of observation is larger than in minirhizotrons, root windows are especially suitable for studying roots in particular soil microsites, interactions with soil micro-organisms (Egli and Kälin 1991), and soil fauna (Wilson et al. 1995). They enable results obtained under controlled conditions by using soil-filled pots or boxes with glass or perspex front plates for root growth studies (for example, LeNoble et al. 1996) to be verified under field conditions.

At present, root windows are not widely used. However, the need to understand in situ root-soil interactions and to study effects of soil micro-organisms on root activity rather than "average" root biomass or root length at a site will foster interest in this inexpensive technique. Root growth along the transparent panel is, of course, not directly comparable to root growth in bulk soil, because three-dimensional root exploration is reduced to two-dimensional growth. The situation at the panel may be similar to, for example, the surface of a smooth flintstone (Kolesnikov 1971; Böhm 1979).

8.3.2 Materials

The same materials used for minirhizotrons (see Sect. 8.2.2.1) or permanent rhizotrons may be used for root windows, but Plexiglas (Häussling et al. 1985,

1991) or glass (Rutherford and Curran 1981; Egli and Kälin 1991) are most common. The disadvantages of the individual materials have not been tested thoroughly. Assumptions that plastic windows induce irregular conditions by electric charges (Voorhees 1976; Rutherford and Curran 1981) have not been substantiated (Reid et al. 1992), but close contact between soil and the plastic plate may be more difficult to obtain with Plexiglas than with glass plates (Taylor and Böhm 1976; Huck and Taylor 1980). The contact between soil and plate must be close for meaningful observations at the window.

Root windows can vary in area according to the specific needs of the investigations. Individual windows for observational purposes are usually as large as 1–3 m^2 (Pearson and Lund 1968; Fordham 1972; Teskey and Hinckley 1981; Tscherning et al. 1995). "Observation pits" (Kolesnikov 1971) or root windows in a size range of 0.7–0.8 m^2 are more suitable for replicated observations and for the study of treatment effects (Egli and Kälin 1991; Häussling et al. 1991). Smaller windows are often in better contact with the soil than larger planes (Rutherford and Curran 1981) and are easier to replicate.

8.3.3 Installation

This description follows the method developed by Häussling et al. (1985, 1991). A similar method is described by Egli and Kälin (1991). For earlier experiences with larger root windows see, for example, Kolesnikov (1971), Böhm (1979), and Rutherford and Curran (1981).

A 50 cm × 40 cm × 0.6 cm Plexiglas plate is put into a laboratory oven. Two sides of the plate are placed on rods and a heavy weight is laid longitudinally in the centre of the plate. The oven is heated to approximately 60 °C and until the plate has become slightly concave. It retains its shape after being removed from the oven and allowed to cool. The root window is then installed by first hammering a straight sharp-edged steel plate (approximately 54 cm wide) into the soil. The steel plate is rammed into the soil at a specific angle (for example, 45 or 60° from the vertical). To obtain a smooth soil profile, the steel plate must be heavy and hammered into soil without deflections. Root windows can also be placed nearly vertical (Böhm 1979) or even with a negative slope (Huck and Taylor 1980). The preferable angle of installation depends on the plant species under investigation, the soil type, and the techniques used for observing root growth at the window.

After the steel plate has been installed, a pit of approximately 55 cm × 55 cm is then dug in front of the steel plate to the depth of installation (see Fig. 8.6). The excavated soil must be removed from the site, to minimise local disturbances of soil chemistry. The soil directly behind the window may not be walked on, as otherwise the physical conditions for root growth may be

Fig. 8.6. A steel plate has been rammed into the soil and a pit has been dug in front of it

altered. The steel plate is then carefully removed from the soil profile. No top soil, not even a small amount, may be transferred down the profile during the installation. If necessary, the soil profile may be smoothed by gently rubbing it horizontally from side to side with dry paper. Any small voids in the profile (for example, from small stones or cutting of woody roots during installation) can be filled with soil from a corresponding soil depth. Large voids, however, cannot usually be filled adequately, and if there are any, the window must be relocated.

The concave-shaped plate is tightly placed with its convex side snugly against the soil profile. If required, a sheet of clear hard plastic can be inserted between the soil profile and the plate, to minimise root damage when the plate is later removed for destructive measurements. To the left and right of the plates, wooden sticks are hammered deep into soil, perpendicular to the plate, so that wedges can be pressed between them and the Plexiglas plate (Fig. 8.7). With the help of the wedges, the sides of the plate are pressed tightly against the soil profile (Fig. 8.8). Because of the concave shape, the Plexiglas is also in close contact with the soil in the centre of the plate. In some soils, it is necessary to press down the wedges further several months or years after installation, in order to maintain the close contact between the Plexiglas plate and the soil profile. For example, soil movement can be appreciable in soils with montmorillonitic clay (Huck and Taylor 1980).

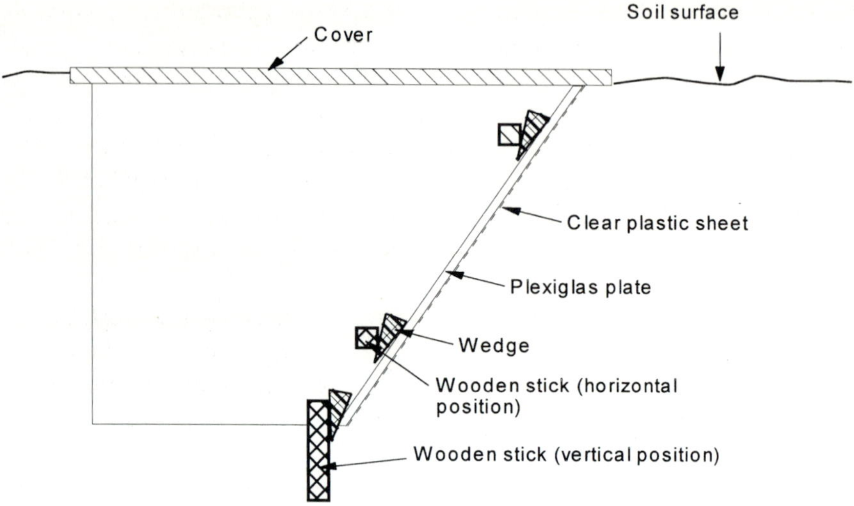

Fig. 8.7. Schematic drawing of a root window installed into soil

Fig. 8.8. The root window (Plexiglas plate) is pressed tightly against the soil profile

Alternatively, jacks can be used to press the plate against the soil profile (Schuurman and Goedewaagen 1971). However, this is more expensive and the jacks reduce the free viewing area of the window. Elaborate systems with spring-loaded windows (Rutherford and Curran 1981) or support frames (Kolesnikov 1971) also bring about a tight soil/plate contact, but require a relatively high technical investment. Large nails or pins are often used to fix the plates directly onto the soil profile, but this may lead to variable soil contact (Böhm 1979) and be appropriate for only a relatively short duration, some soil types, and in certain moisture conditions.

After installation, the soil profile and the pit must be insulated against light and against temperature fluctuations different from the bulk soil (Fig. 8.8).

This can be done by covering the pit with a block of commercially available insulation material (e.g. Styrodur). The block must completely cover the pit to prevent light entering but should not be excessively large. Other materials used for insulation include thick spun glass stapled to plywood covers (Keyes and Grier 1981), expanded polyurethane (Rutherford and Curran 1981), expanded polystyrene (Reid et al. 1992), grass mulch (Fordham 1972), Styrofoam boards (Tscherning et al. 1995), Styrofoam covered with burlap (Hayes and Seastedt 1987) or Styrofoam chips and black plastic foil (Egli and Kälin 1991).

New roots will appear on the window surface soon after installation, depending on vegetation type and time of year. A high number of new roots may originate from old roots cut during installation (for instance when windows are installed into an established forest area or in grassland; Speidel and Weiss 1974). In any case, some time should be allowed after installation before measurements begin. When roots start to appear at the window, it can be seen whether they have grown in soil and are developing their own rhizosphere, or have grown in an air space between soil and the window plate. The window should not be used if the latter situation applies to many roots, because air spaces not only provide room for uninhibited root growth but also allow soil solution to flow freely from the topsoil to deeper soil layers.

In an alternative approach, the plate is fixed a few millimetres from the original profile, and sieved material from the appropriate soil depths is used to reconstruct a soil profile directly behind the window surface (Fordham 1972; Hayes and Seastedt 1987; Reid et al. 1992). Though perhaps providing better control over experimental conditions (Huck and Taylor 1980) this may, of course, lead to unrepresentative root growth in the reconstructed profile (Durrant et al. 1973; Rutherford and Curran 1981). Furthermore, the spatially distinct arrangement of, for example, soil nutrients is lost in this case, as is the original physical condition of the soil. For these reasons, the use of reconstructed soil should be avoided (Zillmann 1956), except when it is impossible to maintain good plate/soil contact with the original soil profile.

8.3.4 Placement

In many cases, root windows have been used to determine root growth dynamics of plant species in relation to their shoot growth cycles. Observations from a few or even a single root window were considered adequate for such basic information. A large number of replicated windows are necessary for more detailed studies or for comparisons of sites or plots. Four root windows per plot were used by Keyes and Grier (1981), whereas Häussling et al. (1991) used ten windows per plot in a Norway spruce forest. Hahn and Marschner (1998) installed 16 windows per plot. Window-to-window variation is greater for some parameters (e.g. total root length) than for others (e.g. weekly root length increment). The number of root windows required per plot therefore depends on the parameter measured, but is probably not much less than in comparable studies with minirhizotrons.

The location of the windows is an important consideration in row crops or in studies with large perennial plants. In forests, the windows are usually placed at a certain distance from and orientation to the stem of a tree (Egli and Kälin 1991; Häussling et al. 1991). Such non-random placement of windows reduces the data variation, but results cannot be extrapolated to the total plot area. Windows placed in different positions with respect to plant rows or individual plants enable the spatial variation of, for example, root growth dynamics or mycorrhizal colonisation of roots to be studied.

8.3.5 Observation Equipment

Almost all measurements that can be taken in permanent rhizotrons (Huck and Taylor 1980) can also be carried out on site at root windows. For example, soil moisture and soil temperature can be measured at different depths and at different distances behind the root window (Fordham 1972; Fernandez and Caldwell 1975; Rutherford and Curran 1981; Teskey and Hinckley 1981). Sensors can easily be installed into the soil through small holes drilled into the plastic windows. These measurements have usually shown that temperature and moisture conditions behind the plate do not differ distinctly from those in the bulk soil (Kolesnikov 1971; Fernandez and Caldwell 1975; Böhm 1979) if the window is well insulated. Additional measurements could include soil O_2 and CO_2 concentrations and soil pH (Taylor et al. 1990; Reid et al. 1992). Novel methodology to sample and analyse very small volumes of soil solution (Göttlein et al. 1996) can be applied to root windows to study ionic concentrations in the rhizosphere and bulk soil solution under natural conditions.

Destructive samplings or measurements can be carried out in smaller additional windows (Rutherford and Curran 1981) or after briefly removing the plate from the soil profile (Häussling et al. 1991). It is possible to remove the plate if the windows are properly constructed, but this interferes with measurements of root growth dynamics because some roots may be affected or even displaced during the removal.

Root length along the plate is usually traced with pens onto acetate sheets placed on the window plane. Different coloured pens can be used to mark new root growth in different time periods (Fordham 1972; Belgrand et al. 1987; Pagès and Serra 1994; Wilson et al. 1995), different root size classes (Neufeld et al. 1989), brown and white roots (Reid et al. 1992) or roots of different plant species

(Tscherning et al. 1995). During the measurements, exposure of roots to light should be minimised. The line lengths on the acetate sheets can later be determined with a grid method, a scanner (Tscherning et al. 1995), or with a digital image analysis system. These techniques can also be used to determine root branching pattern and topology (Belgrand et al. 1987; Colin-Belgrand et al. 1989; Pagès and Serra 1994). A direct scanning system of roots at the transparent interface has been described by Pan et al. (1998).

Depending on plant species and soil type, not all fine roots may be discernible by eye. Microscopic assessment of these roots may be necessary (Rutherford and Curran 1981). Microscopic observations are also helpful or even essential when studying the development of ectomycorrhizal root tips (Egli and Kälin 1991). A hand-lens may be sufficient to study meso- and macroinvertebrate fauna (Wilson et al. 1995). Repeated photography (Wilson et al. 1995) or video imaging of window sections at different size scales can document temporal changes in root and soil properties, similar to in minirhizotron tubes.

8.3.6 Applications

8.3.6.1 Root Morphology, Growth, and Root Turnover

Root length and the number of roots can be measured at the window plane. This results in comparative data sets of, for example, cultivars or species with different drought resistance (Fernandez and Caldwell 1975) or plants in plots with different fertiliser levels (Speidel and Weiss 1974), or plants in competition experiments. This may often be sufficient to test the original hypothesis of an experiment (Wilson et al. 1995). As with minirhizotron data, measurements can be converted into values for bulk soil root density and root numbers or even root biomass (see Sect. 8.2.5.2). These conversions, however, are based on assumptions or empirical calibrations and are usually only of some predictive value (Glinski et al. 1993). No interface method can substitute for destructive soil sampling (see Chap. 6) in the accurate assessment of mean root length densities in the bulk soil. Root windows are better suited than other field methods to determine root branching patterns (Fernandez and Caldwell 1975; Sword 1998). The measurements are similar to those in rhizoboxes in laboratory studies (Glinski et al. 1993; Pagès and Serra 1994; LeNoble et al. 1996), but have the additional value of being carried out on site.

The main advantage of the non-destructive techniques (minirhizotrons and root windows) is, of course, that dynamic changes of root appearance and disappearance, root length and root numbers can be observed. The time between observations is variable, but 2-week intervals are often used (Speidel and Weiss

1974; Fernandez and Caldwell 1975; Keyes and Grier 1981; Häussling et al. 1991; Wilson et al. 1995). The number of roots at the plane may be a more appropriate parameter than root length to measure rooting density (see Sect. 8.2.5.2). However, both parameters may be important for dynamic measurements. For example, in a white oak forest Teskey and Hinckley (1981) observed different peaks during the season for the number of roots and for root elongation. With the help of root windows it can be shown that root elongation in temperate forests is highly positively correlated with soil temperature, except in dry soil (McDougall 1916), but the production of new root tips is correlated with shoot growth events and peaks in spring and autumn. For measurement of root distribution in the soil profile, the number of points in contact with the plate should be assessed (Fernandez and Caldwell 1975), because roots are deflected after their first contact with the plate.

Observations on the appearance and disappearance of new roots can also be used to estimate root turnover (Keyes and Grier 1981; Wilson et al. 1995). It has to be kept in mind, however, that roots at the interface for a relatively long period after installation are comparatively young and not representative of the total root system including older roots (Hayes and Seastedt 1987). Rates of root longevity measured at the interface cannot therefore be used uncritically to calculate turnover of the total root biomass. However, root longevity and root turnover can be assessed with root windows individually for different root types and in relation to measurements of the soil environment.

8.3.6.2 Soil and Rhizosphere Measurements

Soil conditions can be tested non-destructively by inserting the appropriate sensors through the Plexiglas plate. The large spatial heterogeneity over a few centimetres in, for example, soil water potential and NO_3^- concentration or pH in the soil solution can be shown convincingly (Göttlein et al. 1996). Studies carried out conventionally in rhizoboxes or rhizotrons could also be set up with root windows: for example, the study of effects of soil moisture gradients on root growth (Zillmann 1956; Durrant et al. 1973) and mortality (Kosola and Eissenstat 1994). The consequences of high concentrations of mineral elements, pesticides or organochemicals in certain soil zones can also be observed directly (Häussling et al. 1985, 1991; Blaker and MacDonald 1986; Wilson et al. 1995). Processes such as nodulation or pest attacks (Klepper and Kaspar 1994) can be observed on site.

A further advantage of root windows in forests is that the seasonal pattern of mycorrhiza formation and the life span of individual mycorrhizal types can be determined (Egli and Kälin 1991). Small-scale natural disturbances (soil mixing by animals, root severance, local changes of pH) may create additional habitats for mycorrhizal fungi and thus contribute to a high fungal diversity (Bruns 1995). Such hypotheses can be tested with root windows.

It is also possible to manipulate soil moisture and nutrient supply. For example, Larigauderie and Richards (1994) injected small amounts of nutrients into small soil areas. Subsequently, they observed the effect of the local nutrient enrichment on root growth of several arid-land grass species (Larigauderie and Richards 1994).

When the Plexiglas plate is taken off, filter papers or agar sheets can be placed directly on the roots. This allows for the quantification of, for example, soil pH changes in the rhizosphere (Häussling et al. 1985, 1991). Other possible measurements include the determination of phosphatase activity, manganese oxidation or release of phenolics in the rhizosphere (Dinkelaker and Marschner 1992; Dinkelaker et al. 1993).

8.4 Root Screens

Root screens are nylon mesh planes inserted into soil. Some time after installation, the number of roots crossing the mesh is evaluated. Root screens are a new and inexpensive technique for studying rooting intensity in field sites.

To install a root screen, a sharp blade is used to cut a slit into soil, usually at an angle of 45°. Fibreglass screens are then fitted into the slit at different soil depths, the blade removed from soil and the soil faces pressed firmly against the screen. The screens are 5 to 10 cm wide and have a mesh size of 1 mm or more (Fahey and Hughes 1994; Parsons et al. 1994). The sites of screen installation must be marked for later retrieval. In forest soils, screens can also be placed horizontally to investigate roots in the soil surface layer. These screens allow root growth to be observed non-destructively (Fahey and Hughes 1994).

To remove the screens, soil must be lifted from the upward face of the screen. The number of roots penetrating the screen is counted at the site or later in the laboratory. Live and dead roots may be distinguished by visual criteria. After repeated samplings, newly produced root length and even biomass can be calculated from these counts (Fahey and Hughes 1994). As with other interface methods, these calculations require the use of several conversion factors.

Similar to the ingrowth core method (see Chap. 6), the root system is disturbed during installation in plant stands. This may affect root productivity. In contrast to ingrowth cores, no foreign or sieved material needs to be used as a root growth substrate in the root screen method. However, it may be difficult to use root screens in dry soil or at high soil bulk density (Parsons et al. 1994).

The main advantage of root screens is that they are relatively quickly installed and evaluated. This allows observations with a high number of replications at different times of the growing season or in plots with different treatments. The effect of small-scale fertilisation near the screen on root growth can also be studied (Fahey and Hughes 1994). So far, root screens have been used in temperate forests only, but may well be applicable to root studies in agro-ecosystems.

Further Reading

Minirhizotron observation tubes: methods and applications for measuring rhizosphere dynamics. ASA Spec Publ 50: 99–108

Plant Soil 185: 225–258. Selected papers of the minirhizotron workshop, Sweden 17–20 Sept, 1995

References

Andren O, Hansson AC, Vegh K (1993) Barley nutrient uptake, root growth and depth distribution in two soil types in a rhizotron with vertical and horizontal minirhizotrons. Swed J Agric Res 23 (3): 115–126

Bates GH (1937) A device for the observation of root growth in the soil. Nature (Lond) 139: 966–967

Belford RK, Henderson FKG (1985) Measurement of the growth of wheat roots using a TV camera system in the field. NATO Adv Study Inst Ser Ser A Life Sci 86: 99–105

Belgrand M, Dreyer E, Joannes H, Velter C, Scuiller I (1987) A semi-automated data processing system for root growth analysis: application to a growing oak seedling. Tree Physiol 3: 393–404

Beyrouty CA, Wells BR, Norman RJ, Marvel JN, Pillow JA Jr (1987) Characterization of rice roots using a minirhizotron technique. Minirhizotron observation tubes: Methods and applications for measuring rhizosphere dynamics. ASA Spec Publ 50: 99–108

Blaker NS, MacDonald JD (1986) The role of salinity in the development of *Phytophthora* root rot of citrus. Phytopathology 76: 970–975

Bland WL (1993) Cotton and soybean root system growth in 3 soil temperature regimes. Agron J 85: 906–911

Bland WL, Dugas WA (1988) Root length density from minirhizotron observations. Agron J 80: 2, 271–275

Böhm, W (1979) Methods of studying root systems. Ecological Studies 33. Springer, Berlin Heidelberg New York, 188 pp

Box JE, Smucker AJM, Ritchie JT (1989) Minirhizotron installation techniques for investigating root responses to drought and oxygen stresses. Soil Sci Soc Am J 53: 115–118

Bragg PL, Govi G, Cannell RQ (1983) A comparison of methods, included angled and vertical minirhizotrons, for studying root growth and distribution in a spring oat crop. Plant Soil 73: 435–440

Brown DA (1984) Characterization of soybean root distribution. Arkansas Farm Res 33: 3

Brown DA, Upchurch DR (1987) Minirhizotrons: a summary of methods and instruments in current use. Minirhizotron observation tubes: Methods and applications for measuring rhizosphere dynamics. ASA Spec Publ 50: 15–30

Bruns TD (1995) Thoughts on the processes that maintain local species diversity of ectomycorrhizal fungi. Plant Soil 170: 63–73

Buckland ST, Campbell CD, Mackie-Dawson LA, Horgan GW, Duff EI (1993) A method for counting roots observed in minirhizotrons and their theoretical conversion to root length density. Plant Soil 153 (1): 1–9

Campbell CD, Mackie-Dawson LA, Reid EJ, Pratt SM, Duff EI, Buckland ST (1994) Manual recording of minirhizotron data and its application to study the effect of herbicide and nitrogen fertiliser on tree and pasture root growth in a silvopastoral system. Agrofor Syst 26 (2): 75–87

Cheng W, Coleman DC, Box JE Jr (1990) Root dynamics, production and distribution in agroecosystems on the Georgia Piedmont using minirhizotrons. J Appl Ecol 27: 2, 592–604

Cheng W, Coleman DC, Box JE Jr (1991) Measuring root turnover using the minirhizotron technique. Agricult Ecosyst Environ 34: 261–267

Christakos G (1992) Random field models in earth sciences. Academic Press, New York

Colin-Belgrand M, Joannes H, Dreyer E, Pages L (1989) A new data processing system for root growth and ramification analysis: description of methods. Ann Sci For 46 Suppl 305s–309s

De Ruijter F, Veen B, Van Oijen M (1996) A comparison of soil core sampling and minirhizotrons to quantify root development of field-grown potatoes. Plant Soil 182 (2): 301–312

Dinkelaker B, Marschner H (1992) In vivo demonstration of acid phosphatase activity in the rhizosphere of soil-grown plants. Plant Soil 144: 199–205

Dinkelaker B, Hahn G, Marschner H (1993) Non-destructive methods for demonstrating chemical changes in the rhizosphere. II. Application of methods. Plant Soil 155/156: 71–74

Dubach M, Russelle MP (1995) Reducing the cost of estimating root turnover with horizontally installed minirhizotrons. Agron J 87 (2): 258–263

Durrant MJ, Love BJG, Messem AB, Draycott AP (1973) Growth of crop roots in relation to soil moisture extraction. Ann Appl Biol 74: 387–394

Dyer D, Brown DA (1983) Relationship of fluorescent intensity on ion uptake and elongation rate of soybean roots. Plant Soil 72: 127–134

Egli S, Kälin I (1991) Root window technique for in vivo observation of ectomycorrhiza on forest trees. In: Norris JR, Read DJ, Varma AK (eds) Techniques for the study of mycorrhiza. Methods in microbiology, vol 23. Academic Press, London, pp 423–433

Enslin WR, Pregitzer KS, Hendrick RL (1994) MSU ROOTs: a PC-based program to quantify plant roots. Center for Remote Sensing, Michigan State Univerity, East Lansing

Fahey TJ, Hughes JW (1994) Fine root dynamics in a northern hardwood forest ecosystem, Hubbard Brook Experimental Forest, NH. J Ecol 82: 533–548

Ferguson JC, Smucker AJM (1989) Modifications of the minirhizotron video camera system for measuring spatial and temporal root dynamics. Soil Sci Soc Am J 53: 1601–1605

Fernandez OA, Caldwell MM (1975) Phenology and dynamics of root growth of three cool semi-desert shrubs under field conditions. J Ecol 63: 703–714

Fordham R (1972) Observations on the growth of roots and shoots of tea (*Camellia sinensis*, L.) in southern Malawi. J Hortic Sci 47: 221–229

Gijsman AJ, Floris J, Van Noordwijk M, Brouwer M (1991) An inflatable minirhizotron system for root observations with improved soil/tube contact. Plant Soil 134 (2): 261–270

Glinski DS, Karnok KJ, Carrow RN (1993) Comparison of reporting methods for root growth data from transparent-interface measurements. Crop Sci 33: 310–314

Göttlein A, Hell U, Blasek R (1996) A system for microscale tensiometry and lysimetry. Geoderma 69: 147–156

Goulard M (1989) Inference in a coregionalization model. In: Armstrong M (ed) Geostatistics, vol 1. Kluwer, Dordrecht, pp 397–408

Goulard M, Voltz M (1992) Linear coregionalization model: tools for estimation and choice of cross-variogram matrix. Math Geol 24 (3): 269–286

Gregory PJ (1979) A periscope method for observing root growth and distribution in field soil. J Exp Bot 30: 205–214

Hahn G, Marschner H (1998) Effect of acid irrigation and liming on root growth of Norway spruce. Plant Soil 199: 11–22

Hansson AC, Andren O (1987) Root dynamics in barley, lucerne and meadow fescue investigated with a minirhizotron technique. Plant Soil 103: 33–38

Hansson AC, Steen E, Andren O (1992) Root growth of daily irrigated and fertilized barley. Investigation with ingrowth cores, soil cores and minirhizotrons. Swed J Agric Res 22 (4): 141–152

Häussling M, Leisen E, Marschner H, Römheld V (1985) An improved method for non-destructive measurements of the pH at the root-soil interface (rhizosphere). J Plant Physiol 117: 371–375

Häussling M, George E, Lorenz K, Kreutzer K, Marschner H (1991) Einfluss von saurer Beregnung auf Wachstum von Langwurzeln und pH-Werte in der Rhizosphäre von Fichten im Versuch Höglwald. In: Kreutzer K, Göttlein A (eds) Ökosystem-forschung Höglwald: Beiträge zur Wirkung von saurer Beregnung und Kalkung in einem Fichtenbestand. Forstwissenschaftliche Forschungen, Heft 39. Parey, Hamburg, pp 44–48

Hayes DC, Seastedt TR (1987) Root dynamics of tallgrass prairie in wet and dry years. Can J Bot 65: 787–791

Head GC (1966) Estimating seasonal changes in the quantity of white unsuberized root on fruit trees. J Hortic Sci 41: 197–206

Heeraman DA, Juma NG (1993) A comparison of minirhizotron, core and monolith methods for quantifying barley (*Hordeum vulgare* L.) and faba bean (*Vicia faba* L.) root distribution. Plant Soil 148: 29–41

Hendrick RL, Pregitzer KS (1992a) Spatial variation in tree root distribution and growth associated with minirhizotrons. Plant Soil 143 (2): 283–288

Hendrick RL, Pregitzer KS (1992b) The demography of fine roots in a northern hardwood forest. Ecology 73 (3): 1094–1104

Hendrick RL, Pregitzer KS (1993) Patterns of fine root mortality in two sugar maple forests. Nature 361: 59–61

Horgan GW, Buckland ST, Mackie-Dawson LA (1993) Estimating three-dimensional line process densities from tube counts. Biometrics 49 (3): 899–906

Huck MG, Taylor HM (1980) The rhizotron as a tool for root research. Adv Agron 35: 1–35

Hummel JW, Levan MA, Sudduth KA (1989) Minirhizotron installation in heavy soils. Trans ASAE 32: 3, 770–776; Presented as ASAE Paper No 87-1524

Itoh S (1985) In situ measurement of rooting density by micro-rhizotron. Soil Sci Plant Nutr 31 (4): 653–656

Keyes MR, Grier CC (1981) Above- and below-ground net production in 40-year-old Douglas fir stands on low and high productivity sites. Can J For Res 11: 599–605

Klepper B, Kaspar TC (1994) Rhizotrons: their development and use in agricultural research. Agron J 86: 745–753

Kloeppel BD, Gower ST (1995) Construction and installation of acrylic minirhizotron tubes in forest ecosystems. Soil Sci Soc Am J 59 (1): 241–243

Kolesnikov VA (1971) The root system of fruit plants. Mir Publishers, Moscow, 269 pp

Kosola KR, Eissenstat DM (1994) The fate of surface roots of citrus seedlings in dry soil. J Exp Bot 45: 1639–1645

Larigauderie A, Richards JH (1994) Root proliferation characteristics of seven perennial arid-land grasses in nutrient-enriched microsites. Oecologia 99: 102–111

LeNoble ME, Blevins DG, Miles RJ (1996) Prevention of aluminium toxicity with supplemental boron. II. Stimulation of root growth in an acidic, high-aluminium subsoil. Plant Cell Environ 19: 1143–1148

Levan MA, Ycas JW, Hummel JW, Taylor HM (1987) Light leak effects on near-surface soyabean rooting observed with minirhizotrons. Minirhizotron observation tubes: Methods and applications for measuring rhizosphere dynamics. ASA Spec Publ 50: 89–98

Mackie-Dawson LA, Atkinson D (1991) Methodology for the study of roots in field experiments and the interpretation of results. In: Atkinson D (ed) Plant root growth: an ecological perspective. Blackwell, Oxford, pp 25–47

Mackie-Dawson LA, Buckland ST, Duff EI, Pratt SM, Reid EJ, Millard P (1989) The use of in situ techniques for the investigation of root growth. Asp Appl Biol 22: 349–356

Majdi H, Nylund JE (1996) Does liquid fertilization affect fine root dynamics and life span of mycorrhizal short roots? Plant Soil 185: 305–309

Majdi H, Smucker AJM, Persson H (1992) A comparison between minirhizotron and monolith sampling methods for measuring root growth of maize (*Zea mays* L.). Plant Soil 147: 127–134

McDougall WB (1916) The growth of forest tree roots. Am J Bot 3: 384–392

McMichael BL, Taylor HM (1987) Applications and limitations of rhizotrons and minirhizotrons. In: Minirhizotron Observation Tubes: Methods and applications for measuring rhizosphere dynamics. ASA Spec Publ 50: 1–13

McMichael BL, Upchurch DR, Taylor HM (1992) Transparent wall techniques for studying root growth and function in soil. J Plant Nutr 15: 753–762

Meisner CA (1991) Peanut roots, shoot, and yield under water stress. Diss Abstr Int B Sci Eng 52: 1, 3B–4B; Abstr Thesis, University of Georgia, USA, 1990, 125 pp

Melhuish FM, Lang ARG (1968) Quantitative studies of roots in soil. I. Length and diameters of cotton roots in a clay-loam soil by analysis of surface-ground blocks of resin-impregnated soil. Soil Sci 106: 16–22

Merrill SD (1992) Pressurized-wall minirhizotron for field observation of root growth dynamics. Agron J 84 (4): 755–758

Merrill SD, Upchurch DR (1994) Converting root numbers observed at minirhizotrons to equivalent root length density. Soil Sci Soc Am J 58: 1061–1067

Merrill SD, Doering EJ, Reichman GA (1987) Application of a minirhizotron with flexible, pressurized walls to a study of corn root growth. Minirhizotron observation tubes:Methods and applications for measuring rhizosphere dynamics. ASA Spec Publ 50: 131–143

Merrill SD, Upchurch DR, Black AL, Bauer A (1994) Theory of minirhizotron root directionality observation and application to wheat and corn. Soil Sci Soc Am J 58 (3): 664–671

Meyer WS, Barrs HD (1985) Non-destructive measurement of wheat roots in large undisturbed and repacked clay soil cores. Plant Soil 86: 237–247

Murphy JA, Hendricks MG, Rieke PE, Smucker AJM, Branham BE (1994) Turfgrass root systems evaluated using the minirhizotron and video recording methods. Agron J 86 (2): 247–250

Neufeld HS, Durall DM, Rich PM, Tingey DT (1989) A rootbox for quantitative observations on intact entire root systems. Plant Soil 117: 295–298

Pagès L, Serra V (1994) Growth and branching of the taproot of young oak trees – a dynamic study. J Exp Bot 45: 1327–1334

Pan WL, Bolton RP, Lundquist EJ, Hiller LK (1998) Portable rhizotron and color scanner system for monitoring root development. Plant Soil 200: 107–112

Parker CJ, Carr MKV, Jarvis NJ, Puplampu BO, Lee VH (1991) An evaluation of the minirhizotron technique for estimating root distribution in potatoes. J Agric Sci 116 (3): 341–350

Parsons WFJ, Miller SL, Knight DH (1994) Root-gap dynamics in a lodgepole pine forest: ectomycorrhizal and nonmycorrhizal fine root activity after experimental gap formation. Can J For Res 24: 1531–1538

Pearson RW, Lund ZF (1968) Direct observation of cotton root growth under field conditions. Agron J 60: 442–443

Poelman G, Van De Koppel J, Brouwer G (1996) A telescopic method for photographing within 8 × 8 cm minirhizotrons. Plant Soil 185: 163–167

Reid JB, Winfield D, Sorenson I, Petrie RA (1992) Test of the reliability of acrylic plastic windows for studies of kiwifruit roots. N Z J Crop Hortic Sci 20: 463–466

Rutherford MC, Curran B (1981) A root observation chamber for replicated use in a natural plant community. Plant Soil 63: 123–129

Rygiewicz PT, Johnson MG, Ganio LM, Tingey DT, Storm M (1997) Lifetime and temporal occurrence of ectomycorrhizae on ponderosa pine (*Pinus ponderosa* Laws.) seedlings grown under varied atmospheric CO_2 and nitrogen levels. Plant Soil 189: 275–287

Samson BK, Sinclair TR (1994) Soil core and minirhizotron comparison for the determination of root length density. Plant Soil 161 (2): 225–232

Schuurman JJ, Goedewaagen MAJ (1971) Methods for the examination of root systems and roots, 2nd edn. Centre for Agricultural Publishing and Documentation (PUDOC), Wageningen, the Netherlands

Smit AL, Zuin A (1996) Root growth dynamics of Brussels sprouts (*Brassica olearacea var gemmifera*) and leeks (*Allium porrum* L.) as reflected by root length, root colour and UV-fluorescence. Plant Soil 185: 271–280

Smit AL, Groenwold J, Vos J (1994) The Wageningen Rhizolab – a facility to study soil-root-shoot-atmosphere interactions in crops 2. Methods of root observations. Plant Soil 161 (2): 289–298

Smit AL, Booij R, Van der Werf A (1996) The spatial and temporal rooting pattern of Brussels sprouts and leeks. Neth J Agric Sci 44: 57–72

Speidel B, Weiss A (1974) Untersuchungen zur Wurzelaktivität unter einer Goldhaferwiese. Angew Bot 48: 137–154

Sword MA (1998) Seasonal development of loblolly pine lateral roots in response to stand density and fertilization. Plant Soil 200: 21–25

Taylor HM, Böhm W (1976) Use of acrylic plastic as rhizotron windows. Agron J 68: 693–694

Taylor HM, Upchurch DR, McMichael BL (1990) Applications and limitations of rhizotrons and minirhizotrons for root studies. Plant Soil 129: 29–35

Tennant D (1976) A test of a modified line intersect method of estimating root length. J Ecol 63: 995–1001

Teskey RO, Hinckley TM (1981) Influence of temperature and water potential on root growth of white oak. Physiol Plant 52: 363–369

Tscherning K, Leihner DE, Hilger TH, Müller-Sämann KM, El Sharkawy MA (1995) Grass barriers in cassava hillside cultivation: rooting patterns and root growth dynamics. Field Crops Res 43: 131–140

Upchurch DR (1985) Relationship between observations in minirhizotrons and true root length density. PhD Diss Texas Tech Univ, Lubbock (Diss Abstr 85-28594)

Upchurch DR (1987) Conversion of minirhizotron-root intersections to root length density. Minirhizotron observation tubes: Methods and applications for measuring rhizosphere dynamics. ASA Spec Publ 50: 51–65

Upchurch DR, Ritchie J (1983) Root observations using a video recording system in minirhizotrons. Agron J 75: 1009–1015

Upchurch DR, JT Ritchie (1984) Battery-operated color video camera for root observations in minirhizotrons. Agron J 76: 1015–1017

Upchurch DR, McMichael BL, Taylor HM (1988) Use of minirhizotrons to characterize root system orientation. Soil Sci Soc Am J 52: 2, 319–323

Van Noordwijk M (1987) Methods for quantification of root distribution pattern and root dynamics in the field. In: Proc 20th Colloq Int Potash Institute held in Baden bei Wien, pp 263–281

Van Noordwijk M, De Jager A, Floris J (1985) A new dimension to observations in minirhizotrons: a stereoscopic view on root photographs. Plant Soil 86: 447–453

Van Noordwijk M, Brouwer G, Koning H, Meijboom FW, Grzebisz W (1994) Production and decay of structural root material of winter wheat and sugar beet in conventional and integrated cropping systems. Agric Ecosyst Environ 51 (1–2): 99–113

Vogt KA, Grier CC, Gower ST, Sprugel DG, Vogt DJ (1986) Overestimation of net root production: a real or imaginary problem? Ecology 67: 577–579

Volkmar KM (1993) A comparison of minirhizotron techniques for estimating root length density in soils of different bulk densities. Plant Soil 157 (2): 239–245

Voorhees WB (1976) Root elongation along a soil-plastic container interface. Agron J 68: 143

Vos J, Groenwold J (1983) Estimation of root densities by observation tubes and endoscope. Plant Soil 74: 295–300

Vos J, Groenwold J (1987) The relation between root growth along observation tubes and in bulk soil. Minirhizotron observation tubes: Methods and Applications for Measuring Rhizosphere Dynamics. ASA Spec Publ 50: 4, pp 39–49

Wang ZQ, Burch WH, Mou P, Jones RH, Mitchell RJ (1995) Accuracy of visible and ultra violet light for estimating live root proportions with minirhizotrons. Ecology 76 (7): 2330–2334

Wiesler F, Horst WJ (1994) Root growth of maize cultivars under field conditions as studied by the core and minirhizotron method and relationships to shoot growth. Z Pflanzenernähr Bodenkd 157 (5): 351–358

Wilson K, Gunn A, Cherrett JM (1995) The application of a rhizotron to study the subterranean effects of pesticides. Pedobiologia 39: 132–143

Zillmann K-H (1956) Beobachtung des Wurzelwachstums in Feldbeständen. Dtsch Landwirtsch 7: 394–400

CHAPTER 9

The Measurement and Analysis of Fine Root Longevity

J.E. Hooker, R. Hendrick[1], and D. Atkinson[2]

School of Applied Sciences, University of Glamorgan, Pontypridd, Mid-Glamorgan CF37 1DL UK

[1] School of Forest Resources, University of Georgia, Athens, Georgia 30602-2152, USA

[2] SAC, West Mains Rd., Edinburgh, EH9 3JG, UK

CONTENTS

A.L. Smit et al. (Eds.), Root Methods

9.1 Introduction

9.1.1 The Importance of Fine Root Longevity in Ecosystem Processes

A better understanding of ecosystem processes is an important research goal for scientists in the next millennium. It offers the potential to not only make better use of our finite natural resources but also to properly understand the impacts of anthropogenic events such as global climate change. It will also allow us to make more informed assessments of how any deleterious effects may be ameliorated. One of the important research targets for terrestrial ecosystems is to understand biogeochemical cycles, as it is these fluxes (and the processes which drive them), which are fundamental to ecosystem stability.

For many years it has been known that roots play an important part in these processes, firstly by the uptake of nutrients from the soil, and secondly by returning them back into the soil at mortality. However, practical difficulties in quantifying root production, especially in natural ecosystems, have resulted in many estimates of root production and their role in nutrient cycling being derived from above-ground data (Atkinson and Fogel 1997). Recently, more direct methods have become available, and their application has already provided some evidence to suggest that the level of mortality experienced by roots means that their impact on terrestrial biogeochemical cycles may be much greater than previously considered. Essentially, the life of some roots has been shown to be relatively short (Hendrick and Pregitzer 1992, 1993; Hooker et al. 1995; Forbes et al. 1997; Black et al. 1998). Previous estimates based on allometric relationships are thus likely to have been misleading and have underestimated the contribution of roots to system productivity and the cycling of nutrients. However, although this recent data suggests that short lifetimes may be common for roots, more data needs to be collected. Thus, it is hoped that this comprehensive account on appropriate methods will facilitate research.

Carbon Cycling. The need to quantify both root biomass and turnover has become more important as a consequence of interest in global climate change. Cannel et al. 1992, summarizing data from a range of sources, suggested that soil, which contains around 1500 Gt relative to 500 Gt in the terrestrial biosphere, is a major component of the global carbon budget. They acknowledged that, as the part of the biosphere embedded within the soil, plant root systems represented one of the principal means for the movement of carbon from both the atmosphere and the terrestrial biosphere into the soil. However, estimates of the quantity of photosynthate passing to root systems as a whole, and to new root production, and of root turnover in terrestrial ecosystems, are more uncommon than their importance would warrant. This deficiency relates principally to methodological problems in quantifying temporal changes in root production and mortality. Current estimates include those of Atkinson (1985) who, in a study of an established apple plantation, quantified root system production as 2.5 $Mg\,ha^{-1}\,year^{-1}$ and estimated that 75% of this new production turned over within a single year, i.e. an annual root turnover figure of around 1.9 $Mg\,ha^{-1}\,year^{-1}$. Cannel et al. (1992), obtained figures of a similar magnitude when they estimated fine root production in stick spruce, eucalyptus and beech woodland as 2.8, 1.8 and 2.2 $Mg\,ha^{-1}\,year^{-1}$ respectively. More recently, R. Fogel (unpubl. data) quantified the annual production for a *Populus* woodland in northern Michigan, USA as 16.5 Mg and calculated that 5.1 Mg i.e. about 30% was in the form of fine roots. However, neither of the latter studies included estimates of root survival.

Nutrient Cycling. Estimates of the impact of root longevity on nutrient cycling are even less common than those for dry matter production. In one, a study of an established apple orchard, Atkinson (1985) estimated that annual recycling of nitrogen within the root system would amount to approximately 16 $kg\,N\,ha^{-1}\,year^{-1}$. Moreover, variation in new root production during the course of a season suggested that the rate of release of nitrogen as a result of root mortality would also be non-uniform during an annual cycle and that a maximum of 1.5 $kg\,N\,ha^{-1}\,week^{-1}$ was likely to be released as a consequence of root mortality and decomposition. Similar estimates have been made for alfalfa by Andren et al. (1991) who estimated annual nitrogen release due to root decomposition as 32 $kg\,N\,ha^{-1}$. In a more recent study, Goins and Russelle (1996) estimated the annual nitrogen release, as a consequence of fine root turnover in alfalfa, as 60 $kg\,N\,ha^{-1}$. It is thus clear that root decomposition has the potential to influence the size of fertilizer inputs.

Resources Acquisition. Problems in acquiring information on root length in soils has led to many estimates of nutrient inflow to plants that are based upon rather few and infrequent measurements of total root system length. Where the life of roots in soil is long, infrequent measurements of root length will be satisfactory for the calculation of variation in inflow during different parts of the season. However, where root longevity is short, and there is now increasing evidence to show that this is often the case (see above), estimates of nutrient inflow to the root system may be lower than those which will occur in practice. In addition, the length of the root system actually available for nutrient uptake in the soil at any time will be critically influenced by root longevity.

9.1.2 Factors Affecting Fine Root Longevity

Symbiosis. That symbiotic interactions occur between specific groups of microorganisms and plant roots has been known for many years. However, although much research effort has been devoted to the study of these symbioses, studies of their effects on root system development are very limited. It is known that effects on root architecture can occur (e.g. Hooker et al. 1992; Berta et al. 1995; Norman et al. 1996; for review see Hooker and Atkinson 1996). However, the effects of symbiotic interactions on the mortality of roots are almost unknown, with only one report of an appropriate study by Hooker et al. (1995) who used mini-rhizotrons to study mortality of roots in root systems of poplar and identified substantial reductions in the longevity of roots when they were colonized by an arbuscular mycorrhizal fungus (AMF). They also highlighted the importance of this finding for understanding the role of AMF in ecosystem functioning. Further studies are clearly needed.

Soil Temperature. There are only a few reports on the effects of soil temperature on the longevity of plant roots. In the first, Head (1966) studied variation in the longevity of apple roots over a complete season. He found that root longevity was maximal during the winter period (approximately 12 weeks), fell to a minimum value in mid May (approximately 3 weeks) and remained at this rate until autumn. Later, Hendrick and Pregitzer (1993) compared the pattern of fine root mortality in two *Acer saccharum* forests located approximately 80 km apart on a north-south transect. They measured an enhanced rate of root mortality at the southern site and hypothesized that this could be linked to the warmer soil temperatures found at this site. These studies thus both suggested that soil temperature has an effect upon longevity and this was recently confirmed by Forbes et al. (1997) in experiments with *Lolium perenne*. Here they found that increases in temperature resulted in a decrease in the longevity of roots but, because they used soil, could not exclude indirect effects such as changes in soil nutrient levels due to altered microbial activities. However, Forbes et al. (1996) had previously reported the direct effect of temperature on root system architecture of *Plantago lanceolata* in carefully controlled microcosms. The scale of the effects they observed i.e. increased branching with increases in temperature would concur with the data from the later study and (assuming that higher order roots do not survive for as long as lower order roots) strongly suggest that temperature is an important determinant of longevity.

Soil Moisture. Although it is well-known that low soil water potential will adversely effect the rate of root growth (Gregory 1987; Klepper 1987), there have been relatively few studies investigating the effect of soil water potential on root longevity. The potential effect of soil water potential on root longevity will be more complex in perennial crops because of soil water potential effects upon root branching (Atkinson 1993) which in time interacts with root longevity. However, the absence of information on root longevity makes it difficult to assess whether reported effects of low water potentials on root length are a consequence of accelerated root death or impeded new root production, or a combination of both. This question needs to be resolved.

Nutrient Availability. Despite the large number of studies of the effect of nutrient supply on root production and morphology (Drew 1987), there have been relatively few studies of the effect of nutrient supply on root longevity. However, in 1993 Pregitzer et al. assessed the effects of enhanced nitrogen supply on root turnover in *Prunus*. They found that the supply of nutrients greatly enhanced root initiation and that the flush of roots initiated in response to an enhanced nutrient supply had a shorter longevity than those roots which developed in non-enriched patches. It is known that the application of nutrients to a root system can cause a major alteration to root branching (Drew 1987). It is, though, unclear as to whether the response identified by Pregitzer et al. was a direct response in terms of the roots produced, or one consequent upon the production of a number of higher order lateral roots which would themselves survive a shorter period of time than lower order root branches. This is an area which is likely to benefit from future investigation.

Carbon Dioxide. Although elevated levels of CO_2 have been shown to increase root production, e.g. Curtis et al. (1995), Thomas et al. (1996), little information exists about effects on root longevity. In the study by Thomas et al. (1996), increased root production, which varied in magnitude at different times during the season, was accompanied by a change in the timing of the onset of new growth which occurred around 4 weeks earlier in trees growing under elevated CO_2 conditions. Enhanced CO_2 growing conditions were also associated with an increase in CO_2 flux from the soil. However, likely errors associated with measurements of both root production and CO_2 flux were such that it was not possible to use the data of Thomas et al. (1996) to assess whether the roots produced in the high CO_2 atmosphere showed a higher rate of turnover than those produced under ambient conditions. This is another area which would benefit from further investigation.

9.2 Indirect Methods

The use of demographic methods to estimate root production and turnover has been reviewed by Fogel (1991). A number of studies have attempted to estimate root production indirectly by assuming that the ratio of production to mass is similar for root and shoot systems. However, the ratio of root production to shoots is known to vary considerably and has been estimated to be between 20 and 100% of the above ground ratio (Harris et al. 1980). Moreover, assumptions are likely, in most cases, to be invalid and provide misleading data because of spatial and temporal changes in root systems which may not be reflected above-ground. The use of this approach has recently been discussed by Atkinson and Fogel (1997).

An alternative approach is to base estimates upon sequential harvests (see Chap. 8). Production is estimated from measurements of differences in biomass on successive sampling dates with these estimates being corrected for losses (Fogel and Hunt 1979). This process is described by the equation:

$$npp = B_{t+1} - B_t + L,$$

where *npp* is net primary production, B_t is the standing crop at time t and L is the amount lost by rhizodeposition. Estimates obtained using this method are improved if roots are separated into live and dead categories rather than simply working upon the difference between maximum and minimum standing crops for all roots (Fairley and Alexander 1985). This method is usually described as the method of balancing transfers. One problem is that in practice it can be difficult (in material obtained from soil cores) to easily distinguish between live and dead roots. Approaches to resolving these differences have been discussed by Fogel (1985). However, although the harvest method (adjusted by the balancing transfers approach) is helpful in producing relative data for different ecosystems, it does not give sensitive estimates of either production or decomposition; essentially because of production and undetected mortality (e.g. roots that rapidly decompose or are consumed) that may have occurred between sampling dates.

An alternative means of obtaining production estimates is through use of the mesh bag method (e.g. Steen 1991). In this method the amount of root grown within a cylindrical mesh bag filled with a given volume of soil, left in the field for a given period of time, is taken as the rate of root production. The roots, dead and living, found within the bag after a certain period are assumed to have grown into the bag after placement and to represent total production. The method has significant relative value in field trials where experimental treatments are being compared. Again though the derivation of real rates of production are not detectable because of production and undetected mortality that may have occurred between sampling dates.

9.3 Direct, Observation-Based Methods

Until recently, most observational studies of fine root dynamics have reported only net temporal changes in root dry weight or at best total root length or numbers. We now recognize that failing to document changes on an individual root or root-segment basis can render observational techniques (see Chap. 8 on minirhizotron methods) no more informative than traditional coring techniques. Fortunately, observation studies lend themselves to demography-based research, in which the life histories of individual roots are documented from the time of a root's "birth" until its death. This population-based approach enables us to quantify root production, longevity and turnover as distinct processes. Moreover, temporal changes in root dimension and appearance that may be of functional significance can be measured simultaneously.

The analytical techniques of demographers and population ecologists are well-developed, and facilitate relatively sophisticated analyses of dynamic root data. This is particularly true for data derived from studies of root *cohorts*; that is, populations of roots of a known age that were produced during the same, relatively short time period. Cohort studies have a long history in plant, animal and insect ecology, and form the basis of actuarial science. Their history of use in root research is much shorter, but the overall approach shows considerable promise.

9.3.1 Cohort Selection

Cohort-based studies depend upon repeated measures of the same roots, and hence require the ability to return to the same location on either a minirhizotron tube or an observation window (Atkinson 1985). Repeatedly sampling the same area can be accomplished by physically marking sampling locations on tubes or windows, drawing of the root system (Atkinson and Fogel 1997) or via a mechanical guide on the device used to record still or video images. Marking and re-locating individual roots can be accomplished either with computer-aided image analysis programs (e.g. ROOTS and RooTracker) or by hand if using acetate overlays and tracing.

9.3.2 Cohort Size

Cohort sizes should be as large as the time and logistical constraints involved in collecting and processing the data will allow. The degree of precision with which survivorship and mortality rates can be estimated necessarily increases along with cohort size, and accuracy typically increases with sample size as well. Cohort sizes in our own research have ranged from 26 to 245 individual roots. The rooting characteristics peculiar to the plant species and soil environment under study will dictate root densities in observation tubes or windows, and hence the number of roots or root segments encountered per unit sampling area. When compiling cohorts in studies in which sampling areas in the window or rhizotrons are adjacent, it is important that roots growing out of one sampling area and into another are counted only once, or else that root will contribute disproportionately to the data set. As stated, it is logical to conclude that increasing numbers of roots per cohort will result in greater precision of life span estimates, and presumably greater accuracy as well. However, the specific number of roots needed to detect treatment or other differences in life span during some specific period of time depends upon the magnitude of life span difference one wishes to detect (Δ), the choice of significance level with respect to making a Type I error (α, the probability of rejecting a true null hypothesis) and the desired level of the power of the test (1-β, where β = probability of making a Type II error). The number of samples need for a particular combination of Δ, α and β depend upon the distribution of life spans, but values for an assumed exponential distribution with cohorts of equal size have been calculated by George and Desu (1974), some of which are summarized in Table 9.1. Makuch and Simon (1982) present a more generalized approach where the number of treatments is greater than two, and include tabulated data as well as the equation and appropriate constants for various combinations of treatments, α and β. For comparisons of two treatments, their approach also yields the data in Table 9.1, but in addition to making the generalized case for comparisons among more treatments, allows flexibility in the choice of Δ, α and β.

9.3.3 Imaging and Birth Intervals

Given that cohorts are defined as all individual roots produced or "born" at approximately the same time, it is appropriate to ask as to how wide a time inter-

Table 9.1. The number of individuals required in each cohort to detect a desired degree of difference in life spans (Δ) for particular combinations of α and β when treatments are two in number

Level of Δ						
1-β	α	10%	20%	50%	100%	200%
0.95	0.01	3749	951	193	66	17
	0.05	2384	652	132	46	12
0.90	0.01	2870	785	159	55	14
	0.05	1884	515	105	36	9
0.80	0.01	2213	605	123	42	11
	0.05	1360	372	76	26	7

val constitutes "approximately the same time"? In human actuarial studies, periods of one or more years are often considered to constitute a birth interval. However, plant roots may live for as little as a few days to several years, and root longevity varies both within and between species (Nadelhoffer et al. 1985; Hendrick and Pregitzer 1993; Hooker et al. 1995; Black et al. 1998). Rates of root mortality are age-dependent in some species (Hendrick and Pregitzer 1993; Black et al. 1998), and may be considerable within a few days of birth, dropping dramatically thereafter (Pregitzer et al. 1993). Moreover, most plant communities are probably comprised of species with different root longevities, which may complicate studies of root dynamics in mixed-species environments. Equally critical is the choice of the proper interval after which the cohort is resampled. Sampling too infrequently will underestimate actual root birth rates and lead to estimates of survivorship that are too conservative. However, sampling more often than necessary can result in excessive investments of time and resources that result in little or no increase in the quality of data collected.

The age of roots within a cohort are usually defined based upon the date upon which they were first observed. However, the true birth date of a root at the time it is first observed can be anywhere between the day immediately following the previous sampling date and the day immediately preceding the date of first observation. Similarly, the first time a root is observed as dead is unlikely to be the day upon which it died. Without better data from prior experience, one can only assume that the births of new roots and the deaths of old roots are distributed evenly throughout the time period between observations. Therefore, the "average" root is assumed to be born halfway between observations, and similarly to die halfway between observations. Therefore, aging a cohort as zero at the date of first observation and then chronologically from that point forward is numerically equivalent to assuming even distributions of births and deaths between observations. If the distribution of births and deaths is known to be other than even, root ages should be adjusted accordingly.

Clearly, any large-scale, intensive study of root longevity should be based upon some prior knowledge of the life history of roots for that or a similar species or community. Decisions regarding acceptable cohort birth intervals and time periods between sampling will rarely be the same for two investigators or even two successive experiments. However, they should be made with the following criteria in mind:

1. What percentage of the entire cohort born within a given time interval is represented at the time of first observation, and is the likelihood of a root's demise markedly higher in the first few days of existence?
2. How precise must estimates of survivorship and longevity be?
3. In mixed-species experiments, are the observed roots dominated by one or more species with known or suspected root longevities markedly longer or

shorter than other species in the community? Studies in which species-specific root longevity is of interest might require sampling intervals structured around each species.

4. If experimental treatments hypothesized to affect longevity are to be imposed, in what direction (i.e. increased or decreased longevity) are the hypothesized effects to be manifested?
5. To what extent will financial and logistical considerations affect the sampling frequency?

Generally, sampling intensity and frequency will be determined by striving for a balance between the amount of data desired and the amount that can be collected and analyzed. The interval will be generally influenced by the species being studied and the normal rate of its root turnover. We have used intervals of 2 days to 1 month when studying the roots of woody species (e.g. Hendrick and Pregitzer 1992; Pregitzer et al. 1993; Hooker et al. 1995; Black et al. 1998), and Dubach and Russelle (1995) concluded that periods of 2 weeks were appropriate for alfalfa roots. We note that in our experience using video imagery, collecting demographic data adds little to the total cost of the research, and that most of our resources are devoted to extracting and analyzing the data at a later date. It may be best to collect data as frequently as possible, even if there is little chance that all of it can be analyzed by an experiment's end. Partial analysis of frequently collected data can give insight into the amount of information lost, if any, by analyzing data from alternate sampling dates or increasing the interval defining a cohort to increase sample size. For general guidance a sampling protocol is given in Box 9.2.

Finally, before commencing any measurements, it is vital, given the very significant inputs required, that a detailed "cost/benefit analysis" of the protocol be carried out. As for general guidelines for maximizing the return from one's efforts to quantify fine root life span or other demographic processes using observational techniques, we offer the following:

1. Familiarize yourself with the general morphology of the root system from excavations prior to beginning observational studies. How does the appearance (e.g. colour, texture, branching habit, etc.) of the root change as it ages and develops (e.g, from tip to base, or between woody and non-woody roots).
2. What do dead roots look like? Are there abundant obviously dead roots in the soil, or are they rare? Given that we have sometimes witnessed what casually appeared to be dead roots later giving rise to new lateral branches, as well as the fact that many roots disappear altogether between observation intervals due to rapid decomposition, herbivory or a combination of both, it is important that one knows *a priori* something about what to expect during the life history of the roots under study.

Table 9.2. Suggested sampling intervals for observation-based techniques

Vegetation type	Variable	Suggested sampling interval
Annual crops	Overall production and mortality or general life span	1–2 weeks (best); monthly (coarse)
"	Early mortality or precise life span	2–5 days (best); 1–2 weeks (sufficient)
Perennial plants	Overall production and mortality (growing season, annual) or general life span	2 weeks – monthly (best); 2 months (coarse)
"	Overall production and mortality (dormant season) or general life span	monthly – 2 months
#	Early mortality or precise life span	2–7 days (best), 2 weeks or monthly (sufficient or coarse)

3. Where are most of the roots physically located in the soil? Are the roots of primary interest mycorrhizal or those most important for nutrient uptake (likely to be shallow, perhaps within the upper 20–30 cm of the soil), or are deep roots that may be important during periods of high water demand also of interest? The general vertical distribution of roots, as well as those most appropriate to the research question being posed, should guide the selection of minirhizotron length or rhizotron depth.
4. What is the extent of the financial and labour resources, and time, that can be devoted to the study? For example, a 2 m minirhizotron containing perhaps 150 potential images with roots might take only 10 min to image in the field, but image analysis of that tube for a single date's sampling could take from under 30 min to more than 3 h to analyze, depending upon the image analysis system used, the type of data desired and the density of roots along the minirhizotron surface. Moreover, it is our experience that analyzing images in the upper 25–40 cm consumes as much as 75% of analysis time, as most roots are located there. For a minirhizotron 1 m in length and inserted at a 25–45° angle to the soil surface in a perennial plant community (e.g. forest or grassland), perhaps 1.5–2.5 h should be allocated for laboratory image analysis for each sampling date. Time requirements in agricultural systems could be either greater or less, depending upon root densities. Although the specific requirements will vary, completely analyzing images from one sample of 12–15 2 m minirhizotrons, or 24–30 1 m minirhizotrons might take a skilled operator 30–70 h.

5. What is the nature of the soil into which the minirhizotron is to be installed? Soils with a high clay content often cause problems when the clay settles and forms films along the surface of the minirhizotron, thereby obscuring roots. We have partially overcome this problem by hand-digging narrow trenches into which the minirhizotrons are laid, after which soil is hand-packed over the minirhizotrons, rather than sliding the minirhizotrons in a bore-hole.

9.3.4 Censoring

Ideally, cohort studies terminate after all the subjects (i.e. roots) have died. However, a root longevity study will often end prior to the death of all roots under observation. When this occurs, those roots still alive at the study's end are said to be *right censored*. That is, they are known to live for at least the duration of the study, but their exact survival times are unknown. When the roots under study are all comprised of a single cohort, or if cohorts of the same age are being compared, the censoring is said to be *singular*. Another example of singular censoring is when all roots within a cohort are followed until a certain percentage die, at which time the study is terminated.

Progressively censored data are encountered in studies in which roots produced under a single treatment but at different times are still alive at the time of termination. Such roots do not comprise a cohort, but progressively censored data can be analyzed in ways analogous to singularly censored data. We will concentrate primarily on analyses of singularly censored data in our discussions of data analysis below. Technically, most of the roots studied via direct observation will also be *left censored*, in that their exact date of birth is unknown, regardless of their fate during the course of the experiment. Provided that an acceptably short period defining the cohort birth interval has been established, given the investigator's objectives and constraints, this left censoring can be ignored without much consequence.

9.3.5 Survivorship versus Longevity

Before discussing statistical tests appropriate to analyzing root life span data, it is important to distinguish between *survivorship* and *longevity*. Survivorship refers to the proportion of one or more cohorts alive at a particular point in time, while longevity refers to the length of time between birth and death. Survivorship analyses are concerned with the number of root deaths within a particular period of time, while longevity analyses address the length of time until death occurs. With respect to the former, one might wish to compare the root response of two different cultivars to a particular pathogen for a specified number of days after infection, for which statistical contingency tests are appropriate. Alternatively, one might wish to evaluate the effects of two or more levels of soil fertility on root life spans. Statistical tests of treatment effects on longevity quantify differences in the distribution of life spans within each cohort.

Below we recommend some statistical tests appropriate for describing and characterizing the survival and mortality of roots. We also present a number of statistical tests for comparing the longevity of roots subject to two or more different treatments; those treatments might, for example, be soil fertility, irrigation, plant species or time of birth. In our discussion, our objective is not to give either an exhaustive or an abstract treatment of statistical survivorship and life span quantification or analysis. Rather, our intention is to present a broad overview of some of the more common and accessible means of describing the nature of root longevity and mortality, and for testing for differences in root survivorship and longevity, using both real and hypothetical data in selected examples. For more detailed discussions, we refer interested readers to a number of texts and papers addressing the statistical descriptions and comparisons of survival and life span data. Among those we suggest are Gehan (1965), Mantel (1966), Kalbfleisch and Prentice (1980) and Cox and Oakes (1984), Muenchow (1986), Pyke and Thompson (1986), Pollock et al. (1989), Lee (1992). All of these references contain worked examples of one or more statistical applications, and Lee's (1992) is a comprehensive, coherent and highly readable treatise of statistical methods.

9.3.6 Survivorship Comparisons

Comparing the survivorship of two or more cohorts over a particular interval of time is among the simplest of root life span analyses. When only two cohorts are involved, either a simple chi-square test or the *G*-test can be used to determine if the proportion of original roots surviving to some point in time are equivalent. The chi-square test statistic is among the most common of all nonparametric statistical tests, and we will not review it here. The *G*-test is less commonly used, but its use in place of the chi-square test has been encouraged in recent years because of its theoretical appropriateness and greater computational ease (Sokal and Rohlf 1981).

For comparing the survivorship of two cohorts at the end of a specified period of time, the *G* -test takes the following form:

$$G = 2\left(\sum f_c \ln f_c - \sum f_t \ln f_t + n \ln n\right), \qquad (9.1)$$

where the f_c's are the cell frequencies in a 2×2 contingency table of live and dead roots within the two cohorts; the f_t's are the row and column totals, n is the total number of roots in both cohorts, and G is distributed approximately as chi-square with one degree of freedom. A sample calculation for 20 roots in each of two root cohorts that were observed for a period of 1 year is shown in Box 9.1.

BOX 9.1. G-test for Equivalent 1-Year Survivorship of Two Fine Root Cohorts Growing in Hardwood Forests in Michigan, USA

Site	Alive	Dead	Totals
North	14	6	20
South	8	12	20
	–	–	–
Totals	22	18	40

1. Calculate $\Sigma f_c \ln f_c$ for cell frequencies:

$$= 14 \ln 14 + 6 \ln 6 + 8 \ln 8 + 12 \ln 12 = 94.1518$$

2. Calculate $\Sigma f_t \ln f_t$ for row and column totals:

$$= 22 \ln 22 + 18 \ln 18 + 20 \ln 20 + 20 \ln 20 = 239.6586$$

3. Compute $n \ln n$ for total number of roots:

$$= 40 \ln 40 = 147.5552$$

4. $G = 94.1518 - 239.6586 + 147.552 = 2.0452$

9.3.7 Longevity

Given that it is most appropriate to determine the life span of individual roots, this presents both practical and analysis questions. Earlier we explained that the age of the root using any of the currently available direct observational methods will (in practice) be an estimate, with an accuracy related to frequency of observation. However, equally important is how to determine, from observation alone, if a root is alive or not. Discerning live and dead roots is a difficulty long associated with studying root life span and dynamics. A variety of staining techniques have been devised for distinguishing between live and dead roots extracted from the soil, none of which have proved either universally acceptable or consistently accurate. Observational approaches are fraught with similar difficulties. Dead roots are typified by some combination of a dramatic darkening, a loss of "intactness" (e.g blurring of root edges, apparent decomposition, etc.), pronounced shrivelling, colonization by fungal (obviously non-mycorrhizal) hyphae and either gradual or rapid disappearance. However, none of these criteria are necessarily reliably or consistently associated with dead roots. For example, darkening is often associated with root ageing, shrivelling may be due to the collapse of cortical cells in otherwise live but ageing roots, and distinguishing the hyphae of decomposer fungi from mycorrhizal hyphae or hyphae associated with other roots (living or dead) or the bulk soil is difficult. Wang et al. (1995) used a UV light source to distinguish between live and dead roots of eight species of plants, including grasses, forbs and both deciduous and evergreen woody species. The live root assessments of both UV and visible light were similar in their accuracy (0.88 and 0.85, respectively), but they point out that visible light must be used in all cases to make definitive determinations of dead roots. Further testing will be needed to determine whether UV reflectance is a robust and consistent indicator of vitality. It is thus clearly an advantage if detailed anatomical studies have been carried out for the genotype of interest. In apple, for example, MacKenzie (1979) showed that a colour change from white to brown was associated with phellogen development, cortical disintegration, and in roots lacking secondary xylem, death; useful background information for minirhizotron-based investigations of longevity.

In many cases, though, in natural systems roots merely disappear i.e. they will be present on one occasion and absent on the next; presumably due to the activity of saprophytes or predatory insects or invertebrates. The loss of roots can often be seen in images collected a week apart. On other occasions the cortex of a root may be very much reduced and the root will have begun to disintegrate and is clearly not functional; this is usually the case in more controlled laboratory studies where a sterilized soil or sand is often used.

Mortality in roots can usually be determined by slightly different characteristics in species of different types, i.e. annuals, herbaceous perennials and woody perennials. These are detailed in Table 9.3. Woody roots infected by ectomycorrhizas are a special case. Here the long survival of the root in an intact state, following a colour change which would usually precede disintegration and disappearance indicates its ectomycorrhizal status and so it needs to be assessed against a different standard.

In using the minirhizotron method, in conjunction with a database for analyses, e.g. in the program ROOTS (Michigan State University, East Lansing, MI, USA), a series of condition codes can be assigned to roots. Examples are detailed below in Table 9.4.

The data can then be summarized and for simplicity the mean or median values are often calculated. These can be useful statistics but they should be used cautiously and the distribution of the data fully appreciated.

Table 9.3. Characteristics used to determine root mortality

Plant type	Characteristics for root mortality
Annual species (all roots)	Change colour from white to pale brown. Decrease in diameter Roots disappear
Herbaceous perennials ("short roots")	Usually change colour from white to dark brown or black Roots appear white Roots disappear Level of arthropod activity around root increases
Herbaceous perennials ("long roots")	Become dark brown Decrease in diameter or start to appear rough in texture Roots disappear
Woody perennials ("short roots")	Change in colour to mid brown Decrease in diameter Become more attractive to collembolans Roots disappear
Woody perennials ("woody roots")[a]	Change from mid brown to dark brown/black Loose diameter unevenly Uneven decrease in diameter

[a] In ectomycorrhizal species these roots may survive for a period of years.

Table 9.4. Condition codes which can be applied in a database

Code	Description
N	New root. This is assigned only once in a root's life, on the first occasion it appeared.
W	White root. Assigned if the root was white and healthy.
B	Brown root. Assigned if the root was brown and intact.
X	Disintegrating root. Assigned if the root was clearly starting to break up.
A	Absent root. Assigned if the root had disappeared. After a root has disappeared it continues to get an *a* coding on every sample date.

9.3.8 Survivorship and Survival Distributions

Survival analysis is a powerful technique for studying censored data. However, although widely used in medical and industrial studies, it has only recently been applied to root data. Black et al. (1998) recently first used the technique to compare the distribution of root longevities between cohorts. Essentially, a survival distribution is a probability distribution of lifetimes. It includes the survivor function S(t), which defines the probability of mortality for a root at t. Probabilities for an individual surviving beyond time t are then determined by the extent to which they have accumulated hazards and various distributions can then be applied. These include the exponential, Weibull, gamma and lognormal distributions, and each of these should be tested to determine which provides the best fit for the data. For details on how to apply these powerful but reasonably complex techniques we refer readers to Black et al. 1998 (see also Sect. 9.3.10).

9.3.9 Among-Cohort Life Span Comparisons

If none of the roots in a study are censored (that is, all roots have died prior to the completion of the study), standard non-parametric tests for equality of distributions can be used. When only two such distributions are being compared, the Mann-Whitney *U*-test is appropriate (Conover 1980; Lee 1992). The Mann-Whitney *U*-test is a rank test, in which the life spans of two different samples of roots are arranged and ranked in ascending order. The ranks of one sample are then summed (and corrected for rank ties if present), and the resulting test statistic T is compared with the area under the normal distribution at a desired level of α. The two samples need not have the same number of roots. An example is given in Box 9.2.

Censored data are more common than non-censored data in many if not most longevity studies. Lee (1992) has summarized five non-parametric tests appropriate for pair-wise comparisons of fine root life spans in which the data are censored, all of which are also rank tests. These are: Gehan's generalized Wilcoxan test, the Cox-Mantel test, the logrank test, Peto and Peto's generalized Wilcoxan test and Cox's F-test. The first four are appropriate for both singular and progressively censored data. The particular test to use depends upon the size and structure of the data set under consideration (Lee 1992).

The two generalized Wilcoxan tests give relatively more weight to roots that die in the study and are more sensitive to early differences in mortality than the logrank test, which gives a more equal weight to all roots and is more likely to detect differences manifested late in the study. None of these three tests are effective at detecting differences in instances where the mortality or survivorship curves of functions cross; for example, when early mortality is much less in one cohort than the other, but in which a large number of roots die towards the end of the study such that final survivorship is lower. If the sample sizes are less than 50, the data approximate to Weibull or exponential distributions and the data are singly censored, then Cox's F-test has greater statistical power than the Wilcoxan tests. The reader is referred to Lee (1992) for more detail on the strengths and weaknesses of each test.

BOX 9.2. Comparison of Two Non-censored Life Span Distributions for Roots Growing in Soils Receiving Two Different Soil Treatments Using a Mann-Whitney *U*-test

Treatment 1		Treatment 2	
Life span (days)	Rank	Life span (days)	Rank
10	1	18	4
15	2	19	5.5
16	3	24	7
19	5.5	25	9
25	9	25	9
45	13.5	35	11
65	15	37	12
71	18	45	13.5
		69	16
		70	17

Suppose that an experiment designed to asses the impact of two soil treatments on fine root longevity yields the above hypothetical root life spans, and that an α level of 0.10 is to be considered statistically significant.

1. Arrange life spans (scores) in ascending order in each treatment.
2. Assign ranks within each treatment based upon score position within the combined data set. Tied scores receive an average rank that would have been assigned without ties. For example, the 19-day scores (life spans) are the 5th and 6th ordered observations, receiving an average rank of 5.5. Similarly, the 25-day scores would be the 8th, 9th and 10th ordered observations in the absence of ties, receiving an average score of 9.
3. The ranks for one treatment are summed to calculate the test statistic T (treatment 1 in this case, since the number of observations are smaller and hence the calculations simpler).

$$T = 1 + 2 + 3 + 5.5 + 9 + 13.5 + 15 + 18 = 67.0$$

4. In the absence of ties, the above value of T can be used unadjusted as the test statistic. With ties, it must be adjusted by subtracting the mean rank and dividing by the standard deviation using the following formula:

$$T_{adj} = \frac{T - n\dfrac{N+1}{2}}{\left(\dfrac{nm}{N(N-1)}\displaystyle\sum_{i=1}^{N} R_i^2 - \dfrac{nm(N+1)^2}{4(N-1)}\right)^{\frac{1}{2}}},$$

where N is the total number of root life spans, n is the number of life spans in the sample used to calculate T, m is the number of life spans in the other sample, R is the score rank, and ΣR_i^2 is the sum of squares of all N ranks in both samples. In the above example the equation takes the form:

$$T_{adj} = \frac{67 - 8\frac{18+1}{2}}{\left(\frac{(8)(10)}{(18)(17)}(2106) - \frac{(8)(10)(19)^2}{4(17)}\right)^{\frac{1}{2}}} = 0.8022,$$

which corresponds to a p value of approximately 0.221 and indicates that the soil treatments did not have a statistically significant affect on fine root longevity.

We have previously used a variant of Gehan's generalized Wilcoxan test (Hendrick and Pregitzer 1992) to test for life span differences in fine root cohorts observed in two hardwood forests in Michigan, USA. Gehan's test can be used either with individual root life span data, or with observations in which numbers of surviving roots in each cohort are counted at successive observation intervals. In Gehan's test, the life span of each root i in sample x (denoted x_i if dying during the study, or x_i^+ if alive and therefore censored at the study's termination) is compared with the life span of each root in sample y (denoted y_j if dying or y_j^+ if censored). The root in sample x is given a score U_{ij} of 1, 0 or −1 according the following rules:

1. If $x_i > y_j$ or $x_i^+ \geq y_j$, then $U_{ij} = 1$
2. If $x_i = y_j$ or $x_i^+ < y_j$ or $y_j^+ < x_i$ or (x_i^+, y_j) then $U_{ij} = 0$
3. If $x_i < y_j$ or $x_i \leq y_j^+$ then $U_{ij} = -1$

The test statistic W is computed as:

$$W = \sum_{i=1}^{n_1} \sum_{i=1}^{n_2} U_{ij}. \tag{9.2}$$

The variance of W is:

$$\text{Var}\, W = \frac{n_1 n_2}{(n_1 + n_2)(n_1 + n_2)} \sum_{i=1}^{n_1+n_2} (U_i)^2. \tag{9.3}$$

The test statistic W is assumed to be distributed approximately normally with mean and variance of zero. Life span difference between the two samples can be calculated as

$$Z = \frac{W}{(\mathrm{Var}\ W)^{1/2}} \quad (9.4)$$

If the number of roots in each sample is large, calculating W can be laborious. Lee (1992) demonstrates Mantel's (1967) alternative for calculating W that involves combining both samples and then scoring each root based upon its relative life span ranking. In effect, each root's life span is compared with all other roots in the combined sample. Each root's U_i (where $i = 1$ to $n_1 + n_2$) score is then the sum of the number of roots with a life span less than its own, minus the number of roots with longer life spans. In this case, the U_i's comprise a population with a mean of zero, and:

$$W = \sum_{i=1}^{n_1+n_2} U_i. \quad (9.5)$$

If, instead of root life spans, the data are expressed in terms of the number of roots in the original cohorts surviving and dying between each observation, then Pyke and Thompson's (1986) approach combining the methods of both Gehan (1965) and Mantel (1967) in a tabular format can be used, as demonstrated in Box 9.3.

BOX 9.3. Gehan's Generalized Wilcoxan Test for Life Span Differences in Two Root Cohorts

Time (t)	s_1	s_2	S	d_1	d_2	D	$>d_1$	$<d_1$	U_1	d_1U_1
0	149	196	345	10	12	22	323	0	323	3230
1	139	184	323	20	7	27	296	−22	274	5480
2	119	177	296	24	11	35	261	−49	212	5088
3	95	166	261	10	10	20	241	−84	157	1570
4	85	156	241	11	22	33	208	−104	104	1144
5	74	134	208	12	31	43	165	−137	28	336
6	62	103	165	7	6	13	152	−180	−28	−196
7	55	97	152	3	10	13	139	−193	−54	−162
8	46	87	139	6	4	10	129	−206	−77	−462
9	46	83	129	5	5	10	119	−216	−97	−485
10	41	78	119	10	5	15	104	−226	−122	−1220
11	31	73	104	3	9	12	92	−241	−149	−447
12	28[a]	64[a]	92[a]			92[a]	0	−253	−253	−7084
Sum										6972

[a] Censored

The above table summarizes survival (s) and mortality (d) data for two root cohorts ($n_1 = 149$ roots and $n_2 = 196$ roots) from the time of first observation (t_0 = month 0) until the termination of the study 17 months later. Although only 19% of the roots in cohort 1 were alive at the study's end, as opposed to 33% in cohort 2, we need to test for differences in survival distributions to determine if average life spans are the same in both cohorts (H_o). The test statistic W and the determination of a significant cohort difference in root longevity is calculated as follows:

1. For each observation date t, tabulate the number of roots in each sample alive at the time of observation (columns s_1 and s_2 for samples 1 and 2, respectively), and the number of those roots dying between that observation date and the following observation date (columns d_1 and d_2, respectively).
2. For each observation date t, sum columns s_1 and s_2 (= S_t), and d_1 and d_2 (= D_t),
3. For each observation date t, calculate the total number of roots in both cohorts that survive longer than the time period t to $t + 1$ (column $>d_{1,t} = S_t - D_t$), assigning each root a value of 1 according to the rules outlined above. For each observation date t, calculate the total number of roots that died prior to that date during all time periods (column $<d_1 - S_0 - S_t$), assigning each root a value of −1 according to the rules outlined above.
5. Calculate U_i at each time t by adding $>d_{1,t}$ to $<d_{1,t}$,
6. Calculate W using Eq. 9.2 by summing for all t's the product of $d_{1,t}$ and $U_{i,t}$ (column $d_{1,t}\ U_{1,t}$):

$$\begin{aligned} W &= 3230 + 5480 + \ldots - 7084 \\ &= 6972. \end{aligned}$$

7. Calculate the variance according to Eq. 9.3, which equates to the following:

$$\begin{aligned} \text{Var } W &= \frac{n_1 n_2}{(n_1 + n_2)(n_1 + n_2 - 1)} \sum_{t=1} D_t (U_t)^2 \\ &= \frac{(149)(196)}{(149+196)(149+196-1)} 22(323)^2 + 27(274)^2 + \ldots 92(-253)^2 \\ &= 3287278 \end{aligned}$$

8. Calculate Z according to Eq. 9.4:

$$Z - 6972/3287278^{1/2} = 3.7461$$

From a normal distribution table, we find that $p > Z = <0.0001$, and so we can reject the hypothesis that the average life spans in the two cohorts are the same.

When more than two samples or cohorts are to be compared and the data are not censored, the Kruskal-Wallis *H* test may be use in a manner analogous to a one-way analysis of variance F-test. The *H*-test is a rank-based test, in which the life spans in each treatment sample are ranked relative to all other roots in the treatments. *H* is calculated as:

$$H = \frac{12}{N(N+1)} \sum_{j=1}^{K} \frac{R_j^2}{n_j} - 3(N+1), \tag{9.6}$$

where N is the total number of roots in the K samples or treatments, n_j is the number of roots in the j th treatment, and R_j is the sum of all ranks in the j th treatment.

When some of the ranks are tied, H, as computed above, is divided by the value

$$1 - \frac{1}{N^3 - N} \sum_{j=1}^{g} T_j, \tag{9.7}$$

where g is the number of tied groups and $T_j = t_j^3 - t_j$, with t_j equal to the number of tied observations in a tied group (Lee 1992).

The number of roots (n_j) in each sample need not be the same. If the number is greater than or equal to five, H is distributed approximately as χ^2 (Sokal and Rohlf 1981). When the number of roots is less than 5, exact distributions can be found in Kruskal and Wallis (1952). An numerical example with hypothetical root life span data from three different plant species is shown in Box 9.4

When overall significant treatment effects are detected, as in this example, multiple comparisons of the rank sums of each treatment can be made to identify particular pair-wise differences. The particular form of the test depends upon whether the number of roots in each sample are equal, and upon the number of roots in each sample. Lee (1992) has summarized four such tests, one each for both large and small equal-sized samples; and large and small unequal-sized samples.

When more than two (e.g. K) samples need to be compared in which some of the data are censored, an alternative rank test (the K-sample test) can be used in which the ranks are scored as in the Cox-Mantel test. Instead of one set of ranks, two sets, one based upon ascending (R_1) and one based on descending (R_2) scores are calculated, from which the difference between R_1 and R_2 is used to compute a score w_i for each observation i. These values are used to calculate a test X^2, which is assumed to be distributed approximately as chi-square, and is computed as follows:

$$X^2 = \frac{\sum_{j=1}^{K} \frac{S_i^2}{n_j}}{s^2}, \tag{9.8}$$

BOX 9.4. Kruskal-Wallis *H*-test for Species Differences in Uncensored Root Life Span Data

Fine root life spans (days)

Species 1	Rank 1	Species 2	Rank 2	Species 3	Rank 3
45	4.5	21	1	58	8
55	7	27	2	64	10
67	11	32	3	75	12
80	13	45	4.5	87	14
90	15	51	6	91	16.5
91	16.5	62	9	102	18
Sums	67		25.5		78.5

1. Rank the life spans in each species in ascending order, with ranks for n tied scores equal to $1/n$ of the combined ranks, had the scores not been tied.
2. Compute rank sums (R_j) for each species,
3. Compute the correction factor using Eq. 9.7. In the example, the number of tied groups g is 2, and the number of tied scores (t_j) in both groups is 2 and the total number of scores N is 18. The correction factor is then computed as:

$$= 1 - \frac{1}{(18)^3 - 18}\left((2^3 - 2) + (2^3 - 2)\right)$$
$$= 1 - \frac{1}{5814}(12)$$
$$= 0.9979$$

4. Calculate the test statistic H according to Eq. 9.6:

$$H = \frac{\frac{12}{18(19)}\left(\frac{(67)^2}{6} + \frac{(25.5)^2}{6} + \frac{(78.5)^2}{6}\right) - 3(19)}{0.9979}$$
$$= 9.1138$$

The χ^2 value at an α of 0.05 with 2 d.f. is 5.991, and thus we can reject the hypothesis that the root life spans of all the species are the same.

where

$$s^2 = \frac{\sum_{i=1}^{K} w_i^2}{N-1}, \tag{9.9}$$

N is the total number of root life spans in the sample, and w_i^2 are the squared scores for each life span i. The test statistic X^2 is compared with the appropriate chi-square value for $K-1$ degrees of freedom at the desired level of α. The number of life spans in each sample need not be equal. A numerical example using this test is given in Box 9.5.

BOX 9.5. *K*-Sample Test for Equivalent Root Life Spans for Three Plant Species with Censored Observations

Fine root life spans (days)

Sp. 1	$R_{1,i1}$	$R_{2,i1}$	w_{i1}	*Sp.* 2	$R_{1,i2}$	$R_{2,i2}$	w_{i2}	*Sp.* 3	$R_{1,i3}$	$R_{2,i3}$	w_{i3}
45	4	20	−16	21	1	24	−23	58	10	15	−5
55	8	16	−8	27	2	23	−21	64	12	13	−1
67	13	12	1	32	3	22	−18	75	14	10	4
71+	14	1	13	45	4	20	−16	87	15	9	6
90	16	8	8	51	6	19	−13	91+	18	1	17
90	16	6	8	54	7	18	−11	93	19	4	15
91	18	6	12	55	8	16	−8	96	20	2	18
94+	20	1	19	62	11	14	−3	102+	21	1	20
Sums			37				−113				74

Consider the above hypothetical root life span data for three species of plants. We wish to test the hypothesis that the life spans in all three species are the same, with an acceptable α level of 0.05.

1. Rank all non-censored observations in ascending order (= R_1 for each observation i).
2. All censored observations receive a score which is one higher than the adjacent, lower uncensored observation.
3. Reduce the rank scores of tied observations to the lowest value for the score.
4. Rank all observations (both censored and non-censored) in descending order (= R_2 for each observation i).
5. Reduce the rank scores of tied observations to the lowest value for the score.

6. Reduce the rank of all censored observations to 1.
7. Calculate w_{ij} scores by subtracting $R_{2,ij}$ from $R_{1,ij}$ and sum the scores for each sample *j*.
8. Calculate the sum of $w_i^2 = 4352$.
9. Compute the value s^2:

$$= \frac{4532}{(24-1)} = 197.0435.$$

10. Compute the test statistic X^2:

$$= \frac{\frac{(37)^2}{8} + \frac{(-113)^2}{8} + \frac{(74)^2}{8}}{197.0435}$$

$$= 12.4427.$$

The corresponding χ^2 statistic for 2 d.f. and $\alpha = 0.05$ is 5.9915, and we can therefore reject the hypothesis that the root life spans of all three species are the same.

9.3.10 Survival Effects Models

In root longevity studies, the relationship between root life span and some environmental or treatment factor, or an inherent property of the roots themselves, may be of interest. The covariates presumed to be causally related to longevity and survivorship fall into one of two classes: group-based, or individual-based. Examples of the former might include irrigation level, soil temperature, or species, or any other factor to which all roots in a sample are exposed. Individual-based covariates might include root diameter, length or colour. The statistical models used to test these relationships are known as survival effects models. The details of computing both group and individual-based survival effects models are too complex to cover here. Moreover, they are too cumbersome to do by hand. We will thus limit our discussion here to the general principles behind the models' analyses, and some current statistical packages capable of performing such analyses.

Tests of association for individual-based covariates are made with likelihood ratio tests. Likelihood ratio tests are used to test the hypothesis that a model incorporating the effects of one or more individual-based covariates better describes root life spans (in terms of the maximum likelihood estimator λ) than a reduced model in which the covariate(s) has been removed. Since λ can be calculated for exponential, Weibull, lognormal and gamma distributions, and for both censored and non-censored data, likelihood ratio tests can be used with a wide variety of data sets. Among survival analysis programs, SURPH (Center for Quantitative Science, School of Fisheries, University of Washington, Seattle, WA, USA) uses likelihood ratio tests in the construction of survival effects models, and further details of the procedure can be found in the SURPH operating manual and in Hoffmann (1993).

Lee (1992) describes Cox's proportional hazards regression model in which the relationship between group-based covariates and survival times can be examined. The model is non-parametric, and can be used with both censored and non-censored data. The hazard function underlying the survival distribution of the sample serves as the dependent variable, and the covariate effects on the hazard function are proportional to one another; that is, in a sample in which the roots are exposed to one of two covariates, the effect of covariate 2 on root longevity is equal to the effect of covariate 1 multiplied by a constant factor. To estimate the β values associated with each covariate in the regression equation, a maximum likelihood procedure is used where the likelihood function is based upon a conditional probability of failure associated with each covariate. The β's in the model can be added and removed based upon their effect in the presence of other covariates in a step-wise fashion in a manner analogous to multiple linear regression using log-likelihood values.

Two other group-based analyses are available in statistical software programs. The LIFETEST procedure in the SAS (SAS Institute, Cary, NC, USA) uses a rank test of covariate association with life spans that is based upon the rank tests for comparing life span distributions discussed above. Rank statistics are derived from the sum (across all observations) of the product of individual observation rank scores and corresponding covariate values. Ranks are corrected for censoring in one of two ways, depending upon whether the scores are log or Wilcoxan values. Chi-square values for each covariate are generated in the output, as is a stepwise regression model showing the relative improvement in the fit of the model as covariates are added from most to least significant. Further computational and output details can be found in the SAS/STAT user guide. In addition, Allison (1995) has published an excellent manual describing applications and examples of SAS's survival estimation procedures, including suggestions based upon particular data types or desired tests.

The effects of group-based covariates for both single and multiple samples of root life spans are evaluated via analysis of deviance (ANODEV) in the program SURPH. ANODEV is somewhat analogous to an analysis of variance (ANOVA). However, instead of being structured around decomposing sums of squares in order to evaluate treatment effects as with an ANOVA, ANODEV is based upon decomposing the discrepancy between the log-likelihood value attained with a "maximum" model incorporating the effects of all covariates and those models with one or fewer covariates, including the minimum or null model. The deviance of a particular model is defined as twice the difference between the model under consideration and the full model. The fit of an intermediate model is assessed by how much of the discrepancy between the full and null model it explains, accounting for the number of parameters or covariates in the model in question. The deviance is distributed in an asymptotic χ^2 manner. Further details on the theory, computation and presentation of ANODEV can be found in the SURPH users manual.

A newer public-domain program, MARK (Colorado State University, Boulder, CO, USA, http://www.cnr.colostate.edu/~gwhite/mark/mark.htm) can be used to construct models incorporating either group- or individual-based variables. MARK generates survival probabilities for known-fate data (i.e. roots whose fates are followed over time), and can partition variation in life spans into time, age (i.e. different cohorts in the same treatment), treatment and other attributable effects, and their interactions. MARK is based upon Cormack-Jolly-Seber models (Schwarz and Arnason 1996) of survival probabilities, and generates AIC (Akaike's information criterion) scores, which are measures of model parsimony (higher AIC scores = less parsimonious models). Likelihood ratio tests are used to compare models (e.g. time × age × treatment vs. time × age). Details, with an application to the effects of browsing on root lifespan, can be found in Ruess et al. (1998).

9.4 Nutrient Fluxes and Root Longevity

Estimates of fine root longevity, length productivity and length mortality derived from direct observations of fine root life history can be used to estimate the amount of biomass and nutrients associated with these processes. A general estimate of annual biomass and nutrient transfers to (and from) roots on an area basis can be made by multiplying the inverse of fine root longevity expressed in years (years^{-1}) by mean annual root standing biomass or nutrient crop (e.g. $\mathrm{g\,m^{-2}}$). For example, if average fine root longevity is 6 months (0.5 year) and biomass is $250\,\mathrm{g\,m^{-2}}$, annual biomass production (or mortality) can be calculated as:

$$\frac{1}{0.5\ \text{year}} \times 250\ \mathrm{g\,m^{-2}} = 500\ \mathrm{g\,m^{-2}\ year^{-1}}.$$

If mean fine root biomass is stable from year to year, biomass and nutrient allocation to production can be assumed to be equivalent to mortality. However, biomass would not be stable in an immature, developing ecosystem, and the magnitude of inter-annual fluctuations in root biomass and nutrient content in even mature systems are unknown. Even if root biomass and nutrients are stable on an annual basis, several biomass estimates must be made throughout the year, as they can fluctuate widely on an annual basis. Moreover, fine root life spans can vary substantially throughout the year, and can dramatically affect the rate at which the biomass or nutrients associated with any period of productivity is lost to mortality during a particular time interval. For example, sugar maple fine roots produced in the autumn can live for several months longer than roots produced in the spring (Hendrick and Pregitzer 1993). The biomass and nutrients associated with spring root production are therefore cycled much more rapidly than those associated with autumn root senescence.

More accurate measures of biomass and nutrients allocated to root production and returned through mortality for any period of time, or a series of time periods, can be made by direct observations of individual fine roots using some type of rhizotron. The procedure is as follows.

1. With a (mini)rhizotron or other direct observation technique, quantify fine root length at time $t = 0$, and then periodically thereafter.
2. Sample root biomass and/or nutrient content at time $t = 0$ concurrent with the length sampling, and then once more at some subsequent sampling date, preferably when root length is either known or suspected to be greater or less than it was at time $t = 0$.
3. For each successive time period, fine root production and mortality are calculated as a proportion of fine root length at time $t = 0$, as follows:

$$\frac{\text{length production}_{t \text{ to } t=1}}{\text{length}_{t=0}} = \begin{array}{l}\text{proportional root production}\\ \text{for time period } t \text{ to } t+1\end{array}$$

4. Multiply proportional production based upon length by biomass or nutrient content at time $t = 0$.
5. Check accuracy by adding production estimates and subtracting mortality estimates for intervals between initial and final biomass samplings, and by comparing predictions of second biomass observations with actual mass determinations.

Successful application of this approach is based upon several assumptions: (1) the dynamics of fine root biomass and length within a narrow diameter class mimic one another in magnitude, (2) length densities along the (mini)rhizotron have stabilized (i.e. no further post-installation recolonization), and (3) when using the approach to calculate nutrient flux, there is no retranslocation of the nutrient in question from dying roots. Hendrick and Pregitzer (1993) showed that the technique can be used to accurately calculate root biomass dynamics, but the extent to which assumption (3) is true has yet to be proved or disproved conclusively.

9.5 The Future

We believe that the prognosis for continued and improved use of observational techniques to study fine root longevity is good. With respect to data collection, continuing improvements in video technology are constantly being made in consumer products (e.g. video camcorders) and it is these developments that underpin the advances we are able to make in camera-based research tools. More specialized instruments, like the Bartz Technology (Santa Barbara, CA, USA) minirhizotron cameras are also improving. The Bartz cameras are now available in two sizes (2.86 and 5.1 cm), and options include a 100 x zoom function as well as a UV light source. Recent developments by Bartz Technology will allow direct recording of root imagery onto digital media in the field, bypassing the cumbersome and quality-degrading digital-analog-digital conversions associated with image acquisition (digital camera), collection (analog videotape) and analysis (image redigitization).

Further developments and refinements in image analysis tools are also being made. ROOTS (Michigan State University, East Lansing, MI, USA), the first commercially available program developed for analyzing minirhizotron data, is in the early stages of a much-needed update. RooTracker (Duke University, Durham, NC, USA), a Macintosh-based program, also continues to be refined. Scientists working with the Desert Research Institute (Reno, NV, USA) have developed a more automated system (ARCOS) that is also undergoing refinement based upon user feedback. Scientists at the Rhizolab (Wageningen, Netherlands) use a public-domain program (NIH-Image, Macintosh or PC) developed by the US National Institute of Health (http://rsb.info.nih.gov/nih-image). Finally, the development and application of new software programs (e.g. MARK, SURPH) and upgrades of existing software (e.g. SAS) facilitate the design and analysis of more complicated experiments and the extraction of more information about the dynamics of roots and root systems.

The population of scientists studying root longevity and related demographic processes is relatively small, and time and monetary investments into improving our ability to study roots are likely to be also. However, we do stand to benefit tremendously from continuing improvements in consumer electronics and the importance of demographic statistical and numerical computational techniques to a wide array of other disciplines, including forestry, wildlife management, medicine and pharmaceuticals, and the insurance industry. Our task, then, is perhaps not so much to develop novel techniques, but rather to adopt or adapt novel means of applying existing, evolving or new techniques and technology to the study of roots.

References

Allison PD (1995) Survival analysis using the SAS system: a practical guide. SAS Institute, Cary, North Carolima, 292 pp

Andren O, Rajkai K, Katterer T (1991) A non-destructive technique for studies of root distribution in relation to soil moisture. Agric Eco Environ 34: 269–278

Atkinson D (1983) The growth, activity and distribution of the fruit tree root system. Plant Soil 71: 23–37

Atkinson D (1985) Spatial and temporal aspects of root distribution as indicated by the use of a root observation laboratory. In: Fitter AH, Atkinson D, Read DJ, Usher MB (eds) Ecological interactions in soil. Blackwell, Oxford, pp 43–65

Atkinson D (1986) The nutrient requirements of fruit trees: some current considerations. Adv Plant Nutr 2: 93–128

Atkinson D, Fogel R (1997) Roots: measurement, function and dry matter budgets. In: Vangardingen FR, Foody GM, Curran PJ (eds) Scaling-up from cell to landscape. Cambridge University Press, pp 151–172

Berta G, Trotta A, Fusconi A, Hooker JE, Munro M, Atkinson D, Giovannetti M, Morini S, Fortuna P, Tisserant B, Gianinazzi-Pearson V, Gianinazzi S (1995) Arbuscular mycorrhizal induced changes to plant growth and root system morphology in *Prunus cerasifera*. Tree Physiol 5: 281–294

Black KE, Harbron CG, Franklin M, Atkinson D, Hooker JE (1998) Differences in root longevity of some tree species. Tree Physiol 18: 259–264

Cannell MGR, Dewar RC, Thornley JHM (1992) Carbon flux and storage in European forests. In: Teller A, Mathy P, Jeffers JNR (eds) Responses of forest ecosystem to environmental changes. Elsevier, London, pp 256–271

Conover WJ (1980) Practical nonparametric statistics, 2nd edn. John Wiley, New York

Cox DR, Oakes D (1984) Analysis of survival data. Chapman and Hall, London

Curtis PS, Zack DR, Pregitzer KS, Teri JA (1995) Above and below ground response of *Populus grandidentata* to elevated atmospheric CO_2 and soil N availability. Plant Soil 165: 45–54

Drew MC (1987) Function of root tissue in nutrient and water transport. In: Gregory PJ, Lake JV, Rose DA (eds) Root development and function. Cambridge University Press, Cambridge, pp 71–102

Dubach M, Russelle MP (1995) Reducing the cost of estimating root turnover with horizontally installed minirhizotrons. Agron J 87: 258–263

Fairley RI, Alexander IJ (1985) Methods of calculating fine root production in forests. In: Fitter AH, Atkinson D, Read DJ, Usher MB (eds) Ecological interactions in soil. Blackwell, Oxford, pp 37–42

Fogel R (1985) Roots as primary producers in below-ground ecosystems. In: Fitter AH, Atkinson D, Read DJ, Usher MB (eds) Ecological interactions in soil. Blackwell, Oxford, pp 23–36

Fogel R (1991) Root system demography and production in forest ecosystems. In: Atkinson D (ed) Plant root growth. Blackwell, Oxford, pp 89–102

Fogel R, Hunt G (1979) Fungal and arboreal biomass in a western origin Douglas fir ecosystem: distribution patterns and turnover. Can J For Res 9: 245–256

Forbes PJ, Ellison CH, Hooker JE (1996) The impact of arbuscular mycorrhizal fungi and temperature on root system development. Agronomie 159: 617–620

Forbes PJ, Black KE, Hooker JE (1997) Temperature induced alteration to root longevity of *Lolium perenne*. Plant Soil 190: 87–90

Gehan EA (1965) A generalized Wilcoxan test for comparing arbitrarily singly-censored data. Biometrika 52: 650–653

George SL, Desu MM (1974) Planning the size and duration of a clinical trial studying the time to some critical event. J Chron Dis 27: 15–24

Goins GD, Russelle MP (1996) Fine root demography in alfalfa (*Medicago sativa* L). Plant Soil 185: 281–291

Gregory PJ (1987) Development and growth of root systems in plant communities. In: Gregory PJ, Lake JV, Rose DA (eds) Root development and function. Cambridge University Press, Cambridge, pp 147–166

Harris WF, Santantonio D, McGinty D (1980) The dynamic below ground ecosystem. In: Wearing RH (ed) Forests: fresh perspectives from ecosystem analysis. pp 119–129

Head GC (1966) Estimating seasonal changes in the quantity of white unsuberized root on fruit trees. J Hortic Sci 41: 197–206

Hendrick RL, Pregitzer KS (1992) The demography of fine roots in a northern hardwood forest. Ecology 73: 1094–1104

Hendrick RL, Pregitzer KS (1993) Patterns of fine root mortality in two sugar maple forests. Nature 361: 59–61

Hoffman A (1993) Quantifying selection in wild populations using known-fate and mark-recapture designs. PhD Dissertation, University of Washington, Seattle, Washington, 259 pp

Hooker JE, Atkinson D (1996) Arbuscular mycorrhizal fungi-induced alteration to tree-root architecture and longevity. Pflanzenernaehr Bodenkd 159: 229–234

Hooker JE, Munro M, Atkinson D (1992) Vesicular-arbuscular fungi induced alteration in poplar root system morphology. Plant Soil 145: 207–214

Hooker JE, Black KE, Perry RL, Atkinson D (1995) Arbuscular mycorrhizal fungi induced alteration to root longevity of poplar. Plant Soil 172: 327–329

Kalbfleisch JD, Prentice RL (1980) The statistical analysis of failure time data. John Wiley, New York

Klepper B (1987) Origin, branching and distribution of root systems. In: Gregory PJ, Lake JV, Rose DA (eds) Root development and function. Cambridge University Press, Cambridge, pp 103–124

Kruskal WH, Wallis WA (1952) Use of ranks in one-criterion variance analysis. J Am Stat Assoc 47: 583–621

Lee ET (1992) Statistical methods for survival data analysis, 2nd edn. Wiley, New York, 482 pp

MacKenzie KAD (1979) The development of the endodermis and phi layer of apple roots. Protoplasma 100: 21–32

Makuch RW, Simon RM (1982) Sample size requirements for comparing time-to-failure among k treatment groups. J Chron Dis 35: 861–867

Mantel N (1966) Evaluation of survival data and two new rank order statistics arising in its consideration. Cancer Chemother Rep 50: 163–170

Mantel N (1967) Ranking procedures for arbitrarily restricted observations. Biometrics 23: 65–78

Muenchow G (1986) Ecological use of failure time analysis. Ecology 67: 246–250

Nadelhoffer KJ, Aber JD, Melillo JM (1985) Fine roots, net primary production and nitrogen availability: a new hypothesis. Ecology 66: 1377–1390

Norman JR, Atkinson D, Hooker JE (1996) Arbuscular mycorrhizal fungal-induced alteration to root architecture in strawberry and induced resistance to the root pathogen *Phytophthora fragariae*. Plant Soil 185: 191–198

Pollock KH, Winterstein SR, Conroy MJ (1989) Estimation and analysis of survival distributions for radio-tagged animals. Biometrics 45: 99–109

Pregitzer KS, Hendrick RL, Fogel R (1993) The demography of fine roots in response to patches of water and nitrogen. New Phytol 125: 575–580

Pyke DA, Thompson JN (1986) Statistical analysis of survival and removal rate experiments. Ecology 67: 240–245

Ruess RW, Hendrick RL, Bryant JP (1998) Regulation of fine root dynamics by mammalian browsers in early successional Alaskan taiga forests. Ecology 79: 2706–2720

Schwarz CJ, Arnason AN (1996) A general methodology for the analysis of capture-recapture experiments in open populations. Biometrics 52: 860–873

Sokal RR, Rohlf FJ (1981) Biometry, 2nd edn. WH Freeman, New York

Steen E (1991) Usefulness of the mesh bag method in quantitative root studies. In: Atkinson D (ed) Plant root growth. Blackwell, Oxford, pp 75–88

Thomas SM, Whitehead D, Adams JA, Reid JB, Sherlock RR, Leckie AC (1996) Seasonal root distribution and soil surface carbon fluxes for 1-year-old *Pinus radiata* trees growing at ambient and elevated carbon dioxide concentrations. Tree Physiol 16: 1015–1021

Wang Z, Burch WH, Mou P, Jones RH, Mitchell RJ (1995) Accuracy of visible and ultraviolet light for estimating live root proportions with minirhizotrons. Ecology 76: 2330–2334

CHAPTER 10

Root Image Analysis and Interpretation

W. Richner, M. Liedgens[1], H. Bürgi[2], A. Soldati, and P. Stamp[1]

[1] ETH Zürich, Institute of Plant Sciences, ETH Zentrum, LFW A4, 8092 Zürich, Switzerland
[2] Swiss Federal Office of Information Technology, Systems, and Telecommunication (FOITT), Monbijoustrasse 74, 3003 Bern, Switzerland

CONTENTS

A.L. Smit et al. (Eds.), Root Methods

10.1 Introduction

For the description and modelling of root and shoot growth, quantitative information on root morphology and architecture has to be gathered in ecological, agronomic, biological and physiological studies of soil exploration and water and nutrient uptake by root systems. In the past, root length and surface area of washed samples were measured manually, using linear measurement devices (Böhm 1979), solution adsorption methods (Sattelmacher et al. 1983) or line-intercept techniques (Newman 1966; Tennant 1975). In more recent years, image analysis systems have become available for these measurements. Compared with traditional methods, image analysis is faster, more accurate and less prone to human error; with these techniques it is also possible to measure morphological parameters other than the total length of root samples.

Contrary to classical destructive root investigations, the rhizotron and minirhizotron techniques allow the nondestructive monitoring of root dynamics in situ. Due to the complexity of root images recorded through transparent viewing surfaces against the soil, they have been analysed manually so far. However, automated or at least controlled image analysis techniques are necessary for measuring other morphological root parameters on images taken in situ, owing to the large number of pictures that have to be recorded in rhizotron or minirhizotron studies and the limited amount of information that can be obtained by manual counting or measuring.

This chapter will present state-of-the-art image analysis of washed root samples and minirhizotron or rhizotron recordings, describe important steps in the processing of root samples and discuss research needs in these fields.

10.2 Image Analysis of Washed Root Samples

10.2.1 Preparation of Samples

10.2.1.1 Mechanical Separation of Roots from Extraneous Objects

When root samples are washed free from bulk soil, they generally contain soil aggregates, sand grains, plant residues (remains of shoots and roots of previous crops or weeds), bodies and eggs of soil animals, as well as weed and crop seeds. The kind and amount of such unwanted objects left in the root samples depends on the soil type, the crop rotation, the way in which the roots are washed (see Chap. 6) and especially on the mesh size used for collecting washed roots. Automated washing systems allow for smaller mesh sizes. Thus, root samples washed through finer sieves generally contain more fine roots (Livesley et al. 1998), but also more extraneous objects than samples washed manually with coarser sieves. Therefore, mechanical separation of roots from extraneous objects usually precedes measurements of root morphology.

Decanting or flotation techniques (Schuurman and Goedewaagen 1971) can separate roots from unwanted objects that have a specific weight markedly different from that of the roots, such as soil and sand fractions and heavy seeds of weeds. Murphy and Smucker (1995) separated the majority of extraneous objects from roots by repeated decanting of the organic debris that floated more readily than roots. Subsequently, tweezers are used to manually extract some or all of the remaining extraneous objects. If an image analysis system were able to discriminate between roots and extraneous objects according to differences in their shapes, the manual removal process could be less meticulous and could concentrate on removing large extraneous objects, which might cover roots partly or completely, and on residue parts that have a shape similar to that of root objects. If measurement techniques are employed which cannot discriminate between roots and unwanted objects, then the use of tweezers to remove extraneous objects should be done as thoroughly as possible; remaining objects that have a size above the lower limit of spatial resolution of the image acquisition device will be detected as roots, and thus contribute to an overestimation of the total root length. Cleaning roots by hand is very tedious. Thus, image analysis systems should be able to discriminate between roots and unwanted objects (see Sect. 10.2.3.2) if these systems are to be truly advantageous.

10.2.1.2 Staining Roots

The purpose of staining is to enhance the contrast of a root sample. Especially when using simple thresholding techniques (see Box 10.1), a sharp contrast between roots and the background is essential. Because roots are often recorded against a light background, they must usually be stained to allow for sufficiently contrasting images. A variety of dyes of different colours have been used for staining washed roots (Table 10.1). No general recommendations on the use of dyes can be given, because staining depends on root morphological and chemical properties. These may be influenced by the plant species, plant age and the environment in which the plants grew. Because some dyes are toxic to humans, non-toxic dyes should be used or gloves should be worn when handling toxic dyes and stained samples.

BOX 10.1. Image Segmentation

Image segmentation or discrimination is the process of dividing grey-level images into regions or objects of interest for measurements or recognition (Russ 1995), e.g. into plant roots and background. Segmentation produces binary (i.e. black-and-white) images which can be used for further analyses. The most important segmentation technique is grey-level thresholding, but there are numerous alternative, more sophisticated segmentation methods (Fu and Mui 1981; Haralick and Shapiro 1988; Glasbey and Horgan 1995) that are beyond the scope of this chapter.

Grey-level thresholding is very simple and easy to implement. It separates regions of an image into a set of classes, usually two, according to one or two threshold grey-levels. In other words, thresholding establishes which range of grey-levels belongs to the objects and which to the background. When using grey-level thresholding for the segmentation of images, the most critical step is setting the threshold. In an ideal situation, object pixels are restricted to a narrow range of grey-levels, the remaining grey-levels belonging to the background. In this case it is very easy to visually set the threshold value. With real-life images, the brightness histogram of the image can help to reliably determine the threshold value. In the case of a bimodal brightness histogram, the grey-level threshold is chosen from the values between the two peaks (Castleman 1979; Lebowitz 1988). This is, however, not always consistent from one operator to another (Russ 1995), therefore a variety of automated thresholding techniques have been developed, which will not, however, be discussed here.

Roots are usually soaked in the dye solution from a few hours to a few days, depending on the dye, until staining is complete. After staining, roots of all diameter classes should be sufficiently and evenly stained; if not, the staining time may have been too short. Thicker roots sometimes require longer staining, depending on the plant species or dye. It is not always possible to achieve complete staining of the whole sample. Thus, several dyes may have to be tested to determine optimal staining of roots for a given species and plant age. To prevent a degradation of samples, root samples should be placed in a refrigerator or cooling room during staining at temperatures slightly above freezing (usually around 4 °C). Prior to the spreading of the roots for recording, they must be removed from the dye solution with a sieve whose mesh size is smaller than the smallest root objects. Then the dye adhering to the roots in the sieve should be rinsed off with water to avoid staining the scanning tray and the water

Table 10.1. Dyes for staining washed roots to enhance the contrast between roots and a light background

Dye	Authors
Acid fuchsin	Barnett et al. (1987)
Aniline blue	Barnett et al. (1987)
Coomassie Brillant Blue	Tanaka et al. (1995)
Crystal violet	Wilhelm et al. (1983); Kaspar and Ewing (1997)
Dylon stain	Murakami and Yoneyama (1988)
Fluorescein diacetate	McGowan et al. (1983)
Malachite green oxalate	Murphy and Smucker (1995)
Methyl violet	Harris and Campbell (1989)
Methylene blue	Sackville Hamilton et al. (1985); Smit et al. (1994)
Trypan blue	Collins et al. (1987)

in which the roots will be placed. If this were to occur, then the contrast between root boundaries and the background would be strongly reduced.

Contrast can be enhanced by choosing an appropriate background. If white unstained roots are to be measured, a dull dark background is desirable. We found that dull matt black velvet is a suitable background under a transparent scanning tray for recording unstained maize roots, because the rough surface of the velvet does not reflect light to the same extent as smooth objects painted black. On the other hand, stained roots should be measured against a light background. Light boxes or light tables are sometimes placed under a transparent scanning tray; the emitted light is often diffused by means of a milky glass plate, such as white opal glass (Barnett et al. 1987).

10.2.1.3 Spreading of Roots in the Recording Tray

Roots are usually spread out in shallow transparent trays that allow for back-lit camera recording from above or for recording with flatbed scanners from below. Glass trays are generally preferred to plastic materials, because they are more resistant to scratches, which may be resolved as false roots. The trays are usually filled with 2 to 3 mm of water. In this way, roots may be easily separated, and yet are prevented from floating around in the tray. Because the slightest movement of the tray will cause smaller root parts to move around, the final spreading should preferably be done at the place of recording.

Roots should be spread out in the scanning tray so that individual roots do not overlap, thus causing root length to be underestimated and root diameters and branching to be overestimated. In addition, roots, especially neighbour lateral roots, should not stick together after spreading. If the distance between two parallel roots is in the range of a pixel (short for picture element) of the recording device, then the two roots may be detected as one thick root, thus overestimating the mean root diameter and underestimating the total root length (Pan and Bolton 1991) and branching. It can be advantageous to cut longer root segments into shorter pieces to ease spreading and to reduce underdispersion across the recording area (Reicosky et al. 1970) if root samples are large and highly branched. Because many techniques for determining morphological parameters are quite sensitive to orientation of the objects in the images (see Sect. 10.2.3.2), roots should be arranged as randomly as possible to achieve a uniform distribution of orientations of root segments.

Depending on the size and the type of sample, however, spreading can be very time-consuming. Therefore, a compromise between accuracy of spreading and time requirement must be found.

10.2.2 Image Acquisition

10.2.2.1 Recording Devices

The most frequently used devices to record images of washed root samples for image analysis are electronic cameras and scanners (Fig. 10.1), both based on CCD (charge coupled device) sensors. Other devices, such as densitometers (Voorhees et al. 1980) or photocopiers (Collins et al. 1987), are no longer very important because of generally limited spatial resolutions or their intricate

Fig. 10.1. Stained roots spread out in a transparent tray on top of a desktop scanner. (Photo with permission from U. Zimmermann, Institute of Plant Sciences, ETH Zürich)

Table 10.2. Sizes of black-and-white, 8-bit grey-level and 24-bit colour images of A4 dimension (21.0 × 29.7 cm) scanned at resolutions of 150, 300, 600 and 1200 dpi[a]

Image type	Resolution (dpi)			
	150	300	600	1200
	File size (MB[b])			
Black-and-white (single-bit)	0.3	1.0	4.1	16.6
256 grey-levels (8-bit)	2.1	8.3	33.2	132.8
16.7 million colours (24-bit)	6.2	24.9	99.6	398.3

[a] Dots per inch.
[b] Megabytes.

handling. Imaging devices should produce images of roots which contrast strongly with the background at a spatial resolution high enough to resolve the thinnest roots in a sample. Thus, the critical aspects of image acquisition devices are their radiometric and spatial resolutions (Glasbey and Horgan 1995).

The radiometric resolution is the maximum number of brightness levels that may be recorded. In an image, a value of brightness is associated with each pixel. For grey-scale images this value frequently ranges from 0 to 255 (8-bit grey scale images); 0 usually stands for black and 255 for white. For most biological applications, between 16 and 256 grey-levels are considered sufficient (Glasbey and Horgan 1995). Most of the modern CCD cameras and scanners allow the recording of coloured images, but almost all present root image analysis systems still rely on monochrome information only, owing to the large file sizes of digitised colour images (Table 10.2).

The spatial resolution is given by the number of pixels per unit scanning area, in the case of scanners usually expressed as dpi (dots per inch). It determines the size of the smallest objects that can be recorded. Recent research has shown that 56% of the total root length of field-grown maize consisted of roots thinner than 0.175 mm (Pallant et al. 1993). Roots thinner than 0.5 mm may constitute up to 80% of the total root length of a barley root system (Hackett 1968). Due to the significant proportion of fine roots in most root systems, the maximum spatial resolution of a recording device should allow for the recording of the thinnest roots in a sample. Roots with diameters close to one pixel length may cause problems in thresholding (see Box 10.1) of root images (Smit et al. 1994), i.e. the division of them into roots and background. As a conservative rule of thumb, the diameter of the thinnest roots in a sample should be at

Table 10.3. Smallest root diameters that can be theoretically and practically resolved at scanning resolutions of 150, 300, 600 and 1200 dpi.[a] The calculation of the practically smallest root widths is based on minimal root widths of three pixels, which is required by most image analysis algorithms for reliably measuring root diameters

	Scanning resolution (dpi)			
	150	300	600	1200
	Resolved root diameter (μm)			
Theoretical	169	85	42	21
Practical	508	254	127	64

[a] Dots per inch.

least about three times the pixel size of the recording device to ensure a reliable detection of roots and an accurate measurement of root diameter (see Sect. 10.2.3.2). Thus, the practical maximum resolution of an imaging device is somewhat lower than the theoretical maximum resolution (Table 10.3).

Cameras. Electronic cameras turn analogue voltage signals into digital signals using analogous-to-digital (A/D) converters, such as computer frame grabbers, whereas digital cameras and scanners do not require such converters. The main advantage of cameras is their short scanning time and, in the case of CCD cameras, their high spatial resolution. While the outdated tube cameras often had optical resolutions too limited to detect higher-order lateral roots (Voorhees et al. 1980; Lebowitz 1988; Cunningham et al. 1989; Harris and Campbell 1989), newer CCD cameras usually have typical spatial resolutions of about 700 × 500 pixels. More expensive CCD cameras have superior resolutions of up to 5000 × 5000 pixels (Krumenaker 1994).

By keeping scanning areas relatively small, however, the necessary spatial resolutions for recording fine roots may also be achieved with cheaper cameras. Especially with low-resolution cameras, however, this may lead to very small recording areas. Because preparing and spreading roots in a number of small scanning trays is more time-consuming than using a larger tray covering the same area, recording samples may be optimised by using robotic cameras that are mounted on an X–Y moving gantry (Murphy and Smucker 1995). This allows a large scanning area to be divided into several sub-images of constant dimension, which are recorded automatically one after the other. Thus, spatial resolution may be kept sufficiently large for the individual sub-images. Additionally, optical distortion at the boundaries of the recording area is reduced when scanning areas are smaller compared to recording larger areas with a greater distance between the camera and the tray. However, some of the recorded objects may be truncated by the sub-image borders, thus dividing a root into two parts. This is a potential problem when roots are discriminated from extraneous objects by shape indices (see Sect. 10.2.3.2), because the truncated root segments may no longer meet the shape requirements of roots, and thus would be eliminated as extraneous objects, leading to an underestimation of total root length and area.

A general problem with camera-recording is achieving uniform lighting of the scanning area. Samples are lit from the top and/or sides using incandescent or special lamps, or from the back using light trays. Lighting from above using lamps may produce shadows, especially of thick roots, and extraneous light may influence the recording. Therefore, it is usually preferable to back-light the objects. In this way roots can be recorded as dark objects before a light background. A milky glass plate is usually used with light-boxes to produce diffused light. To prevent reflections from the room lighting or sunlight from windows, shields or boxes are sometimes used to prevent extraneous light from reaching the scanning area. To reduce a further source of error, Russ (1995) suggests avoiding automatic gain or brightness circuits that are available in some types of standard cameras, because they may change the image contrast or linearity in response to bright or dark regions that do not even lie within the digitised portion of the image.

Scanners. Like CCD cameras, scanners use a solid-state detector array for recording images, but in contrast to cameras, this array is linear only and moves along the scanning area for recording (Russ 1995). Scanners have been used for recording washed root samples that are spread in transparent trays (Krstansky and Henderson 1989; Pan and Bolton 1991; Smit et al. 1994). These are generally standard flatbed scanners, which are primarily designed to scan flat objects, but there are also special 3-D scanners that have a depth of resolution of several centimetres (Smit et al. 1994). Roots are usually spread in 2 to 3 mm of water; therefore, the depth of resolution of standard flatbed scanners is usually sufficient for accurate recording of root samples (Kirchhof and Pendar 1993).

Desktop scanners provide high quality optical characteristics for recording images, thus enabling accurate and precise estimates of root length and other morphological root parameters (Pan and Bolton 1991). A major advantage is their high spatial resolution, which is up to 600 dpi even with mid-level scanners, which is often better than the resolution of cameras (Smit et al. 1994). Only the *optical* resolution of scanners should be considered, because the higher *interpolated* resolutions that are given by many scanner manufacturers do not really mean a higher spatial resolution. In addition, lighting problems are reduced, compared with camera-based systems, although light intensity may be lower near the boundaries of the scanning area (Russ 1995). To reduce uneven lighting, the warm-up time of the scanner should be long enough for the light source to become stable (Russ 1995). Scanners may also cast shadows around the roots, although usually to a lesser extent than when a light source is placed above the root tray for the recording of roots with a top-mounted camera. To reduce this problem, scanned root samples can be simultaneously lit from below and from above (Bauhus and Messier 1999) using a transparency adapter unit in place of the scanner lid.

A disadvantage is that scanners are much slower at capturing an image than are CCD cameras with a fixed scanning area (Berntson 1992), but the time required for recording with a scanner is still much shorter than for spreading a root sample. On the other hand, camera-based systems mounted on an X–Y moving gantry may take even longer to record a scanning area that is divided into a large number of sub-images, although its total size is comparable to that of a desktop scanner.

Using conventional mid-level desktop scanners with an optical resolution of 600 dpi, roots as thin as about 130 μm can be resolved (Table 10.3), provided that the contrast between roots and the background is sharp and uniform. This resolution results in image file sizes from 4 to over 30 megabytes (MB) when scanning A4-sized black-and-white or 256-grey-levels images, respectively (Table 10.2). If lateral roots with diameters of about 50 μm are to be measured, which is in the range of the thinnest reported roots (Pallant et al. 1993), scanners with a spatial resolution of at least 1200 dpi must be used (Table 10.3). This, however, results in huge image files, e.g. more than 130 MB with 256 grey-levels (Table 10.2). This is beyond the processing capability of most PC-based image analysis programs for measuring roots. To overcome this limitation, it can be advantageous to split up such large images after recording into several smaller images or, alternatively, only record parts of the whole scanning area at once.

10.2.2.2 Storage of Images

While older root image analysis systems (e.g. "Delta-T Mk II", Delta-T Devices, Burwell, UK; Table 10.4) were designed for real-time processing of root samples, with neither necessity nor possibility of storing digital images, the majority of newer systems require the creation of digital images prior to processing. For most projects involving measurements of morphological traits of washed root samples, it is worthwhile storing the recorded images at least until the end of the project so that samples can be remeasured if necessary. Furthermore,

Table 10.4. Overview of widely-used root image analysis systems

System	Platform	Principle	Individual object measurements	Major root parameters provided	Manufacturer/ programmer
Delta–T Mark II	Modified area meter	Line–intercept method (Harris and Campbell 1989)	No	Length, area, mean diameter (derived from length and area measurements)	Delta–T Devices Ltd., Burwell, Cambridge, UK
Delta–T SCAN	DOS	Modified length–intercept method (Kirchhof and Pendar 1993)	Yes, but not in length-measurement mode	Length (grouped into user-definable diameter classes), area, number of root tips	Delta–T Devices Ltd., Burwell, Cambridge, UK (*http://www.delta-t.co.uk*)
NIH Image 1.62	Macintosh[a]	Macro–based multipurpose system (Rasband and Bright 1995); used for root studies by several groups (e.g. Smit et al. 1994; Dowdy et al. 1995; Tanaka et al. 1995)	Yes	Multipurpose system	W. Rasband, National Institutes of Health (NIH), Bethesda, MD, USA (public domain software; *http://rsb.info.nih.gov/nih-image*)
ROOTEDGE 2.3	DOS	Edge chord algorithm (Ewing and Kaspar 1995; Kaspar and Ewing 1997)	Yes	Length, width, perimeter	R.P. Ewing and T.C. Kaspar, USDA/ARS National Soil Tilth Laboratory, Ames, IA, USA (public domain software: *http://www.nstl.gov/software/rootedge/*)
WinRHIZO/ MacRHIZO	Windows/ Macintosh	Skeletonisation method (Bauhus and Messier 1999)	Yes	Length (grouped into user-definable diameter classes), area, diameter, number of root tips and root forks, topology and fractal analysis, colour analysis (see Chap. 4)	Régent Instruments Inc., Quebec, Canada (*http://www.regent.qc.ca*)

[a] "Scion Image", a Windows version of "NIH Image", is available from Scion Corporation, Frederick, MD, USA (http://www.scioncorp.com).

images can be analysed for additional morphological parameters later when new algorithms become available.

If images have to be analysed using several computer programs, or if they are exchanged with co-operating laboratories, a compatible image format should be chosen. There are a variety of image file formats, but only a few of them are supported by various computer platforms and programs. The most widely supported bitmap file format is the TIFF (tagged image file format) format. TIFF is capable of handling grey-level and colour data and has the ability to store data in a compressed form. There are, however, a variety of options for TIFF files, so that a TIFF file created by one program may not be readable by another (Russ 1995). On the other hand, there are inexpensive programs for both personal computers (PCs) and Macintosh computers for easily converting images between different file formats.

As shown in Table 10.2, increasing the spatial and radiometric resolutions leads to large images, which require more space for storage. This can amount to several gigabytes of storage capacity, depending on the size of a research project. Image compression can reduce the amount of storage space, and, ultimately, the cost, but a discussion of compression techniques is beyond the scope of this handbook. See Jaïn (1989) for more information on image compression.

Data are usually stored on floppy disks, computer hard disks, tapes and various optical storage devices. Because of their limited capacity, genuine floppy disks with a capacity of below 2 MB are not suited for storing images. Although hard disks with a capacity of 30 gigabytes and more have become much less expensive, they are still too expensive for the long-term storage of large amounts of stored information as compared with other media. Hard disks have the advantage of the fastest access times compared with other storage media. On the other hand, the storage cost per MB is still too great and the lifetime of data is short compared with other storage media. Although somewhat slower than hard disks, optical storage media still have relatively short access times, but like hard disks they are comparatively expensive per amount of stored information. Compact disc writers are now less expensive and offer a good way of storing up to 650 MB of data permanently on inexpensive CD-ROM or temporarily on CD-RW media. Tapes of a variety of standards seem to be an easy and inexpensive storage medium for large numbers of digital root images. The disadvantage is that the access time for individual files is much longer than for hard disks because of sequential access of data. Thus, images should be grouped logically prior to storage on tape, according to the way they will be retrieved later for analysis.

Similar to backup procedures for all kinds of data, it should be kept in mind that image files must always be saved on a second medium, which should be stored at a place other than the original data. In this way, the risk of losing images, and ultimately data, can be kept at a minimum.

10.2.3 Image Analysis Techniques

What is required of an image analysis system for analysing washed root samples will depend on the type of samples to be analysed and on the parameters to be measured. If the samples contain roots only and if length is the only parameter to be measured, then automated line-intercept techniques may be sufficient. If the root sample contains extraneous objects, and more specific measurements (e.g. diameter, branching) are required, then more sophisticated analysis techniques will be needed.

10.2.3.1 Segmentation of Images of Washed Roots

In image analysis, segmentation (see Box 10.1) of objects from the background is a necessary task before being able to assess morphological parameters of the objects. Although the background of washed root samples is relatively uniform, segmentation is crucial for the automated analysis of root images (Smucker et al. 1987). This section will discuss grey-level thresholding, which is most often used for the segmentation of washed root samples, and alternative segmentation techniques.

Grey-Level Thresholding. For the segmentation of washed root samples, the grey-level thresholding technique is used most frequently. When using thresholding, the most critical step is setting the grey-level threshold (see Box 10.1). In a few cases, two threshold grey-levels are used instead of just one (Smit et al. 1994; Kaspar and Ewing 1997). The two threshold levels define the upper and the lower limit of the range of grey-levels belonging to the roots.

Contrast is often rather weak in root samples, which usually contain roots in the range of tens of micrometres (Pallant et al. 1993) to several millimetres, and in which lighting is not uniform across the whole scanning area. When decreasing the threshold grey-level to resolve the thinnest roots in a sample, more background pixels along the root boundaries are classified as object pixels and, thus, the contours of thicker roots will blur and expand (Fig. 10.2). In addition, a great deal of noise may occur in the image as a result of a too low grey-level threshold. On the other hand, when optimising the threshold value for thicker roots, the thinnest roots will not be resolved. Figure 10.2 shows that there may be drastic changes in resulting image quality when changing the threshold by just one grey-level.

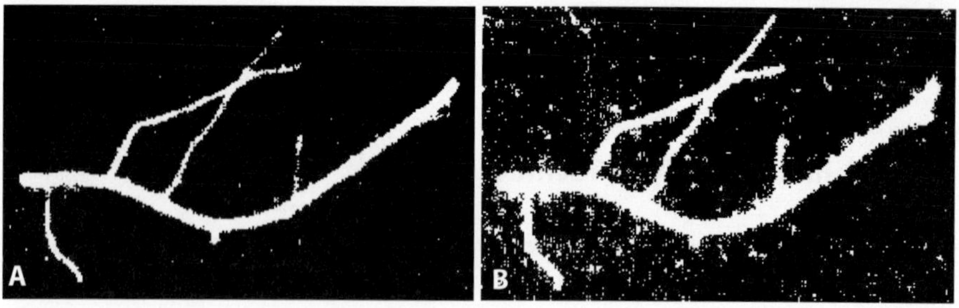

Fig. 10.2. Effect of choosing different brightness levels in grey-level thresholding on the quality of the resulting binary image showing an unstained maize root segment. **A** threshold grey-level: 221; **B** threshold grey-level: 222 (original image: 256 grey-levels, spatial resolution: 300 dpi)

Staining roots may eliminate some of the problems involved in thresholding, though not completely (see Sect. 10.2.1.2). As a rule of thumb, Kaspar and Ewing (1997) suggest selecting the lowest possible threshold setting that does not change background pixels into object pixels for images of the recording tray filled with water but without roots. Because it is difficult to consistently set the threshold level in a series of images or when different people are recording images in turns, a standard threshold level is chosen for all images.

In summary, simple grey-level thresholding is appropriate only if there is good contrast between roots and background and a uniform lighting of the whole scanning area. Because these requirements are most often not met, it is difficult to visually determine the optimal threshold value (Tollner et al. 1994).

Alternative Segmentation Techniques. When processing root images recorded by cameras or scanners, the segmentation by thresholding usually leads to objects with irregular contours producing spurious laterals, and it is difficult to consistently set the grey-level threshold for a series of samples. Therefore, attempts have been made to overcome these limitations using alternative segmentation techniques. Apart from grey-level thresholding, however, only a very limited number of more sophisticated segmentation techniques have been applied in the analysis of washed root systems. Smit et al. (1994) first estimated the background, using local maximum and minimum filters. Before the final histogram-based grey-level thresholding, they applied a Laplacian-like sharpening filter. Smucker et al. (1987) used a combination of edge enhancement and grey-level thresholding. Koller et al. (1995) proposed a new algorithm for the segmentation and local description of elongated, symmetric line-like structures using a non-linear combination of linear filters. We adapted and implemented

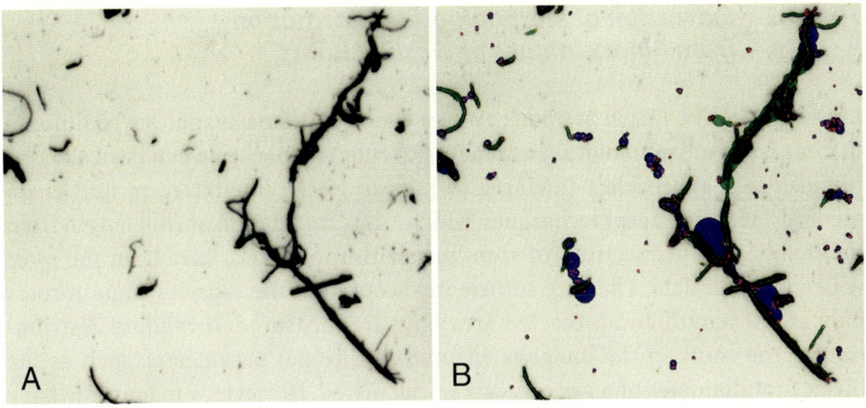

Fig. 10.3. Original 256-grey-scale image (**A**) of washed white clover (*Trifolium pratense L.*) roots, and resulting image (**B**) after image segmentation and length analysis. *Green lines* are root objects, *red crosses* are ending points of individual root segments and *blue circles* are extraneous objects discriminated from roots using a 4:1 length-to-diameter ratio. (Photos with permission from U. Zimmermann, Institute of Plant Sciences, ETH Zürich)

this algorithm (Fig. 10.3) into a program which provides local shape attributes (position, width and direction of the centre line) that are necessary for the morphological description of root segments directly during the detection of curvilinear objects, i.e. roots.

Editing of Segmented Images. The segmentation of images of low contrast or images that contain very thin roots may result in root objects that are not completely resolved. If these fragmented roots are measured, length and area is underestimated. In addition, some of these root fragments may not satisfy the length-to-width criteria that are used to eliminate extraneous objects from the image (see Sect. 10.2.3.2). If root fragments are excluded from the measurements in this way, measurement errors even increase. Therefore, image-processing algorithms (Russ 1990) can be applied to improve the segmented images prior to morphological measurements. Kaspar and Ewing (1997) used *closing* to reduce fragmentation of root objects and to smooth object edges without deleting small object fragments. *Closing* involves *dilation* (i.e. changing any background pixel into an object pixel if at least one of its neighbour pixels is an object pixel) followed by *erosion*, which is the reverse of dilation.

10.2.3.2 Extraction of Morphological Information from Binary Images of Washed Roots

Line-Intercept vs. Chain Methods. When classifying image analysis techniques that measure morphological parameters of root samples, one can basically distinguish two approaches (Ewing and Kaspar 1995): line-intercept and chain methods. Line-intercept techniques, which allow calculation of root length from the number of intersections of scan lines with root pixels, have been the most widely used to date. They are sufficiently accurate if the samples contain roots only and if length and projected area only are measured. If random distribution of the roots in the image is ensured, additional parameters, such as the mean root diameter of a sample, can be calculated. However, a major drawback of such line-intercept approaches is that individual objects within a sample cannot be described morphologically. Thus, line-intercept techniques produce only limited information about the sample. Generally, it is of interest to determine additional parameters:

1. Distribution of root diameters instead of the mean root diameter of the sample;
2. Branching parameters;
3. Shape parameters of individual root objects.

In contrast to line-intercept techniques, chain methods are able to measure these parameters by identifying and analysing individual objects in the image instead of establishing global image features. In principle, chain methods, which are based on Freeman (1970) chain coding, sum the length of pixels in a chain, with length correction for pixels connected diagonally (Kaspar and Ewing 1997). Chain algorithms primarily provide the object perimeter by counting the pixels at the object's border, but they can also be used to measure the length of skeletonised objects of one pixel width (see below).

Determination of Morphological and Architectural Root Parameters. Root length is the primary root trait to be measured. The measurement of length of root samples allows the determination of root length per plant in pot studies or root length densities in different soil layers in field studies. Root measurements often include projected root area, from which the root surface area can be calculated by multiplying it by π, assuming that roots are cylindrical (Kokko et al. 1993). From total length and projected area, the mean diameter of a sample is often calculated. Root length has been rarely classified according to specified diameter classes (but see Pallant et al. 1993; Dowdy et al. 1995; Pietola and Smucker 1998), and branching patterns (Zoon and Van Tienderen 1990) or plant root architecture (Berntson 1992) have seldom been determined.

Root length has been used more often to describe root systems than other parameters (Berntson 1992). In the early days of root research, root length was determined using linear measurement devices such as callipers, rulers or opisometers. These techniques are generally accurate and precise (Reicosky et al. 1970), but prone to human error. Additionally, they are too time-consuming for studies involving large numbers of root samples, but for special purposes, such as determining the lengths of individual main axes or laterals, these methods are still applied.

Based on the principle of Newman (1966), later modified by Marsh (1971) and Tennant (1975), line-intercept methods were widely used for measuring root length. According to Tennant (1975), root length L is equal to:

$$L = (\pi \cdot N_L \cdot D_L)/2, \tag{10.1}$$

where N_L is the number of root intersections with parallel lines, and D_L is the distance between these lines.

Initially, intersects of roots with parallel, regularly spaced lines were counted manually, but this was tedious work and prone to human error (Bland and Mesarch 1990). In the 1970s, photoelectric detection was used increasingly to determine root intersects with regularly spaced scan lines (Rowse and Phillips 1974; Goubran and Richards 1979; Richards et al. 1979; Wilhelm et al. 1983; Ottman and Timm 1984; Collins et al. 1987; Morita et al. 1988). Such photoelectric systems, based on line-intercept principles, were reasonably accurate at relatively high resolutions, but root parameters other than length could not be measured. Furthermore, they were rather expensive and complex (Harris and Campbell 1989). Later, line-intercept methods were developed to process root images recorded with video cameras (Voorhees et al. 1980; Barnett et al. 1987; Burke and LeBlanc 1988; Cunningham et al. 1989; Harris and Campbell 1989) or flatbed scanners (Krstansky and Henderson 1989; Kirchhof 1992). By establishing a regression function between the number of pixels and root length using samples of known size, the total root length in a sample can be estimated. This "grid-intercept" approach is an improvement over the manual line-intercept method; the digitised image represents a very fine pixel grid whose intersections with roots are counted.

Line-intercept based systems are much faster than manual measurement techniques. Accuracy and precision depend on the root size in a sample, orientation of the root objects and root overlap (Melhuish 1968). Random orientation of root objects is a requirement of the line-intercept calculation of root length; non-random placement results in lower accuracy and precision. To prevent negative effects of non-random root orientation, roots can be repositioned and recorded for a second time after rotating the scanning trays by 90° (Harris and Campbell 1989). Spreading of roots, however, is very time-consuming. The line-incercept method is prone to errors due to root overlap at higher densities of root objects in the scanning area (Barnett et al. 1987). Several authors (Harrris and Campbell 1989; Kirchhof 1992; Kirchhof and Pendar 1993) provided correction factors for compensating for the underestimation of root length due to root overlap.

Apart from line-intercept methods, chain methods are used for measuring root length. Using chain algorithms, object length can be measured directly after thinning the objects to a skeleton (see Box 10.2), i.e. a line of one pixel width, or by utilising algorithms to calculate length from perimeter and area.

BOX 10.2. Skeletonisation

Apart from perimeter measurements (Ewing and Kaspar 1995), objects in segmented binary images usually must be reduced to a skeleton or centre line, i.e. a line of one pixel width, for their morphological description (Russ 1990). This thinning operation is most often accomplished by an iterative erosion technique (Russ 1990). The skeleton contains both metric and topological information (e.g. the number of end points or the number of nodes); it has approximately the same length and topology as the original object.

Thinning algorithms are needed to create skeletons by eroding pixels step-by-step from the border to the centre of an object. For accurate measurements of objects, the thinning process must guarantee connectivity (avoid removing border pixels and thus causing discontinuities) and avoid end shortening (avoid removing pixels at the ends of the line). Alternatively, medial axis transformations (Blum 1967) might be applied instead of skeletonisation to produce a central line on which each pixel is at an equal distance from at least two border pixels. The medial axis transformation has the advantage over the thinned skeleton in that not only can the length of the object be measured by counting the pixels on the medial axis, but also the diameter of the object can be measured at each pixel of the medial axis simultaneously and accurately. Although medial axis transformation has the potential for accurate and economic determinations of length and diameter (Russ 1995), it has not been used for root image analysis so far.

The skeletonisation approach has been most often used for the image analysis of washed root samples (e.g. Lebowitz 1988; Zoon and Van Tienderen 1990; Travis et al. 1993; Smit et al. 1994; Tanaka et al. 1995; Pietola and Smucker 1998). Prior to measurements, a thinning algorithm is applied to create root skeletons. The success of measurements based on thinning algorithms depends on the quality of the previous segmentation process. If the borders of the objects are irregular, skeletonisation may produce spurious lateral branches (Russ 1995), thus leading to an unrealistic description of roots, especially to an overestimation of length and branching. After finishing the thinning process, the length of the object can be measured as a function of the number of vertically, horizontally and diagonally adjacent pairs of pixels in the skeleton (Lebowitz 1988). It is important to take into account diagonally adjacent pairs of pixels, because the distance between them is theoretically greater by a factor of $\sqrt{2}$ than the distance between vertical or horizontal pairs of pixels. Thus, length L can be calculated as (Lebowitz 1988):

$$L = D_P \cdot ((H+V) + \sqrt{2} \cdot X), \tag{10.2}$$

where D_P is the distance between any horizontally or vertically adjacent pixels, and H, V, X are the number of horizontally, vertically and diagonally adjacent pairs of pixels in the skeleton.

However, Smit et al. (1994) showed that the calculation of Lebowitz (1988) does not give a correct weighting factor for pairs of adjacent pixels when averaged over all possible directions of root segments. Instead of using correction factors of 1 and $\sqrt{2}$ for horizontal/vertical and diagonal pairs of pixels, respectively, Smit et al. (1994) derived an overall correction factor of 1.12. For a more accurate calculation of the length of randomly oriented lines, Beckers and Smeulders (1992) replaced the strictly geometric distances of 1.0 and 1.414 (1.21 on average) between orthogonal and diagonal pixel neighbours by 0.948 and 1.340 (1.14 on average), which is quite close to the correction factor derived by Smit et al. (1994). Thus, calculation of root length L is simplified as followed (Smit et al. 1994):

$$L = N \cdot C / R, \tag{10.3}$$

where N is the number of root skeleton pixels, C the correction factor (1.12), and R the spatial resolution of the imaging device in pixels per centimetre.

Compared with the direct measurement of root skeleton length, root length is less often derived from perimeter measurements without a preceding skeletonisation of root objects (Pan and Bolton 1991; Kokko et al. 1993; Dowdy et al. 1995; Ewing and Kaspar 1995); this approach is less demanding on computer memory and time than the above-mentioned skeletonisation process (Ewing and Kaspar 1995). To overcome the above-mentioned problems of non-random orientation of objects, Ewing and Kaspar (1995) developed an edge chord algorithm (Fig. 10.4) that calculates the length of lines drawn between the edges of steps along the perimeter of the digitised objects. Its advantages are the small memory requirement and the fact that the orientation of root objects has only a minimal effect on measured length. The calculation of object length from area

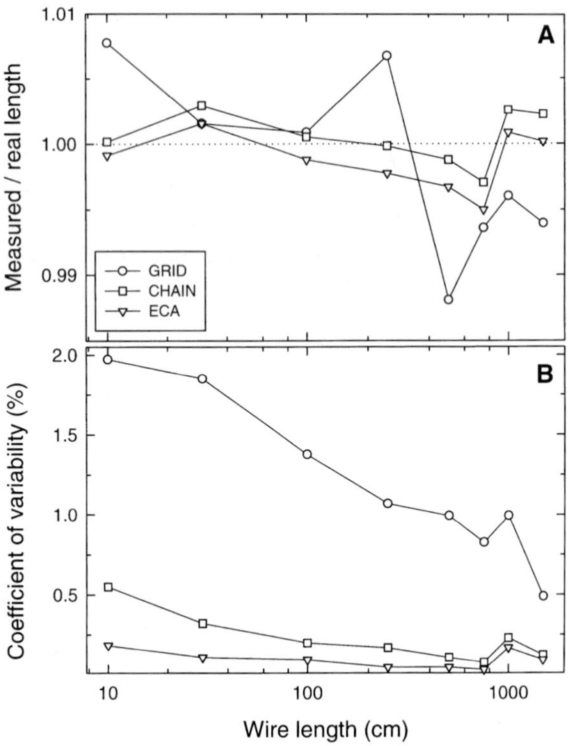

Fig. 10.4. Relative length (measured/real) of randomly oriented wire pieces (**A**) and coefficients of variability of length measurements (**B**) by the grid-intercept method (*GRID*), a chain method according to Freeman (1970) as corrected by Kulpa's (1977) estimate of the theoretical error (*CHAIN*) and the edge chord algorithm by Ewing and Kaspar (1995; *ECA*). Wire length is shown on a log scale. (Modified from Ewing and Kaspar (1995), with permission from the Journal of Computer-Assisted Microscopy, Volume 7, Plenum Press)

and perimeter, which was measured by a chain-type method and the edge chord algorithm by Ewing and Kaspar (1995), has been shown to be more accurate (Fig. 10.4A) and more repeatable (Fig. 10.4B) than length measurements by the grid-intercept method.

There is abundant information on comparisons of true lengths or manual line-intercept measurements with digital length measurements of root samples or calibration objects. Although experimental conditions, imaging devices and image analysis approaches varied widely, most of the reported coefficients of determination (r^2) of the association between manual and digital length measurements were greater than 0.95 (e.g. Barnett et al. 1987; Morita et al. 1988; Farrell et al. 1993; Shuman et al. 1993), and, in some cases, nearly 1 (e.g. Wilhelm et al. 1983; Harris and Campbell 1989; Zoon and Van Tienderen 1990; Kirchhof 1992; Smit et al. 1994; Dowdy et al. 1995, 1998; Tanaka et al. 1995). There were no clear-cut differences in the coefficients of determination between digital line-intercept techniques and chain methods. However, results of a direct comparison of algorithms by Ewing and Kaspar (1995) suggest that chain-type algorithms may be more accurate and precise than the line-intercept method.

Farrell et al. (1993) and Murphy and Smucker (1995) attributed weaker relationships between manually and digitally measured root lengths to the lower spatial resolution of the imaging devices compared with the human eye, leading to an underestimation of digital measurements. In the study of Murphy and Smucker (1995), this was only relevant for perennial ryegrass (*Lolium perenne* L.), whose root system is characterised by a large portion of very fine roots, but not for the coarser-rooted alfalfa (*Medicago sativa* L.). This underlines the need for high-resolution imaging devices that allow reliable recording of the thinnest roots in a sample (see Sect. 10.2.2.1).

Average coefficients of variation for repeated measurements of root length and for different operators are usually larger for the manual line-intercept technique than for digital methods (Farrell et al. 1993). Thus, digital image analysis techniques may provide more accurate and reproducible measurements of root length than manual methods.

Projected root area was initially determined with photoelectric devices designed to measure leaf surface areas (Morrison and Armson 1968; Armson 1972). Because the spatial resolution of these devices was rather poor, this technique was problematic if too many fine roots were present in the samples (Kemph 1976). These systems usually required calibration for measuring root area as was the case when measuring leaf area.

With more recent systems, projected root area is usually measured as the sum of all root pixels, which is converted to the appropriate real units of measure; total root surface area can then be calculated by multiplying the projected root area of a sample by π, assuming that roots are cylindrical (Kokko et al. 1993). Systems that yield length and mean diameter of roots allow calculation of root surface area by multiplying the product of length and mean diameter by π. However, this method may be inaccurate when root diameters in the sample are not evenly distributed above and below the mean diameter (Kirchhof 1992). Thus, the alternative calculation of total root surface area as the sum of root surfaces of groups of roots belonging to predefined complementing diameter classes is preferable. The preferred method is to directly measure the root area of individual root segments or to calculate it from their lengths and diameters; total root surface area is the sum of the areas of single root segments. In this way, surface area measurements are expected to be more accurate, and large, compact extraneous objects, which falsely increase the area of a sample even more than the length, may be eliminated from the measurement according to shape parameters.

In line-intercept systems, **mean root diameters** can be calculated by dividing projected root area by the total length of a sample, provided that the system allows for area measurements. In contrast, chain methods using skeletonisation algorithms enable the direct measurement of the root diameter of individual objects (Lebowitz 1988). The diameter is, for each point on the object's skeleton, directly proportional to the number of iterations necessary in the erosion process that leads to the skeleton. This approach is more accurate than the calculation of mean root diameter using the line-intercept method.

As a consequence of the thinning process, however, diameter measurements are unreliable if the diameter of the thinnest roots is close to the lower limit of spatial resolution (Smucker et al. 1987). If the ratio of root diameter to the pixel size of the recording device is low, overestimates of root width may occur because pixels at the root boundary that are only partially occupied by the root contribute their entire length to the measurement of root diameter (Pan and Bolton 1991). With increasing root diameters, this overestimation of root width becomes less important. Therefore, Smucker et al. (1987) suggest that root segments should be at least four pixels wide for accurate diameter measurements. This should be considered when selecting scanning resolutions (see Sect. 10.2.2.1).

Methods that require skeletonisation for length measurements also enable the determination of **branching parameters**, which was previously done manually (e.g. Costigan et al. 1982). Image analysis allows the automatic determination of branching, as shown by Smucker et al. (1987) and Zoon and van Tienderen (1990) for washed root segments. The thinned skeleton can be used to determine the degree of branching of a sample. Basically, the number of root branches can be determined by counting the number of pixels that have three neighbouring pixels in a 3×3 pixel region. In this way, Zoon and van Tienderen (1990) determined a branching ratio, by dividing the difference between branch pixels and intersection pixels by root length.

For entire seedling root systems, **root architecture** (Berntson 1992) can be determined by image analysis. Large and complex root systems may be divided into multiple parts, e.g. single main axes, which are recorded and analysed individually. There are two basic requirements for measurements of root architecture: first, the root system has to be thinned to a skeleton, represented by a chain of connected root pixels; second, no loops enclosing background pixels should occur, and no roots should cross (Berntson 1992). Thus, samples for such analyses require even more careful spreading than samples for length measurements. After converting digitised root images into a binary tree data structure, it has to be checked visually for deviations from the original image, and, if needed, the image can be edited using a pixel editor program (Berntson 1992). Berntson's (1992) method provides a measurement of the components of root system architecture (Fitter 1987). The output of this program can be stored and used in models of the 3-D architecture of plant root systems (Berntson 1994). Theory and measurement of root architecture are described further in Chapter 4.

Recently, methods have become available for the computer-assisted description of **fractal geometry** (Berntson et al. 1998). The reader is referred to Chapter 4 for more information on theory and measurement of fractal geometry.

Discrimination of Roots from Extraneous Objects. Most root image analysis systems do not provide morphological parameters of individual objects, so there is no way to discriminate extraneous objects from roots. With systems that provide single-object traits, the classical way of differentiation is to use simple shape indices.

Murphy and Smucker (1995) used a length-to-width ratio of 4:1 as a criterion for discriminating roots from unwanted objects; objects with length-to-width ratios greater than 4 were assumed to be roots, and objects with ratios less than 4 were excluded from root measurements. Dowdy et al. (1998) recently introduced a length-to-width ratio of 15:1 for discriminating between extraneous objects and roots, which are assumed to have much greater length-to-width ratios than unwanted objects. The 15:1 ratio was derived from corn root samples through iterations. For hand-picked root samples with no extraneous objects, an r^2 of 0.97 was found for the correlation between length measurements by image analysis, using the 15:1 ratio, and a manual line-intercept method (Dowdy et al. 1998). The correlation of root lengths of samples with extraneous objects measured with a 15:1 and a 4:1 discrimination ratio gave an r^2 of 0.80 (Dowdy et al. 1998).

More complex shape indices derived from morphological parameters of objects have not yet been used to discriminate roots from extraneous objects. Because roots are usually several times longer than they are wide, they may be discriminated from extraneous objects, which usually have a more compact shape than roots, by various kinds of morphological parameters, such as the spherical index (Glasbey and Horgan 1995), describing the roundness of objects. Colour may be a useful additional criterion for discrimination, because unwanted objects are often darker and more opaque than roots. Collins et al. (1987) achieved some differential staining of extraneous organic matter with trypan blue, which suggests that there may be some potential for discriminating roots and unwanted organic objects by selective staining.

The user has to be aware that the above-mentioned discrimination approaches are never perfect. Although a given length-to-width ratio may exclude the majority of extraneous objets from morphological measurements, an unknown number of unwanted objects will not be eliminated and an unknown amount of root objects will be falsely removed.

Differentiation Between Viable and Dead Roots. An accurate, automated discrimination between living and dead roots would be highly desirable because it is essential to know the ratio of living to dead roots in samples for studying

Table 10.5. Dyes used for vital-staining of plant roots

Dye	Authors
2,3,5-triphenyltetrazolium chloride	Goedewaagen (1954); Sator and Bommer (1971); Crapo and Coleman (1972); Knievel (1973); Clemensson-Lindell (1994)
2,3,5-triphenyltetrazolium bromide	Jacques and Schwass (1956)
Congo red	Ward et al. (1978)

root function and turnover (Knievel 1973). However, differentiation is technically difficult (Schuurman and Goedewaagen 1971), and basic knowledge of objective differentiation criteria is scarce (see Chap. 9).

Because certain dyes are used to discriminate between dead and living roots (Table 10.5), they could be used for a selective measurement of previously live roots in a sample if the contrast between roots stained in this way and non-stained roots is great enough to reliably discriminate between both groups. Ottman and Timm (1984), using an image analyser, measured the length of viable roots that were effectively differentiated from non-viable roots and organic residues by preferential staining with trypan blue. As an alternative to staining, differences in colour and/or texture information might be used for discriminating between living and dead roots, but more information on relationships between these parameters and root viability is needed first.

10.2.4 Root Image Analysis Systems

Besides a variety of custom-built systems, there are a few recently developed systems that have been widely used (Table 10.4). Except for the "NIH Image" (National Institutes of Health (NIH), Bethesda, MD, USA) and "MacRHIZO" (Régent Instruments Inc., Quebec, Canada) programs for Macintosh computers and the "Delta-T Mk II" (Delta-T Devices, Burwell, UK), which is not computer-based, all the other programs listed in Table 10.4 run on PCs. While "Delta-T SCAN" and "ROOTEDGE" (National Soil Tilth Laboratory, Ames, IA, USA) are designed for the DOS operating system, "WinRHIZO" (Régent Instruments Inc., Quebec, Canada) runs under the Windows environment. There are public domain programs for Macintosh computers ("NIH Image") and PCs ("ROOTEDGE"), but the majority of systems are distributed commercially, often in combination with imaging devices such as flat-bed scanners ("Delta-T SCAN", "MacRHIZO" and "WinRHIZO"). The outdated "Delta-T Mark II" is no longer commercially available, but was widely used in recent years. While "NIH Image" is a general purpose program for which task-specific macros have to be written, the other systems have no option for user-written extensions.

All these systems are designed for measuring root length and, in most cases, root surface area. Some of the programs, e.g. "Delta-T SCAN" (Kirchhof 1992; Kirchhof and Pendar 1993), and "MacRHIZO"/"WinRHIZO" (Arsenault et al. 1995; Bauhus and Messier 1999), have algorithms for correcting errors in the measurement of root length and area due to root overlaps. In addition to measurements of root length, area and diameter, some of the systems can be used for evaluations of other morphological parameters. Among other parameters, "Delta-T SCAN" allows for object counting and root tip counting; "MacRHIZO"/"WinRHIZO" for morphological measurements using user-definable colour classes, root tip counting and root topology analyses.

If roots washed from soil containing unwanted organic objects are to be measured, then the systems should allow for individual object measurements and, thus, the possibility of discriminating between roots and extraneous objects. The "MacRHIZO"/"WinRHIZO" and the "ROOTEDGE" (Kaspar and Ewing 1997) programs and the system described by Murphy and Smucker (1995) have this feature. However, if only root samples from hydroponics or solid substrates without soil are measured, then the systems that do not allow for object-based measurements can also give accurate results because such samples usually contain no or only few extraneous objects. The "Delta-T SCAN" programs stands between these two groups, because it allows for single-object measurements of various parameters but not of root length.

No information can be given yet on the comparative accuracy of the different systems, because there are no data on comparisons between image analysis systems or programs. It is not possible to give general recommendations on the use of root image analysis systems, because their options and drawbacks vary greatly, and the requirements of scientists studying roots are diverse.

10.2.5 Priorities for Image Analysis of Washed Roots

The major priority in developing more sophisticated image analysis techniques is to find more accurate image segmentation techniques that do not rely on simple grey-level thresholding. A second priority is the reliable differentiation between roots and extraneous objects based on shape parameters and/or colour. In both cases, new algorithms should be developed together by biologists, agronomists and ecologists on the one hand and image analysis specialists on the other. The adaptation of algorithms from other fields also may be applicable.

Computer equipment for handling large colour image files is available today, and, thus, the use of colour information in the analysis of washed samples may be beneficial in several ways. The inclusion of colour may help to differentiate between living and dead roots, between roots and extraneous objects, and in assessing the portion of root length or area infected by root diseases, as shown by Kokko et al. (1993) for subcrown internode discoloration in wheat caused by common root rot.

None of the image analysis techniques described have provided an overall solution for the fast and accurate measurement of morphological and topological root parameters. A special solution for each situation has seemed to be more suitable to date. Possibly, the ongoing dramatic development of computer resources will enable scientists to fully exploit the potential of image analysis techniques in the morphological description of root samples in the future.

10.3 Image Analysis of Minirhizotron Images

The minirhizotron technique allows for non-destructive measurements and minimises point-to-point variation because repeated root observations are possible at the same positions in the soil. The analysis of minirhizotron images, however, is tedious and prone to human error. Images are usually analysed by manually counting the number of roots or root elements (Upchurch 1987; see Chap. 8), which may be quite closely correlated with root length on the same images (Box and Ramseur 1993). Only in a few cases has root length in minirhizotron images been measured manually by tracing roots on a TV or computer monitor (Cheng et al. 1990; Hendrick and Pregitzer 1992) or by means of the modified line-intercept method (Hirasawa et al. 1995). These attempts to directly measure root lengths yielded more accurate information on root morphology, but required even more time for analysing images compared with manual counting, especially for tracing of roots. Because minirhizotron investigations usually require the recording and evaluation of a large number of images, there is a need for replacing the above-mentioned approaches by semi- or fully-automated analyses of images.

10.3.1 Segmentation of Minirhizotron Images

Image quality is an important prerequisite for the successful image analysis of video-recorded roots (Smucker et al. 1987). There are several factors that may contribute to the poor quality of minirhizotron images. The non-uniform diffusion of light due to tube curvature may result in inconsistencies in brightness observed in digitised images (Brown and Upchurch 1987). Non-perpendicular viewing or camera-tube non-linearity may lead to geometrically distorted images (Heeraman et al. 1993). Low battery power can also contribute to photometric distortion due to poor illumination by the incandescent bulbs of the minirhizotron camera (Heeraman et al. 1993).

The major problem in the automated analysis of minirhizotron images is the much more complex background compared with washed samples, both in colour and in texture. A variety of objects, such as reflective soil particles, droplets of condensation, scratches on the tubes, electronic noise and soil fauna, may have similar brightness values as the roots (Smucker et al. 1987; Casarin et al. 1991; Heeraman et al. 1993; Andrén et al. 1996). Furthermore, the brightness of roots may differ due to variable luminance along the same root, to loss of root contact with the tube wall, to roots being out of focus (leads to greater apparent diameters of roots or to a more reflective background) and to the structure of small roots, which appear to be fragmented (Casarin et al. 1991).

Additionally, the background of a given minirhizotron image position may not be constant over time due to varying soil moisture contents, settling of soil (Böhm 1979), diurnal variations in root diameters (Huck et al. 1970), reorientation of soil particles during root elongation (Goss 1991), the displacement of soil particles into pores adjacent to the tube wall by percolating water, movement of the minirhizotron tube (Heeraman et al. 1993) and activity of soil animals. As a consequence, the major problem in segmentation (see Box 10.1) of minirhizotron images is the discrimination of roots from a very complex, changing background.

10.3.1.1 Enhancement of Minirhizotron Images

Several techniques of image enhancement may have the potential to facilitate the subsequent segmentation of minirhizotron images. Contrary to the majority of systems reported for the analysis of washed roots, enhancing contrast has usually been used prior to the analysis of minirhizotron images. Smucker et al. (1987) applied low-pass filters to eliminate high-frequency noise from the image, and high-frequency filters averaged the intensity values of each pixel. Casarin et al. (1991) increased the contrast for roots of small to medium diameter and reduced the noise to some extent on uniform luminosity areas of the background using linear convolution, a technique which is effective for low dynamic images. Heeraman et al. (1993) used linear contrast stretch enhancement. Andrén et al. (1996) reduced background noise, which was created by

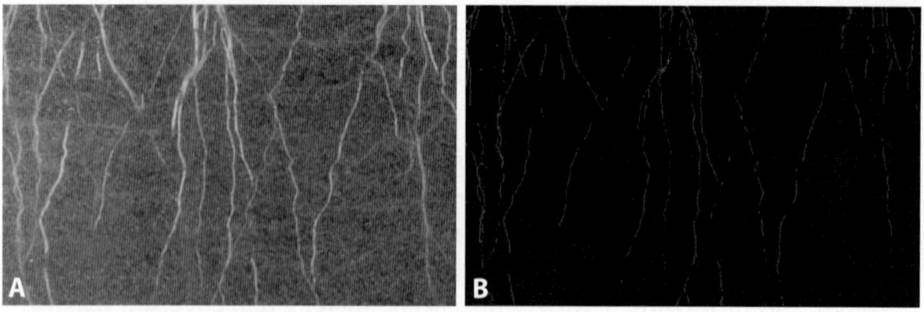

Fig. 10.5. Original minirhizotron image of meadow fescue (*Festuca pratensis* Huds.) roots (**A**) and roots identified by median and convolution filtering, multiple grey-level thresholding and skeletonisation procedures according to Andrén et al. (1996; **B**). (Photos with permission from E Sindhøj, Department of Soil Science, Swedish University of Agricultural Sciences, Uppsala)

reflections from sand particles, by two-step median filtering using a 5 × 5 matrix.

10.3.1.2 Segmentation of Minirhizotron Images

Global techniques, such as grey-scale thresholding, can only be used for the segmentation (see Box 10.1) of images if the objects to be recognised are consistently lighter or darker than the background. Roots in minirhizotron images are usually lighter than the background, but the resulting light intensity histograms are not bimodal (Nater et al. 1992). Andrén et al. (1996) applied grey-level thresholding using a fixed brightness level for differentiating roots from the soil background (Fig. 10.5). To avoid background scatter, they chose a conservative (i.e. high) threshold level; as a consequence, only the lightest parts of roots were detected (Andrén et al. 1996). Thus, the discrimination of roots from the soil background by grey-level thresholding is not reliable, especially in the case of thin secondary and higher order roots (Casarin et al. 1991). In addition, the manual selection of a threshold level is dependent on the operator's judgement and, thus, may lead to error. To overcome this limitation, Smucker et al. (1987) applied grey-level thresholding, with the mean intensity value of all the pixels along previously tracked root edges automatically set as the threshold. This approach gave good results when the roots had a similar and constant intensity. Although roots exhibit large, localised changes in light intensity, other features such as pores and cracks may exhibit the same rate of change as do roots (Nater et al. 1992). Thus, segmentation techniques based on thresholding or edge detection only are inappropriate.

Because global segmentation techniques are not feasible, several groups applied local techniques. Casarin et al. (1991) used linear flat-hat transforms

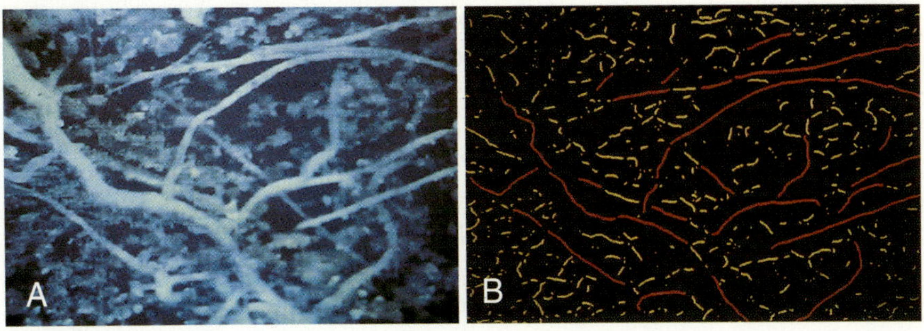

Fig. 10.6. Original minirhizotron image of alfalfa (*Medicago sativa* L.) roots (**A**) and the centre lines of roots resulting from a ridge detection approach (**B**). (Photos from Smucker (1993), with permission from the Annual Review of Phytopathology, Volume 31, Annual Reviews)

(Meyer 1979) at different diameters for segmenting images. With this approach, the extraction of medium and large roots presented few problems (Casarin et al. 1991). Other approaches for identifying roots in minirhizotron images are ridge detection by identifying the parallel edges of each root segment (Ferguson et al. 1990), allowing an identification of 94–96% of the roots, depending on the complexity of images (Fig. 10.6). Smucker (1993) reported token-based algorithms, which are an example of a knowledge-based feedback approach, for improved identification of roots on a complex background. A further development of Nater et al. (1992) tested several artificial neural system architectures for segmentation. Inputs to their models were horizontal and vertical derivative images produced from the raw image by smoothing and differentiation with least square procedures. A training image was created by hand-editing the original image. After about 40 iterations, most pixels in the training image were correctly classified, whereas accuracy was much reduced when processing other images for which the system had not been trained. The ability to adjust classification, even in the presence of noise, is a major advantage over classical segmentation approaches (Nater et al. 1992).

In addition to the above-mentioned contour-based approaches, a region growing approach was recently reported by Jankowski et al. (1995). After the initial detection of seeds, i.e. significant root regions, these were expanded to optimal root contours by a heuristic search for parallel root structures. Jankowski et al. (1995) detected at least one seed on each root segment in 28 out of 32 tested images, but no seeds could be detected if the inner part of root segments were strongly structured. Results of the segmentation technique used by Jankowski et al. (1995) were better than when processing the same test images with the technique reported by Smucker et al. (1987).

10.3.2 Extraction of Information from Binary Minirhizotron Images

10.3.2.1 Determination of Morphological and Architectural Root Parameters

After segmentation of images, morphological root traits can be determined in basically the same way as for washed roots (see Sect. 10.2.3.2). The primary root parameter to be measured is also root length. Root length was measured in all reported attempts to analyse minirhizotron images. In some cases, root length was categorised by diameter classes (Smucker et al. 1987; Casarin et al. 1991). Measurements of length from rhizotron and minirhizotron video images according to the technique by Casarin et al. (1991) showed less than 10% difference from values determined manually with Newman's method (Girardin et al. 1991). Smucker et al. (1987) reported a difference of 7.6% between manual line-intercept and digital analysis techniques. On average, Andrén et al. (1996) observed that root lengths measured using image analysis were 9% smaller than manually traced root lengths.

Projected root surface area is easily determined by measuring the image area occupied by the roots relative to the total image area (Heeraman et al. 1993), but area measurements probably do not very precisely reflect the true root surface area, because visible root boundaries are usually not smooth and contours of roots are often partially covered by soil particles. Smucker et al. (1987) found a difference of 6.2% between manual and digital measurements of projected area.

Acrylic rhizotron walls do not seem to disturb secondary root ramification (Jordan 1992), and, thus, branching indices could be determined in minirhizotron images as suggested by Smucker et al. (1987) and Casarin et al. (1991). However, there are no data available to demonstrate whether root branching can be determined reliably using the relatively small minirhizotron pictures only. Before measuring branching indices, thinning of the roots is required (see Box 10.2).

The development of morphological root parameters over time was evaluated by Casarin et al. (1991) and Heeraman et al. (1993). The latter detected changes in root appearance and disappearance with time by assigning a unique primary colour, i.e. red, green or blue, to selected growth stages and then overlaying the images of a position in soil taken at the different recording times. Areas of no change appeared white, whereas changed areas appeared as one of the primary colours or a complementary colour as a result of the mixing of two primary colours. However, the resulting colour composites were interpreted only visually. Although the automated detection of temporal changes in root parameters is very desirable, such analyses would be prone to differences in root detection due to the possibility of a changing background and slight variations in the position of the camera, despite the use of camera indexing handles as described by Ferguson and Smucker (1989).

A differentiation between living and dead roots is desirable. Different fluorescence of living and dead roots may have potential for discriminating between them by means of image analysis, provided that viable roots fluoresce more intensely than dead ones. Brown and Scott (1984) showed that actively growing parts of soybeans fluoresce strongly under ultraviolet light and that fluorescence decreases with root age and ceases at the beginning of suberisation. Wang et al. (1995) found that an ultraviolet light source was more accurate than visible light for estimating live proportions of roots of woody species in a minirhizotron, whereas the opposite was true for grass species.

10.3.2.2 Discrimination of Roots from Extraneous Objects

There is a variety of extraneous objects potentially visible in minirhizotron images. Extraneous objects, typical of minirhizotron images, are scratches on tube walls, water drops or water vapour, and soil animals. Such extraneous objects may be excluded by shape parameters (see Sect. 10.2.3.2). Smucker et al. (1987) used a maximum and a minimum root length and a length-to-width ratio of 4:1 as criteria for this discrimination. One problem with this approach is that root segments may quite often have shorter length-to-width ratios than 4:1, because root segments are either partially covered by soil particles or other extraneous objects, such as organic residues, or because segmentation results in fragmented root segments that are separated by gaps. Scratches, water droplets and certain mineral particles are often isolated and strongly reflect light, making them easier to eliminate (Casarin et al. 1991).

10.3.3 Priorities for Image Analysis of Minirhizotron Images

Image quality is crucial to the analysis of minirhizotron images and must be improved. Nater et al. (1992) suggest testing different wavelength filters for the minirhizotron light source for increased contrast between roots and background.

Minirhizotron images are usually evaluated by manual counting or tracing of roots. To enable a more efficient method of analysis, all the recorded images must be made available in digital form. This would make the visual evaluation more flexible, eliminating the necessity of observing the samples in the order in which they were recorded, thus making a digital analysis possible (Dubach and Russelle 1995). With digital images, standard image enhancement techniques may be applied to increase the contrast for easier visual or automated evaluation of images. There are two possible approaches to automatically digitising minirhizotron images. First, images could be directly digitised during recording, thus eliminating the need to temporarily store them on video tapes. Second, image code information as provided, for example, by a screen marker system (Bartz Technology, Santa Barbara, California) could be used to capture video frames from a video tape and to select an individual frame to be saved for each tube position according to unique coded information corresponding to this tube position.

A major conclusion of previous attempts to process minirhizotron images is that segmentation techniques based only on conventional grey-level thresholding are inappropriate. Thus, more sophisticated segmentation techniques specifically adapted to discriminate roots from a complex background have to be developed. An artificial neural systems approach and a contour-based technique have already shown some potential for automatic analysis, at least of images with relatively good, uniform contrast between roots and the soil background. This requirement, however, is usually not met for the majority of minirhizotron images. Thus, semi-automated systems that allow for user-input, as proposed, for example, by Daneels et al. (1993) for identifying and outlining objects in images, should be evaluated as intermediate steps in the image analysis of minirhizotron images, especially when dealing with a very complex background or plant species with very thin or translucent roots.

Results from the digital analysis of images that cover a wide range of soil conditions, plant species and growth stages should be compared with those from manual measurements of the same images. To date, such comparisons have only been made with a very limited number of images. Thus, a conclusive assessment of existing approaches for the digital processing of minirhizotron images is not yet possible.

Further Reading

Glasbey CA, Horgan GW (1995) Image analysis for the biological sciences. John Wiley, Chichester

Price TV, Osborne CF (1990) Computer imaging and its applications to some problems in agriculture and plant science. Crit Rev Plant Sci 9: 235–266

Russ JC (1990) Computer-assisted microscopy: the measurement and analysis of images. Plenum Press, New York

Russ JC (1995) The image processing handbook. 2nd edn. CRC Press, Boca Raton, Florida

Smucker AJM (1993) Soil environmental modifications of root dynamics and measurement. Annu Rev Phytopathol 31: 191–216

Tillett RD (1991) Image analysis for agricultural processes: a review of potential opportunities. J Agric Engin Res 50: 247–258

References

Andrén O, Elmquist H, Hansson AC (1996) Recording, processing and analysis of grass root images from a rhizotron. Plant Soil 185: 259–264

Armson KA (1972) Distribution of conifer seedling roots in a nursery soil. For Chron 48: 141–143

Arsenault JL, Pouleur S, Messier C, Guay R (1995) WinRHIZO™, a root measuring system with a unique overlap correction method. HortScience 30: 906 (Abstr)

Barnett CE, White RA, Petrovic AM, Good GL (1987) An automated apparatus for measuring root length. HortScience 22: 140–144

Bauhus J, Messier C (1999) Evaluation of line root length and diameter measurements obtained with RHIZO image analysis. Agron J 91: 142–147

Beckers ALD, Smeulders AWM (1992) Optimal length and distance estimation in three dimensions. Comput Vision Graph Image Process 55: 296–306

Berntson GM (1992) A computer program for characterizing root system branching patterns. Plant Soil 140: 145–149

Berntson GM (1994) Modelling root architecture: are there tradeoffs between efficiency and potential of resource acquisition? New Phytol 127: 483–493

Berntson GM, Lynch J, Snapp S (1998) Fractal geometry and the description of plant root systems: current perspectives and future applications. In: Baveye P, Parlange JY, Stewart BA (eds) Fractals in soil science. CRC Press, Boca Raton, Florida, pp 113–152

Bland WL, Mesarch MA (1990) Counting error in the line-intercept method of measuring root length. Plant Soil 125: 155–157

Blum H (1967) A transformation for extracting new descriptors of shape. In: Whaten-Dunn W (ed) Models for the perception of speech and visual form. MIT Press, Cambridge, Massachusetts, pp 153–171

Böhm W (1979) Methods of studying root systems. Springer, Berlin Heidelberg New York

Box JE Jr, Ramseur EL (1993) Minirhizotron wheat root data – comparisons to soil core root data. Agron J 85: 1058–1060

Brown DA, Scott HD (1984) Dependence of crop growth and yield on root development and activity. In: Barber SA, Bouldin DR (eds) Roots, nutrient and water influx, and plant growth. Soil Science Society of America (SSSA), Madison, Wisconsin, pp 101–136

Brown DA, Upchurch DR (1987) Minirhizotrons: a summary of methods and instruments in current use. In: Taylor HM (ed) Minirhizotron observation tubes: methods and applications for measuring rhizosphere dynamics. ASA Spec Publ 50: 15–30

Burke MK, LeBlanc DC (1988) Rapid measurement of fine root length using photoelectronic image analysis. Ecology 69: 1286–1289

Casarin M, Jacquey S, Fouere A, Girardin P (1991) Digital picture processing applied to the evaluation of plant root dynamics. In: McMichael BL, Persson H (eds) Plant roots and their environment. Proc Symp Int Soc of Root Research (ISRR), 21–26 August, 1988, Uppsala, Sweden. Elsevier, Amsterdam, pp 570–575

Castleman KR (1979) Digital image processing. Prentice-Hall, Englewood Cliffs, New Jersey

Cheng W, Coleman DC, Box JE Jr (1990) Root dynamics, production and distribution in agroecosystems on the Georgia Piedmont using minirhizotrons. J Appl Ecol 27: 592–604

Clemensson-Lindell A (1994) Triphenyltetrazolium chloride as an indicator of fine-root vitality and environmental stress in coniferous forest stands – applications and limitations. Plant Soil 159: 297–300

Collins RP, Gregory PJ, Rowse HR, Morgan A, Lancashire B (1987) Improved methods of estimating root length using a photocopier, a light box and a bar code reader. Plant Soil 103: 277–280

Costigan PA, Rose JA, McBurney T (1982) A microcomputer-based method for the rapid and detailed measurement of seedling root systems. Plant Soil 69: 305–309

Crapo NL, Coleman DC (1972) Root distribution and respiration in a Carolina old field. Oikos 23: 137–139

Cunningham M, Adams MB, Luxmoore RJ, Post WM, DeAngelis DL (1989) Quick estimates of root length, using a video image analyzer. Can J For Res 19: 335–340

Daneels D, van Campenhout D, Niblack W, Equitz W, Barber R, Bellon E, Fierens F (1993) Interactive outlining: an improved approach using active contours. In: Niblack W (ed) Storage and retrieval for image and video databases. Proc SPIE 1908. SPIE, Bellingham, Washington, pp 226–233

Dowdy RH, Nater EA, Dolan MS (1995) Quantification of the length and diameter of root segments with public domain software. Commun Soil Sci Plant Anal 26: 459–468

Dowdy RH, Smucker AJM, Dolan MS, Ferguson JC (1998) Automated image analyses for separating plant roots from soil debris elutriated from soil cores. Plant Soil 200: 91–94

Dubach M, Russelle MP (1995) Reducing the cost of estimating root turnover with horizontally installed minirhizotrons. Agron J 87: 258–263

Ewing RP, Kaspar TC (1995) Accurate perimeter and length measurement using an edge chord algorithm. J Comput Assist Microsc 7: 99–100

Farrell RE, Walley FL, Lukey AP, Germida JJ (1993) Manual and digital line-intercept methods of measuring root length – a comparison. Agron J 85: 1233–1237

Ferguson JC, Smucker AJM (1989) Modifications of the minirhizotron video camera system for measuring spatial and temporal root dynamics. Soil Sci Soc Am J 53: 1601–1605

Ferguson JC, Chang SL, Smucker AJM (1990) Automated segmentation of minirhizotron images based on the coherence of intensity gradients. In: Abstr Conf on analytical methods for quantifying root and soil dynamics. St Louis, Missouri, 13–14 August, 7pp

Fitter AH (1987) An architectural approach to the comparative ecology of plant root systems. New Phytol 106: 61–77

Freeman H (1970) Boundary encoding and processing. In: Lipkin BS, Rosenfeld A (eds) Picture processing and psychopictorics. Academic Press, New York, pp 241–266

Fu KS, Mui JK (1981) A survey on image segmentation. Pattern Recogn 13: 3–16

Girardin P, Trendel R, Jacquey S, Casarin M, Goss M, Fouere A (1991) Comparison of in situ root observations with rhizoscope, video recording and endoscope. In: McMichael BL, Persson H (eds) Plant roots and their environment. Proc Symp Int Soc of Root Research (ISRR), 21–26 August, 1988, Uppsala, Sweden. Elsevier, Amsterdam, pp 603–607

Glasbey CA, Horgan GW (1995) Image analysis for the biological sciences. John Wiley, Chichester

Goedewaagen MAJ (1954) Beoordeling van de vitaliteit van het wortelstelsel. Landbouwkd Tijdschr 66: 553–555

Goss MJ (1991) Consequences of the activity of roots on soil. In: Atkinson D (ed) Plant root growth: an ecological perspective. British Ecological Society Special Publication No 10. Blackwell, Oxford, pp 171–186

Goubran FH, Richards D (1979) The estimation of root length in samples and sub-samples: comparison of a visual and an automatic method. Plant Soil 52: 77–83

Hackett C (1968) A study of the root system of barley. 1. Effects of nutrition on two varieties. New Phytol 67: 287–299

Haralick RM, Shapiro LG (1988) Segmentation and its place in machine vision. Scanning Microsc (Suppl) 2: 39–54

Harris GA, Campbell GS (1989) Automated quantification of roots using a simple image analyzer. Agron J 81: 935–938

Heeraman DA, Crown PH, Juma NG (1993) A color composite technique for detecting root dynamics of barley (*Hordeum vulgare* L.) from minirhizotron images. Plant Soil 157: 275–287

Hendrick RL, Pregitzer KS (1992) The demography of fine roots in a northern hardwood forest. Ecology 73: 1094–1104

Hirasawa T, Masahide T, Ishihara K (1995) A minirhizotron method for measuring root systems of soybean plants growing in the field. Jpn J Crop Sci 64: 78–85

Huck MG, Klepper B, Taylor HM (1970) Diurnal variations in root diameter. Plant Physiol 45: 529–530

Jacques WA, Schwass RH (1956) Root development in some common New Zealand pasture plants. 7. Seasonal root replacement in perennial ryegrass (*Lolium perenne*), Italian ryegrass (*L. multiflorum*) and tall fescue (*Festuca arundinacea*). N Z J Sci Technol Sect A 37: 569–583

Jain AK (1989) Fundamentals of digital image processing. Prentice-Hall, Englewood Cliffs, New Jersey

Jankowski M, Breckle S-W, Posch S, Sagerer G, Veste M (1995) Automatische Detektion von Wurzelsystemen in Minirhizotron-Bildern. In: Sagerer G, Posch S, Kummert F (eds) Mustererkennung 1995 - Verstehen akustischer und visueller Informationen. 17. DAGM-Symposium, Bielefeld, 13–15 September 1995. Springer, Berlin, Heidelbery New York, pp 176–185

Jordan MO (1992) Can rhizotrons be used for the study of corn (*Zea mays* L.) root ramification? Agronomie 12: 3–14

Kaspar TC, Ewing RP (1997) ROOTEDGE: software for measuring root length from desktop scanner images. Agron J 89: 932–940

Kemph GS (1976) Measuring fibrous roots with a leaf area meter. J Range Manage 29: 85–86

Kirchhof G (1992) Measurement of root length and thickness using a hand-held computer scanner. Field Crops Res 29: 79–88

Kirchhof G, Pendar K (1993) Delta-T SCAN® user manual. Delta-T Devices, Burwell, Cambridge, UK

Knievel DP (1973) Procedure for estimating ratio of live to dead root dry matter in root core samples. Crop Sci 13: 124–126

Kokko EG, Conner RL, Kozub GC, Lee B (1993) Quantification by image analysis of subcrown internode discoloration in wheat caused by common root rot. Phytopathology 83: 976–981

Koller TM, Gerig G, Székely G, Dettwiler D (1995) Multiscale detection of curvilinear structures in 2-D and 3-D image data. Proc 5th Int Conf on Computer Vision, 20–23 June, Cambridge, Massachusetts. IEEE Computer Society Press, Los Alamitos, California, pp 864–869

Krstansky JJ, Henderson GS (1989) Computerised measurement of fine root length using a desktop image scanner. In: Agronomy abstracts. American Society of Agronomy (ASA), Madison, Wisconsin, 307 pp

Krumenaker L (1994) Hardware, software advances brighten image-analysis systems picture. Scientist 8: 17–19

Kulpa Z (1977) Area and perimeter measurement of blobs in discrete binary pictures. Comput Graph Image Process 6: 434–451

Lebowitz RJ (1988) Digital image analysis measurement of root length and diameter. Environ Exp Bot 28: 267–273

Livesley SJ, Stacey CL, Gregory PJ, Buresh RJ (1998) Sieve size effects on root length and biomass measurements of maize (*Zea mays*) and *Grevillea robusta*. Plant Soil 207: 183–193

Marsh B (1971) Measurement of length in random arrangements of lines. J Appl Ecol 8: 265–267

McGowan M, Armstrong MJ, Corrie JA (1983) A rapid fluorescent-dye technique for measuring root length. Exp Agric 19: 209–216

Melhuish FM (1968) A precise technique for measurement of roots and root distribution in soils. Ann Bot 32: 15–22

Meyer F (1979) Cytologie quantitative et morphologie mathématique. Thesis, École des Mines de Paris, Paris

Morita STS, Haruki Y, Yamazaki K (1988) Morphological characters of rice roots estimated with a root length scanner. Jpn J Crop Sci 57: 371–376

Morrison JK, Armson KA (1968) The rhizometer – a device for measuring roots of tree seedlings. For Chron 44: 21–23

Murakami T, Yoneyama T (1988) Comparison of root length of two rice (*Oryza sativa* L.) varieties by using an image analyzer. Plant Soil 105: 287–289

Murphy SL, Smucker AJM (1995) Evaluation of video image analysis and line-intercept methods for measuring root systems of alfalfa and ryegrass. Agron J 87: 865–868

Nater EA, Nater KD, Baker JM (1992) Application of artificial neural system algorithms to image analysis of roots in soil. 1. Initial results. Geoderma 53: 237–253

Newman EI (1966) A method of estimating the total length of roots in a sample. J Appl Ecol 3: 139–145

Ottman MJ, Timm H (1984) Measurement of viable plant roots with the image analyzing computer. Agron J 77: 1018–1020

Pallant E, Holmgren RA, Schuler GE, McCracken KL, Drbal B (1993) Using a fine root extraction device to quantify small diameter corn roots ($\leq$ 0.025 mm) in field soils. Plant Soil 153: 273–279

Pan WL, Bolton RP (1991) Root quantification by edge discrimination using a desktop scanner. Agron J 83: 1047–1052

Pietola L, Smucker AJM (1998) Fibrous carrot root responses to irrigation and compaction of sandy and organic soils. Plant Soil 200: 95–105

Rasband WS, Bright DS (1995) NIH Image: a public domain image processing program for the Macintosh. Microbeam Anal Soc J 4: 137–149

Reicosky DC, Millington RJ, Peters DB (1970) A comparison of three methods for estimating root length. Agron J 62: 451–453

Richards D, Goubran FH, Garwoli WN, Daly MW (1979) A machine for determining root length. Plant Soil 52: 69–76

Rowse HR, Phillips DA (1974) An instrument for estimating the total length of root in a sample. J Appl Ecol 11: 309–314

Russ JC (1990) Computer-assisted microscopy: the measurement and analysis of images. Plenum Press, New York

Russ JC (1995) The image processing handbook, 2nd edn. CRC Press, Boca Raton, Florida

Sackville-Hamilton NR, Jones M, Harper JL (1985) Automated analysis of roots in bulk soil. In: Atkinson D (ed) Plant root growth: an ecological perspective. British Ecological Society Special Publication No 10. Blackwell, Oxford, pp 291–301

Sator C, Bommer D (1971) Methodological studies to distinguish functional from non-functional roots of grassland plants. In: Ellenberg H (ed) Integrated experimental ecology: methods and results of ecosystem research in the German Solling project. Springer, Berlin Heidelberg New York, pp 72–74

Sattelmacher B, Klotz F, Marschner H (1983) Comparison of two non-destructive methods for determination of root surface areas. J Plant Nutr Soil Sci 146: 449–459

Schuurman JJ, Goedewaagen MAJ (1971) Methods for the examination of root systems and roots, 2nd edn. Centre for Agricultural Publishing and Documentation, Wageningen, the Netherlands

Shuman LM, Ramseur EL, Wilson DO (1993) Video image method compared to a hand method for determining root lengths. J Plant Nutr 16: 563–571

Smit AL, Sprangers JFCM, Sablik PW, Groenwold J (1994) Automated measurement of root length with a three-dimensional high-resolution scanner and image analysis. Plant Soil 158: 145–149

Smucker AJM (1993) Soil environmental modifications of root dynamics and measurement. Annu Rev Phytopathol 31: 191–216

Smucker AJM, Ferguson JC, DeBruyn WP, Belford RK, Ritchie JT (1987) Image analysis of video-recorded plant root systems. In: Taylor HM (ed) Minirhizotron observation tubes: methods and applications for measuring rhizosphere dynamics. ASA Spec Publ 50: 67–80

Tanaka S, Yamauchi A, Kono Y (1995) Easily accessible method for root length measurement using an image analysis system. Jpn J Crop Sci 64: 144–147

Tennant D (1975) A test of a modified line intersect method of estimating root length. J Ecol 63: 995–1001

Tollner EW, Ramseur EL, Murphy C (1994) Techniques and approaches for documenting plant root development with X-ray computed tomography. In: Anderson SH, Hopmans JW (eds) Tomography of soil-water-root processes. Proc Symp of the Soil Science Society of America (SSSA), Minneapolis, Minnesota, 4 Nov 1992. SSSA Special Publication No 36. Soil Science Society of America (SSSA), Madison, Wisconsin, pp 115–133

Travis AJ, Murison SD, Chesson A (1993) Estimation of plant-cell wall thickness and cell-size by image skeletonization. J Agric Sci 120: 279–287

Upchurch DR (1987) Conversion of minirhizotron-root intersections to root length density. In: Taylor HM (ed) Minirhizotron observation tubes: methods and applications for measuring rhizosphere dynamics. ASA Spec Publ 50: 51–65

Voorhees WB, Carlson VA, Hallauer EA (1980) Root length measurement with a computer-controlled digital scanning microdensitometer. Agron J 72: 847–851

Wang ZQ, Burch WH, Mou P, Jones RH, Mitchell RJ (1995) Accuracy of visible and ultraviolet light for estimating live root proportions with minirhizotrons. Ecology 76: 2330–2334

Ward KJ, Klepper B, Rickman RW, Allmaras RR (1978) Quantitative estimation of living wheat-root lengths in soil cores. Agron J 70: 675–677

Wilhelm WW, Norman JM, Newell RL (1983) Semiautomated X-Y-plotter-based method for measuring root lengths. Agron J 75: 149–151

Zoon FC, Van Tienderen PH (1990) A rapid quantitative measurement of root length and root branching by microcomputer image analysis. Plant Soil 126: 301–308

CHAPTER 11

Computer-Assisted Tomography and Magnetic Resonance Imaging

S. Asseng[1], L.A.G. Aylmore[2], J.S. MacFall[3], J.W. Hopmans[4] and P.J. Gregory[5]

[1] CSIRO, Division of Plant Industry, Private Bag, PO, Wembley, Western Australia 6014, Australia
[2] The University of Western Australia, Soil Science and Plant Nutrition, Nedlands, Western Australia 6907, Australia
[3] School of the Environment, Duke University, Durham, North Carolina 27706, USA
[4] University of California, Dept. Land, Air and Water Resources, Hydrology Program, 123 Veihmeyer Hall, Davis, California 95616, USA
[5] Department of Soil Science, The University of Reading, Whiteknights, PO Box 233, Reading, RG6 6DW, UK

CONTENTS

11.1 Introduction

Until recently, techniques for direct measurements on roots and soil water in the rooting zone have been largely destructive, or simply lacked the spatial resolution necessary for meaningful definition of root and water content distrib-

A.L. Smit et al. (Eds.), Root Methods

utions. With advances in X-ray physics, detector technology and mathematical reconstruction theory, a solution to this problem was essentially achieved by Hounsfield (1972), who developed the technique known as computer-assisted tomography (CAT) or computed tomography (CT). In addition to X- and γ-rays (photons), tomographic imaging is also applicable to other types of energy such as electrons, protons, alpha particles, lasers, radar, seismic and nuclear magnetic resonance (NMR).

Of particular value is the ability of CT scanners to characterise the internal structure and the nature of the components present in the soil. For example, knowledge of the size and distribution of soil pores is important to an understanding of soil processes, including water flow and redistribution, aeration, and root penetration. In particular, the ability to monitor root proliferation and distribution by a non-invasive technique is essential for detailed studies of water uptake by plant roots.

In conventional radiography the transmission of radiation through a three-dimensional object is used to produce a two-dimensional image of the internal features of the object on a radiation-sensitive film. Attenuation occurs because the photons of the incident beam may be absorbed by the material (photoelectric absorption), or may be deflected (Compton scattering), both of which lead to a decrease in the detected radiation intensity (Fig. 11.1). The final radiographic image represents the spatial variation of radiation attenuation of the object, perpendicular to the beam. However, it does not provide information on attenuation variation along the beam direction.

In CT, multiple scans from different angles in a given plane provide a large number of ray-sums or projections. Using a back projection technique, a two-dimensional image of a horizontal slice is reconstructed numerically to yield the distribution of attenuation coefficients at discrete points within the slice along the beam path (Aylmore 1993). In the simplest scanning systems these are obtained using both linear and rotational movements of the source/detector

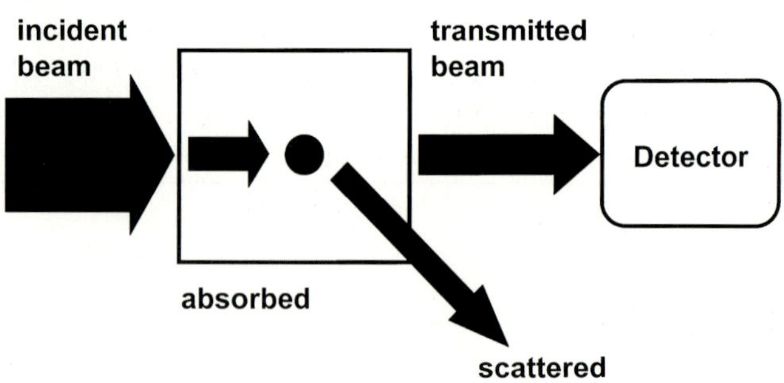

Fig. 11.1. Attenuation of a narrow beam of radiation by absorption and scattering. (After Aylmore 1993)

system. A complete set of parallel ray-sums represents a single projection for that view of the cross-sectional layer of the object. By either moving the object or the source plus detector, multiple parallel slices yield three-dimensional images. Alternatively, cone beam imaging in combination with a two-dimensional detector array allows three-dimensional imaging using a single object rotation.

Application of CT to X- and γ-ray attenuation and nuclear magnetic resonance has provided exciting new methods for the non-destructive imaging of roots and soils to study root behaviour and soil-plant-water relations in space and time. In this chapter we discuss the principles of X- and γ-ray CT scanning, the use of dual-energy measurements and nuclear magnetic resonance imaging and their application and limitations in soil and root studies.

11.2 X-ray Computed Tomography

X-ray beams, generated by electrons, impinge upon a suitable target material. The beam is polychromatic in nature with a photon energy typically normally distributed. The subsequent attenuation of an X-ray beam as it passes through a medium depends on its electron density and packing density, and the energy of the radiation. Since X-ray CT systems usually do not exceed 140 kV excitation, photoelectric absorption is the main mechanism of absorption in soil materials.

The ability of the scanner to resolve voids and objects of different size and spacing (spatial resolution) is a function of the picture element size (pixel or voxel dimensions). The array of attenuation values is graphically displayed as an image on a high-resolution monitor. Although some scanners allow voxel sizes from $1.0 \times 1.0 \times 8.0\,mm^3$ to as small as $0.1 \times 0.1 \times 0.5\,mm^3$, even smaller spatial resolutions of 10–25 μm can be achieved using smaller source beam diameters and recent developments in detection technology (Steude et al. 1994). Attenuation coefficients are converted to an internationally standardised scale, which is expressed in Hounsfield units (HU), where the attenuation of the scanned medium is measured relative to that of pure water. Optimal contrast is achieved while maximising the difference in HU-values between the features in the image, e.g., such as occurs between a water-containing root growing in dry soil material.

11.2.1 Applications

Numerous workers (Petrovic et al. 1982; Hainsworth and Aylmore 1983, 1986; Crestana et al. 1985; Anderson et al. 1988; Tollner et al. 1987, 1994; Steude et al.

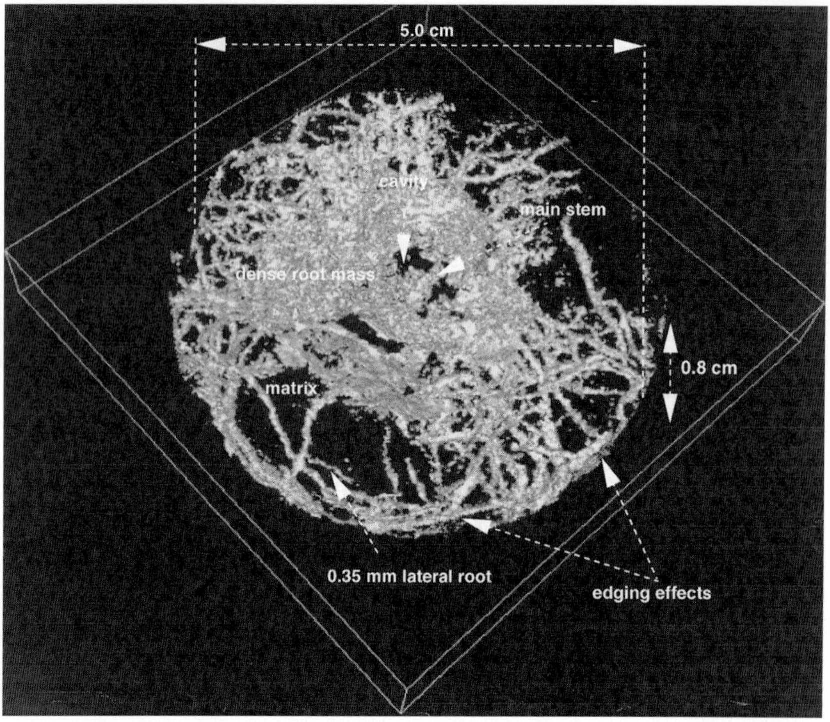

Fig. 11.2. An image of 14-day-old bush bean (*Phaseolus vulgaris* L.) roots in situ, derived from a high-energy industrial X-ray CT. (After Heeraman et al. 1997; with kind permission from Kluwer Academic Publisher)

1989; Vaz et al. 1989; Hopmans et al. 1992) have demonstrated that conventional X-ray medical scanners can provide information otherwise not obtainable on the spatial distributions of bulk density and water content in soil columns and in particular that associated with plant roots (Hainsworth and Aylmore 1983, 1986; Aylmore and Hamza 1990; Hamza and Aylmore 1991, 1992a, b).

Total root length has been estimated from an image of a plant root system in situ (Fig. 11.2) with a resolution of 0.16 mm by Heeraman et al. (1997; Box 11.1). The image was derived with a high-energy industrial X-ray CT system. Although alternative CT techniques using, for example, cone beam geometry in combination with an areal detector array, could potentially improve the spatial resolution to 5–20 μm, it would not necessarily make possible the imaging of very fine roots of this size. First, it would greatly reduce the maximum sample size to 1 cm diameter or smaller. Secondly, when reducing the voxel size further, it would become much more difficult to separate roots from the soil matrix, as roots and water-filled pores have similar attenuation values.

BOX 11.1. Root Length Estimate with a High-energy Industrial X-ray CT System

In an experiment by Heeraman et al. (1997), bush bean plants were transplanted to 5-cm-diameter PVC columns containing a fine sandy soil maintained at field capacity, after pre-germinating the seeds on paper towel. Plants were allowed to grow for 14 days in a controlled growth chamber. Three plants were destructively sampled and analyzed for root growth parameters, including root length, root dry weight, specific root length and root volume. One plant was used for the CT measurements. The stem of the plant was excised at its base, and the root system was imaged with a high-energy (420 keV) third generation industrial X-ray CT scanner (Steude et al. 1994). The scanner was equipped with a linear detector array of 125 detectors with each detector being 0.2109×0.2109 mm in size. Final spatial resolution after reconstruction was 0.16 mm. Horizontal slices (tomograms) were obtained by rotating the object; the time to scan a single slice was 6.2 min. Forty slices were obtained for a total rooting depth of 8 mm (1.2–2 cm depth below the soil surface). Three-dimensional images were obtained by combining a series of 40 individual tomograms into a three-dimensional data set of 8 mm height. The three-dimensional data set was visualized using a modified version of VIS-5D version 4.0 (Fig. 11.2). The volumetric rendering image of roots at the 1.2–2 cm depth uses this imaging software to construct a three-dimensional isosurface of attenuation coefficients. Total root length was estimated from root volume, assuming that the roots are cylindrical. Total root volume as estimated from CT was 2.97 cm^3, whereas destructive root measurements yielded a total root volume of 2.2 cm^3 for the same depth interval. Total root length as estimated from destructive and CT measurements were 8.65 and 11.7 m for the 1.2–2.0 cm depth. The overestimation of root volume using CT can be caused by signal noise and volume averaging assumptions.

An alternative means of improving the spatial resolution while still retaining the ability to scan large samples is to use a microfocal X-ray beam and to employ "area of interest" scanning techniques. Columns can be scanned rapidly at several angles to identify areas where roots are likely to be present and then those areas can be scanned in more detail. This procedure has the advantages of reducing scanning time and data storage while improving resolution.

Another example of the application of X-ray CT in root research is presented in Fig. 11.3 from Hamza and Aylmore (1992b) which shows a cross section through a lupin plant root and surrounding soil. The total number of pixels, mean, standard deviation and root mean square deviation of pixel values

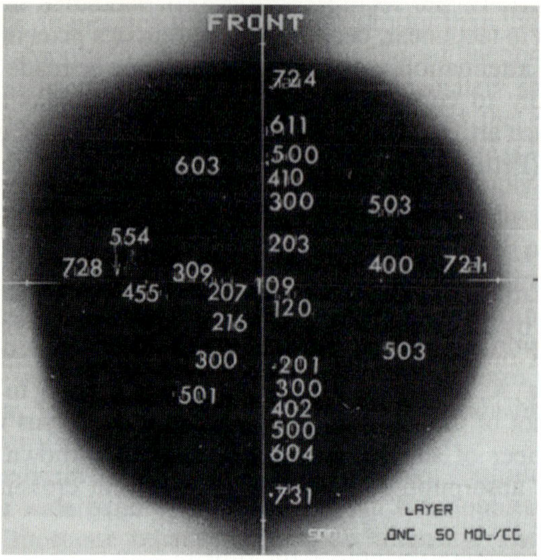

Fig. 11.3. X-ray CT scan showing a cross-section through a lupin plant root and surrounding soil. Hounsfield unit values vary throughout the section of the root. The root diameter is about 4 mm. (After Hamza and Aylmore 1992b)

in the selected region, its area and volume, and the distance between two points are instantly available.

In root water uptake studies by Hainsworth and Aylmore (1983, 1986, 1989), changes in volumetric water associated with single radish roots in a non-swelling 15% kaolinite clay and 85% sand mixture provided the first detailed observations of this type. Uptake of water along the radish roots was clearly shown to be non-uniform with depth, with the roots first removing water from the top, and later drawing water at greater depths, as soil hydraulic resistance became a major limiting factor in the upper soil layer (Hainsworth and Aylmore 1989). Subsequently, Aylmore and Hamza (1990) and Hamza and Aylmore (1991, 1992a, b) used a combination of CT scanning and a Na-ion specific microelectrode technique to measure concomitantly the spatial distribution of soil water content and Na^+ ion concentration in close proximity to lupin and radish plant roots (Box 11.2). Figure 11.4 illustrates the detail and accuracy with which the CT technique was able to resolve the changes in water content by root water uptake at different positions along the root resulting from differences in water content and electrolyte. Recently, Hamza et al. (1996) have used CT to study the spatial distribution of water content in the vicinity of interacting multiple radish roots and the effects of water stress on root shrinkage and recovery.

11.2.2 Limitations

A number of factors impose limitations on the applicability of X-ray CT which are related to spatial resolution, energy absorption, and changing soil bulk density. For example, the true spatial resolution value in the study by Warner

BOX 11.2. Monitoring Soil Water Content and Na^+ Concentrations Near Plant Roots

A combination of computer-assisted tomography applied to X- and γ-ray attenuation and Na^+ liquid ion-exchanger microelectrodes was used by Hamza and Aylmore (1992a,b) to monitor soil water contents and Na^+ ion concentrations at root depths of 3, 6 and 9 cm and at zero, 2, 4, 6, and 8 h intervals from the diurnal commencement of transpiration. The plants were subjected to two levels of transpirational demand and five Na^+ soil solution concentration levels.

The scan times took upwards of 1.4 s. The image was displayed immediately following the scan on a video screen in a 512 × 512 matrix of 2 × 2 mm pixels (Fig. 11.3). The software provided with the scanner (Somatom DR/H scanner with SCINTILLARC 700) was able to magnify any area in the scanned slice and pinpoint any pixel (from the root center to the bulk soil) to determine its water content automatically through determining its attenuation coefficient. This could be done in very small distance intervals (i.e. 0.1 mm).

The software was also able to perform some statistical functions to calculate the total water content (through Hounsfield units) of the whole or part of the scanned slice, thus making it easy to calculate water uptake by the plant root for any given section and time. Measurable differences in water content of 0.006 g/cm^3 can be readily detected by both X- or γ-ray CAT scanning (Hainsworth and Aylmore 1983, 1988).

In combination with the monitoring of soil water content, Na^+-LIX microelectrodes were used, in situ, to determine Na^+ concentrations at the root surface. For this purpose soil pots were specifically designed and constructed to allow accurate positioning of the microelectrode tips at the root surface (Hamza and Aylmore 1991). To ensure root/electrode contact, a 0.1 mm-diameter hole was made from the surface of the soil to the bottom to encourage the root subsequently planted, to follow. The microelectrode shaft was manoeuvred inside a slightly larger plastic tube. Special attention had to be paid to prevent the microelectrode tip from penetrating the root surface where it would read the cortex or xylem concentration instead of that at the root surface. The microelectrode pots were amenable to CAT scanning so that the water drawdown in any layer, and in particular that associated with the microelectrode, could be monitored continuously.

et al. (1989) was about twice the spatial resolution determined by the scanner. This results from the volume averaging that occurs if the interface of two or more materials with different attenuation values are present within a voxel. For

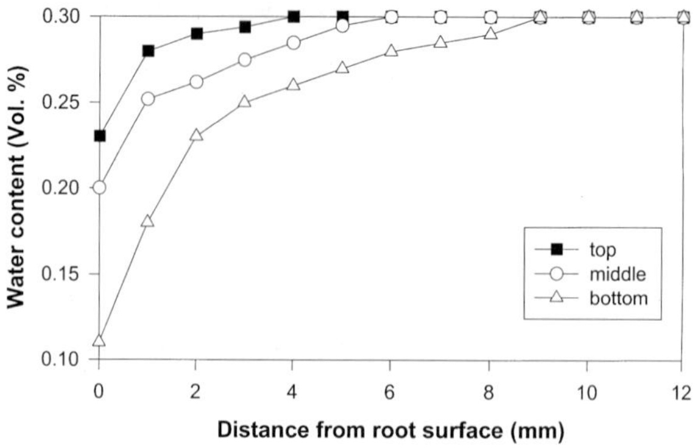

Fig. 11.4. Water drawdowns near top, middle, and bottom sections of a single radish root after 2 h of transpiration measured by CT scanning. (After Hamza and Aylmore 1992a)

example, if two materials each occupy half of the volume of a voxel (e.g., a root surrounded by a soil matrix), then the attenuation of the voxel will be the arithmetic average of the two. Consequently, studies have been largely limited to soils having macropores of a dimension equal to or larger than the voxel size.

Another limitation of X-ray CT in soils arises from the selective absorption of low-energy X-rays. The energy spectrum of X-ray scanners is polychromatic, and preferential absorption of the lower energy X-rays occurs as the beam traverses the soil. This process is referred to as "beam hardening" (Herman 1980). Physical or mathematical filters can partially correct for beam hardening (Brooks and Di Chiro 1976), but such artefacts will undoubtedly occur in soils when using relatively low-energy X-rays.

When only very small differences in attenuation exist between an object and the surrounding soil, the ability to accurately resolve the object (imaging contrast) is also reduced. While X-ray CT provides excellent resolution for some studies, its usefulness in soil systems has invariably been restricted by its inability to distinguish between changes in water content and bulk density in swelling soils. Consequently, the applications of X-ray CT to quantitative soil-water studies in swelling soils have been largely limited to statistical assessments of macroporosity distributions before and after complete wetting and drying cycles (Phogat and Aylmore 1989), and the measurement of water uptake in proximity to plant roots in non-swelling soils (Hainsworth and Aylmore 1986; Aylmore and Hamza 1990; Hamza and Aylmore 1992a, b). Even in non-swelling soils, however, localised changes in bulk density may occur close to roots caused by the root displacing particles when elongating. These changes are not detected by current X-ray systems and may lead to errors in estimates of water uptake.

Their quantitative limitations, great expense and low accessibility have seriously limited the use of medical X-ray scanners for soil-water-plant studies. Moreover, these scanners are designed to monitor patients in a horizontal position, and their construction is not generally suitable for studies involving plants growing in vertically-positioned soil columns. Finally, the proprietary nature of these commercial systems makes software modifications difficult.

11.3 Gamma-Ray Computer-Assisted Tomography

Because of the quantitative limitations and restricted accessibility of X-ray systems, work in several laboratories has concentrated on potentially more suitable and less expensive CT systems using γ-ray sources (Hainsworth and Aylmore 1983, 1988; Crestana et al. 1986; Brown et al. 1993). Gamma-ray sources range in strength from a few millicuries to 500 Ci and have been used to obtain useful tomographic images in acceptable scanning times, even with a source brightness which is several orders of magnitude less than that produced by X-ray sources (MacCuaig et al. 1986). The use of γ-sources where the energy spectrum is monochromatic, eliminates the beam hardening and absorption edge problem. Gamma-ray sources also offer additional advantages compared with X-ray tubes, including lower cost, compactness, portability and easy access to a very wide range of photon energies.

A block diagram illustrating the operation of a γ-CT system (Aylmore 1993) is shown in Fig. 11.5. The main body consists essentially of two platforms, with one allowing for vertical motion of the source/detector system, and the other providing linear translation and rotational motions of the object to be scanned.

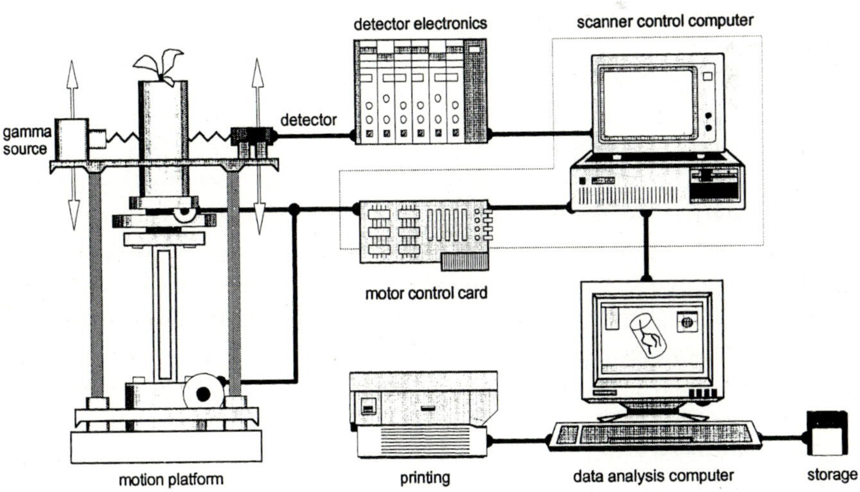

Fig. 11.5. Block diagram of γ-ray CT scanning system constructed at the University of Western Australia, consisting of three subsystems: computer, data acquisition, and motion control systems. (After Aylmore 1993)

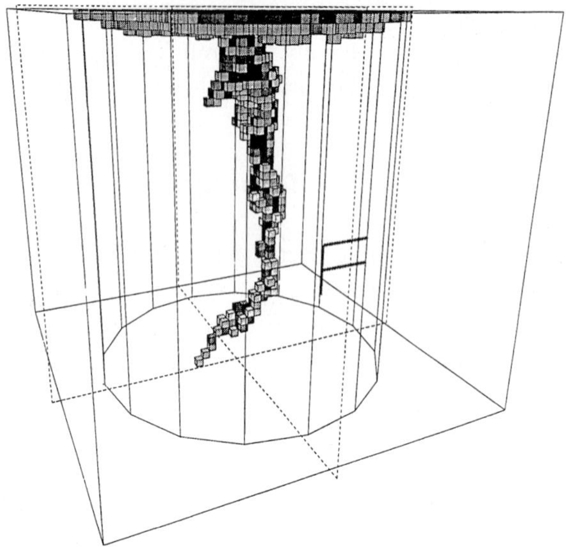

Fig. 11.6. Lupin root in a soil column. Three-dimensional image reconstruction obtained by subtractive imaging using a γ-ray CT. (After Aylmore 1993)

The main platform supports the lead shielding for the source and scintillation detector. The CT scan platform is placed at a fixed level in an opening in the middle of the γ-platform, so that the γ-platform can move independently of the CT scan platform. This system is capable of scanning columns up to 10 cm in diameter and 150 cm in length. The systems developed by Crestana et al. (1986) and Brown et al. (1993) use a similar parallel beam approach.

11.3.1 Applications

Gamma attenuation measurements have been used for many years to monitor changes in average bulk density and water content in sections through soil columns (Gurr 1962; Groenvelt et al. 1969; Ryhiner and Pankow 1969). The application of CT to γ-attenuation is illustrated in Fig. 11.6, showing how the development of a lupin tap root in a soil column can be accurately defined by subtractive imaging, thereby eliminating the soil's contribution to the image.

11.3.2 Limitations

The major disadvantage of γ-scanners is their low source brightness (Aylmore 1994). The relatively low photon emission from γ-sources as compared with X-ray sources requires much longer scanning times and larger voxel sizes ($5\,mm^3$,

Brown et al. 1993), and measurements are largely restricted to slowly changing or steady state processes. As with X-ray CT, measurements of the spatial distribution of water content in soils by a single γ-source is limited to uniform and non-swelling soils.

11.4 Dual-Energy Scanning

A major limitation of the use of single energy X- or γ-ray CT scanning has been the necessary assumption of uniform or constant bulk density. As soil attenuation is a function of both bulk density and volumetric water content, an accurate determination of water content in soils by these methods is impossible if the bulk density changes during the experiment (Petrovic et al. 1982; Hainsworth and Aylmore 1983; Anderson et al. 1988; Phogat and Aylmore 1989). Even in non-swelling soils, the general non-uniformity of packed soil columns makes accurate measurement of water content distribution near plant roots difficult, since the technique requires exact positioning of the soil column in a sequence of scans.

To monitor changes in the spatial distribution of both bulk density and water content in swelling and shrinking soils, independent estimates of attenuation associated with both bulk density and water content with different energies can be combined. In dual energy γ-CT scanning, a low-energy and a high-energy γ-source have been used (e.g., Cs-137 and Am-235 or Yb-169; Aylmore 1993).

11.4.1 Applications

Effective studies of plant root development and water uptake demand the ability to accurately and simultaneously monitor changes in water content and bulk density in close proximity to the roots. Phogat et al. (1991) demonstrated that, using dual-source γ-CT scanning, it is possible to measure the spatial distributions of bulk density and water content in swelling soils simultaneously and non-destructively with a satisfactory level of precision.

11.4.2 Limitations

Unfortunately, the relatively long scanning time required has limited the efficient application of γ-CT to plant root studies. For example, the study of Phogat et al. (1991) showed that about 169 s was needed for an individual ray-sum. In their study, 112 h were needed to complete a dual-source scan, which resulted in an average standard deviation in pixel water content values of 0.025 cm^3. Such large counting times limit its applicability to steady-state systems only.

11.5 Magnetic Resonance Imaging

An alternative approach to studying soil water extraction by plants uses proton (^1H) nuclear magnetic resonance imaging (MRI), as presented in Woods et al. (1989). Nuclei suitable for MRI are characterised by a non-zero spin quantum number I and include ^1H, ^2H, ^{13}C, ^{31}P, ^{17}O, ^{19}F, ^{23}Na, ^{27}Al, and ^{31}P. Of the NMR detectable nuclei, ^1H is intrinsically the most sensitive to detection. It is for this reason that most imaging experiments focus on ^1H, with biological and/or soil water being the most abundant detectable pool of nuclei. In practice, the sample is surrounded with a radiofrequency (*rf*) coil, and placed into the bore of a superconducting magnet. In a uniform magnetic field all protons resonate at the same specific frequency (Larmor frequency), which is determined by magnetic field strength. Following placement in the static magnetic field, the *rf* transmitter is used to apply a pulse of radio waves at the Larmor frequency. This generates a magnetic field, designated B1, which can be sensed by the nuclei, causing them to "tip sideways". The degree of tipping is dependent on the strength and duration of the applied *rf* pulse (Fig. 11.7). When the *rf* radiation is turned off, the spins realign with the static magnetic field. As they re-align, energy from the transverse magnetisation is detected as an AC current by a receiver coil placed near the specimen. Detected signals are then amplified, filtered and sent to a computer for digitisation and recording of the signal frequency, amplitude and decay. If a magnetic field gradient is applied, protons at different locations will resonate at different frequencies. If three orthogonal

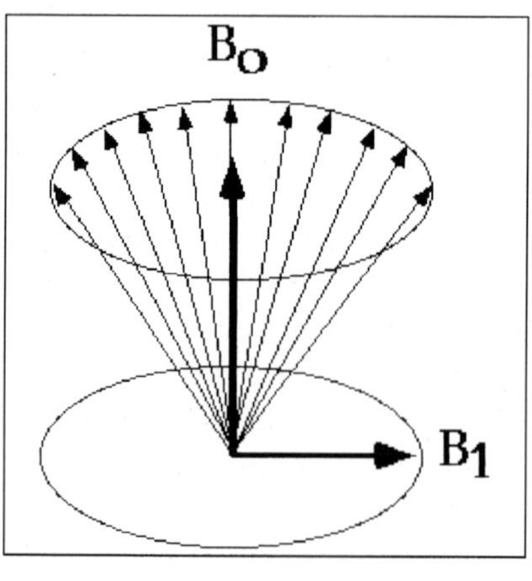

Fig. 11.7. Vector diagram showing the orientation of the primary magnetic field (*Bo*) which is created by the superconducting magnet, and the perpendicular field created by the radio frequency pulse (*B1*). Following placement in the magnet bore, nuclei align themselves so that they process in alignment with Bo

gradients are applied, analysis of the different frequencies can be used to generate a proton density map or image. By changing the three gradients it is possible to control such variables as slice position, slice thickness and resolution, thus producing a magnetic resonance image. These differences in detected signals provide "contrast" to the image, or the ability to distinguish different physiological, biochemical or anatomical properties within a specimen, including the water distribution in soil surrounding roots as well as roots themselves.

There are basically five fundamental parameters in nuclear magnetic resonance (NMR) studies which will determine signals detected from a localised region of tissue, the voxel. The fundamental parameters are spin density (how many detectable nuclei are present within a voxel, representing volumetric water content), spin-lattice relaxation (T1), spin-spin relaxation (T2), experimental spin-spin relaxation (T2*), and movement (e.g. diffusion and flow). Spin-lattice relaxation and spin-spin relaxation are functions of the degree of interaction of the water with its external molecular environment. For example, soil type affects the binding energy of the water at the liquid-solid interface which may depend on the chemical composition of the soil and soil solution as well. The wide range of properties which determine detection ability and image contrast makes MRI an extremely flexible and powerful technique, but they also greatly complicate the image interpretation. Choice of image acquisition protocol will greatly change the appearance of an image and provides a much wider range of applications than X-ray CT scans, γ-detection or dual-energy scanning.

11.5.1 Applications

It has been demonstrated that large roots (Omasa et al. 1985) and seedlings (Rogers et al. 1985) can be detected in soil by MRI. In an extensive examination of the MRI properties of soil, plant roots were imaged when grown in 30 different agricultural and 8 artificial soils. Image quality and the ability to detect roots suffered significantly when the ferromagnetic particle content was larger than 4%. Below 4%, there was still significant variation in the suitability of a soil as a root imaging substrate. Seven native soils, however, presented excellent delineation of the root systems, suggesting that MRI holds great promise for the non-destructive study of roots within a soil matrix (Bottomley et al. 1986; Rogers and Bottomley 1987). Magnetic resonance imaging has also been demonstrated to give information on the pore structure and wettability of natural sandstone, providing visualisation of wetting heterogeneities (Guillot et al. 1994). The T1 relaxation time of natural sandstone was shown to increase significantly with salinization, changing the wettability of the surface from moderately hydrophilic to hydrophobic. Similarly, the binding properties of water, as reflected by the relationship between relaxation time and water content, have been studied for samples of fine sand. When moistened with distilled water, the T1 values increased from 472 ms at 5% (w/w) to 1265 ms at 25% (w/w) water. The increase was linear between 5 and 20%, with a slight reduction in slope between 20 and 25%. The water potential measured by the pressure plate method showed the water potential ranged from 0.01 to 0.0 MPa (J.S. MacFall 1990, unpubl. data), indicating water was not tightly held by the sand grains. Further studies with a variety of sand samples of the same approximate particle size showed a range of T1 relaxation times, ranging from 565 to 1474 ms when the sample contained 25% water (MacFall and Johnson 1994a). In an early study, MacFall et al. (1991) used a series of reference tubes (sand phantoms) filled with acid-washed sand at various water contents to provide a rapid reference calibration curve. This was used to monitor water uptake by Loblolly pine seedlings from a fine sand and to quantitatively compare the relative efficiencies of fine, lateral and taproots in water uptake.

Similar experiments can be conducted for other plants, examining effects of stress, cultivar comparisons, competition, etc. Image acquisition protocols have also been developed to enhance signals detected from functional roots, suppressing signals from water in the surrounding soil/sand. This approach allows enhanced visualisation of the root system, potentially providing new tools to study root water content, conductance, growth, distribution, architecture and turnover. In an early work, Bottomley et al. (1986) used a spatial resolution of 0.6 mm with an unidentified slice definition to observe the movement of a dilute solution of $CuSO_4$ (a contrast agent which enhanced the detected signal) into and through roots of *Vicia faba*. Brown et al. (1986) were able to differentiate anatomical regions of the *Pelargonium hortorum* roots with a spatial resolution of 0.1 mm × 0.1 mm × 1.2 mm without the use of contrast agents. In other work, a synthetic soil mix was developed which gave a small magnetic resonance detectable signal while supporting plant root growth (Brown et al. 1991).

These image acquisition protocols have been extended from only two-dimensional image sets to three-dimensional acquisition, reconstruction and views of intact roots in a soil container (MacFall and Johnson 1994a). Images acquired with specially tailored pulse sequences provide sufficient contrast between roots and surrounding soil or sand to allow digital extraction of the root system or segments from the data set. Roots can be extracted either through use of a seeding algorithm or simply by adjusting the viewing opacity and window/level during rendering of the three-dimensional volume (MacFall and Johnson 1994b; Fig. 11.8). This approach has allowed not only visualisation but quantitative measurement of changes in root volume, surface area, and total

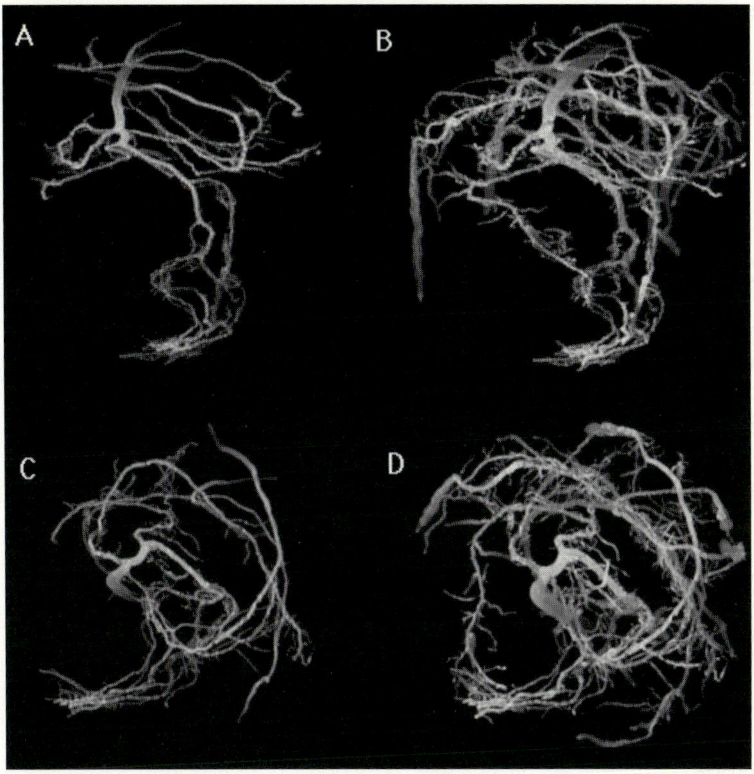

Fig. 11.8. Magnetic resonance images acquired of a pine seedling root system at two points in time. Views **A** (side view) and **C** (top view) were acquired soon after plants were transplanted into fine sand. Views **B** (side view) and **D** (top view) were acquired 6 weeks after the initial imaging experiments and clearly show the extent of root growth over this period. (J.S. MacFall, unpubl. data)

root length of the root systems of potted Loblolly pine seedlings over a period of several months. Similar studies can be done with other plants, allowing study of many topics such as pathogen effects, root turnover, comparative growth rates, etc. More recent work with pulse sequence parameter optimisation has demonstrated that three-dimensional root images can be acquired, both of intact, entire root systems within a container and of roots within intact field soil cores, further enhancing the potential utility of the technique. Roots can also be studied in a solid, synthetic matrix. Root decay was observed by three-dimensional imaging techniques of detopped root systems embedded in plaster of paris. As roots decayed, preferential patterns of water flow could be observed following the root channels (Liu et al. 1994).

11.5.2 Limitations

The presence of paramagnetic elements such as Fe^{2+}, Fe^{3+}, Mn^{2+} and Cu^{2+} at levels commonly found in soils may cause interference, thus requiring plants to be grown in porous media other than soil or in synthetic soil mixes (Cassel and Nielsen 1994).

One major difficulty is in the adsorption of water to soil particles such as clays. The signal becomes non-detectable when water becomes tightly bound to a solid substrate and less mobile, creating difficulties in quantitatively measuring water which is physically bound within the soil matrix (MacFall et al. 1991). In many cases, however, when the water content becomes low enough that detection becomes limiting, the water potential is also so low that the remaining water would be unavailable for plant uptake. Preliminary work has shown a correlation between water content (and relaxation time) with soil water potential (J.S. MacFall, unpubl.), therefore, from the perspective of water flux, the inability to detect bound water may simply reflect a limiting soil water potential.

Another constraint to the application of MRI is the available scanner time. For the acquisition of a three-dimensional data set, the imaging time required is the product of the time for each slice and the number of slices in the three-dimensional image set. For measurement of water depletion zones, the acquisition time may be up to 8 h, and if the plant is rapidly transpiring, there may be substantial changes in the shape of the depletion zone in this period. Nevertheless, as rapid pulse sequences (e.g. fast spin echo) are developed for plant imaging applications, much shorter acquisition periods will be required.

Similar to X- and γ-ray CT, most existing measuring systems are non-portable and plants have to be grown in soil cores or taken as core samples from the field. The size of MRI systems limits the size of the sample, with sample sizes ranging from <1 cm to 80 cm diameter depending on the opening size of the magnet and field strength. Generally, the larger the sample which can be accommodated, the larger the voxel sizes comprising the image. A limitation of MRI is also the high cost of the system.

11.6 Future Prospects

Industrial CT systems designed for object size/weight/density flexibility and high sensitivity, and not restricted by object motion or radiation dose level, have recently been developed [e.g., at Advanced Research and Applications Corporation (ARACOR), Sunnyvale, California, USA, and Surrey Medical Imaging Systems (SMIS), Surrey, UK]. These include systems capable of inspecting objects up to 2.4 m in diameter by 5 m long and weighing up to 49 500 kg.

Reduction in scanning times to allow more rapid monitoring of changes in soil water content would seem a priority for soil and plant studies. Recently, to reduce scanning time, fourth generation fixed-detector and rotating source scanners have been developed, in which a large number of detectors are mounted on a fixed ring and an X-ray tube inside this ring continually rotates around the object. Furthermore, improved dimensional resolution will enhance structural definition in soil systems. X-ray CT systems have been developed recently to provide 1 μm spatial resolution [e.g., synchrotron X-ray computed macrotomography (CMT), Brookhaven National Synchrotron Light Source X26 beam line, Spanne et al. 1994]. However, current pixel dimensions of the order of 0.5 to 1 mm are quite adequate to allow meaningful resolution of many of the controversies associated with water extraction by plant roots. CT scanning has the potential to provide detailed volume distributions of water content and potential throughout a complete plant root system and to allow the efficiency of different parts of the root system in extracting water and interaction between adjacent roots to be examined. Unfortunately, attempts in this direction (L.A.G. Aylmore and P.J. Gregory, unpubl.; A. Reid, R.D. Schuller and L.A.G. Aylmore, unpubl.) have, as yet, been largely confounded by difficulties in obtaining uniform packing, compensating for variations in soil bulk density and accuracy of repositioning of the column in the scanner. Future portable CT systems will allow scans under field conditions. For example, Onoe et al. (1983) described the use of a portable X-ray CT scanner for measuring annual growth rings of live trees.

In view of their substantially lower cost and some superior quantitative characteristics, such as the absence of beam hardening and absorption edge problems, γ-ray tomographic systems are likely to become a useful tool for soil and plant studies.

Although MRI has its limitations as well, particularly in the study of high Fe-containing soils, it clearly has enormous potential in the study of plant physiological processes and water uptake by plant roots. The flexibility of the imaging protocols provides the tools to study not only root distributions and anatomy, but to probe physiological function. Low cost, portable units for veterinary applications are currently in development, and these may open the doors for greater utility in field applications and to a wider range of users. The development of broad-line MRI techniques may also allow sub-millimetre spatial resolution of water distribution in soils and plants.

11.6.1 Conclusions

Possible applications, advantages and present constraints of X-ray, γ-ray, dual energy systems and MRI for root related studies are summarised in Table 11.1.

Table 11.1. Summarised applications, advantages and present constrains of X-ray, γ-ray, dual energy systems and MRI for root related studies. Note that all applications refer to studies with soil columns

System	Applications	Advantages	Limitations
X-ray CT	Dynamics of roots, soil water or bulk density	High resolution Fast scanning	Beam hardening and edge absorption Requires uniform soil density Requires non-swelling soils
Gamma-ray CT	Steady state roots, soil water or soil density	No beam hardening and edge absorption problems Relative low costs	Requires uniform soil density Requires non-swelling soils Low resolution Slow scanning
Dual energy scanning	Soil water content and bulk density	Can simultaneously quantify changes in soil water content and bulk density (swelling soils)	Very slow scanning
MRI	Dynamics of root structure and soil water	High resolution No soil density effects	Requires absence of paramagnetic elements and high clay contents Slow scanning

Despite some of the current limitations there is no doubt that the application of these innovative techniques will, with further developments, provide major tools for soil and plant scientists and have the potential to resolve major controversies with respect to the dynamics of plant root systems and water uptake by plant roots.

Acknowledgements. We thank Drs D. Rolin and A. Perry for helpful discussions, Drs K. Shackel and A. McNeill for comments on the manuscript and the Grains Research and Development Corporation for financial support.

References

Anderson SH, Hopmans JW (eds) (1994) Tomography of soil-water-root processes. Proc Symp sponsored by Division S-1 and S-6 of the Soil Science Society of America in Minneapolis,

Minnesota, 4 November 1992. SSSA Special Publication No 36. American Society of Agronomy, Soil Science Society of America, Madison, Wisconsin

Anderson SH, Gantzer CJ, Boone JM, Tully RJ (1988) Rapid non-destructive bulk density and soil water content determination by computed tomography. Soil Sci Soc Am J 52: 35–40

Aylmore LAG (1993) Use of computer-assisted tomography in studying water movement around plant roots. Adv Agron 49: 1–54

Aylmore LAG (1994) Application of computer-assisted tomography to soil-plant-water studies – an overview. In: Anderson SH, Hopmans JW (eds) Tomography of soil-water-root processes. Proc Symp sponsored by Division S-1 and S-6 of the Soil Science Society of America in Minneapolis, Minnesota, 4 November 1992. SSSA Special Publication No 36. American Society of Agronomy, Soil Science Society of America, Madison, Wisconsin, pp 7–16

Aylmore LAG, Hamza M (1990) Water and solute movement to plant roots. In: Transactions 14th Congr of Soil Science. Kyoto, Japan, 12–18 Aug. ISSS, Wageningen, pp 124–129

Bottomley PA, Rogers HH, Foster TH (1986) NMR imaging shows water distribution and transport in plant root systems in situ. Proc Natl Acad Sci 83: 87–89

Brooks RA, Di Chiro G (1976) Principles of computer-assisted tomography (CAT) in radiographic and radioisotopic imaging. Phys Med Biol 21: 689–732

Brown DP, Pratum TK, Bledsoe C, Ford ED, Cothern JS, Perry D (1991) Noninvasive studies of conifer roots: nuclear magnetic resonance (NMR) imaging of Douglas-fir seedlings. Can J For Res 21: 1559–1566

Brown GO, Stone ML, Gazin JE (1993) Accuracy of gamma ray computerised tomography in porous media. Water Resour Res 29: 479–486

Brown JM, Johnson GA, Kramer PJ (1986) In vitro magnetic resonance microscopy of changing water content in *Pelargonium hortorum* roots. Plant Physiol 82: 1158–1160

Cassel DK, Nielsen DR (1994) The realization of a dream. In: Anderson SH, Hopmans JW (eds) Tomography of soil-water-root processes. Proc Symp sponsored by Division S-1 and S-6 of the Soil Science Society of America in Minneapolis, Minnesota, 4 November 1992. SSSA Special Publication No 36. American Society of Agronomy, Soil Science Society of America, Madison, Wisconsin, pp 1–6

Crestana S, Mascarenhas S, Pozzi-Mucelli RS (1985) Static and dynamic three-dimensional studies of water in soil using computed tomographic scanning. Soil Sci 140: 326–332

Crestana S, Cesareo R, Mascarenhas S (1986) Using a computer-assisted tomography miniscanner in soil science. Soil Sci 142: 56–61

Groenvelt PH, De Swart JG, Cisler J (1969) Water content measurement with 60 keV gamma ray attenuation. Bull Int Assoc Sci Hydrol 14: 67–78

Guillot G, Chardaire-Riviere C, Bobroff S, Le Roux A, Roussel JC, Cuiec L (1994) Characterisation of wetting heterogeneities in sandstone rocks by MRI. Magn Reson Imaging 12: 365–368

Gurr CG (1962) Use of gamma rays in measuring water content and permeability in unsaturated columns of soils. Soil Sci 94: 224–229

Hainsworth JM, Aylmore LAG (1983) The use of computer-assisted tomography to determine spatial distribution of soil water content. Aust J Soil Res 21: 435–443

Hainsworth JM, Aylmore LAG (1986) Water extraction by single plant roots. Soil Sci Soc Am J 50: 841–848

Hainsworth JM, Aylmore LAG (1988) Application of computer-assisted tomography (CAT) to gamma attenuation measurement of soil water content. Aust J Soil Res 26: 105–110

Hainsworth JM, Aylmore LAG (1989) Non uniform soil water extraction by plant roots. Plant Soil 113: 121–124

Hamza M, Aylmore LAG (1991) Liquid ion exchanger microelectrodes used to study soil solute concentrations near roots. Soil Sci Soc Am J 55: 954–958

Hamza M, Aylmore LAG (1992a) Soil solute concentration and water uptake by single lupin and radish plant roots. I. Water extraction and solute accumulation. Plant Soil 145: 187–196

Hamza M, Aylmore LAG (1992b) Soil solute concentration and water uptake by single lupin and radish plant roots. II. Driving forces and resistances. Plant Soil 145: 197–206

Hamza MA, Anderson SH, Aylmore LAG (1996) Studies of root function using computer-assisted tomography. Proc ASSSI and NZSSS National Soils Conf, Melbourne, pp 97–98

Heeraman DA, Hopmans JW, Clausnitzer V (1997) Three-dimensional imaging of plant roots in situ with X-ray computed tomography. Plant Soil 189: 167–179

Herman GT (1980) Image reconstruction from projections: the fundamentals of computed tomography. Academic Press, New York

Hopmans JW, Vogel T, Koblik PD (1992) X-ray tomography of soil water distribution in one-step outflow experiments. Soil Sci Soc Am J 56: 355–362

Hounsfield GN (1972) A method of an apparatus for examination of a body by radiation such as X- or gamma-radiation. British Patent No. 1283915. British Patent Office, London

Liu IY, Waldron LJ, Wong STS (1994) The application of nuclear magnetic resonance imaging to study preferential water flow through root channels. In: Anderson SH, Hopmans JW (eds) Tomography of soil-water-root processes. Proc Symp sponsored by Division S-1 and S-6 of the Soil Science Society of America in Minneapolis, Minnesota, 4 November 1992. SSSA Special Publication No 36. American Society of Agronomy, Soil Science Society of America, Madison, Wisconsin, pp 135–148

MacCuaig N, Tajuddin AA, Gilboy WB (1986) Industrial tomography using a position-sensitive carbon-fibre anode proportional counter. Nucl Instrum Methods 242: 620–625

MacFall JS, Johnson GA (1994a) Use of magnetic resonance imaging in the study of plants and soils. In: Anderson SH, Hopmans JW (eds) Tomography of soil-water-root processes. Proc Symp sponsored by Division S-1 and S-6 of the Soil Science Society of America in Minneapolis, Minnesota, 4 November 1992. SSSA Special Publication No 36. American Society of Agronomy, Soil Science Society of America, Madison, Wisconsin, pp 99–114

MacFall JS, Johnson GA (1994b) The architecture of plant vasculature and transport as seen with magnetic resonance microscopy. Can J Bot 72: 1561–1573

MacFall JS, Johnson GA, Kramer PJ (1991) Comparative water uptake by roots of different ages in seedlings of Loblolly pine (*Pinus taeda* L.) New Phytol 119: 551–560

Omasa K, Onoe M, Yamada H (1985) NMR imaging for measuring root system and soil water content. Environ Control Biol 23: 99–102

Onoe M, Tsao JM, Nakamura H, Kogure J, Kawamura H, Yoshimatsu M (1983) Computed tomography for measuring annual rings of live trees. Proc IEEE 71: 907–908

Petrovic AM, Siebert JE, Rieke PE (1982) Soil bulk density analysis in three dimensions by computed tomographic scanning. Soil Sci Soc Am J 46: 445–450

Phogat VK, Aylmore LAG (1989) Evaluation of soil structure by using computer-assisted tomography. Aust J Soil Res 27: 313–323

Phogat VK, Alymore LAG, Schuller RD (1991) Simultaneous measurement of the spatial distribution of soil water content and bulk density. Soil Sci Soc Am J 55: 908–915

Rogers HH, Bottomley PA (1987) In situ nuclear magnetic resonance imaging of roots: influence of soil type, ferromagnetic particle content, and soil water. Agron J 79: 957–965

Rogers HH, Bottomley PA, Foster TH (1985) Application of nuclear magnetic resonance imaging to plant root studies. Proc Int Conf Soil Dynam 5: 1152–1157

Ryhiner AH, Pankow J (1969) Soil moisture measurement by the gamma transmission method. J Hydrol 9: 194–205

Spanne P, Jones KW, Prunty L, Anderson SH (1994) Potential applications of synchrotron computed microtomography to soil science. In: Anderson SH, Hopmans JW (eds) Tomography of soil-water-root processes. Proc Symp sponsored by Division S-1 and S-6 of the Soil Science Society of America in Minneapolis, Minnesota, 4 November 1992. SSSA Special Publication No 36. American Society of Agronomy, Soil Science Society of America, Madison, Wisconsin, pp 43–58

Steude JS, Hopkins F, Anders JE (1994) Industrial X-ray computed tomography applied to soil research. In: Anderson SH, Hopmans JW (eds) Tomography of soil-water-root processes. Proc Symp sponsored by Division S-1 and S-6 of the Soil Science Society of America in Minneapolis, Minnesota, 4 November 1992. SSSA Special Publication No 36. American Society of Agronomy, Soil Science Society of America, Madison, Wisconsin, pp 29–42

Steude JS, Sullivan CP, Chaney RC, McKee M, O'Shea LE (1989) Tracer experiments in saturated sand by computed tomography and inductively coupled plasma. In: Hoddinott K, Lamb RO (eds) Proceeding of the Symposium on Physio-Chemical Properties of Soil, Rock, and Related Materials. St Louis, Missouri 29 June 1989. American Society of Testing and Materials, Philadelphia, pp 171–184

Tollner EW, Verma BD, Cheshire JM (1987) Observing soil-root-interactions and soil organisms using X-ray computer tomography. Trans Am Soc Agric Engin 30: 1605–1610

Tollner EW, Murphy C, Ramseur EL (1994) Techniques and approaches for documenting plant root development with X-ray computed tomography. In: Anderson SH, Hopmans JW (eds) Tomography of soil-water-root processes. Proc Symp sponsored by Division S-1 and S-6 of the Soil Science Society of America in Minneapolis, Minnesota, 4 November 1992. SSSA Special Publication No 36. American Society of Agronomy, Soil Science Society of America, Madison, Wisconsin, pp 115–134

Vaz CMP, Crestana S, Mascarenhas S, Cruvinel PE, Reichardt K, Stolf R (1989) Using a computed tomography miniscanner for studying tillage-induced soil compaction. Soil Technol 2: 313–321

Warner GS, Nieber JL, Moore ID, Geise RA (1989) Characterizing macropores in soil by computed tomography. Soil Sci Soc Am J 53: 653–660

Woods RT, Hennessy JJ, Kwok E, Hammer BE (1989) NMR microscopy – A new biological tool. Bio Techniques 7: 616–622

CHAPTER 12

Isotope Techniques

I.J. Bingham[1], A.D.M. Glass[2], H.J. Kronzucker[2], D. Robinson[3*], and C.M. Scrimgeour[4]

[1] Scottish Agricultural College, Craibstone Estate, Aberdeen AB21 9YA, UK
[2] Department of Botany, University of British Columbia, Vancouver, V6T 1Z4, Canada
[3] Department of Plant and Soil Science, University of Aberdeen, Aberdeen AB24 3UU, UK
[4] Scottish Crop Research Institute, Invergowrie, Dundee DD2 5DA, UK

CONTENTS

* Principal author

A.L. Smit et al. (Eds.), Root Methods

12.1 Introduction: Why Use Isotopes in Root Studies?

The short answer to this question is *detection*. Isotopes allow materials in certain pools (e.g. roots) to be distinguished from those in the rest of an experimental system (e.g. soil). This is useful when studying the detailed locations (Baldwin and Tinker 1972; Brownlee et al. 1983) or C dynamics of roots in soil (Milchunas et al. 1985; Swinnen et al. 1994), both of which are almost impossible to accomplish using non-isotopic techniques. The effectiveness with which roots access certain nutrient pools or water in soil can be estimated using isotopically labelled nutrient (e.g. Powlson et al. 1986; Hodge et al. 1998) and water sources (Caldwell and Richards 1989). The isotope dynamics of certain pools can provide useful information about the structure and function of the system. For example, the efflux kinetics of radioactivity from labelled roots into an unlabelled bathing solution reflects the sizes and permeabilities of the various compartments in the root – apoplast, symplast, vacuole, etc. (Elliot et al. 1984; Kronzucker et al. 1997). These are just some of many of the possible applications of isotopes in root studies; others are listed in Box 12.1 (see also Appendix).

Most such applications involve introducing an isotopically distinct (but chemically indistinguishable) *tracer* into an experimental system. This approach has been used widely in laboratory and pot-scale studies and field experiments (Box 12.1). Experiments from <1 h up to 1 year are possible with this approach. Longer experiments are impracticable as tracers are eventually diluted beyond detection (but see, e.g. Clark 1977). It is sometimes possible, and more convenient (especially at field scales), to use distinct natural abundances of isotopes as *natural tracers*, provided that isotopic abundances in different parts of the system of interest are distinguishable and do not change appreciably during transformation and transport processes. The appearance of an isotopic abundance characteristic of a certain source pool in a sink pool may indicate transfer of material between pools. For example, Watkins et al. (1996) used differences in the natural abundance of ^{13}C in C_3 and C_4 plants to measure

BOX 12.1. Selected Examples of Root Research Utilising Isotope Techniques

Research area	*References*
Analysis of transport pathways	Dick and ap Rees (1975); Farrar (1985)
Biosynthetic rate determination	Robbins and Rost (1987); Lee and Lewis (1994)
Carbon partitioning	Caldwell et al. (1977); Atkinson and Farrar (1983); Farrar (1985); Gordon et al. (1985); Svejcar et al. (1990)
Carbon transport between plants via mycorrhizal networks	Watkins et al. (1996)
Cell cycle kinetics	Clowes (1971); Berta et al. (1991)
Compartmental analysis of ions	Cram (1968); Behl and Jeschke (1982); Jeschke (1982); Fox et al. (1992); Kronzucker et al. (1995a,b)
Fertiliser recovery by crops	Powlson et al. (1986)
Fluxes through competing metabolic pathways	Bryce and ap Rees (1985); Dieuaide-Noubhani et al. (1995)
Hydraulic lift	Caldwell and Richards (1989)
Identifying locations of ion uptake along and within roots	Clarkson et al. (1968); Kashirad et al. (1973); Marschner and Richter (1973); Kochian and Lucas (1983); Kraus et al. (1987); Lazof et al. (1992)
Influx and efflux of ions and organic molecules	Elliott et al. (1984); Aslam et al. (1994); Jones and Darrah (1992, 1994, 1996)
Inter-root competition for nutrients	Caldwell et al. (1985, 1987)
Kinetics of ion uptake	Deane-Drummond and Glass (1983); Glass et al. (1985); Siddiqi et al. (1989, 1990); Clarkson et al. (1996)
Mobilisation of ions in the rhizosphere	Marschner et al. (1989); Treeby et al. (1989)
Mycorrhizal contribution to nutrient and water capture	Brownlee et al. (1983); Pfeffer et al. (1997)
N_2 fixation in the field	Androssoff et al. (1994)
Nutrient uptake from non-uniformly distributed sources	Drew et al. (1969); Drew and Nye (1970); Robinson et al. (1994)
Rhizodeposition of carbon	Keith et al. (1986); Gregory and Atwell (1991); Mary et al. (1992); Cheng et al. (1994); Swinnen et al. (1994); Lichtfouse et al. (1995)
Root growth of competing plants	Ludlow et al. (1976); Wong and Osmond (1991); Polley et al. (1992)
Root-microbe competition for nutrients	Jackson et al. (1989)
Root turnover	Milchunas et al. (1985)
Shoot to root O_2 transport	Yoshida and Eguchi (1994)

Shoot to root recycling of N	Clark (1977); Cooper and Clarkson (1989)
Spatial location of roots in soil	Baldwin and Tinker (1972)
Turnover of metabolite pools	Farrar (1991)
Visualisation of ion depletion zones around roots	Bhat and Nye (1973); Krauss et al. (1987)
Water uptake by and loss from roots	Shone and Flood (1980); Bacic and Ratcovic (1987); Dalton (1989); Dawson (1993); Thorburn and Ehleringer (1995)
Xenobiotic metabolism and transport in roots	Ezra et al. (1985); d'Arcy-Lameta and Bompiex (1991)

transfer of carbon between such plants linked by a common mycorrhizal network. Natural tracers may be used over time-scales from <1 day to >10 years (as with "bomb"-^{14}C: Harrison et al. 1990). It is impractical to do very short (<1 h) experiments with natural tracers.

It is important to note two things about natural isotope abundances. First, using natural abundances to trace elements assumes that isotope fractionations are negligible (Box 12.2). Second, isotope fractionations – when they occur – can provide valuable insights into the operation of complex biological systems (e.g. Robinson et al. 1998), but that approach does not use the same ideas and interpretive tools as tracer studies; it is vital that the two are not confused. "Fractionation"-based approaches are in their infancy in root studies, and so will not be considered further, but see Handley and Scrimgeour (1997) and Scrimgeour and Robinson (2001) for further information.

Many considerations dictate which isotope to use in a root study. These include the availability of isotope and instrumentation, complexity of sample preparation, detection limits and precision, half-life (for radioisotopes), costs and safety. Often there may be little or no choice because only one type of isotope or instrumentation is available, but several elements have more than one suitable isotope. For example, C partitioning to roots, and between roots and the rhizosphere, can be measured with either ^{13}C or ^{14}C (Warembourg and Kummerow 1991; Boutton 1991b), or N uptake by roots can be measured using ^{15}N or ^{13}N (Clarkson et al. 1996).

Stable isotopes have the advantage that they are non-hazardous and can be used without special precautions. These lend themselves well to field and large pot experiments, unlike radioisotopes where containment is usually required and large quantities of contaminated soil must be disposed of. On the other hand, radioisotopes can generally be detected in smaller quantities than stable isotopes. For example, the theoretical limits of detection for ^{13}N are as low as 5×10^{-19} mol compared with 10^{-9} mol for ^{15}N (Meeks 1993).

Another major advantage of radioisotopes is that visualisation of the isotope is relatively straightforward using autoradiography or electronic imaging systems (see Sect. 12.3.1.5), enabling the distribution of the tracer to

BOX 12.2. Isotopic Fractionation

The natural abundance of isotopes varies in different materials. These variations reflect isotopic *fractionations* between the heavier and lighter isotope in a pair such as $^{15/14}N$, $^{18/16}O$, etc. Fractionations are caused by processes such as evaporation, condensation, and diffusion, and by chemical reactions.

Equilibrium fractionations may occur in certain chemical equilibria; e.g., $NH_3 + H^+ \leftrightarrow NH_4^+$, ^{15}N is ca. 20‰ more concentrated in NH_4^+ than in NH_3. Isotopes also fractionate in systems which are not at equilibrium, because chemical reactions may occur at different (usually slower) rates for molecules containing the heavier isotope of a pair compared with those containing the lighter isotope. These are known as *kinetic fractionations*. For example, when water evaporates, residual liquid is enriched in D (by up to ca. 19‰) and in ^{18}O (by up to ca. 3‰) and the vapour is correspondingly depleted in these isotopes. When tracing the uptake of isotopically different water sources by deep-rooted vegetation (White et al. 1985) it is essential to measure $\delta^{18}O$ and δD in xylem water, not leaf water, as the latter is subject to evaporative enrichments which fractionate the respective isotopes.

Isotopically enriched tracers should label pools such that the influence of fractionations becomes trivialised. Thus, for example, to distinguish root-derived C from soil-derived C, plants should be labelled with ^{13}C to be at least 20‰ different from soil C, ideally more. Otherwise, the $^{13/12}C$ fractionations which may occur during respiration (up to ca. 5‰; Boutton 1991a) may mask genuine differences between the two C sources. It is often impossible to use $\delta^{15}N$ measurements to trace N sources in the environment partly because of $^{15/14}N$ fractionations which can occur as an N source (e.g., soil organic N) is converted to plant-available forms and assimilated by the plant(s) of interest (Handley and Scrimgeour 1997; see Sect. 12.4.3).

be mapped in root cells, tissues or systems. Imaging of stable isotopes is possible, but it requires sophisticated instrumentation, i.e. a secondary ion mass spectrometer (Grignon et al. 1992; see Sect. 12.3.1.2). For some applications with radioisotopes, the element used may be less important than its other properties such as energy, penetrating power and half-life (Appendix).

Our aim is to provide the basic information needed to use stable isotopes and radioisotopes in root studies. Specific methodologies for studying root structure and function, e.g. nutrient uptake, water transport and rhizodeposition, are dealt with in other chapters (e.g. Chaps. 13, 14) and technical details are given in the original publications cited in the text (including Appendix).

Useful further reading includes: Vose (1980), Slater (1990), Smith (1991a) and Smith and Cresser (2001) for radioisotopes (although the first is now rather dated); Lajtha and Michener (1994), Boutton and Yamasaki (1996) and Scrimgeour and Robinson (2001) for stable isotopes; Coleman and Fry (1991) and Knowles and Blackburn (1993) for C and N isotopes, respectively; and Rorison and Robinson (1986) and Caldwell and Virginia (1989) specifically for some isotope applications in root studies.

All work with radioisotopes is potentially hazardous. It must be done according to the health and safety regulations of your institution, and only after you have consulted the appropriate safety officer. The information given here is for guidance only.

12.2 Isotope Notation and Terminology

12.2.1 Stable Isotopes

Stable isotope abundances are quantified as ratios of heavy to light isotopes in a sample. Suppose we have a sample of N in which 5 out of every 100 atoms are ^{15}N, and the rest ^{14}N; the *isotope ratio* (usually abbreviated to IR and given the symbol R) of the sample is 5/95 = 0.0526.

Another way to express the abundance of the heavier isotope in an isotope pair is as the fraction of the total amount of that element. This is the *atom fraction*, or, because atom fractions are often small, the *atom percent abundance* (A), which is 100 times the atom fraction. In the example for ^{15}N given above, A is $[5/(5 + 95)] \times 100 = 5$ atom %. The *mass* per cent of ^{15}N in the sample is not, however, 5%. It is calculated by multiplying A of each isotope by its atomic mass: $(5 \times 15)/[(5 \times 15) + (95 \times 14)] \times 100 = 5.338\%$. In a tracer experiment, even controls to which no tracer has been added contain background amounts of the isotope. To find the amount of isotope in the sample that originated from the tracer, deduct the control's A from the sample's. The difference is the sample's *atom percent enrichment* or *excess*.

The A notation is convenient for tracer work. But if we wish to use natural variations in isotopic abundance, those data sometimes vary in only the 4th or 5th decimal place when expressed as percentages. We express natural abundances as the relative difference between the isotopic ratio in the sample and that in a standard, in parts per thousand or "per mil" (‰); this is the "δ" notation. δ_{sample} is defined as:

$$\delta_{sample} = [(R_{sample} - R_{standard})/R_{standard}] \times 1000. \tag{12.1}$$

R_{sample} and $R_{standard}$ are the IRs of sample and standard, respectively (see above). Values of $R_{standard}$ are listed in Table 12.1. Laboratory working standards are

Table 12.1. Isotope ratios (IR) in the international standard materials for the analysis of D, ^{13}C, ^{15}N, ^{18}O and ^{34}S

Standard material	Isotope pair	IR, $R_{standard}$
Vienna Standard Mean Ocean Water (V-SMOW)	D/H	0.00015576
Vienna Standard Mean Ocean Water (V-SMOW)	$^{18}O/^{16}O$	0.00200520
Vienna PeeDee Belemnite (V-PDB)	$^{18}O/^{16}O$	0.0020671
Vienna PeeDee Belemnite (V-PDB)	$^{13}C/^{12}C$	0.0112372
Atmospheric N_2	$^{15}N/^{14}N$	0.0036765
Canyon Diablo Troilite	$^{34}S/^{35}S$	0.0450045

calibrated against these primary standards using materials supplied by the International Atomic Energy Agency (Austria), the Los Alamos National Laboratory (USA), and other agencies. Suppose we had a sample whose $^{13/12}C$ ratio was 0.0111372, i.e. it differed by only –0.0001 from the standard. In δ notation, this difference is –8.90‰. By definition, all standards have a δ of zero. IRs in samples may be either positive or negative on a δ scale. Positive values mean that the heavier isotope is more abundant compared with the standard; negative values, that the heavier isotope is less abundant than in the standard.

If two source materials have isotopic compositions that are (a) constant, and (b) different, we can deduce how much of each is present in a common sink using the latter's isotopic composition (*if* there is little isotopic fractionation during transfer from source to sink: Box 12.2), relative to the difference in IR between the sources, using the notion of *mixing ratios*. If an isotope has two sources, "A" and "B", the percentage derived from "A" in a common sink is:

$$f_A = [(\delta_{sample} - \delta_B)/(\delta_A - \delta_B)] \times 100. \qquad (12.2)$$

δ_{sample} is the δ of the sink. δ_A and δ_B, the δ of each source, are the *end members* of the line along which all isotopic mixtures of "A" and "B" must lie (e.g. Lajtha and Michener 1994, p. 18). The precision of f_A depends on the difference between δ_A and δ_B relative to the precision with which δ can be measured.

If this precision is 1‰, a difference of 5‰ between δ_A and δ_B limits the precision of f_A to 20%, at best; i.e. we can say only that 0, 20, 40, 60, 80, or 100% of the isotope came from source "B". In practice, the precision will be even less than this, as the analytical error will also apply to the δ of each end member. If there is a difference of 20‰ between δ_A and δ_B, and the analytical precision is 0.1‰, then we can estimate f_A to within 1%, good enough for most practical purposes.

12.2.2 Radioisotopes

Some of the most commonly used radioisotopes (e.g. ^{3}H, ^{14}C, ^{35}S, ^{32}P) decay by emitting negatively charged *beta* particles (β^-). Others emit *positrons* (β^+) produced when a proton decays to a neutron. Positrons are unstable and are annihilated by interacting with electrons, liberating *gamma* (γ) radiation. The measurement of γ radiation can be used to detect β^+ emitters (e.g. ^{13}N: see Sect. 12.5.3). γ rays may also be released from some isotopes after the emission of β^- or alpha (α) particles. α particles consist of two protons and two neutrons, i.e. they are He nuclei ($^{4}He^{2+}$); α emitters are used rarely in biological work. More detailed accounts of the chemistry of radioisotopes are given in Slater (1990) and Vose (1980).

Radioactive decay is random. The number of decaying nuclei at a given time is proportional to the number of atoms present. So, the decay of an isotope over time is exponential. Decay rate is characteristic of the isotope and is described by its *half life* ($t_{1/2}$). This is the time taken for the radioactivity present to fall by half. $t_{1/2}$ varies widely among isotopes: see Appendix. When conducting experiments with short $t_{1/2}$ isotopes such as ^{42}K and ^{13}N (Behl and Jeschke 1982; Glass et al. 1985), a large proportion of the activity initially present in a root or soil sample may be lost through decay before the sample is counted. Thus experiments must be kept short ($\leq 6 \times t_{1/2}$) and the final measurements adjusted to account for the decay occurring during the experiment. High initial activities must be used to ensure that there is sufficient remaining at the end of the experiment to detect accurately. One of the major reasons for using the analogue ^{86}Rb instead of ^{42}K in early studies of K^+ uptake kinetics was its more convenient $t_{1/2}$ (Kochian and Lucas 1982, 1983; Appendix).

The energy associated with radioactive decay is also characteristic of the isotope. This is measured in units of keV ($1\ eV = 1.6 \times 10^{-16}$ J). The traditional unit of *activity* was the Curie (Ci). $1\ Ci \equiv 3.7 \times 10^{10}$ disintegrations per second (dps) or 2.22×10^{12} disintegrations per minute (dpm). Its definition was based on the amount of radioactivity in 1 g of ^{226}Ra. Although still widely used, the Ci has been replaced by the SI unit, the Bequerel (Bq): 1 Bq = 1 dps; i.e. $1\ Bq \equiv 2.7 \times 10^{-11}$ Ci.

Specific activity is the amount of radioactivity (Bq) of a substance per unit amount of material. A particular radio-labelled compound can sometimes be purchased at a range of different specific activities. It is then usually diluted to the required specific activity with a known quantity of a *carrier*, comprising the *stable* isotope(s) of the element. An accurate knowledge of the specific activity is not necessary in applications where only the presence or absence of activity in a sample is of interest, such as mapping the location of roots in the soil (Baldwin and Tinker 1972). But the specific activity of the source, and in some cases individual metabolite pools, must be known when quantifying nutrient fluxes, translocation or compartmentation of ions and metabolites in roots (Behl and Jeschke 1982; Atkinson and Farrar 1983; Chap. 13).

12.3 Instrumentation and Laboratory Facilities for Isotope Work

12.3.1 Instrumentation

12.3.1.1 Mass Spectrometry

Most root studies which involve stable isotopes will require the measurement of IRs using a *continuous-flow isotope ratio mass spectrometer* (CF-IRMS); see Scrimgeour and Robinson (2001). This requires relatively large samples (μg to mg dry wt.) and pre-conversion to a suitable form for measurement. An IRMS measures the IR of only a few simple gases (H_2, N_2, CO_2, SO_2). A CF-IRMS is usually a bench-top system requiring relatively few services and costs ca. $US 150 000.

Solid samples (e.g. plant and soil material) are oven-dried and ground to powder in a ball-mill to ensure homogeneity. Sub-samples of 1 to 10 mg are weighed into tin-foil cups and loaded onto an auto-sampler, a standard feature in modern CF-IRMS. It is good practice to re-dry stored, previously dried samples before weighing to remove any adsorbed moisture. Carry-over of samples in successive grindings must be avoided. Scrupulous cleaning of the ball-mill capsule and avoidance of excessively dusty working conditions are essential, especially if measuring natural isotopic abundances.

Liquid samples (e.g. soil and hydroponic solutions) often have to be pre-concentrated (e.g. by micro-diffusion or ion exchange) so that a small aliquot can be pipetted into a tin cup, before being freeze dried in situ. If extracted fractions or individual metabolites are to be analysed, they must be highly purified and free of any solvents or reagents containing the element of isotopic interest.

Gas samples (e.g. CO_2 from root respiration studies) can be introduced directly into the MS via an auto-sampler if many samples are to be run. Trace gases require collection over long periods to ensure that enough sample is available for analysis.

Standards are included between every 10–20 samples to calibrate the analysis. Calibration is checked by including a standard sample (e.g. plant material of a known isotopic composition) within each batch of samples.

12.3.1.2 High-Sensitivity Mass Spectrometry

Gas-chromatograph-MS (GC-MS) or *high performance liquid chromatograph-MS* (HPLC-MS) combine separation of minute amounts (pg to ng) of complex mixtures with IR measurement, but sample enrichments of ca. 1 atom % or more are required for reliable quantitation. GC-MS or HPLC-MS can measure only percentage differences in IR. These instruments are useful in metabolic studies, e.g. to determine the fate of N assimilated by mycorrhizal roots (Martin and Botton 1993).

IRMS is not sensitive enough to quantify isotopes with small natural abundances, e.g. ^{14}C, ^{41}K, and with very dilute tracers. Greater sensitivity is achieved by adding a second mass acceleration/separation step, as in *tandem MS-MS*. MS-MS requires little or no sample preparation of even complex mixtures and tissues, and analysis is faster due to the elimination of a lengthy chromatographic separation step required in GC-MS or HPLC-MS.

Other high-sensitivity MS instruments include *secondary ion mass spectrometry* (SIMS). This uses high-energy (MeV) "primary" ion beams to sputter "secondary" (negatively charged) ions from specimen surfaces. The sensitivity of SIMS is usually limited to 10^{-7} to 10^{-8} minor atoms per major atom. One of its strengths is the ability to image the spatial distributions of isotopes within tissue sections (Grignon et al. 1992; Lazof et al. 1994). Modified versions of SIMS – often referred to as *accelerator mass spectrometry* (AMS; Trumbore 1996) – can reach detection limits of 10^{-15} atoms atom^{-1}. Nepstad et al. (1994) related root biomass to changes in ^{14}C natural abundance with soil depth, ^{14}C being measured by AMS.

12.3.1.3 Scintillation Counting

Scintillation counting is the most widely used technique for measuring radioactivity. Its advantages over other techniques include its sensitivity, versatility and ease of use. The basic principles are the same whether determining α, β, or γ radiation.

The kinetic energy associated with α, β or γ radiation is transferred to a fluorescent molecule (scintillator or fluor). This transfer is either direct (solid scintillation or γ counting) or indirect via a solvent (liquid scintillation counting). The resulting excitation of the fluor causes it to emit photons ("scintillation"). These are then detected by photomultiplier (PM) tubes and counted electronically. Radioactivity is reported as counts per second or minute (cps or cpm). Not all disintegrations are measured and so it is essential to know the counting efficiency to convert cps to dps. All instruments will register background counts in the absence of a radioactive sample. This must be subtracted from the sample's cps. In most instruments this is done automatically. See Simonnet (1990) and Goulding and Slater (1994) for more detailed accounts of the principles of scintillation counting.

Liquid scintillation counting (LSC) is used mostly in biological work for measuring weak or medium energy β^- emitters. A sample is placed into a vial with a *scintillation cocktail*. This cocktail consists of an organic solvent and one or more fluors. Careful sample preparation is critical for maximising counting efficiency with LSC (Coupland 1986).

Quenching is a major cause of poor counting efficiency. *Colour quenching* occurs when pigmented samples are counted. Problems are most frequently encountered with leaf extracts containing chlorophyll. Although less of a problem with roots, colouration may occur with some extraction procedures. Emitted fluorescence is absorbed or scattered by the pigments before it can be detected by the PM tubes. Colour quenching can be reduced by bleaching the sample with, e.g. H_2O_2 prior to counting (Coupland 1986). However, in some scintillation cocktails, this may lead to a different problem of *chemiluminescence* (see below). An alternative strategy is to reduce the sample volume relative to that of the cocktail, effectively "diluting out" the colour. But take care that there is sufficient energy in the vial to register cps significantly above background. *Chemical quenching* refers to interference with the energy transfer from solvent to fluor by components of the sample. If chemical or colour quenching occurs, a correction must be applied to calculate the true dps in the sample. Several techniques are available for quench correction including *internal standardisation and external standardisation* (for details see Simonnet 1990; Goulding and Slater 1994). Modern LSC instruments have automatic external standardisation. For ease of use and accuracy, this is usually the method of choice. Some instruments also have software for determining dps directly from an analysis of the full energy spectrum of the isotope without the need for quench correction (e.g. Direct DPM, Canberra Packard).

Two other important sources of interference in LSC are chemiluminescence and *photoluminescence.* Chemiluminescence can result from reactions among components of the sample and the scintillation cocktail that generate weak light, but are independent of radioactive decay events. Storing samples before counting is advisable to allow any chemiluminescence to decay. Photoluminescence occurs when components of the sample and the vial absorb light and re-emit it at wavelengths detectable by the PM tubes. It is mostly a problem with pigmented samples. Sample storage in the dark prior to counting can help overcome the problem. Some instruments have a facility for identifying and correcting for luminescence, although this should not be regarded as a substitute for careful sample preparation.

Since β^- particles travel only short distances, it is essential that the sample is distributed homogeneously in the scintillation cocktail. LSC is best suited to counting *liquid* samples. This would include samples of soil solution, hydroponics or root extracts (Gordon et al. 1985; Jones and Darrah 1994). Radioactive *gases* can be counted as solutions after dissolving them in an appropriate solvent; e.g. $^{14}CO_2$ respired from roots or soil micro-organisms, or that liberated from oxidation of root tissue (see below), can be trapped in NaOH or an organic amine (e.g. phenylethylamine or ethanolamine) and counted

(Swinnen et al. 1994). Organic amines are preferable to inorganic bases as the former have a greater trapping capacity, and produce less quenching and chemiluminescence (Coupland 1986).

Often, the radioactive sample to be counted is a *solid*. This may be excised roots, or an insoluble residue after extraction, e.g. cell wall or DNA (Robbins and Rost 1987). Although these may be counted directly (Farrar 1985, 1993), there are two potential problems. First, much radiation may be absorbed by the sample itself or its solid support, e.g. glass fibre filter (*self-absorption*). This leads to a loss in counting efficiency. This is most pronounced with weak β emitters, e.g. ^{3}H. Second, counting efficiency will be influenced greatly by the orientation of the sample in the scintillation vial relative to the PMs. Efficiency losses through this are unpredictable and cannot be corrected by standardisation. Homogeneity can be increased by first grinding the sample then adding water and a gel-forming scintillation solution. The sample is dispersed by shaking. A gel forms after the shaking stops, holding the sample in suspension for counting.

The best way to treat more bulky material is to solubilize or oxidise it before counting. Plant samples can be solubilized using strong alkalis such as quaternary amines (Zocchi and Hanson 1982; Coupland 1986; Bingham et al. 1998). This procedure is straightforward. Its main drawback is that only small samples can be treated. Solubilization can lead to chemiluminescence with some cocktails, overcome by selecting a compatible cocktail. Mineral elements bound to or incorporated in structural material can be counted after acid digestion (Jensen and Kylin 1980). Dalal (1979) and Amato (1983) have described wet oxidation techniques for determining ^{14}C in root and soil samples; organic ^{14}C is oxidised by strong acid and trapped as $^{14}CO_2$.

For ^{3}H and ^{14}C, an alternative to solubilization and wet oxidation is dry sample oxidation. Dedicated equipment is available for this. The sample is combusted in an atmosphere of O_2. The H_2O and CO_2 which are produced are trapped, and the remaining oxides (e.g. SO_3 or NO_x) released as waste. Radioactivity in the trapped samples is then counted by LSC. Dry oxidation has several advantages over other forms of sample preparation. It is quick, large samples can be handled, and it greatly reduces chemiluminescence and quenching. Sequential extraction of soluble C fractions can precede sample oxidation (e.g. Gordon et al. 1985).

A possible alternative to LSC is Cerenkov counting. When high-energy charged particles pass through a transparent medium (e.g. H_2O) at a speed faster than light in that medium, blue-white light is emitted which can be measured directly in an LS counter (Vose 1980, p. 122). Cerenkov counting is possible for γ emitters and hard β emitters with a particle energy greater than 0.5 MeV (the *Cerenkov limit*). This includes many isotopes used in studies of ion transport in roots; ^{32}P, ^{36}Cl, ^{24}Na, ^{42}K, ^{86}Rb and ^{13}N. Samples, including root extracts or excised roots, are placed in water and counted directly (Behl and Jeschke 1982; Lefebvre and Glass 1982; Glass et al. 1985). It is a simple, cost-effective method of counting because no organic cocktail is required; it is not affected by chemical quenching, although colour quenching can occur. Self-absorption can be a problem with excised roots. To overcome this, establish a calibration curve by varying the root mass in the vial (Glass et al. 1985). Counting efficiencies range from 0.1 to 0.8 depending on the isotope. Efficiency increases with increasing E_{max} above the Cerenkov limit (Vose 1980; Goulding and Slater 1994).

12.3.1.4 *γ Counting*

γ rays are more penetrating than most β^- particles, and so tend to pass through a scintillation cocktail without transferring energy to a solvent or fluor. Standard γ counters use a cylindrical, solid crystal fluor (e.g. NaI laced with thallium) into which a well has been bored in the centre. Sample vials are transferred to the well for counting (Smith 1991b, p. 350).

Because the sample is external to the fluor, γ counting does not suffer in the same way as LSC from problems with quenching and luminescence. Both liquid and solid samples can be counted, but counting efficiency depends on the sample geometry. The top of the sample must be positioned deep in the crystal's well and ideally a series of samples should have a constant height. In root studies, the lack of quenching problems is particularly attractive since virtually no sample preparation is required other than excision of the parts of interest (cf. Glass et al. 1985; Kronzucker et al. 1995a,b,1997). Most instruments have calibration procedures that determine automatically the resolution, background and counting efficiency of each detector (see Simonnet 1990 for principles).

Many solid materials, other than the NaI crystals usually employed in γ counters, can be excited to fluorescence by impacting radiation. These have potential applications in root research. Minchin et al. (1994) used a series of scintillation detectors positioned against leaves and two halves of a split-root system to measure ^{11}C transport within intact barley plants.

12.3.1.5 *Autoradiography*

Autoradiography is an extremely useful technique for determining the distribution of a radioactive tracer within a sample.

Macro-autoradiographs of whole root systems, and similar structures, can be made by taping the plant material to a piece of stiff card within an *X*-ray film cassette or similar holder. To prevent redistribution of the isotope, the roots are oven or freeze dried before a sheet of *X*-ray film is clamped tightly against the tissue and exposed for a period of hours to days (Vose 1980, p. 212 et seq.; Farrar 1993). An alternative to drying is to secure the film over fresh roots (protecting the emulsion with a layer of thin cellophane) and freezing at –20 °C during exposure. The exposure time required will depend on the isotope used, its activity, the type of film and the temperature. As in all types of autoradiography, a certain amount of experience and trial and error is required to judge the period accurately. If replicate samples are available, try developing these for different times to check for adequate exposure. Typical applications include identifying major sinks for assimilates in roots and root-mycorrhizal associations (e.g. Brownlee et al. 1983) and determining the systemic activity of root-applied pesticides (d'Arcy-Lameta and Bompeix 1991).

In situ autoradiography has been useful in mapping the distribution of roots growing in soil. This has been done in two ways. Soil is loaded with a radioisotope, roots are grown though the soil, the soil and roots are embedded in resin and then sectioned horizontally. The cut surface is placed in contact with *X*-ray film for a period of time and developed (Nye and Tinker 1977 p. 141). This technique also allows ion depletion zones around roots to be visualized (Kraus et al. 1987). Second, the tops of plants are injected with radioisotopes of readily translocated elements. Sheets of *X*-ray film are driven into the soil through the root system, or placed against the window of a root observation box against which roots are growing. The distribution of labelled roots contacting the plane of the film is indicated by the appearance of dark spots on the autoradiograph. Suitable isotopes are those which are phloem mobile, have medium to high energy to overcome absorption by adhering soil, and a $t_{1/2}$ of several days or more (e.g. ^{14}C, ^{32}P, ^{35}S; *not* ^{45}Ca or ^{3}H). Baldwin and Tinker (1972) visualised the distributions of the inter-penetrating root systems of two plants. Each plant was injected with a β emitter differing in E_{max} (see Appendix): ^{35}S or ^{33}P in one plant, and ^{32}P in the other. The difference in E_{max} allowed the smaller energy to be screened out by a sheet of Melinex covering the film. Baldwin and Tinker (1972) give some useful hints about the practical application of this technique.

For studies of cellular and sub-cellular localisation (micro-autoradiography), liquid emulsion is used to provide the close contact needed between the section and film (Laskey 1990). After development, the slide can be viewed under bright or dark field optics to reveal, respectively, black or white grains where the isotope is located. Isotopes which are not incorporated into structural material during labelling of the roots may become redistributed during slide preparation. This can be avoided by freeze-substitution of the root before embedding (e.g., Kochian and Lucas 1983). Micro-autoradiography is best suited to the weak β^- emitters ^{3}H, ^{14}C and ^{35}S. Their energy is soon dissipated within the emulsion producing a discrete, high resolution image. Higher energy emitters (e.g. ^{32}P) generally produce insufficient resolution for microscopic examination.

Comparing the autoradiograph with a photograph of the same structure allows the distribution of the tracer to be determined (e.g. Brownlee et al. 1983). The autoradiograph can be quantified by densitometry or image analysis (Kochian and Lucas 1983) against standards of known activity. But note that the density of image is proportional to the logarithm of exposure and that, above a threshold exposure, further exposure does not lead to an increase in density (Farrar 1993).

Electronic systems are now available for imaging radioactivity (e.g. InstantImager, Packard). By combining direct nuclear counting with image analysis, the distribution of radioactivity in samples can be quantified in real time without the need for lengthy exposure and development of film. Its advantages over standard autoradiography are its speed, ease of use, sensitivity and linearity of response. Because results are generated within 10–30 min, autoradiography of short $t_{1/2}$ isotopes is possible. Techniques for autoradiography in biochemical and molecular work, e.g. visualising ^{32}P probes, are covered by Laskey (1990).

12.3.1.6 *Magnetic Resonance Techniques*

Magnetic resonance (MR) involves subjecting molecules to an oscillating electromagnetic field and measuring the frequency at which they come to resonate with that field. Most MR work involves detecting nuclear resonance, so it is known as nuclear MR (NMR). The advantage of NMR is that it is that it is highly compound-specific. It gives intra-molecular, site-specific information, often with a minimum of sample preparation, providing that the compound(s) of interest are sufficiently concentrated and isotopically enriched in the sample. NMR usually requires at least 10 atom % labelling for quantitation and 1–100 mg samples, and uses ca. 100% labelled substrates. Nuclei of only certain isotopes can be detected and quantified by NMR: e.g. ^{1}H, D, ^{13}C, ^{15}N, ^{17}O and ^{31}P. NMR spectra do not usually give IRs (or specific activities) directly (unlike MS), but these may be deduced from, e.g. the extent of ^{1}H-^{13}C couplings in the spectrum (Dieuaide-Noubhani et al. 1995). Examples of NMR spectroscopy as applied to nutrient transport in mycorrhizal root systems are described by Pfeffer et al. (1997).

In root studies, NMR has been used exclusively for metabolic investigations. The work of Dieuaide-Noubhani et al. (1995) is a superb example of its use to clarify pathways of C metabolism in root tips, using both ^{13}C and ^{14}C labelling. Bacic and Ratkovic (1987) used NMR to monitor water exchange within roots after labelling with D_2O. For further information on the use of NMR in plant nutrition studies see Lee and Ratcliffe (1993).

12.3.2 Laboratory Facilities

No laboratory is designed or equipped ideally for all isotope applications. What follows is a brief outline of some of the facilities that allow isotope work to be done safely and reliably.

12.3.2.1 General

It is a good idea to keep experimental and sample preparation areas separate from those containing the analytical instruments. It is also sensible, if possible, to separate areas and equipment (e.g. grinders) in which isotopically labelled and unlabelled materials are used. Accidental contamination by even a small amount of a tracer can ruin natural abundance measurements. All working areas should be kept clean and uncluttered. This applies as much to the areas where soil is prepared as to the analytical laboratory.

12.3.2.2 Radioisotopes

Basic safety points are described fully by Vose (1980) and Prime and Frith (1990). Most importantly, all radioisotope work must conform to acceptable health and safety standards in addition to those that are in force to ensure good general laboratory practices. These differ from country to country, although the basic rules apply everywhere. All workers with radioisotopes should be properly trained in their use, monitoring and disposal.

Radioisotopes pose an internal hazard if material is ingested, inhaled or absorbed through the skin and an external hazard from the radiation field surrounding an exposed source. The external hazard should be minimised by using appropriate shielding and by maximising the distance between the operator and the radioactive source. Access to "hot" areas should be restricted. If possible, glasshouses or growth rooms should be designated specially for radioisotope work. At the very least, the presence of radioactive material should be displayed clearly and shielded from other laboratory areas. When using high energy sources, the external hazard should be monitored during the experiment and the acute and cumulative exposure kept as low as possible and certainly within the legal safety limits. Operators may be required to wear personal dosimeters or film badges. However, these provide an indication of the level of exposure only *after* the dose has been received.

Laboratories should have work surfaces and floors that are as continuous as possible to minimise the extent to which radioactive materials can lodge and which can be cleaned and decontaminated easily. This includes walls with washable paintwork.

Operations that generate dust, e.g. grinding, sieving and sectioning of plant and soil material should be carried out in a negative-pressure glove box or a fume hood. Particular care is necessary when conducting experiments using e.g. ^{14}C and ^{35}S to avoid accidental release of $^{14}CO_2$, $^{35}SO_2$ or $H_2{}^{35}S$ into a confined atmosphere. Degeneration of ^{35}S-amino acids during storage can release radioactive volatile compounds. $^{14}CO_2$ may be respired from plants and microbes labelled with ^{14}C. Where accumulations of radioactively labelled gases are likely, apparatus should be vented in a fume cupboard. Alternatively $^{14}CO_2$ may be trapped by e.g. self-indicating soda lime or KOH, and the radioactive material disposed of as solid or liquid waste respectively.

The *minimum* containment for solid or liquid materials is a tray large enough to contain any spillage on an open bench in a designated laboratory. The minimum personal protection should be laboratory coat, safety spectacles or goggles and vinyl or latex gloves. Handle all radioactive materials with forceps. Before, during and after each experiment or operation involving radioisotopes, monitor the working area and laboratory clothes for contamination. This is usually done with swab counts or calibrated count rate meters suitable for the isotope in use.

The use of radioisotopes in the field is possible (e.g. ^{14}C to measure rhizodeposition of C; Swinnen et al. 1994), but presents some practical difficulties in terms of containment, control of public access, long-term contamination and waste disposal. A complete risk assessment of the fate of the radioisotope in the environment is required. As the Chernobyl incident demonstrated all too well, the uncontrolled introduction of radioisotopes to large areas of land can have unforeseen long-term consequences. Where possible, the use of *stable* isotopes should be explored for field experiments.

12.3.2.3 Stable Isotopes

Fewer specialised facilities are needed to handle and use stable isotopes. An important one that is essential for natural abundance $\delta^{13}C$ experiments is to grow plants in a well-ventilated glasshouse. This avoids unacceptable fluctuations in the $\delta^{13}C$ of source CO_2 which might confound later analyses of $\delta^{13}C$. For this reason, $\delta^{13}C$ experiments in controlled environment chambers should be avoided, unless those chambers have accurate and reliable control of the internal $[CO_2]$ and $\delta^{13}C$: few do.

12.4 Stable Isotopes

12.4.1 ^2H (D) and ^{18}O

Deuterium (^2H or D) and ^{18}O are considered together as these isotopes feature jointly in many applications, particularly on water uptake by roots, but they differ in how they are prepared for IRMS. Generally, ^{18}O, when used as a tracer, is expensive but easy to measure, while D is cheap and more difficult to measure. Suitable working MS standards for D and ^{18}O in H_2O are a local tap-water and tap-water with a little $D_2O/H_2{}^{18}O$ added. Both are calibrated against V-SMOW (Table 12.1).

Water itself is not suitable for IRMS analysis because of memory effects (i.e. water tends to remain adsorbed on surfaces inside the MS inlet). It must first be converted to, or equilibrated with, more suitable gases: CO_2 for ^{18}O; H_2 for ^2H analysis. ^{18}O is measured by equilibration of H_2O with CO_2. The O atoms in water and CO_2 exchange, and the ^{18}O content of the CO_2 reflects that of the water. This is done readily on large sample sets, with the CO_2 being sampled and measured automatically. For D analysis, water is converted to H_2, usually by reduction with hot Zn or U. Prosser and Scrimgeour (1995) combined an alternative equilibration method (i.e. using H_2 and a Pt catalyst) with automated CF-IRMS of H_2.

Additional steps are required to measure D or ^{18}O in H_2O within matrices such as soil or plant tissue. Existing methods rely mainly on azeotropic distillation to extract H_2O. This involves distilling the H_2O in the material with a solvent such as toluene. Each distillation takes several hours, requires solvent and expensive glassware, and needs samples containing at least 5 ml of H_2O. Direct equilibration of such samples avoids the extraction step and is more suited to large sample sets (Scrimgeour 1995). ^{18}O in H_2O in plant stems, woody twigs and soil equilibrates with added CO_2 at a similar rate to ^{18}O in free H_2O. This method gives accuracy and precision comparable to azeotropic extraction. Preparation time is shorter and smaller samples (containing as little as 100 μl) can be used. With CF-IRMS of the equilibrated CO_2, up to 100 samples can be analysed daily, making ecological-scale studies a practical proposition. Direct equilibration with H_2 can also be used to analyse H_2O in intact plants. However, complete equilibration does not occur unless the sample is first heated to 100 °C under vacuum. Presumably, this releases tightly bound H_2O that exchanges slowly with other H_2O in the system.

Given sufficient concentration and isotopic enrichment, D is quantifiable by NMR as Bacic and Ratkovic (1987) showed in their study of H_2O exchanges in maize roots.

12.4.2 ^{13}C

^{13}C tracers are readily available as $^{13}CO_2$ or as a wide range of labelled organic compounds up to 99 atom %. Most commercially available compressed CO_2 is isotopically distinct from atmospheric CO_2 (Schnyder 1992). Whereas the $\delta^{13}C$ of the latter is ca. –8‰, CO_2 produced (industrially in, e.g. distilleries, breweries, petrochemical plants) has $\delta^{13}C$ of –25 to –50‰. A working standard for natural abundance determination of $\delta^{13}C$ by CF-IRMS is a leucine/citric acid mixture (Handley et al. 1993). An internal standard "plant" material of commercially available flour is always included in each batch of samples. To measure $\delta^{13}C$ of CO_2 produced by roots and soil microbial respiration, Robinson and Scrimgeour (1995) used as a standard 5% CO_2 in N_2, calibrated against V-PDB by comparison with other gas samples (Table 12.1).

Natural variations in $\delta^{13}C$ are potentially very useful tracers of C. The difference in $\delta^{13}C$ between C_3 and C_4 species may be used to detect C from one source against a background of C from the other, e.g. root C of maize (C_4) in soil containing soil organic matter derived only from C_3 vegetation (Balesdent and Mariotti 1996).

Within plants, different C-containing molecules have different $\delta^{13}C$ values. These arise because of $^{13/12}C$ fractionations that occur during their enzyme-catalysed synthesis. Compared with hexose, lignin is up to 4‰ more depleted in ^{13}C; cellulose is ca. 2‰ more enriched; and chitin – not a constituent of plants, but an important one of pathogenic and mycorrhizal fungi – is up to 4‰ more enriched (Gleixner et al. 1993). Protein and, particularly, lipids are depleted in ^{13}C relative to carbohydrate. Knowing the ^{13}C composition of individual compounds allows deeper interpretations to be made of, e.g. the fate of decomposing root residues (Lichtfouse et al. 1995) or of the metabolic interactions between roots and associated fungi (Gleixner et al. 1993).

^{13}C is also quantifiable by NMR given sufficient sample size and isotopic enrichment (Dieuaide-Noubhani et al. 1995; Sect. 12.3.1.6).

12.4.3 ^{15}N

^{15}N has been used most often as a field- or pot-scale tracer for N, in "fertiliser recovery" trials. ^{15}N enrichments of ca. 1 atom % are usually adequate in the fertiliser. ^{15}N-labelled inorganic salts and organic compounds are readily available with ^{15}N abundances of up to 99 atom %. ^{15}N-enriched N_2 is also available. Less widely known, but of potential use in solution culture experiments, is ^{15}N-*depleted* material. This is produced as a by-product of the ^{15}N enrichment process (industrial $^{15/14}N$ fractionation), and it is much cheaper than the ^{15}N-enriched fraction. It is useful for growing plants at a very small background of

^{15}N as the ^{15}N depleted material can have a ^{15}N abundance of ca. only 0.005 atom % (Vose 1980, p. 165–7).

^{15}N is quantified by CF-IRMS as for $\delta^{13}C$ (see above). For natural abundance measurements of both $\delta^{13}C$ and $\delta^{15}N$ on the same sample, % N is measured first during a run optimised for $\delta^{13}C$. This information is then used to calculate how much sample is required to give enough N in a subsequent analysis optimised for $\delta^{15}N$. This can greatly improve the accuracy of the $\delta^{15}N$ determination. Working standards are as for $\delta^{13}C$, calibrated against atmospheric N_2.

When ^{15}N is used as a tracer in soil, data interpretation may be problematical if the mixing of ^{15}N labelled materials with native N pools in the soil is incomplete or, in some way, influences N turnover. This can cause the so-called added N interactions that Jenkinson et al. (1985) and Powlson and Barraclough (1993) explain fully.

Natural variations in $\delta^{15}N$ are not as useful as are those of $\delta^{13}C$ to trace the element. Variations in whole-plant $\delta^{15}N$ can be greater than that of the $\delta^{15}N$ of putative external N sources (be they NO_3^-, NH_4^+ or N_2: Handley and Scrimgeour 1997). This means that it is rarely possible to attribute changes in plant $\delta^{15}N$ to the uptake of different sources of N unless corroboratory evidence is available (preferably complete analyses of the amounts and $\delta^{15}N$ of all N sources, something which is not yet technically feasible). Many $\delta^{15}N$ studies in the literature should be read with a critical eye and not taken at face value. No methods are yet available to routinely measure the $\delta^{15}N$ values of soil N constituents such as NO_3^- reliably (see Scrimgeour and Robinson 2001). The isolation of NO_3^- from soil by diffusion or distillation is not specific to NO_3^- and "NO_3^-" thus isolated may contain contaminants which contribute to the apparent $\delta^{15}N$ value of the sample. Samples must be screened for the presence of potential contaminants. This is done most easily by comparing mass amounts of N measured colorimetrically (which is specific for NO_3^-) with those measured by CF-IRMS (which are not specific). If the two amounts do not agree to within 10%, say, reject the data (Robinson and Conroy 1999).

NMR may quantifiably distinguish different ^{15}N-labelled fractions in root tissues or extracts (Fox et al. 1992; Sect. 12.3.1.6).

12.4.4 ^{34}S

There are only a few sources of ^{34}S labelled tracers, available mainly in inorganic forms; e.g., $Ca^{34}SO_4$ which costs around $US 4000 per g isotope. Naturally, dissimilatory reduction of SO_4^{2-} by microbes depletes ^{34}S in the sulfides that are produced by up to 50‰ (see Lajtha and Michener 1994, p. 59).

A major reason why S-containing compounds in the environment *may* differ in $\delta^{34}S$ is, however, not because of $^{34/32}S$ fractionations, but because of pollution. Atmospheric inputs of S-compounds produced from, e.g., coal burning, can be detected in soils and waters if their $\delta^{34}S$ values are sufficiently distinct from non-anthropogenic sources. However, Kirchmarr et al. (1996) concluded that "shifts in soil $\delta^{34}S$ cannot be used as a tool for quantitative determinations of S turnover" in soil, nor, presumably to identify S sources absorbed by roots in situ.

The determination of $\delta^{34}S$ is more difficult than for the other stable isotopes described so far. Analysis is usually done on SO_2, but reliable conversion of samples to SO_2 is tricky. Scrimgeour and Robinson (2001) describe the latest analytical developments which deal with such problems.

12.5 Radioisotopes

12.5.1 ^3H

^3H (tritium) is usually used in root studies to measure uptake, transport and metabolism of organic compounds. A wide range of ^3H-labelled molecules is available. For example, amino acid uptake by maize roots was studied by measuring the depletion of ^3H-aspartic acid from culture solution (Jones and Darrah 1994). Rates of biosynthetic reactions can be determined by measuring the incorporation of ^3H-labelled precursors into macromolecules. This approach has been used to investigate the effects of sulfonylurea herbicides on DNA synthesis in pea root tips (Robbins and Rost 1987).

Greater resolution can be achieved in micro-autoradiography using ^3H rather than ^{14}C-labelled material, although it is less suitable for macro-autoradiography because of self-absorption by root tissue (but see d'Arcy-Lameta and Bompeix 1991). A standard technique for measuring cell proliferation in root meristems involves incubating roots in ^3H-thymidine to label newly synthesised DNA and then microautoradiography of root squashes or sections to follow the progression of the labelled DNA through the cell cycle (Clowes 1971; Berta et al. 1991). Visualising the location of inhibitors in roots has provided valuable information on the spatial patterns of ion uptake. Microautoradiography of ^3H-NEM, a potent inhibitor of K^+ uptake, showed that NEM binding occurred almost exclusively in the peripheral layers of the maize root cortex (Kochian and Lucas 1983).

Another advantage of ^3H over ^{14}C is the greater specific activity of material that can be obtained. Radio-chemicals containing a single ^3H atom per molecule have a maximum specific activity of $1.07 \times 10^6\,GBq\,mol^{-1}$ compared with $2.31 \times 10^3\,GBq\,mol^{-1}$ for ^{14}C. In metabolic studies where small amounts of isotope incorporation are anticipated, or where small concentrations of source compound are being used, ^3H may be the preferred isotope because of the better sensitivity of detection that its greater specific activity allows.

A potential limitation of ^3H is isotope exchange. ^3H attached to -COOH, -OH and -NH_2 groups may exchange with H_2O, thus compromising the interpretation of tracer experiments (Vose 1980, p. 148). Labelling should therefore be on parts of the molecule that do not exchange (d'Arcy-Lameta and Bompeix 1991).

12.5.2 ^{14}C and ^{11}C

Some of the major questions being addressed in studies of root and rhizosphere function concern fluxes and partitioning of C within the plant-soil-microbe system. ^{14}C is widely used as a tracer in such work. Since root systems derive most of their C from photosynthesis, $^{14}CO_2$ is often used to label photosynthate to measure the flux of C to the root system and its partitioning between root and rhizosphere. For details of experimental techniques and numerical analyses see Keith et al. (1986), Gregory and Atwell (1991) and Swinnen et al. (1994). General features of $^{14}CO_2$ feeding systems are similar in most laboratories, differing only in their level of sophistication (Warembourg and Kummerow 1991; Farrar 1993). Whole plants or individual leaves may be sealed in a feeding-chamber into which $^{14}CO_2$ is released by the addition of acid to $^{14}CO_3^-$ or $H^{14}CO_3^-$. Uniformity of labelling is improved by ensuring adequate mixing of air within the chamber or by using a flow-through system.

These techniques are suitable for short-term experiments to pulse-label plant C pools, e.g., when measuring the time courses of C partitioning between roots and rhizosphere (Meharg 1994). Longer-term feeding (>24 h), during which a uniform supply of $^{14}CO_2$ is maintained, is necessary to label soluble C pools to constant specific activity (Farrar 1985). Here a syringe pump can be used to add, at a constant rate, a $^{14}CO_3^-$ solution to acid, and the $^{14}CO_2$ produced is swept by an air stream through a leaf chamber or whole-plant cuvette (Warembourg and Kummerow 1991; Farrar 1993). Releasing $^{14}CO_2$ into the air will raise the concentration of CO_2 above ambient. When control of the CO_2 concentration is required, it may be necessary to scrub CO_2 from the airstream before adding it back at a known dose-rate. Diluting the $^{14}CO_3^-$ with carrier CO_2 allows both the desired specific activity and total CO_2 concentration to be generated. Continuous feeding over extended periods is required to uniformly label root structural (i.e., cell wall) material. The relative merits of pulse-labelling and continuous labelling for measuring rhizodeposition of C have been discussed by Meharg (1994).

Translocation of specific ^{14}C-labelled organic compounds (e.g. phloem-mobile herbicides) can be measured by applying solutions directly to abraded or non-abraded leaves, or by incubating cut ends of otherwise intact leaves in ^{14}C-solutions (Thompson et al. 1986). This technique, when used with ^{14}C-sucrose, provides an alternative to $^{14}CO_2$ for labelling root carbohydrate pools. In an investigation of the pathways of phloem unloading in pea root tips, Dick and ap Rees (1975) applied ^{14}C-sucrose to the base of roots whose cortex had been removed to expose the vascular tissue. For studies of root physiology the simplest, and often most effective way, of supplying ^{14}C-labelled organic compounds to roots is via solution in hydroponics. Farrar (1991) pulse-labelled barley roots with ^{14}C-sucrose to measure rates of starch and fructan turnover. Jones and Darrah (1992, 1996) supplied maize roots with ^{14}C-sugars and organic acids to study mechanisms of C exudation and re-sorption. However, consideration should be given to whether normal endogenous C pools are labelled by this method.

An extensive range of ^{14}C-labelled organic molecules is available in addition to $^{14}CO_3^-$ and $H^{14}CO_3^-$ salts and $^{14}CO_2$. Organic products may be labelled at specific positions in the molecule (e.g. [1-^{14}C] glucose in which the C-1 atom is the ^{14}C isotope, the rest being ^{12}C). These are useful if the metabolic fate of a particular fraction of the molecule is to be followed or relative fluxes through different metabolic pathways (e.g., glycolysis versus oxidative pentose-phosphate pathway) are measured (Bryce and ap Rees 1985; Dieuaide-Noubhani et al. 1995). In uniformly labelled products (e.g. [U-^{14}C] glucose), each C atom has an equal chance of being ^{14}C. These are less expensive and are appropriate for studies of uptake and metabolism where the metabolic fate of the whole molecule (or of the C in it) is of interest. Suppliers may even be willing to synthesise specific products that are not normally available (but at a price!).

Although ubiquitous in atmospheric CO_2, most atmospheric ^{14}C is not strictly "natural" in occurrence. It was produced largely from thermonuclear weapons testing in the late 1950s and early 1960s (Goh 1991). The abundance of atmospheric ^{14}C is changing in a predictable (and, for us, a useful) way as it decays. We can use this decay to distinguish C fixed by plants recently (i.e., since ca. 1960) and introduced to soil via their roots from C fixed earlier (Harrison et al. 1990; Nepstad et al. 1994).

^{14}C can be detected by LSC or autoradiography. Its extremely long $t_{1/2}$ (Appendix) means that any contamination is serious, particularly if liquids or solids find their way into cracks in floors or benches, because the activity will not quickly decay to safe levels. Detection by AMS is possible (Nepstad et al. 1994), as is ^{14}C-NMR (Dieuaide-Noubhani et al. 1995), given appropriate sample sizes and specific activities.

In certain types of work, the long $t_{1/2}$ of ^{14}C can be a limitation. ^{11}C, in contrast, has a $t_{1/2}$ of 20.3 min which means that after administering a pulse of $^{11}CO_2$ most of its activity will have decayed within 1 h. Measured activity then represents mainly *current* assimilate, rather than material in storage pools (unlike

^{14}C). By re-supplying $^{11}CO_2$ at set intervals, repeat or continuous labelling experiments are possible on the *same plant*, to investigate short-term changes in translocation from leaves to roots, roots to rhizosphere, or between competing roots (Spence and Sharpe 1991; Minchin et al. 1994; Minchin and Thorpe 1996). ^{11}C in intact tissues may be detected continuously using scintillation detectors coupled to a ratemeter. Many of the precautions required for ^{11}C experiments are the same as those which apply to ^{13}N (see Sect. 12.5.3).

12.5.3 ^{13}N

The opportunities and practicalities of using high energy, short $t_{1/2}$ isotopes in root studies are illustrated by ^{13}N. The greater sensitivity of detection possible with ^{13}N compared with ^{15}N has allowed low capacity constitutive N transport systems in roots (Lee and Drew 1986; Kronzucker et al. 1995a), N influx (Siddiqi et al. 1990; Kronzucker et al. 1995a) and efflux (Lee and Clarkson 1986; Kronzucker et al. 1995a,b) to be analysed in detail. The ability to measure efflux directly and sensitively, makes it possible to estimate N exchange times for subcellular compartments. Then, using labelling times $<t_{1/2}$ for exchange of a compartment, it is possible to measure the unidirectional flux across a particular membrane rather than the net N flux (balance between influx and efflux). For example, knowing that $t_{1/2}$ for exchange of the cell wall and cytoplasm was ca. 20 to 30 s, and ca. 6 to 8 min, respectively, Kronzucker et al. (1995a,b) designed protocols to measure plasma membrane influx in roots of *Picea glauca* (white spruce), rather than the "quasi-steady" flux to the vacuole (Cram 1968). It is impossible to distinguish the component fluxes (influx and efflux) by ion depletion (Chap. 13).

As with ^{11}C, the high energy of the ^{13}N γ emission (0.5 MeV maximum emission energy) permits non-invasive translocation studies using whole plants and repeated observations on the same plant after allowing for isotopic decay (Caldwell et al. 1984). Yet, experiments lasting as long as 75 to 90 min are feasible if coupled to methods such as HPLC, to investigate the metabolic fates of ^{13}N tracer (Meeks 1993). However, because of the short $t_{1/2}$, longer-term studies (>>1 h, e.g., to resolve vacuolar characteristics by compartmental analysis), are impossible. In this case, ^{15}N becomes the isotope of choice.

Special problems associated with the use of ^{13}N fall into two categories; both arise from its short $t_{1/2}$ (9.96 min). First, rapid decay dictates that much more radioactivity (>100×) be delivered to the laboratory than is required for longer-lived tracers, to compensate for decay during the experiment. Second, all procedures must be rigorously pre-planned to avoid needless time losses, and experiments performed close to a cyclotron where the ^{13}N is produced.

The high energies of the positron emission and γ radiation (Appendix) demand extensive shielding of the experimental area with lead bricks as well as the development of automated methods to minimise exposure to this radiation. Radiation dosimeters are mandatory as well as Pb aprons and eye shielding. Precautions adopted for the use of ^{13}N at the University of British Columbia are as follows. Generation of $^{13}NO_3^-$ by proton irradiation of water (Meeks 1993) is controlled remotely from outside a "hot" cell lined with 50 mm Pb bricks. Transfer from the cyclotron to the U.B.C. campus hospital is in an underground pneumatic tube, the 2.5 km distance being covered in 2 min. Transfer of ^{13}N from the collection point to the laboratory (0.5 km) is by car, the tracer being contained inside a Pb cylinder. The cylinder is carried on a specially constructed trolley to distance the experimenter as far as possible from the ^{13}N. Contaminants ($^{13}NO_2^-$, $^{13}NH_4^+$, ^{18}F) are removed in a fume hood lined on all sides with 50 mm Pb bricks (Lee and Clarkson 1986; Lee and Drew 1986; Siddiqi et al. 1989), where conversions to $^{13}NO_2^-$, or $^{13}NH_4^+$ are done as required. It is important to consider the possibility of exposing individuals in adjacent laboratories or offices. A sheet of 15 mm Pb-impregnated glass (180 × 270 mm) mounted on a sliding track over the open fume hood, provides a "safe" viewing window. Within the hood itself is a large mirror so that direct viewing of the isotope is avoided. Isotope spills decay within the hour leaving behind only trace amounts of ^{13}N.

12.5.4 ^{32}P

^{32}P is one of the most widely used radioisotopes in root studies. It has been used to measure P uptake, to visualise root locations in soil, and P depletion zones around roots. Some of the most interesting applications have been in conjunction with its sister isotope, ^{33}P, such as interspecific root competition for P (Caldwell et al. 1987) and distinguishing between P influx and its simultaneous efflux (Elliot et al. 1984).

The usual aim of adding ^{32}P to soil is to label the plant-available pool(s) without increasing P availability significantly. But labelling soil with ^{32}P presents problems due to the exchange of inorganic P between solid and liquid phases. Drew and Nye (1970) incubated ^{32}P-labelled soil for up to 2 months before using it in a P uptake experiment to allow isotopic equilibration to occur. In such applications, the ^{32}P content of the soil solution should, ideally, be measured sequentially to check the approach of the soil to isotopic equilibrium. Phosphate does not move greatly in soil (unlike NO_3^-, for example) and considerable spatial variability of [P] can occur over short distances (<10 cm). If a uniform distribution of ^{32}P with depth is required, inject it at several depths rather than rely on downward percolation of solution which produces very uneven distributions of ^{32}P (Jupp et al. 1987).

The high energy β^- particles represent a significant external hazard when dispensing isotope stocks and conducting experiments in solution culture. However, field experiments are possible because the activity is greatly reduced when P becomes adsorbed in the soil. The risk of leaching is low, and the short $t_{1/2}$ means that activity rapidly decays. Vose (1980, p. 268 et seq.) discusses the practicalities of using ^{32}P in field experiments. If ^{32}P is to be used in a P-fertiliser experiment, uniformly labelled material cannot be produced by adding carrier-free ^{32}P to a unlabelled P fertiliser; it must be purchased ready-labelled (Vose 1980, p. 278).

Jupp et al. (1987) described a neat, non-destructive, semi-quantitative method to measure ^{32}P uptake by plants in soil. They measured the activity of ^{32}P in a leaf to which a scintillation probe had been attached. This allowed the time-course, but not the amount, of ^{32}P uptake to be obtained.

12.5.5 ^{86}Rb and ^{42}K

Many physiological studies of K uptake by roots have actually used an analogue, ^{86}Rb. This is because ^{86}Rb has a more convenient $t_{1/2}$ compared with that of the obvious K isotope, ^{42}K (Appendix 12.1). ^{86}Rb has been used extensively in studies of uptake rate-ion concentration relations in the laboratory (e.g. Kochian and Lucas 1982, 1983). ^{86}Rb has also been used to estimate root activity at specific locations in soil (Vose 1980, p. 291). However, a number of studies demonstrate that discrimination between Rb and K may occur during influx and efflux across root membranes; the effects depend on species and experimental conditions (Jensen and Kylin 1980; Behl and Jeschke 1982; Drew and Saker 1984). Experiments of this type using ^{86}Rb must, therefore, be interpreted with caution; at best they provide a qualitative indication of possible treatment effects on K uptake and transport.

^{86}Rb is usually quantified by Cerenkov counting, and ^{42}K by Cerenkov or γ counting (Behl and Jeschke 1982; Jeschke 1982), although other techniques have been used, e.g. LSC and end window Geiger counting (Jensen and Kylin 1980; Drew and Saker 1984).

12.5.6 ^{35}S

This is the isotope of choice in studies of S uptake by roots. But as suitable IRMS systems become available, we may see greater use of ^{34}S in the future. Robinson et al. (1994) supplied a fraction of a wheat root system in soil with ^{35}S labelled K_2SO_4 solution at a [S] of 2 mol m^{-3} and ^{35}S activity of ca. 460 kBq cm^{-3}; this was more than adequate to allow the isotope to be quantified in shoots 7 days later. Other examples of the use of ^{35}S with hydroponically grown plants are in Lee (1982), Holobradá and Kubica (1989) and Sunarpi and Anderson (1996).

^{35}S is also widely used in studies of root metabolism. The effects of treatments on current protein synthesis can be determined by incubating roots in ^{35}S-methionine and measuring its incorporation into proteins. Quantitative autoradiography of ^{35}S-labelled proteins on two-dimensional gels has shown that synthesis of specific proteins may be regulated by the carbohydrate supply to the root (Baysdorfer and VanDerWoude 1988).

12.5.7 Other Radioisotopes

Convenient isotopes are available for a number of other elements of interest to root scientists including some important plant micro-nutrients, e.g. Mn (^{54}Mn) Zn (^{65}Zn), Mo (^{99}Mo), Fe (^{59}Fe), Cu (^{64}Cu), Cl (^{36}Cl), plus Ca (^{45}Ca) and Na (^{22}Na). In most cases they have been used as tracers for measuring ion uptake, translocation and compartmentation. For example, Lasat and Kochian (1997) used ^{65}Zn to study the mechanisms of Zn hyper-accumulation in *Thlaspi caerulescens*, a candidate species for phyto-remediation of Zn-contaminated soils. Similarly, ^{59}Fe has been used to investigate the role of phyto-siderophores (non-proteinogenic amino acids) in Fe mobilisation in the rhizospheres of and uptake by Fe-deficient grasses (Marschner et al. 1989; Treeby et al. 1989), and the sites of Fe uptake by maize roots (Kashirad et al. 1973). Details of radiation type, E_{max}, and $t_{1/2}$ can be found in Vose (1980).

12.6 Concluding Remarks

This chapter has illustrated the breadth of applications that stable and radioisotopes have in root studies. As instrumentation evolves, this range is sure to increase, e.g. we expect to see in the near future many uses of electronic autoradiography to show how roots work in real time. In some areas, further developments are needed in basic chemistry, rather than in isotope chemistry per se, e.g. in the isolation of plant and soil N species for reliable $\delta^{15}N$ determinations. If that particular development occurs, there will be a parallel need for improved theories to usefully interpret such $\delta^{15}N$ data. In physiology, too, certain breakthroughs are required, e.g. micro-sampling of plant tissues and organelles for the measurement of their isotopic composition to address still-unanswered questions about metabolite compartmentation. Whatever the developments, the use of isotopic (or any other) techniques should be driven by biological questions, and not the other way around.

Acknowledgments. The Scottish Crop Research Institute and the Scottish Agricultural College receive grant-in-aid from the Scottish Executive Rural Affairs Department.

References

Amato M (1983) Determination of carbon ^{12}C and ^{14}C in plant and soil. Soil Biol Biochem 15: 611–612

Androssoff GL, van Kessel C, Pennock DJ (1994) Landscape-scale estimates of dinitrogen fixation by *Pisum sativum* by nitrogen-15 natural abundance and enriched isotope dilution. Biol Fertil Soils 20: 33–40

Aslam M, Travis RL, Huffaker RC (1994) Stimulation of nitrate and nitrate efflux by ammonium in barley (*Hordeum vulgare* L.) seedlings. Plant Physiol 106: 1293–1301

Atkinson CJ, Farrar JF (1983) Allocation of photosynthetically-fixed carbon in *Festuca ovina* L. and *Nardus stricta* L. New Phytol 95: 519–531

Bacic G, Ratkovic S (1987) NMR studies of radial exchange and distribution of water in maize roots: the relevance of modelling and exchange kinetics. J Exp Bot 38: 1284–1297

Baldwin JP, Tinker PB (1972) A method for estimating the lengths and spatial patterns of two interpenetrating root systems. Plant Soil 37: 209–213

Balesdent J, Mariotti A (1996) Measurement of soil organic matter turnover using ^{13}C natural abundance. In: Boutton TW, Yamasaki S (eds) Mass spectrometry of soils. Dekker, New York, pp 83–111

Baysdorfer C, VanDerWoude WJ (1988) Carbohydrate responsive proteins in the roots of *Pennisetum americanum*. Plant Physiol 87: 566–570

Behl R, Jeschke WD (1982) Potassium fluxes in excised barley roots. J Exp Bot 33: 584–600

Bengtsson B (1982) Uptake and translocation of calcium in cucumber. Physiol Plant 54: 107–111

Berta G, Tagliasacchi AM, Fusconi A, Gerlero D, Trotta A, Scannerini S (1991) The mitotic cycle in root apical meristems of *Allium porrum* L. is controlled by the endomycorrhizal fungus *Glomus* sp. strain E_3. Protoplasma 161: 12–16

Bhat KKS, Nye PH (1973) Diffusion of phosphate to plant roots in soil. I. Quantitative autoradiography of the depletion zone. Plant Soil 38: 161–175

Bingham IJ, Blackwood JM, Stevenson EA (1998) Relationship between tissue sugar content, phloem import and lateral root initiation in wheat. Physiol Plant 103: 107–113

Blackburn TH, Knowles R (1993) Introduction. In: Knowles R, Blackburn TH (eds) Nitrogen isotope techniques. Academic Press, London, pp 1–10

Boutton TW (1991a) Stable carbon isotope ratios of natural materials: II. Atmospheric, terrestrial, marine and freshwater environments. In: Coleman DC, Fry B (eds) Carbon isotope techniques. Academic Press, London, pp 173–185

Boutton TW (1991b) Stable carbon isotope ratios of natural materials: I. Sample preparation and mass spectrometric analysis. In: Coleman DC, Fry B (eds) Carbon isotope techniques. Academic Press, London, pp 155–171

Boutton TW, Yamasaki S (eds) (1996) Mass spectrometry of soils. Marcel Dekker, New York

Brownlee C, Duddridge JA, Malibari A, Read DJ (1983) The structure and function of mycelial systems of ectomycorrhizal roots with special reference to their rôle in forming inter-plant connections and providing pathways for assimilate and water transport. Plant Soil 71: 433–443

Bryce JH, ap Rees T (1985) Comparison of the respiratory metabolism of *Plantago lanceolata* L. and *Plantago major* L. J Exp Bot 36: 1559–1565

Caldwell CD, Fensom DS, Bordeleau L, Thompson RG, Drouin R, Didsburg R (1984) Translocation of ^{13}N and ^{11}C between nodulated roots and leaves in alfalfa seedlings. J Exp Bot 35: 431–443

Caldwell MM, Richards JH (1989) Hydraulic lift: water efflux from upper roots improves effectiveness of water uptake by deep roots. Oecologia 79: 1–5

Caldwell MM, Virginia RA (1989) Root systems. In: Pearcy RW, Ehleringer J, Mooney HA, Rundel PW (eds) Plant physiological ecology: field methods and instrumentation. Chapman and Hall, London, pp 367–398

Caldwell MM, White RS, Moore, Camp LB (1977) Carbon balance, productivity and water use of cold-winter desert shrub communities dominated by C_3 and C_4 species. Oecologia 29: 275–300

Caldwell MM, Eissenstat DM, Richards JH, Allen MH (1985) Competition for phosphorus: differential uptake from dual-isotope labeled soil interspaces between shrub and grass. Science 229: 384–386

Caldwell MM, Richards JH, Manwaring JH, Eissenstat DM (1987) Rapid shifts in phosphate acquisition show direct competition between neighbouring plants. Nature 327: 615–616

Cheng W, Coleman DC, Carroll CR, Hoffman CA (1994) Investigating short-term carbon flows in the rhizospheres of different plant species using isotopic trapping. Agron J 86: 782–788

Clark FE (1977) Internal cycling of ^{15}nitrogen in a shortgrass prairie. Ecology 58: 1322–1333

Clarkson DT, Sanderson J, Russell RS (1968) Ion uptake and root age. Nature 220: 805–806

Clarkson DT, Gojon A, Saker LR, Wiersema PK, Purves JV, Tillard P, Arnold GM, Paans AJM, Vaalburg W, Stulen I (1996) Nitrate and ammonium influxes in soybean (*Glycine max*) roots – direct comparison of ^{13}N and ^{15}N tracing. Plant Cell Environ 19: 859–868

Clowes FAL (1971) The proportion of cells that divide in root meristems of *Zea mays* L. Ann Bot 35: 249–261

Coleman DC, Fry B (eds) (1991) Carbon isotope techniques. Academic Press, London

Cooper HD, Clarkson DT (1989) Cycling of amino-nitrogen and other nutrients between shoots and roots in cereals – a possible mechanism integrating shoot and root in the regulation of nutrient uptake. J Exp Bot 40: 753–762

Coupland D (1986) Sample preparation for liquid scintillation counting. Aspects Appl Biol 11: 55–66

Cram WJ (1968) Compartmentation and exchange of chloride in carrot root tissue. Biochim Biophys Acta 163: 339–353

Dalal RC (1979) Simple procedure for the determination of total carbon and its radioactivity in soils and plant materials. Analyst 104: 151–154

Dalton FN (1989) Plant root water extraction studies using stable isotopes. In: Loughman BC (ed) Structural and functional aspects of transport in roots. Kluwer, Dordrecht, pp 151–155

d'Arcy-Lameta A, Bompeix G (1991) Systemic transport of tritiated phosphonate in tomato plantlets (*Lycopersicon esculentum* Mill). Pestic Sci 32: 7–14

Dawson TE (1993) Water sources of plants as determined from xylem-water isotopic composition: perspective on plant competition, distribution and water relations. In: Ehleringer JR, Hall AE, Farquhar GD (eds) Stable isotopes and plant carbon-water relations. Academic Press, London, pp 465–496

Deane-Drummond CE, Glass ADM (1983) Compensatory adjustments in ion fluxes into barley (*Hordeum vulgare* L. cv. Betzes) seedlings in response to differential root/shoot growth temperature. J Exp Bot 34: 1711–1719

Dick PS, ap Rees T (1975) The pathway of sugar transport in roots of *Pisum sativum*. J Exp Bot 26: 305–314

Dieuaide-Noubhani M, Raffard G, Canion P, Pradet A, Raymond P (1995) Quantification of compartmented metabolic fluxes in maize root tips using isotope distribution from ^{13}C- or ^{14}C-labelled glucose. J Biol Chem 270: 13147–13159

Drew MC, Nye PH (1970) The supply of nutrient ions by diffusion to plant roots in soil. III. Uptake of phosphate by roots of onion, leek and rye-grass. Plant Soil 33: 545–563

Drew MC, Saker LR (1984) Uptake and long distance transport of phosphate, potassium and chloride in relation to internal ion concentrations in barley: evidence of non-allosteric regulation. Planta 160: 500–507

Drew MC, Nye PH, Vaidyanathan LV (1969) The supply of nutrient ions by diffusion to plant roots in soil. I. Absorption of potassium by cylindrical roots of onion and leek. Plant Soil 30: 252–270

Ehleringer JR (1991) $^{13}C/^{12}C$ fractionation and its utility in terrestrial plant studies. In: Coleman DC, Fry B (eds) Carbon isotope techniques. Academic Press, London, pp 187–200

Elliot GC, Lynch J, Läuchli A (1984) Influx and efflux of P in roots of intact maize plants. Plant Physiol 76: 336–341

Ezra G, Rusness DG, Lamoureux GL, Stephenson GR (1985) The effect of CDAA (*N,N,*-Diallyl-2-chloroactetamide) pretreatments on subsequent CDD injury to corn (*Zea mays* L.). Pestic Biochem Physiol 23: 108–115

Farrar JF (1985) Fluxes of carbon in roots of barley plants. New Phytol 99: 57–69

Farrar JF (1991) Starch turnover: its role in source-sink relations and a comparison with the role of sucrose. In: Bonnemain JL, Delrot S, Lucas WJ, Dainty J (eds) Recent advances in phloem transport and assimilate compartmentation. Quest Editions, Nantes, pp 213–223

Farrar JF (1993) Carbon partitioning. In: Hall DO, Scurlock JMO, Bolhàr-Nordenkampf HR, Leegood RC, Long SP (eds) Photosynthesis and production in a changing environment: a field and laboratory manual. Chapman and Hall, London, pp 232–246

Fox GG, Ratcliffe RG, Robinson SA, Slade AP, Stewart GR (1992) Detection of ^{15}N-labelled ammonium. In vivo ^{15}N NMR versus mass spectrometry. J Magnet Reson 96: 146–153

Glass ADM, Thompson RG, Bordeleau L (1985) Regulation of NO_3^- influx in barley. Studies using $^{13}NO_3^-$. Plant Physiol 77: 379–381

Gleixner G, Danier H-J, Werner RA, Schmidt H-L (1993) Correlations between the ^{13}C content of primary and secondary plant products in different cell compartments and that in decomposing basdiomycetes. Plant Physiol 102: 1287–1290

Goh KM (1991) Bomb carbon. In: Coleman DC, Fry B (eds) Carbon isotope techniques. Academic Press, London, pp 147–151

Gordon AL, Ryle GJA, Mitchell DF, Powell CE (1985) The flux of ^{14}C-labelled photosynthate through soyabean root nodules during N_2 fixation. J Exp Bot 36: 756–769

Goulding KH, Slater RJ (1994) Radioisotope techniques. In: Wilson K, Walker J (eds) Principles and techniques of practical biochemistry. Cambridge University Press, Cambridge, pp 227–274

Gregory PJ, Atwell BJ (1991) The fate of carbon in pulse-labelled crops of barley and wheat. Plant Soil 136: 205–213

Grignon N, Halpern S, Gojon A, Fragu P (1992) ^{14}N and ^{15}N imaging by SIMS microscopy in soybean leaves. Biol Cell 74: 143–146

Handley LL, Raven JA (1992) The use of natural abundance of nitrogen isotopes in plant physiology and ecology. Plant Cell Environ 15: 965–985

Handley LL, Scrimgeour CM (1997) Terrestrial plant ecology and ^{15}N natural abundance; the present limits to interpretation for uncultivated systems with original data from a Scottish old-field. Adv Ecol Res 27: 133–212

Handley LL, Daft MJ, Wilson J, Scrimgeour CM, Ingleby K, Sattar MA (1993) Effects of the ecto- and VA-mycorrhizal fungi *Hydnagium carneum* and *Glomus clarum* in the $\delta^{15}N$ and $\delta^{13}C$ values of *Eucalyptus globulus* and *Ricinus communis*. Plant Cell Environ 16: 375–382

Harris D, Paul EA (1991) Techniques for examining the carbon relationships of plant-microbial symbioses. In: Coleman DC, Fry B (eds) Carbon isotope techniques. Academic Press, London, pp 39–52

Harrison AF, Harkness DD, Bacon PJ (1990) The use of bomb-^{14}C for studying organic matter and N and P dynamics in a woodland soil. In: Harrison AF, Ineson P, Heal OW (eds) Nutrient cycling in terrestrial ecosystems: field methods, application and interpretation. Elsevier, London, pp 246–258

Hodge A, Stewart J, Robinson D, Griffiths BS, Fitter AH (1998) Root proliferation, soil fauna and plant nitrogen capture from nutrient-rich patches in soil. New Phytol 139: 479–494

Holobradá M, Kubica S (1989) The role of maize root tissues in sulphate absorption and radial transport. In: Loughman BC (ed) Structural and functional aspects of transport in roots. Kluwer, Dordrecht, pp 101–105

Jackson LE, Schimel JP, Firestone MK (1989) Short-term partitioning of ammonium and nitrate between plants and microbes in a annual grassland. Soil Biol Biochem 21: 409–415

Jenkinson DS, Fox RH, Rayner JH (1985) Interactions between fertilizer nitrogen and soil nitrogen – the so called 'priming' effect. J Soil Sci 36: 425–444

Jensen P, Kylin A (1980) Effects of ionic strength and relative humidity on the efflux of K^+ (^{86}Rb) and Ca^{2+} (^{45}Ca) from roots of intact seedlings of cucumber, oat and wheat. Physiol Plant 50: 199–207

Jeschke WD (1982) Shoot-dependent regulation of sodium and potassium fluxes in roots of whole barley seedlings. J Exp Bot 33: 601–618

Jones DL, Darrah PR (1992) Re-sorption of organic components by roots of *Zea mays* L. and its consequences in the rhizosphere I. Re-sorption of ^{14}C labelled glucose, mannose and citric acid. Plant Soil 143: 259–266

Jones DL, Darrah PR (1994) Amino-acid influx at the soil-root interface of *Zea mays* L. and its implications in the rhizosphere. Plant Soil 163: 1–12

Jones DL, Darrah PR (1996) Re-sorption of organic components by roots of *Zea mays* L. and its consequences in the rhizosphere III. Characteristics of sugar influx and efflux. Plant Soil 178: 153–160

Jupp AP, Newman EI, Ritz K (1987) Phosphorus turnover in soil and its uptake by established *Lolium perenne* plants. J Appl Ecol 24: 969–978

Kashirad A, Marschner H, Richter C (1973) Absorption and translocation of ^{59}Fe from various parts of the corn root. Z Pflanzenernähr Bodenk 134: 136–147

Keith H, Oades JM, Martin JK (1986) Input of carbon to soil from wheat plants. Soil Biol Biochem 18: 445–449

Kirchmarr H, Pichlmayer F, Gerzabek MH (1996) Sulfur balances and sulfur-34 abundance in a long-term fertilizer experiment. Soil Biol Biochem 59: 174–178

Knowles R, Blackburn TH (eds) (1993) Nitrogen isotope techniques. Academic Press, London

Kochian LV, Lucas WJ (1982) Potassium transport in corn roots. I. Resolution of kinetics into a saturable and linear component. Plant Physiol 70: 1723–1731

Kochian LV, Lucas WJ (1983) Potassium transport in corn roots. II. The significance of the root periphery. Plant Physiol 73: 208–215

Kraus M, Fusseder A, Beck E (1987) Development and replenishment of the P-depletion zone around the primary root of maize during the vegetative period. Plant Soil 101: 247–255

Kronzucker HJ, Siddiqi MY, Glass ADM (1995a) Compartmentation and flux characteristics of nitrate in spruce. Planta 196: 674–682

Kronzucker HJ, Siddiqi MY, Glass ADM (1995b) Compartmentation and flux characteristics of ammonium in spruce. Planta 196: 691–698

Kronzucker HJ, Siddiqi MY, Glass ADM (1997) Conifer root discrimination against soil nitrate and the ecology of forest succession. Nature 385: 59–61

Kuhn AJ, Bauch J, Schröder WH (1995) Monitoring uptake and contents of Mg, Ca and K in Norway spruce as influenced by pH and Al, using microprobe analysis and stable isotope labelling. Plant Soil 168–169: 135–150

Lasat MM, Kochian LV (1997) Physiological basis for Zn hyperaccumulation in *Thlaspi caerulescens*. In: Flores HE, Lynch JP, Eissenstat D (eds) Radical biology: advances and perspectives on the function of plant roots. American Society of Plant Physiologists, Rockville, pp 139–149

Laskey RA (1990) Radioisotope detection using X-ray film. In: Slater RJ (ed) Radioisotopes in biology: a practical approach. Oxford University Press, Oxford, pp 87–107

Lajtha K, Michener RH (eds) (1994) Stable isotopes in ecology and environmental science. Blackwell, Oxford

Lazof DB, Rufty TW, Redinbaugh MG (1992) Localization of nitrate absorption and translocation within morphological regions of the corn root. Plant Physiol 100: 1251–1258

Lazof DB, Goldsmith JKG, Rufty TW, Suggs C, Linton RW (1994) The preparation of cryosections from plant tissue: an alternative method appropriate for secondary ion mass spectrometry studies of nutrient tracers and trace metals. J Microsc 176: 99–109

Lee RB (1982) Selectivity and kinetics of ion uptake by barley plants following nutrient deficiency. Ann Bot 50: 429–449

Lee RB, Clarkson DT (1986) Nitrogen-13 studies of nitrate fluxes in barley roots. I. Compartmental analysis from measurements of ^{13}N efflux. J Exp Bot 37: 1753–1767

Lee RB, Drew MC (1986) Nitrogen-13 studies of nitrate fluxes in barley roots. II. Effect of plant N-status on the kinetic parameters of nitrate influx. J Exp Bot 37: 1753–1767

Lee RB, Lewis MJ (1994) Synthesis and ^{15}N-labelling of glutamine and glutamate in maize roots during early stages of ^{15}N-ammonium assimilation. J Exp Bot 45: 767–777

Lee RB, Ratcliffe RG (1993) Nuclear magnetic resonance studies of the location and function of plant nutrients in vivo. Plant Soil 155–156: 45–55

Lefebvre DD, Glass ADM (1982) Regulation of phosphate influx in barley roots: effects of phosphate deprivation and reduction of influx with provision of orthophosphate. Physiol Plant 54: 199–206

Lichtfouse É, Dou S, Girardin C, Grably M, Balesdent J, Behar F, Vandenbroucke M (1995) Unexpected ^{13}C-enrichment of organic components from wheat crop soils: evidence for the in situ origin of soil organic matter. Org Geochem 23: 865–868

Ludlow MM, Troughton JH, Jones RJ (1976) A technique for determining the proportion of C3 and C4 species in plant samples using stable natural isotopes of carbon. J Agric Sci 87: 625–632

Macklon AES, Sim A (1981) Cortical cell fluxes and transport to the stele in excised root segments of *Allium cepa* L. 4. Calcium as affected by its external concentration. Planta 152: 381–387

Macko SA, Ostrom NE (1994) Pollution studies using stable isotopes. In: Lajtha K, Michener RH (eds) Stable isotopes in ecology and environmental science. Blackwell, Oxford, pp 45–62

Marschner H, Richter C (1973) Akkumulation und Translokation von K^+, Na^+ und Ca^{2+} bei Angebot zu einzelnen Wurzelzonen von Maiskeimpflanzen. Z Pflanzenernähr Bodenkd 135: 1–15

Marschner H, Treeby M, Römheld V (1989) Role of root-induced changes in the rhizosphere for iron acquisition in higher plants. Z Pflanzenernähr Bodenkd 152: 237–240

Martin F, Botton B (1993) Nitrogen metabolism of ectomycorrhizal fungi and ectomycorrhiza. Adv Plant Pathol 9: 83–102

Mary B, Mariotti A, Morel JL (1992) Use of ^{13}C variations at natural abundance for studying the biodegradation of root mucilage, roots and glucose in soil. Soil Biol Biochem 24: 1065–1072

McLaughlin MJ, Alston AM, Martin JK (1987) Transformations and movement of P in the rhizosphere. Plant Soil 97: 391–399

Meeks JC (1993) ^{13}N techniques. In: Knowles R, Blackburn TH (eds) Nitrogen isotope techniques. Academic Press, London, pp 273–303

Meharg AA (1994) A critical review of labeling techniques used to quantify rhizosphere carbon-flow. Plant Soil 166: 55–62

Milchunas DG, Lauenroth WK, Singh JS, Cole CV, Hunt HW (1985) Root turnover and production by ^{14}C dilution: implications of carbon partitioning in plants. Plant Soil 88: 353–365

Minchin PEH, Thorpe MR (1996) What determines carbon partitioning between competing sinks? J Exp Bot 47: 1293–1296

Minchin PEH, Thorpe MR, Farrar JF (1994) Short-term control of root:shoot partitioning. J Exp Bot 45: 615–622

Mordacq L, Mousseau M, Deléens E (1986) A C-13 method of estimation of carbon allocation to roots in a young chestnut coppice. Plant Cell Environ 9: 735–739

Nepstad DC, de Carvalho CR, Davidson EA, Jipp PH, Lefebvre PA, Negreiros GH, da Silva ED, Stone TA, Trumbore SE, Viera S (1994) The rôle of deep roots in the hydrological and carbon cycles of Amazonian forests and pastures. Nature 372: 666–669

Nye PH, Tinker PB (1977) Solute movement in the soil-root system. Oxford: Blackwell Scientific Publications

Pfeffer PE, Shachar-Hill Y, Bécard G, Rolin D, Douds DD (1997) Nutrient transport and metabolism in the life cycle of arbuscular mycorrhizae as examined by NMR spectroscopy. In: Flores HE, Lynch JP, Eissenstat D (eds) Radical biology: advances and perspectives on the function of plant roots. American Society of Plant Physiologists, Rockville, pp 187–209

Polley HW, Johnson HB, Mayeux HS (1992) Determination of root biomasses of three species grown in a mixture using stable isotopes of carbon and nitrogen. Plant Soil 142: 97–106

Powlson DS, Barraclough D (1993) Mineralization and assimilation in soil-plant systems. In: Knowles R, Blackburn TH (eds) Nitrogen isotope techniques. Academic Press, London, pp 209–242

Powlson DS, Hart PBS, Pruden G, Jenkinson DS (1986) Recovery of ^{15}N-labelled fertilizer applied in autumn to winter wheat at four sites in eastern England. J Agric Sci 107: 611–620

Prime D, Frith B (1990) Radioisotope use. In: Slater RJ (ed) Radioisotopes in biology: a practical approach. Oxford University Press, Oxford, pp 7–30

Prosser SJ, Scrimgeour CM (1995) High-precision determination of $^2H/^1H$ in H_2 and H_2O by continuous-flow isotope ratio mass spectrometry. Anal Chem 34: 1992–1997

Raney F, Vaadia Y (1965) Movement of tritiated water in the root system of *Helianthus annuus* in the presence and absence of transpiration. Plant Physiol 40: 378–388

Robbins J, Rost TL (1987) Chlorsulfuron inhibition of cell cycle progression and the recovery of G_1 arrested cells by Ile and Val. J Plant Growth Regul 6: 67–74

Robinson D, Conroy JP (1999) A possible plant mediated feedback between elevated CO_2, denitrification and the enhanced greenhouse effect. Soil Biol Biochem 31: 43–53.

Robinson D, Millard P (1986) Short-term uptake rates and partitioning of nitrogen in a potato crop. J Exp Bot 38: 551–557

Robinson D, Scrimgeour CM (1995) The contribution of plant C to soil CO_2 measured using $\delta^{13}C$. Soil Biol Biochem 27: 1653–1656

Robinson D, Smith KA (1991) Analysis of nitrogen isotope ratios by mass spectrometry. In: Smith KA (ed) Soil analysis, 2nd edn. Dekker, New York, pp 465–503

Robinson D, Linehan DJ, Gordon DC (1994) Capture of nitrate from soil by wheat in relation to root length, nitrogen inflow and availability. New Phytol 128: 297–305

Robinson D, Handley LL, Scrimgeour CM (1998) A theory for $^{15}N/^{14}N$ fractionation in nitrate-grown vascular plants. Planta 205: 397–406

Rorison IH, Robinson D (1986) Mineral nutrition. In: Moore PD, Chapman SB (eds) Methods in plant ecology, 2nd edn. Blackwell, Oxford, pp 145–213

Schnyder H (1992) Long-term steady-state labelling of wheat plants by use of natural $^{13}CO_2/^{12}CO_2$ mixtures in an open, rapidly turned-over system. Planta 187: 128–135

Scrimgeour CM (1995) Measurement of plant and soil water isotope composition by direct equilibration methods. J Hydrol 172: 261–274

Scrimgeour CM, Robinson D (2001) Stable isotope analyses and applications in soil and environmental science. In: Smith KA, Cresser MS (eds) Soil and environmental analysis. Dekker, New York

Shone MGT, Flood AV (1980) Studies on the uptake and loss of water by barley roots in relation to changes in root resistance. J Exp Bot 31: 1147–1159

Siddiqi MY, Glass ADM, Ruth TJ, Fernando M (1989) Studies of the regulation of nitrate influx by barley seedlings using $^{13}NO_3^-$. Plant Physiol 90: 806–813

Siddiqi MY, Glass ADM, Ruth TJ, Rufty TW (1990) Studies of the uptake of nitrate in barley 1. Kinetics of $^{13}NO_3^-$ influx. Plant Physiol 93: 1426–1432

Simonnet G (1990) The scintillation counter. In: Slater RJ (ed) Radioisotopes in biology: a practical approach. Oxford University Press, Oxford, pp 31–85

Slater RJ (ed) (1990) Radioisotopes in biology: a practical approach. Oxford University Press, Oxford

Smith KA (ed) (1991a) Soil analysis, 2nd edn. Dekker, New York

Smith KA (1991b) Nuclear and radio-chemical analysis. In: Smith KA (ed) Soil analysis, 2nd edn. Dekker, New York, pp 325–375

Smith KA, Cresser MS (eds) (2001) Soil and environmental analysis. Dekker, New York

Spence RD, Sharpe PJH (1991) Uses and procedures for ^{11}C. In: Coleman DC, Fry B (eds) Carbon isotope techniques. Academic Press, London, pp 245–258

Sunarpi, Anderson JW (1996) Distribution and redistribution of sulfur supplied as [^{35}S] sulfate to roots during vegetative growth of soybean. Plant Physiol 110: 1151–1157

Svejcar T, Boutton TW (1985) The use of stable carbon isotope analysis in rooting studies. Oecologia 67: 205–208

Svejcar T, Boutton TW, Trent JD (1990) Assessment of carbon allocation with stable isotope labelling. Agron J 82: 18–21

Swinnen J, van Veen JA, Merckx R (1994) ^{14}C pulse-labelling of field-grown spring wheat: an evaluation of its use in carbon budget estimations. Soil Biol Biochem 26: 161–170

Thompson LML, Sanders GE, Pallett KE (1986) Experimental studies into the uptake and translocation of foliage applied herbicides. Aspects Appl Biol 11: 45–54

Thorburn PJ, Ehleringer J (1995) Root water uptake of field-growing plants indicated by measurements of natural-abundance deuterium. Plant Soil 177: 225–233

Treeby M, Marschner H, Römheld V (1989) Mobilization of iron and other micronutrients from a calcareous soil by plant-borne, microbial, and synthetic metal chelators. Plant Soil 114: 217–226

Trumbore SE (1996) Applications of accelerator mass spectrometry to soil science. In: Boutton TW, Yamasaki S (eds) Mass spectrometry of soils. Dekker, New York, pp 311–340

Voroney RP, Winter JP, Gregorich EG (1991) Microbe/plant/soil interactions. In: Coleman DC, Fry B (eds) Carbon isotope techniques. Academic Press, London, pp 77–99

Vose PB (1980) Introduction to nuclear techniques in agronomy and plant biology. Pergamon Press, Oxford

Warembourg FR, Kummerow J (1991) Photosynthesis/translocation studies in terrestrial ecosystems. In: Coleman DC, Fry B (eds) Carbon isotope techniques. Academic Press, London, pp 11–37

Wassenaar LI (1995) Evaluation of the origin and fate of nitrate in the Abbotsford Aquifer using the isotopes of ^{15}N and ^{18}O in NO_3^-. Appl Geochem 10: 391–405

Watkins NK, Fitter AH, Graves JD, Robinson D (1996) Quantification using stable carbon isotopes of carbon transfer between C_3 and C_4 plants linked by a common mycorrhizal network. Soil Biol Biochem 28: 471–477

White JWC, Cook ER, Lawrence JR, Broecker WS (1985) The D/H ratios of sap in trees: implications for water sources and tree ring D/H ratios. Geochim Cosmochim Acta 49: 237–246

White PJ, Clarkson DT, Earnshaw MJ (1987) Acclimation of potassium influx in rye (*Secale cereale*) to low root temperatures. Planta 171: 377–385

Wong SC, Osmond CB (1991) Elevated atmospheric partial pressure of CO_2 and plant growth. III. Interactions between *Triticum aestivum* (C_3) and *Echinochloa frumentacea* (C_4) during growth in mixed culture under different CO_2, N nutrition and irradiance, with emphasis on below-ground responses using $\delta^{13}C$ value of root biomass. Aust J Plant Physiol 18: 137–152

Yoshida S, Eguchi H (1994) Environmental analysis of aerial O_2 transport through leaves for root respiration in relation to water uptake in cucumber plants (*Cucumis sativus* L.) in O_2-deficient nutrient solution. J Exp Bot 45: 187–192

Zocchi G, Hanson JB (1982) Calcium influx into corn roots as a result of cold shock. Plant Physiol 70: 318–319

Appendix. Commonly Used Isotopes in Root Studies (Based on Vose, 1980, pp. 143–144)[a]

Element	Isotope	E_{max} (MeV) β	E_{max} (MeV) γ	$t_{1/2}$	Method	Selected examples of applications in root studies
Calcium	^{40}Ca			s[b]		
	^{45}Ca	0.254		162 days	LSC; AR	T: Marschner and Richter (1973); Jensen and Kylin (1980); Macklon and Sim (1981); Bengtsson (1982); Zocchi and Hanson (1992)
Carbon	^{11}C		0.9600	20.5 min (c.p.)	γ counting	GR: Spence and Sharpe (1991) T: Minchin et al. (1994); Minchin and Thorpe (1996)
	^{12}C			s[b]		
	^{13}C			s	MS	GR: Boutton (1991b); Ehleringer (1991) T: Svejcar et al. (1990); Mordacq et al. (1986); Lichtfouse et al. (1995); Hodge et al. (1998) NT: Ludlow et al. (1976); Svejcar and Boutton (1985); Wong and Osmond (1991); Mary et al. (1992); Polley et al. (1992); Robinson and Scrimgeour (1995); Watkins et al. (1996)
	^{14}C	0.156		5730 years	LSC; AR; MS (for NT); NMR	GR: Warembourg and Kummerow (1991); Harris and Paul (1991); Voroney et al. (1991); Goh (1991); Meharg (1994) T: Caldwell et al. (1977); Swinnen et al. (1994); NT: Harrison et al. (1990); Nepstad et al. (1994) AR: Brownlee et al. (1983)

Hydrogen	^{1}H			s[b]		
	^{2}H (D)			s	MS; NMR	GR: Dawson (1993) T: Caldwell and Richards (1989) NT: White et al. (1985); Thorburn and Ehleringer (1995); NA: Dalton (1989) AR: Kochian and Lucas (1983)
	^{3}H (T)	0.0181		12.26 years	LSC: AR	GR: Vose (1980) T: Raney and Vaadia (1965); Shone and Flood (1980)
Nitrogen	^{13}N	1.19	0.51	9.7 min (c.p.)	γ counting	GR: Meeks (1993) T: Lee and Clarkson (1986); Siddiqi et al. (1989, 1990); Kronzucker et al. (1995a, b); Clarkson et al. (1996)
	^{14}N			s[b]		
	^{15}N			s	MS; NMR	GR: Jenkinson et al. (1985); Robinson and Smith (1991); Handley and Raven (1992); Blackburn and Knowles (1993) T: Clark (1977); Powlson et al. (1986); Robinson and Millard (1986); Jackson et al. (1989); Grignon et al. (1992); Clarkson et al. (1996) NT: Androssoff et al. (1994)
Oxygen	^{16}O			s[b]		
	^{18}O			s	MS	T: Yoshida and Eguchi (1994) NT: Wassenaar (1995)
Phosphorus	^{31}P			s[b]		
	^{32}P	1.71		14.3 days	Cerenkov; AR	GR: Vose (1980) T: Drew and Nye (1970); Elliott et al. (1984); Caldwell et al. (1985); Jupp et al. (1987); McLaughlin et al. (1987) AR: Bhat and Nye (1973)

Appendix. *(cont.)*

Element	Isotope	E_{max} (MeV)		$t_{1/2}$	Method	Selected examples of applications in root studies
		β	γ			
	^{33}P	0.25		25 days	LSC; AR	T: Elliott et al. (1984); Caldwell et al. (1985); McLaughlin et al. (1987) AR: Kraus et al. (1987)
Potassium	^{39}K			s[b]		
	^{41}K			s	MS	T: Kuhn et al. (1995)
	^{42}K	3.55, 2.0	1.52	12.4 h	γ counting; Cerenkov	T: Marschner and Richter (1973); Behl and Jeschke (1982)
Rubidium	^{85}Rb			s[b]		
	^{86}Rb	1.77, 0.7	1.08	18.7 days	γ counting; Cerenkov	T: Clarkson et al. (1968); Drew et al. (1969); Kochian and Lucas (1982, 1983); White et al. (1987)
Sulphur	^{32}S			s[b]		
	^{34}S			s	MS	GR: Macko and Ostrom (1994) NT: Kirchmarr et al. (1996)
	^{35}S	0.168		86.7 days	LSC; AR	T: Lee (1982); Holobradá and Kubica (1989); Robinson et al. (1994); Sunarpi and Anderson (1996) AR: Baldwin and Tinker (1972); Baysdorfer and Vanderwoude (1988)

[a] Abbreviations are as follows: AR autoradiography, c.p. cyclotron produced, GR general reference, LSC liquid scintillation counting, MS mass spectrometry, NMR nuclear magnetic resonance, NT natural tracer, s stable, T tracer.

[b] most abundant isotope.

CHAPTER 13

Assessing the Ability of Roots for Nutrient Acquisition

Ch. Engels[1], G. Neumann[2], T.S. Gahoonia[3], E. George[2], and M. Schenk[4]

[1] Agrarökologie, Universität Bayreuth, 95440 Bayreuth, Germany
[2] Institut für Pflanzenernährung (330), Universität Hohenheim, 70593 Stuttgart, Germany
[3] Department of Agricultural Science, The Royal Veterinary & Agricultural University, 1871 Frederiksberg C, Copenhagen, Denmark
[4] Institut für Pflanzenernährung, Universität Hannover, 30419 Hannover, Germany

CONTENTS

Dedicated to Professor Dr. Dres. h.c. H. Marschner

A.L. Smit et al. (Eds.), Root Methods

13.1 Introduction

Nutrient acquisition by roots from soil is a complex process which is dependent on several root features: (1) morphological root characteristics, including mycorrhizal associations, which determine the extent of the interface between plant and soil (2) ability to modify the nutrient availability in the rhizosphere, and (3) ability for nutrient uptake through the plasma membranes (for reviews see Barber 1984; Clarkson 1985; Marschner 1995). The relative importance of these factors for nutrient acquisition is dependent on environmental conditions and the specific nutrient, particularly its chemical availability and mobility in the soil.

The most convenient way to estimate nutrient acquisition by roots is to measure the nutrient accumulation rate in the whole plant between two harvest dates. Such a measurement, however, gives no information about the contribution of the various root characteristics mentioned above. Thus, statements e.g. on the ability for nutrient acquisition of a particular genotype, or on the consequence of a specific agronomic measure for nutrient uptake, can be drawn only for the specific environmental conditions under which the plants have been cultured. Furthermore, accumulation rate is difficult to measure in trees which are several years old. Therefore, it is often necessary to determine separately the contributions of the various root characteristics involved in nutrient acquisition. The measurement of morphological root characteristics has been described in detail in the preceding chapters. In this chapter an overview is given on the principal methods of measuring or estimating root-induced modifications in the rhizosphere, the ability of roots to take up nutrients, and the contribution of mycorrhizal fungi to these processes.

13.2 Modification of Nutrient Availability in the Rhizosphere

Roots may modify the physical, chemical and biochemical characteristics of the soil in their vicinity (rhizosphere) by uptake and release of different compounds (Grayston et al. 1996). These alterations lead to modifications of the biomass and species composition of microorganisms which in turn change rhizosphere conditions. In view of the large diversity of substances which can modify the rhizosphere soil, this chapter does not cover analytical methods for measurement of the various compounds. It is confined to techniques for collection of root exudates directly and for sampling rhizosphere, as opposed to bulk, soil for subsequent extraction and analysis of rhizosphere compounds.

13.2.1 Collection of Root Exudates

Various techniques have been used for the collection of root exudates in a solution surrounding the roots (Fig. 13.1). Of course only water soluble, diffusible

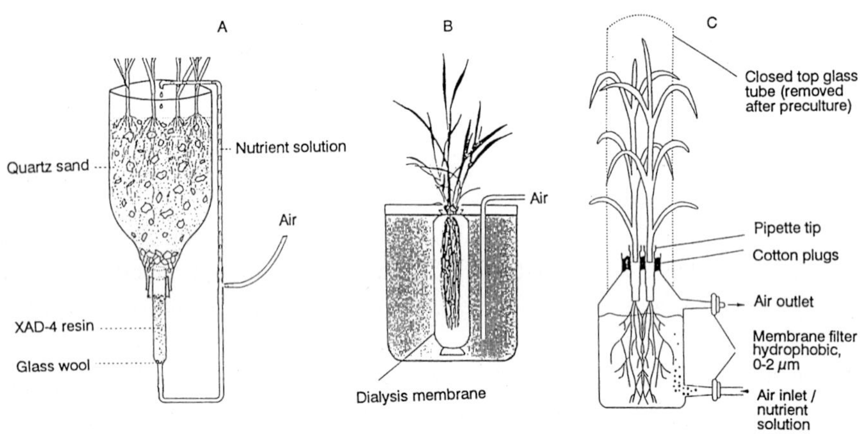

Fig. 13.1. Examples of plant culture systems used for the collection of root exudates. **A** Plant culture in solid medium (quartz sand); root exudates are eluted from the sand with the circulating nutrient solution and may be selectively retained in exchange resin at the bottom of the system (Tang and Young 1982, reprinted with permission of ASPP). **B** Plant culture in nutrient solution; root exudates are concentrated within a dialysis membrane around the roots (Tadano and Sakai 1991, reprinted with permission of Japanese Society of Soil Science and Plant Nutrition). **C** Hydroponic system for plant culture and collection of root exudates under axenic culture; for further details see Box 13.1 (von Wirén et al. 1995, reprinted with permission of The New Phytologist)

exudates can be measured in the solution, whereas exudates adhering to the root surface (e.g. mucilage) are not or only partially sampled by these techniques.

13.2.1.1 Collection in Trap Solutions

Aerated Trap Solutions. Root exudates may be collected by transferring the roots from the culture medium into an aerated trap solution. By this technique root exudation may be assessed during a well-defined period of time. However, its application should be restricted to plants grown in nutrient solution. Attempts to apply the technique to culture systems using solid media (soil, sand) almost certainly cause root injury and thus introduce artefacts. On the other hand, the physical nature of the medium surrounding the roots affects the total amount and composition of substances released by the roots (Boeuf-Tremblay et al. 1995), and exudation is stimulated by the mechanical impedance during growth in solid media. Simulation of the mechanical forces imposed on roots growing in soil may be achieved by addition of glass ballotini to the nutrient solution (Barber and Gunn 1974; Schönwitz and Ziegler 1982) or by growing the plants in sand culture using percolation techniques for exudate sampling (Fig. 13.1A).

For the collection of high molecular weight compounds, roots may be enclosed in dialysis membranes which are non permeable to compounds exceeding a certain molecular weight, to increase the concentration of these compounds in the vicinity of the roots, and thus, to facilitate their analytical detection (Fig. 13.1B; Sakai and Tadano 1993). In open (flowing culture) systems (Fig. 13.1A) the exudates are continuously displaced and may be removed from the circulating nutrient solution e.g. by exchange resins (Petersen and Böttger 1991).

Percolation Techniques. Collection of exudates from undisturbed root systems of plants grown in solid media (e.g. sand, vermiculite) may be achieved by percolating the culture vessels with the trap solution. Prior to collection, rhizosphere products accumulated during the culture period due to chemical and microbial modification of root exudates, should be removed by repeated washings with the trap solution. Thereafter, the vessels are incubated for a short period (e.g. 30–60 min), or repeatedly percolated with the trap solution, and exudates released during this time are subsequently collected with the percolate (Miyasaka et al. 1991; Hülster and Marschner 1994; Shepherd and Davies 1994a; Johnson et al. 1996). For quantitative evaluation of rhizodeposition of organic compounds, isotopic labelling techniques using $^{14}CO_2$ (Keith et al. 1986; Hodge et al. 1996; Johnson et al. 1996) or $^{15}NH_3$ (Janzen 1990; Reining et al. 1995) may be employed before percolation.

Control samples should be collected from culture vessels without plants. Since sorption of certain exudate compounds to the solid matrix of the culture medium cannot be excluded, recovery experiments and comparison with results obtained from experiments in nutrient solution are essential.

Composition of Trap Solutions and Sample Processing. Generally, trap solutions used for collection of root exudates are nutrient solutions of the same composition as the culture medium (Miyasaka et al. 1991; Shepherd and Davies 1994a; Johnson et al. 1996), solutions of 0.5–2 mM $CaSO_4$ or $CaCl_2$ (Ohwaki and Hirata 1990), or distilled water (Lipton et al. 1987; Hülster and Marschner 1994; von Wirén et al. 1994). The nature of the trap solution may influence exudation rates of certain compounds and subsequent sample preparation as well. Since the osmotic strength of commonly used nutrient solutions is generally low, osmotic effects of the trap solution on the membrane permeability are not to be expected even if distilled water is used as the trap solution. However, exudation of low molecular weight compounds may be affected by the ionic composition of the medium surrounding the roots, providing counter ions for transport mechanisms. For example, efflux of amino acids and sugars from plant roots seems to occur by passive diffusion, but effective resorption by active transport mechanisms has been reported (Jones and Darrah 1993, 1994) which can be affected by the ionic composition of trap solutions. Release of organic acids from plant roots, which is probably mediated by active efflux mechanisms (Dinkelaker et al. 1989; Jones and Darrah 1995), may be affected in a similar way. However, comparison of either nutrient solution, or water, or 20 mM KCl as trap solutions for amino acids released from roots of *Brassica napus* L., showed no differences in exudation during collection periods between 0.5 and 6 h (Shepherd and Davies 1994b). In order to avoid nutrient depletion in the roots during collection of exudates, water or $CaSO_4/CaCl_2$ solutions should only be used if collection is confined to short time periods.

Collection of root exudates in trap solutions generally yields a very diluted solution of exudates. Before analysis, it is often necessary to increase the exudate concentration by decreasing the volume of the trap solution e.g. by freeze drying or rotavaporation, depending on the stability of the compounds of interest. Depending on the composition of the trap solution, volume reduction may lead to high salt concentrations which may interfere with subsequent analysis or even result in precipitation of exudates (e.g. Ca-citrate, Ca-oxalate, proteins). Thus, if possible, interfering salts should be removed by use of ion exchange resins prior to volume reduction. Exudates may be also enriched and prefractionated by techniques of solid phase extraction (ion exchange resins for charged compounds, reversed phase matrices for lipophilic compounds). High molecular weight constituents (polysaccharides, proteins) can be precipitated by addition of organic solvents (e.g. 80% v/v) or separated by ultrafiltration.

13.2.1.2 Localized Sampling

Exudation is not uniform over the whole root system, and considerable longitudinal gradients may exist in exudation rates along the roots. For example, enhanced release of organic acids in P-deficient rape (Hoffland et al. 1989) or of phytosiderophores in Fe-deficient maize (Römheld 1991) is largely confined to the root tip (apical cm). As the density of microorganisms which may catabolize exudates is higher in basal root zones (Uren and Reisenauer 1988), the localization of enhanced exudation in the root tip is expected to increase the efficiency of exudates in mobilizing P and Fe in the rhizosphere. Moreover, a distinct diurnal rhythm exists for the release of phytosiderophores from Fe-deficient plants (Marschner et al. 1986), which may also increase the probability that phytosiderophores mobilize Fe in the rhizosphere before being degraded by microorganisms. For the assessment of the ecological relevance of exudation for nutrient acquisition, it may therefore be important to measure the variability of exudation in space and time.

The variability of exudation in time can be measured well with the above mentioned methods, e.g. by frequent exchange of the solution surrounding the roots. There are also several methods to evaluate the variability of exudation along the root axis. For example, Hoffland et al. (1989) spread roots of plants grown in nutrient solution on a glass plate, and placed small plastic rings (diameter 1.2 cm) over the root zone from which exudation was to be measured. After covering the roots outside the rings with agar solution, a small volume of nutrient solution (200 μl) was pipetted into the rings. The solution in the rings was collected after incubation for 2 h at room temperature and high humidity, and analyzed for exudates. Delhaize et al. (1993) transferred solution-grown plants to large Petri dishes filled with nutrient solution, and sealed the plastic rings for separation of distinct root zones with vacuum grease. Alternatively, the whole root system may be covered with a thin prefixed layer of 1% agarose (low electroendosmosis type; Marschner et al. 1987). During incubation the agarose layer should be moistened with distilled water to prevent desiccation. After incubation the agarose is cut into pieces according to the position of the root segments below, from which exudation is to be measured, and subsequently analysed for exudates (Marschner et al. 1987). However, exact localization of the diffusion zones of exudates is difficult when using agarose sheets for exudate collection. It is also possible to spread the root system on a layer of fleece which is covered by a sheet of filter paper and moistened with water. Roots selected for exudate sampling are underlaid by small sheets of polyethylene foil, and different root zones are placed between two discs of moist chromatography paper (5 mm in diameter, preparative quality) which was previously washed with methanol and distilled water. The rest of the root system is covered with filter paper, and moistened with nutrient solution to avoid desiccation. During incubation the filter discs are occasionally remoistened with 10 μl of distilled water. After 3 h the filter discs are removed and exudates are extracted in a small volume (100 μl per three discs) of distilled water by centrifugation (10 min) in a microliter centrifuge. In the case of organic acids, the samples can be analyzed directly by HPLC, and quantitative determinations of exudates originating from single root tips or root segments with a spatial resolution of 5 mm even from plants with low exudation rates are possible (Neumann, unpublished results). Similarly, spatial variation of flavonoids released from roots of soybean seedlings has been examined by placing the roots between two layers of cellulose acetate filters as a medium for sorption of root exudates (Kape et al. 1992).

13.2.1.3 Collection from Soil-Grown Plants

For collection of root exudates from soil-grown plants, as described above, pieces of absorbent material (e.g. chromatography paper, glass fibre filters, solid agar sheets, polyacrylamide gels) may be placed on the roots for several hours. During collection the absorbent material has to be pressed onto the roots to ensure proper contact with them. To avoid root injury during uncovering from the soil, exudates can be collected from roots growing along the internal surface of root observation windows (rhizoboxes or windows installed in the field, see Chap. 8). This also allows localized collection from well-defined root segments. To prevent roots sticking to the plastic lid when the window is opened, a transparent plastic sheet is placed between the front lid and the soil. Using small sheets of filter paper as an absorbent material, temporal changes have been monitored in the concentration of organic acids in the soil solution obtained from the rhizosphere of proteoid roots of *Hakea undulata* (Neumann et al. 1995; Dinkelaker et al. 1996). When using agarose gels for exudate collection, gel shrinking sometimes occurs due to diffusion of water into the soil matrix.

In principle, in soil-grown plants the solution surrounding the roots can also be collected and compared with the solution from the bulk soil. Recently, Göttlein et al. (1996) described the construction of miniaturized suction cups which allow the microscale collection of soil solution from the rhizosphere. The solution is sampled in micro ceramic cells (outer diameter 1 mm, length 5 mm) which are connected to a vacuum collecting device. Also, extraction of root exudates from rhizosphere soil (see Sect. 13.2.2), collected from proteoid roots of *Banksia integrifolia L.* (Grierson 1992) and *Lupinus albus* L. (Dinkelaker et al. 1989; Gerke et al. 1994) has been reported.

For the techniques of exudate sampling from soil-grown plants described above it has to be considered that exudates may be degraded by microorganisms and adsorbed to the soil matrix. Thus, for collection of root exudates, sterile absorbent material should be used (e.g. agar sheets, methanol-washed filter paper), and the collection should be terminated before substantial microbial colonization of the absorbent material occurs. Furthermore, recovery experiments may be helpful to assess microbial degradation or sorption of exudates to the soil matrix under the specific experimental conditions. In such experiments, well-defined amounts of exudate compounds are added to experimental soil and recovery of these compounds is measured after incubation under sterile (e.g. chloroform atmosphere) and non-sterile conditions.

13.2.1.4 Collection of Mucilage

Mucilage is a gelatinous, root-produced material mainly consisting of polysaccharides and polyuronic acids, and is deposited primarily in the region around the root tip. Possible functions of mucilage include protection of the root from desiccation (Leiser 1968), reduction of friction between the growing root tip and the soil, improvement of the root-soil contact, and complexation of toxic minerals such as aluminium (Horst et al. 1982). For collection, it may be removed from the root surface with a soft brush and transferred to cellulose acetate filters (Horst et al. 1982), or it may be simply removed by use of forceps (Dinkelaker et al. 1996). Mucilage formed on nodal roots of maize after immersion into distilled water can be collected by vacuum suction (Morel et al. 1986). Also, ultrasonication treatments have been described to remove exudate compounds adhering to the root surface without visible damage or influence on subsequent plant growth (Barber and Gunn 1974).

13.2.1.5 Factors Affecting Recovery of Root Exudates

External Concentration. The total amount of substances released by the roots is dependent on their concentration in the medium surrounding the roots. Low external concentrations increase the transmembrane concentration gradient and enhance the release, and decrease the reabsorption (retrieval) of exudates. Therefore, it may be expected that exudate collection in systems with circulating nutrient solution or frequent replacement of the nutrient solution in static systems may lead to unrealistically high exudation rates (Prikryl and Vancura 1980; Jones and Darrah 1993). On the other hand, in closed (static) systems (Fig. 13.1B,C) the concentration of exudates in the nutrient solution is increasing with time, leading to higher retrieval into the roots as was demonstrated for sugars and amino acids in root exudates of maize (Jones and Darrah 1993, 1994). Estimates of exudation for the two techniques may differ by an order of magnitude and it is not clear which method most closely resembles exudation in soil (Darrah 1996).

Effects of Microorganisms. Microorganisms can modify and degrade low molecular weight exudates. This process can be diminished in open systems when exudates may be protected from microbial degradation by sorption to exchange resins (Fig. 13.1A). Alternatively, plants can be grown in axenic culture (Fig. 13.1C). Various systems have been described to grow sterile plants in hydroponic culture (e.g. Lipton et al. 1987; Jones and Darrah 1992; von Wirén et al. 1995; Box 13.1). It has to be kept in mind, however, that microorganisms affect both the quality and quantity of exudates released from the roots. In culture systems where the microorganisms are spatially separated from the roots by a Millipore membrane ("microcosm", see Meharg and Killham 1991), it was demonstrated that exudation may not only be affected through root infection but also by microbial metabolites. Depending on the microbial inoculum, exudation varied in the range of 3–34% of recently assimilated carbon, compared with 1% in the absence of microorganisms (Meharg and Killham 1995).

BOX 13.1. Example of a Hydroponic System for Sterile Plant Culture (Wirén et al. 1995, Fig. 13.1C)

Prior to germination, surface sterilization of seeds is performed by immersion for 1–5 min into 96% ethanol and 25 min into 18% (v/v) H_2O_2, followed by rinsing the seeds five times in sterile 2.5 mM $CaSO_4$ solution. To select sterile seedlings, the seeds are germinated on sterile agar plates containing half-strength casein-peptone/soymeal-peptone nutrient broth (Merck No. 5459) for 3 days at 25 °C in the dark. Uninfected seedlings are inserted into pipette tips in autoclaved culture vessels containing 600 ml autoclaved nutrient solution and the shoot is covered with a closed top-glass tube for a preculture period of 9–10 days. Thereafter the top glass tube is removed and the pipette tips are sealed with 45 °C melted paraffin. Sterilized cotton plugs are inserted between the pipette tips to maintain axenic conditions in the culture vessel. The nutrient solution is continuously aerated through 0.2 μM membrane filters and the air inlet filter is also used to replace the nutrient solution. Aliquots of the nutrient solution are regularly checked for sterility via incubation for 2 days at 28 °C on nutrient agar.

Several methods are available for soil sterilization such as sterilization by heat, gamma-irradiation and fumigation with toxic chemicals (Johnson and Curl 1972; Box 13.2). However, soil sterilization procedures can change chemical soil properties and may lead to increased plant availability of nutrients such as nitrogen, phosphorus and toxic elements (Johnson and Curl 1972). Furthermore, it is very difficult to prevent re-colonization of sterilized soil by microorganisms, particularly during long-term cultivation of plants.

BOX 13.2. Methods for Soil Sterilization

Sterilization by Heat

Dry heat pasteurization may be used to rid soil of most fungal propagules. Sieved soil samples are heated twice to 70–80 °C for 24 h, each followed by incubation at room temperature for 24 h. Steaming at 83 °C for 30 min results in a better heat distribution in the soil and kills most pathogenic fungi, nematodes, insects and weed seeds, but some bacteria and viruses may survive dry or moist heat treatment. The soil structure seems not to be significantly affected by these methods (Johnson and Curl 1972). Autoclaving of homogenized soil samples (120 °C, 1.1 atm, 30–60 min) destroys most microorganisms but the soil structure is altered significantly, and may result in the liberation of nutrients and toxic elements (Alef 1995).

Chemical Sterilization

Soil samples (250 g or more) can be disinfected by fumigation in a large desiccator containing a beaker of alcohol-free chloroform. The desiccator is evacuated and incubated in the dark for 18–24 h at 25 °C. Thereafter re-evacuation is repeated three times for 3 min to remove the chloroform. Fumigation with chloroform destroys all microorganisms but may also significantly alter the soil structure (Powlson and Jenkinson 1976a). Fumigation with other chemicals such as ethylene oxide (explosive) or methyl bromide can be used with similar results (Johnson and Curl 1972).

Gamma Irradiation

Soil samples sealed in polyethylene bags are exposed to an irradiation dose of 2.5 Mrad at a rate of about 2 Mrad h^{-1}. The procedure eliminates all microorganisms and there seems to be only little effect on soil structure. However, enzyme activities such as those of phosphatase and urease are still detectable after the sterilization treatment (Johnson and Curl 1972).

To prevent microbial degradation of exudate compounds, various antibiotics such as Rifampicin/Tetracyclin (50/25 ppm; Schwab et al. 1983; Ohwaki and Hirata 1992), Cefatoxim/Trimethaprim (30/20 ppm; Azaizeh et al. 1995) or "Micropur" (Amann and Amberger 1989) have been added to trap solutions, or agar sheets used as collection media for root exudates, or washing solutions used for root pretreatment prior to exudate collection. However, antibiotics may have phytotoxic effects depending on plant species and dosage (Smart et al. 1995). Reducing the time period of collection of root exudates from plants grown under non-sterile conditions (Tagaki et al. 1984; Cakmak and Marschner

1988) is another strategy to overcome problems of microbial degradation of exudate compounds.

Root Injury. During transfer of root systems into trap solutions, as well as during application of absorbent materials or preparation of root systems for exudate collection, roots may be exposed to stress due to mechanical damage (e.g. rupture of roots hairs, epidermal cells) or rapid change in the environmental conditions (e.g. temperature, pH, oxygen availability), which may affect exudation rates. The general influence of handling during exudate collection on plant growth may be assessed by comparing plants either subjected or not subjected to the collection procedure (Barber and Gunn 1974). The effects of the routine procedure of handling on root exudation may roughly be estimated by comparison of the qualitative and quantitative changes in patterns of exudation of plants subjected to handling procedures of different intensity.

13.2.2 Separation of Rhizosphere from Bulk Soil

Shaking Off Soil Adhering to Roots. This simple method for separation of rhizosphere and bulk soil involves careful excavation of plant roots. The soil loosely adhering to the roots (regarded as bulk soil) is collected by gently shaking the roots until most soil has dropped onto clean paper. Separation of soil from roots is facilitated by a short period of air drying, which dries the soil without making the roots brittle. The more tightly adhering soil (regarded as rhizosphere soil) is then shaken off separately on another clean paper. Alternatively, tightly adhering soil can be washed off from the roots and air dried. All visible root debris should be removed using forceps or by sieving to avoid contamination of the rhizosphere soil with root residues during analysis. The method can be applied to plants grown in pots or in the field (Smiley 1974). The method was first described by Starkey (1931) and was used to study e.g. the pH (Riley and Barber 1971; Smiley 1974), or the concentrations of various mineral elements in the rhizosphere (Sihna and Singh 1976; Hendriks and Jungk 1981).

Using this method, rhizosphere soil can be obtained from plants growing under natural conditions, both in pot and field experiments, without the need for technical equipment. The main disadvantage of the method, however, is the low spatial resolution of sampling not only along the root axis but also in a radial direction. Depending on the sand and clay content and thus the formation of aggregates, different amounts of soil are removed from the roots by shaking (Hendriks and Jungk 1981). Therefore, the method is applicable mainly for qualitative assessment of modifications in the rhizosphere, particularly if large differences exist between rhizosphere and bulk soil.

Collection of Soil at Well-Defined Distances from Roots. Papavizas and Davey (1961) designed a microsampler for obtaining soil at increasing distances (3 mm increments) from the surface of individual lupin roots. The sample collector consisted of seven thin-walled steel tubes, 50 mm long with an inside diameter of 3 mm, soldered adjacent to each other in a plane with an overall width of 22 mm. To collect the soil samples, the collector was inserted vertically into the soil immediately adjacent and parallel to the primary root of lupin seedlings. Another possibility for obtaining soil at well defined distances from the roots is to confine the roots to certain soil zones by nets with a mesh width preventing penetration by roots or root hairs but allowing free exchange of exudates and soil solution. Boero and Thien (1979) used a system of concentric cylinders made of a nylon netting and confined the root system to the central core. Helal and Sauerbeck (1983) divided pots filled with soil by vertical stainless steel screens (pore diameter 31.5 μm, open area 25%) into a central root zone, an adjacent soil zone of 10 mm width which could be penetrated by root hairs and an outer soil zone which which was free from roots and root hairs. The principle of this method has also been used to study phosphorus depletion and pH changes at the soil-hyphae interface of plants infected with mycorrhiza (Li et al. 1991). Instead of dividing the pots vertically into different soil zones, horizontal separation can also be achieved (Kuchenbuch and Jungk 1982, see below).

Thin Slicing of Rhizosphere Soil. Many of the processes involved in nutrient acquisition occur in the soil zone very close to the roots. Therefore, it is necessary to investigate this zone with high spatial resolution. This can be achieved by slicing the rhizosphere soil into thin layers with a microtome. Farr et al. (1969) held onion roots between two moist soil blocks. Onion roots do not have many root hairs and do not branch. Therefore, they did not grow into the soil blocks, and a defined root-soil interface was obtained. After 12 days, soil and plant roots were separated. The soil was frozen in liquid nitrogen, sliced into thin sections using a freezing microtome and analyzed for K, Ca and pH. Kuchenbuch and Jungk (1982) separated roots from soil columns horizontally by using a fine mesh (30 μm) nylon screen allowing the penetration of root hairs but not that of roots. At harvest, the roots had formed a dense root mat on the screen, so that the whole screen could be regarded as a root surface. The soil column was separated from the screen (root surface) with a knife, quickly frozen into liquid nitrogen and sliced at −4 °C into 0.06 mm layers with a freezing microtome. The method has been applied to study the rhizosphere concentrations of potassium (Kuchenbuch and Jungk 1984), magnesium (Seggewiss and Jungk 1988), phosphorus (Gahoonia et al. 1992), and aluminium (Gahoonia 1993) and phosphatase activity (Tarafdar and Jungk 1987). Gahoonia and Nielsen (1991, 1992) developed an improved method for studying the rhizosphere of plants at more advanced stages of development under controlled nutritional conditions (Fig. 13.2). Plants were precultured for about 10 days in PVC tubes filled with vermiculite as illustrated in Fig. 13.2A. The bottoms of the tubes were closed with fine nylon cloth impervious to roots. The plants were supplied with a nutrient solution via two wicks covered with black PVC foil. When a root mat had developed on the nylon cloth after about 9 days, the nylon was removed and the PVC tube with plants transferred to the test soil columns (Fig. 13.2B). A nylon screen of inner mesh size 53 μm at the bottom of the upper soil column prevented penetration of roots to the lower column. The roots grew quickly into the 1 cm soil layer of the lower column and formed a new root mat on top of the nylon screen. To maintain defined soil moisture content, the soil columns were placed on a small cup-shaped sand bath, connected to a reservoir of distilled water with a wick. To obtain rhizosphere soil of defined distance from the roots, the columns were separated from the root mats, quickly frozen in liquid nitrogen and sliced into thin layers using a freezing microtome.

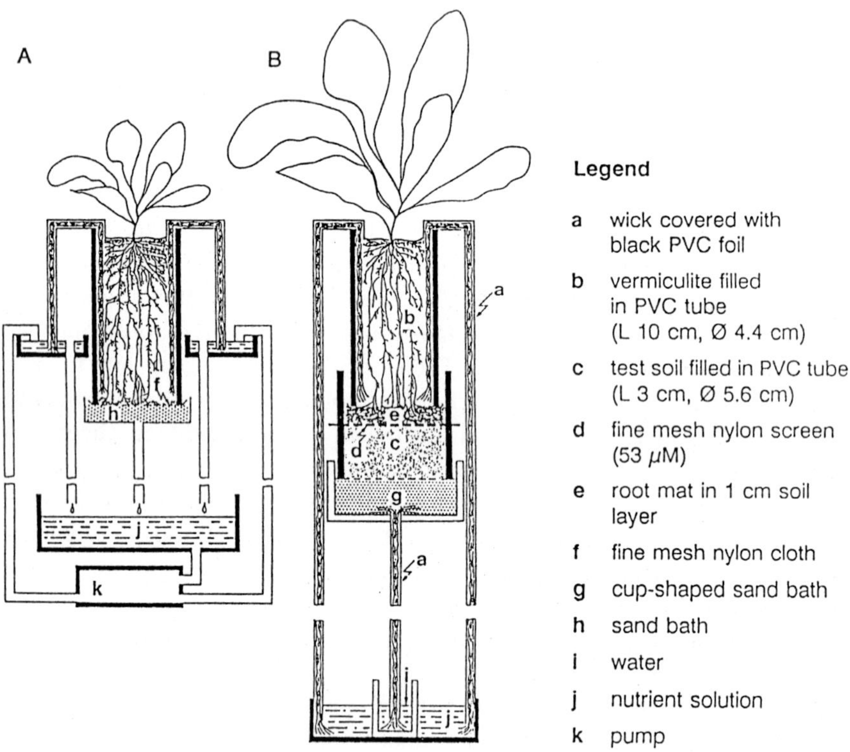

Fig. 13.2. Example of a plant culture system for obtaining rhizosphere soil at defined distance from the root surface (root mat). **A** Pre-experimental system. **B** Experimental system. For further explanation see text (Gahoonia and Nielsen 1992; with kind permission from Kluwer Academic Publishers)

13.2.3 In Situ Measurements in the Rhizosphere

Various non-destructive staining techniques have been developed to demonstrate chemical changes in the rhizosphere (Fig. 13.3; Box 13.3; Box 13.4). Generally, sheets of agar, agarose, polyacrylamide or paper are used as carrier matrices for indicators and substrates for enzyme or colour reactions. After application to the surface of root systems of plants grown in hydroponic or soil culture (rhizoboxes, root windows; see Chap. 8), qualitative, semiquantitative or in some cases even quantitative evaluation of changes in the rhizosphere is possible.

BOX 13.3. Methods for Visualization of Chemical Changes in the Rhizosphere

Rhizosphere pH

Root systems are embedded in 0.75% (w/v) agar solutions (35 °C, pH 6) containing nutrient solution or $CaSO_4$ [1 mM], and 0.006% (w/v) bromokresol-purple as a pH-indicator. After an incubation period of between 15 min and 2 h, pH-induced colour changes along the roots become visible. For sensitive detection of pH changes, the use of agar with low buffering capacity (e.g. Merck No. 1614) or even agarose (1% w/v) is recommended. A stock solution of 1% (w/v) bromokresol-purple is prepared, dissolving the pH indicator in distilled water by dropwise addition of 1 N NaOH during 20–40 min. Complete dissolution is indicated by a constant pH. Thereafter, the pH of the indicator solution is adjusted to 6.0 by addition 1 N H_2SO_4.

Reduction of Fe(III)

A 0.75% (w/v) agar solution containing 100 μM Fe(III)EDTA and 300 μM BPDS as chelator for Fe(II) can be used for embedding of the root systems (Marschner et al. 1982) or for application of agar sheets (3 mm) onto the root surface (Dinkelaker et al. 1993a). Reduction of Fe(III) is indicated by formation of a red complex after 30 min to several hours. Alternatively, the agar solution may be prepared with 100 μM Fe(III)EDTA, 10 mM Mes-buffer pH 4–4.6, and 200 μM Ferrozin, resulting in the formation of a violet complex of Fe(II)-Ferrozin.

Reduction of Mn (IV)

$KMnO_4$ [1 mM] is added to the agar solution (0.75% w/v) pH 6.0, and kept at 50 °C for 2 h to induce formation of MnO_2. After application to plant roots, MnO_2 reduction is indicated by decoloration in the brown agar layer during an incubation period of between one and several days (Marschner et al. 1982). Alternatively, filter papers (e.g. MN-260;

Macchery and Nagel, Düren, Germany) are impregnated in a solution of 10 mM $KMnO_4$ for 7 h, forming MnO_2 (brown) by oxidation of cellulose in the paper. After washing (distilled water) and drying, the papers can be stored for months without loss in colour intensity. The impregnated paper sheets are applied with proper contact to the root surface (e.g. rhizoboxes, root windows) and fixed with pins. Decoloration can be observed after a period of 30 min to several hours.

Complexation of Aluminium
Gels of polyacrylamide (1 mm, 10% T); agar (3 mm, 0.75% w/v; Dinkelaker et al. 1993a), or agarose (3 mm, 1% w/v; Dinkelaker et al. 1997) are used as a carrier matrix for red-coloured Al-Aluminon complexes. For gel preparation, a solution of $Al(NO_3)_3$ [250 μM], Aluminon [88 μM] and ascorbic acid [1.4 μM] adjusted to pH 4.2 is used for either incubation (2 h, polyacrylamide gels) or direct addition to the solutions of agar or agarose after cooling to approximately 50 °C. Thereafter, gel sheets are placed onto the root surface and covered with plastic foil. Complexation of Al by chelators with a higher affinity to Al compared with Aluminon (e.g. organic acids in root exudates) is indicated by zones of decoloration around the root, developing within a few hours.

13.2.3.1 Rhizosphere pH

Root-induced changes of pH in the rhizosphere are mainly due to an imbalance in cation-anion uptake which depends particularly on the nitrogen source, or due to a deficiency in nutrients e.g. in phosphate or iron (Marschner and Römheld 1983). For demonstration of pH changes in the rhizosphere, root systems are embedded in agar solutions containing nutrient solution or $CaSO_4$, and bromocresol-purple as a pH-indicator (Box 13.3). After an incubation period of between 15 min and 2 h, pH-induced colour changes along the roots become visible (Fig. 13.3A), a yellow colour indicating acidification, and purple or violet colours indicating alkalization. The technique is applicable for plants grown in nutrient solution (Marschner et al. 1982) and in rhizoboxes in soil (Marschner and Römheld 1983). As an alternative to embedding root systems in agar solution, the roots may be covered with solid agar sheets, permitting repeated measurements with the same plants. Quantitative measurements of the pH changes in the rhizosphere are possible by insertion of microelectrodes into the various pH zones qualitatively indicated by the colour changes in the agar layer (Häussling et al. 1985; Gollany and Schumacher 1993). Recently, computer controlled videodensitometry has been used to quantify pH changes in the rhizosphere after visual demonstration using the agar technique described above (Jaillard et al. 1996).

BOX 13.4. Methods for Estimation of Enzyme Activity in the Rhizosphere

Peroxidase

Peroxidase released from roots of sterile grown *Festuca rubra* has been photometrically detected in the nutrient solution by addition of an equal volume of 100 mM citrate-phosphate buffer, pH 5.5, using 100 mM H_2O_2 as a substrate and phenolic acids or guaiacol [10 mM] as hydrogen donors for the enzymatic reaction (Vaughan et al. 1994).

Acid Phosphatase and Protease

For determination of acid phosphatase activity, the production of p-nitrophenol by hydrolysis of p-nitrophenyl phosphate is measured photometrically (Bartlett and Lewis 1973). Since in many cases acid phosphatase was found to be predominantly associated with rhizodermal cell walls (Bieleski and Johnson 1972) or released in the mucilage (Felipe et al. 1979), whole root systems or root cuttings are incubated in the reaction buffer at 20 °C for 20–60 min. (Bielski and Johnson 1972).

A non-destructive method for visualization of phosphatase activity in soil-grown plants (rhizoboxes, root windows) has been described by Dinkelaker and Marschner (1992) based on the enzymatic hydrolysis of 1-naphtylphosphate by root acid phosphatase, yielding 1-naphtol as a reaction product which forms a red complex with Fast Red TR (diazotized 2-amino-5-chloro-toluene 1,5-naphtalene disulfonate). Filter papers impregnated with the reagents are placed onto the root surface of plants grown in rhizoboxes and activity of acid phosphatase is indicated by development of red colour zones within approximately 2 h (Dinkelaker and Marschner 1992; Dinkelaker et al. 1996). Similarly, protease activity was visualized by decoloration of polyacrylamide gels (10% T) containing gelatine (0.5%), dye-labelled by Remazol-Brilliant-Blue, which were placed onto the root surface (B. Dinkelaker et al. unpubl.).

13.2.3.2 Redox Processes

Plant roots are able to influence the redox state of the rhizosphere soil either by reduction or oxidation. Reduction of iron or manganese oxides due to the activity of plasma membrane-bound reductases (Bienfait et al. 1983) or by release of reducing substances (Marschner et al. 1986) such as phenolic acids or carboxylic acids (Hether et al. 1984; Dinkelaker et al. 1993b, 1996) can increase the availability of iron and manganese especially in soils of neutral or alkaline pH (Cairney and Ashford 1989). In contrast, in marsh plants such as rice, growing in wet or flooded soils, excessive uptake of reduced iron, man-

Fig. 13.3. Examples for the demonstration of chemical and biochemical modifications of the rhizosphere; further details in the text, and Boxes 13.3 and 13.4. **a** Formation of yellow colour in agar containing bromocresol-purple, induced by decline of pH around roots of P-deficient buckwheat (V. Römheld, unpubl.). **b** Formation of a red complex in agar containing BPDS, induced by the reduction of Fe^{3+} to Fe^{2+} in Fe-deficient groundnut (*left*; V. Römheld, unpubl.). **c** Decolorization in agar containing brown MnO_2, induced by the reduction of Mn^{4+} to Mn^{2+} (Dinkelaker et al. 1993b; with kind permission from Kluwer Academic Publishers). **d** Oxidation power of roots of *Phragmites australis* (Armstrong et al. 1992, reprinted with permission of The New Phytologist). *Above*: formation of blue colour in agar containing colourless leuco methylene blue, induced by oxidation to methylene blue. *Below*: iron oxidation also becomes visible as an orange-brown precipitate in the rhizosphere of soil-grown plants. **e** Decoloration in agar containing red Al-aluminon complexes induced by root exudates of *Lupinus luteus* forming more stable complexes with Al (B. Dinkelaker, unpubl.). **f** Formation of redish-brown colour on filter papers soaked with 1-naphtyl phosphate and Fast Red TR, induced by the formation of 1-naphtol through phosphatases (B. Dinkelaker, unpubl.)

ganese and sulfides is avoided by an increase in the redox potential of the rhizosphere relative to the bulk soil. In these plant species an internal channel system (aerenchyma) allows diffusion of oxygen from the shoot to the roots and subsequent release into the rhizosphere (Trolldenier 1988). For demonstration of redox changes in the rhizosphere agar gels or filter papers impregnated with redox indicators are applied to the root surface (Box 13.3).

Reduction of Fe(III). The method most frequently used for demonstration of Fe(III) reduction is an agar technique where reduction of Fe(III) is indicated by the formation of a red complex between Fe(II) and BPDS (bathophenanthroline-disulfonic acid; Marschner et al. 1982; Fig. 13.3B). Instead of BPDS, $K_3Fe(CN)_6$ has also been used as a redox indicator (Brown and Ambler 1974). For quantitative evaluation of the Fe(III) reducing capacity of roots from plants grown in nutrient solution, the formation of a red complex between Fe(II) and Ferrozine (3-(2-pyridyl)-5,6-diphenyl-1,2,4-triazine-p,p′-disulfonic acid) has been measured spectrophotometrically at 562 nm in nutrient solution (pH 5.4) containing 100 μM Fe(III)EDTA and 300 μM Ferrozine (Horst et al. 1992).

Reduction of Mn(IV). Reduction of Mn(IV) in the rhizosphere has also been demonstrated by use of the agar technique described above (Box 13.3). After application of agar sheets with MnO_2 to plant roots, MnO_2 reduction is indicated by decoloration in the brown agar layer (Marschner et al. 1982; Fig. 13.3C). A modification of the method to increase its sensitivity and applicability for soil-grown plants has been described by Dinkelaker et al. (1993a). Based on the method of Uren (1981), MnO^{2-}-impregnated paper sheets are applied to the surface of roots growing e.g. in rhizoboxes (Box 13.3).

Redox State in General. For quantitative measurements of redox changes in the rhizosphere, platin microelectrodes have been employed (Schaller and Fischer 1985; Flessa and Fischer 1992) which are inserted into the rhizosphere soil. Leuco methylene blue (Methylene blue [20–25 mg l^{-1}] reduced to the colourless form by addition of $Na_2S_2O_4$ [200 mg l^{-1}]) has been frequently used in combination with agar techniques to visualize oxidative changes in the rhizosphere (Trolldenier 1988; Armstrong et al. 1992; Fig. 13.3D), whereas reductive changes were demonstrated by use of methylene blue (Karpov and Potapov 1975).

An agar medium which was blackened by precipitated ferrous sulfide was used to demonstrate the oxidizing power of rice roots (Trolldenier 1988). FeS precipitation was obtained by reduction of $FeSO_4$ [5 mM] with Na_2S [2–4 mM] which was added to the agar solution at 50 °C. FeS oxidation was indicated by transparent zones in the rhizosphere of rice plants cultivated for 11 days in the semi-solid agar medium containing FeS and nutrient solution.

13.2.3.3 Complexation of Aluminium

Aluminium (Al) toxicity is a major factor limiting plant growth in strongly acidic soils. Depression of root elongation is the most obvious toxic effect of Al on plant growth (Delhaize and Ryan 1995). Plant species and cultivars vary widely in their resistance to Al toxicity. One possible cause for Al tolerance is detoxification of Al by complexation either within the plant or in the rhizosphere (Miyasaka et al. 1991; Delhaize and Ryan 1995).

Complexation of Al in the mucilage covering the root apex has been demonstrated by hematoxylin staining (Horst et al. 1982). Also, root-derived organic acids such as malic, citric and oxalic acids may complex and thus detoxify Al in the rhizosphere (Miyasaka et al. 1991; Delhaize et al. 1993; Pellet et al. 1995). A technique to demonstrate Al complexation in the rhizosphere of soil-grown plants by applying agar containing red-coloured Al-aluminon complexes to the roots has been described by Dinkelaker et al. (1993a,b, 1997; Box 13.3). Complexation of Al by chelators with a higher affinity to Al compared with aluminon (e.g. organic acids in root exudates) is indicated by zones of decoloration around the root, developing within a few hours (Fig. 13.3E). Control experiments should be carried out, measuring Al concentrations in the different gel zones, to make sure that decoloration is really a result of Al complexation and not of root uptake of the water-soluble Al-aluminon complex.

13.2.3.4 Detection of Low Molecular Weight Rhizosphere Compounds

Exudation of citric acid from proteoid roots of white lupin (*Lupinus albus* L.) grown in a calcareous soil has been detected by eye due to precipitation of crystals of calcium citrate in the rhizosphere (Dinkelaker et al. 1989). Low molecular weight compounds may be sampled on absorbent materials such as chromatography paper, agarose gels (Dinkelaker et al. 1996) or cellulose acetate filters (Kape et al. 1992). For example, phenolic compounds in the rhizosphere of several tree species grown in rhizoboxes have been visualized after a 4-h collection period with chromatography paper (MN 260, Macchery and Nagel, Düren, Germany) or agarose gels (1% w/v, 2 mm), and subsequent spraying with Folin Ciocalteau reagent (Merck, Darmstadt, Germany) and 20% w/v Na_2CO_3 (Dinkelaker et al. 1996). Similarly, other spray reagents commonly used for detection of different groups of compounds in paper- or thinlayer chromatography may be suitable for visualization of rhizosphere compounds (e.g. ninhydrin for amino acids, anilinphtalate for sugars). Moreover, many phenolics become visible by examination of the carrier media in UV-light (360 nm, 254 nm) due to intense autofluorescence (Kape et al. 1992).

13.2.3.5 Detection of Enzyme Activities

Activities of various enzymes, such as acid phosphatase (Bieleski and Johnson 1972; Bartlett and Lewis 1973) peroxidases (Vaughan et al. 1994), invertase, cellobiase, adenosinetriphosphatase, pyrophosphatase and nuclease (Chang and Bandurski 1964) have been detected in root bathing solutions of different plant species. For measurements of enzyme activities, the solutions are generally mixed with an appropriate buffer and a natural or artificial substrate of the enzymatic reaction (Box 13.4). Reaction products are detected by eye (Fig. 13.3F) or spectrophotometrically.

13.2.3.6 Assessment of Nutrient Distribution in the Rhizosphere by Autoradiograpy

Autoradiography has been used to study the nutrient concentration profile in soil in the vicinity of plant roots. More details are given in Chapter 12. In principle, plants are grown in soil labelled with radioactive tracers. After a few days, X-ray film is placed on the soil around roots, and tracer concentration is estimated from blackening of the film.

Plant Culture. Most frequently, plants are grown in flat rhizoboxes (depth of the soil layer 0.5 cm) in which the roots grow along the removable front cover. A thin (e.g. 8 μm) plastic foil or mylar film can be placed between the front cover and labelled soil, on which the film is placed at the end of the experimental period (Bhat and Nye 1973; Claassen et al. 1981). As this technique is non-destructive, several autoradiograms can be made with the same plants during their development. Ernst et al. (1989) cultured the plants in rhizoboxes with non-labelled soil. After different growing periods, the front cover was removed and the soil-root interface covered with a prefixed sheet (2.1 mm thickness) of 0.75% (w/v) agar containing ^{32}P. After an uptake period of 24 h the soil-root interface including the agar sheet was covered by X-ray film for autoradiography. The above mentioned methods of plant culture have been criticized because nutrient flow towards the root takes place only in the direction of the plane but not radially from all around the root as would be the case under "natural" conditions (Kraus et al. 1987). To overcome this problem, plants were grown in cylindrical pots containing an additional small cylinder in the centre, which was filled with labelled soil. The top of the inner cylinder was covered with a nylon net. A small maize seedling was transplanted to the experimental device with its primary root forced into a small hole which was bored vertically through the nylon and the first 2–3 cm of the central cylinder. The rest of the roots were allowed to grow in the non-labelled soil outside the central cylinder. At the end of the experimental period the soil core from the inner cylinder was separated and frozen in liquid nitrogen. The whole soil core was cut into disks of about 0.5 cm thickness using a stone cutting machine and these disks were used for autoradiography (Kraus et al. 1987).

Soil Labelling, Preparation and Evaluation of the Radioautogram. Various radioactive tracers have been used for autoradiography, including ^{32}P and ^{33}P (Bhat and Nye 1973; Claassen et al. 1981; Kraus et al. 1987), ^{86}Rb (Claassen et al. 1981), ^{33}K (Claassen and Jungk 1982), ^{45}Ca (Barber and Ozanne 1970; Wilkinson et al. 1968a), ^{99}Mo (Lavy and Barber 1964), ^{65}Zn (Wilkinson et al. 1968b) and ^{35}S (Sanders 1971). The tracer has to be thoroughly mixed with the soil. The initial specific activity needed is dependent on the half life of the isotope and the duration of the experiment, and is normally in the range of 1–10 $\mu Ci\,g^{-1}$ soil. For the radioautogram, X-ray film is brought into close contact with the foil covering the soil and exposed to the radiation e.g. for 4 h or 1 day either at root temperature (Claassen et al. 1981) or in a freezer (Kraus et al. 1987).

For quantitative evaluation of the radioautograms, blackening of the film is measured densitometrically (Claassen et al. 1981; Kraus et al. 1987). For standardization of the densiometer readings, radioautograms have to be prepared from soil samples containing various amounts of tracer. Further details of quantitative autoradiography are given e.g. by Passioura (1972).

13.3 Nutrient Uptake

13.3.1 Principles of Nutrient Uptake

Mechanisms for Nutrient Uptake. In the soil, nutrients are transported to the root surface by massflow and diffusion. Transport of nutrients by these physical processes is restricted within the apoplast of roots by hydrophobic suberin deposits in the endodermis, and in some plant species also in the exodermis. So, even if membrane transport does not occur at the root periphery (root hairs, epidermis), nutrients still have to cross the plasma membranes at the endodermis before they can reach the symplasm of the roots. The plasma membrane is an effective barrier to diffusion of nutrients and is the site where selection and metabolic control of nutrient uptake takes place. Across the plasma membrane between the external solution and the cytoplasm, an electrical potential difference of 100–200 mV (inside negative) exists, which is generated by the activity of the membrane-bound H^+-ATPase, pumping H^+ from the cytoplasm into the apoplasm. The electrochemical potential gradient of H^+, generated by this enzyme, is used for the membrane transport of nutrients via transport proteins. The electric potential drives ions through channels. The activity of both transporters and ion channels is metabolically controlled. The direction of ion flow through an open channel is determined by the electrochemical gradient across the plasma membrane. Therefore, ion channels are very likely to provide the mechanism for ion efflux, i.e. release of ions from the cytoplasm to the external solution. In principle, one can envisage that net uptake of ions into the cell may be controlled by both influx and efflux (Deane-Drummond 1990; White et al. 1992; Lee 1993).

Transporters. At low external concentrations (below 1 mM) the membrane transport of nutrients is mediated mainly by specific high affinity transport proteins (Glass et al. 1992; Maathuis and Sanders 1996; Miller and Smith 1996), which show Michaelis-Menten kinetics, i.e. saturation at high external concentrations. At high external concentrations a second set of quite distinct low affinity transporters come to dominate the overall absorption of ions. Uptake by low affinity transporters often linearly increases with increasing external nutrient concentrations. Thus, depending on the external nutrient concentration in an experimental set up, the activity of different uptake systems may be assessed. In soil, the concentration of most nutrients at the root surface is expected to be low, so that membrane transport is mainly mediated by high affinity uptake systems.

At low nutrient concentrations in the external solution the outermost cell layers of the roots (rhizodermis) absorb most of the nutrients delivered by diffusion in the soil solution (Rufty et al. 1986; Clarkson 1996). At high external concentrations a nutrient-rich solution may enter the apoplast of the root cortex, and all cells of the cortex may contribute to nutrient uptake. Thus, depending on the nutrient concentration in an experimental set up, different tissues or locations within the roots might be assessed for their ability for nutrient uptake.

Metabolic Control. Nutrient transport across plasma membranes of root cells may be modified in the short-term, i.e. within seconds to minutes, by a change in the transmembrane potential which is established by the membrane-bound H^+ ATPase. Rapid changes of the root environmental conditions, e.g. in temperature (Minorsky and Spanswick 1989), or mechanical disturbance of the roots occuring even when roots are gently transferred from one nutrient solution to another (Miller 1981; Bloom and Sukrapanna 1990) may lead to a sudden decrease in nutrient uptake which is presumably caused by a decrease in the transmembrane potential. If the roots are not seriously injured nutrient uptake recovers within minutes.

In the long-term (hours to days), nutrient uptake may also be regulated by the concentrations of mineral nutrients and their metabolites in the roots, or by various shoot signals translocated to the roots in the phloem, including sugars, malate, nutrients and their metabolites (Marschner 1995). Excision of roots stops transfer of theses signals. Therefore, measurements on nutrient uptake in excised roots should be completed before interruption of signalling from the shoot becomes effective, i.e. within minutes to a few hours.

Parameters for Estimating the Ability of Roots to Take Up Nutrients. The ability of roots for nutrient uptake and/or translocation to the shoot can be estimated by various parameters which assess different root characteristics. The actual activity of the transport systems mediating membrane transport of ions from the external solution into the cells may be estimated by measurement of ion influx, i.e. the unidirectional flux of ions from the external solution into the cells. Influx can only be measured under laboratory conditions (see Sect. 13.3.6), but the roots used for influx studies can be sampled from the field (Chapin et al. 1986).

The ability of roots to absorb nutrients from the external solution (soil solution or nutrient solution) may be estimated by measurement of net uptake (net influx) of ions into roots. Net uptake of an ion is the result of mechanisms mediating influx and efflux. It can be determined for roots from field-grown plants in situ (see Sect. 13.3.2) or under laboratory conditions (see Sect. 13.3.3).

If ion influx is mediated by saturable transport mechanisms, the relationship between influx and external concentration obeys Michaelis-Menten kinetics and may be characterized by I_{max} (maximum influx at external ion concentrations saturating the activity of the transport system) and K_m (Michaelis-Menten constant, external concentration where influx = $I_{max}/2$). If the relationship between net uptake and external concentration obeys Michaelis-Menten kinetics, the potential ability of roots for net uptake can also be characterized by I_{max}, K_m, and c_{min}, where c_{min} is the minimum concentration to which roots may deplete the external solution, i.e. the concentration where influx = efflux. The kinetic parameters for influx and net uptake can only be determined under laboratory conditions (see Sect. 13.3.6).

The ability of roots to translocate nutrients to the shoot may be assessed from measurement of net accumulation rates of ions in the shoot (net translocation rate), or from ion fluxes in the exudate of decapitated plants (gross translocation rate). Both net translocation rates and xylem fluxes can be assessed in plants grown in nutrient solution or soil (see Sect. 13.3.4).

13.3.2 Nutrient Uptake into Intact Plants

13.3.2.1 Nutrient Accumulation in Plants Under In Situ Conditions (Williams Formula)

Nutrient accumulation in plants may be determined from plant biomass increment between sequential harvests, and nutrient concentrations in the biomass (see Sect. 13.3.7 for methods to measure nutrient concentrations in plant tissue). Depending on the root parameters which are measured, nutrient accumulation between two harvest dates can be based e.g. on root biomass, root length, or root surface area. According to Williams (1948) net uptake rates (U) can be calculated with the following formula:

$$U = (W_2 - W_1)/(t_2 - t_1) \times \ln(R_2/R_1)/(R_2 - R_1),$$

whereby W_1 and W_2 are the nutrient contents (biomass x concentration) of the plants at times (t) 1 and 2, and R_1 and R_2 are the root parameters (e.g. root biomass, length) measured at t_1 and t_2. For the above formula it is assumed that root growth is exponential between the two harvest dates. For the assumption of linear root growth the formula is modified to

$$U = (W_2 - W_1)/(t_2 - t_1)/(R_2 + R_1)/2.$$

U is a parameter to estimate mean uptake rates of the roots between the harvests. The time interval between sequential harvests ($t_2 - t_1$) may vary from a few days (Engels and Marschner 1992) to several months (Headley et al. 1985). The lower limits of the time interval are given by the need for a significant increment in either biomass or nutrient concentration between two harvest dates. With long intervals between harvests, U, as calculated by the Williams formula, is only a rough parameter for estimating root activity, and may be falsified by (1) biomass losses e.g. through herbivores or leaf fall, (2) nutrient losses e.g. by leaching or retranslocation to the roots and subsequent release into the rhizosphere, and (3) nutrient uptake or release via aerial parts of the plant, which may be considerable for nitrogen and sulfur. Furthermore, it is obvious that increasing the time interval between harvests complicates the assessment of the mean root biomass that has contributed to the observed nutrient uptake, because root growth under field conditions is neither linear nor exponential but is strongly dependent on plant age and environmental conditions in the shoot and root zone. If uptake rates are based not on root biomass but on root length or surface area, additional sources of error may arise because specific root length and root diameter may also vary depending on plant age and soil conditions (e.g. temperature, mechanical impedance, moisture).

In the field it is difficult to obtain data for root biomass. Furthermore, considerable nutrient losses from roots may occur during washing out from soil (Grzebisz et al. 1989). Therefore, often only nutrient accumulation rates in the shoots are considered in the Williams formula, i.e. net translocation rates from the roots to the shoot. In large perennial plant species like trees, it is not easy to gain data for the nutrient accumulation in the above-ground organs of the plant. From data on nutrient accumulation rates in individual plant parts, e.g. young growing leaves, conclusions on root activity can be drawn only if tracers are applied to the soil, because increment of nutrient content in the organs may be the result of both uptake by roots or remobilization from other plant organs, e.g. roots or stems.

The importance of adequate experimental design for the statistical analysis of data was emphasized by Chapin and van Cleve (1989). For example, a comparison of the effect of two treatments on nutrient uptake requires that a block containing sufficient individuals of both treatments for all harvest dates is established. This block constitutes an individual replicate and should be replicated sufficiently to provide the necessary sample size. At each harvest date a plot is randomly selected from within each block, and individuals of the two treatments are harvested from that plot. At the next harvest date another plot is randomly selected from each replicate block etc. An example of an experimental design for four individuals per treatment, tree harvest dates and five individual replicates is given in Fig. 13.4.

In conclusion, growth analysis and calculation of the net uptake rate by the Williams formula estimate the in situ root activity of intact plants growing under specific environmental conditions, as an average of the whole root system and the time interval between sequential harvest dates. The main advantage of this method is that there is no need for any manipulations of the plants or soil surrounding the roots during the uptake study, thus excluding the risk of measuring artefacts. Disadvantages of the method include the poor resolution of nutrient uptake in time and space and, at least for field-grown plants, the difficulty of estimating the root size on which net uptake is to be based. Furthermore, from nutrient accumulation rates in field-grown plants it is not possible to differentiate whether observed variations in nutrient uptake between treatments are due to differences in the ability of the roots to take up nutrients or to modifications of nutrient availability in the rhizosphere.

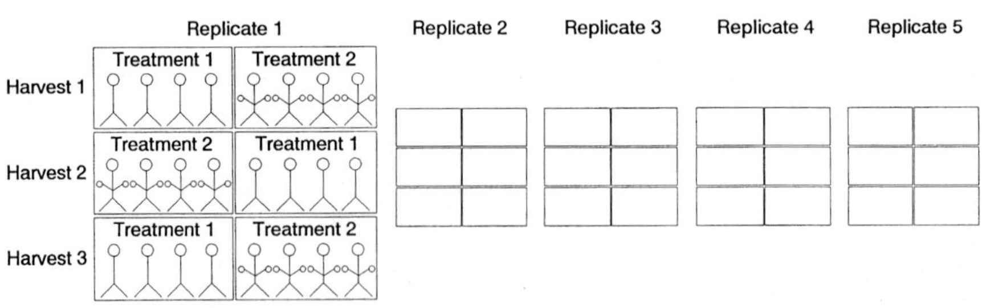

Fig. 13.4. Experimental design for comparing net uptake rates of treatments 1 and 2 during two time intervals, using the Williams formula

13.3.2.2 Methods to Assess the Variation of Nutrient Uptake Within an Intact Root System

Nutrient uptake within a root system can vary depending on e.g. root age (Marschner and Richter 1973; Ernst et al. 1989) or local nutrient availability (Drew and Saker 1975; van Vuuren et al. 1996). In principle, differences between various root regions in their ability to take up nutrients can be assessed using excised roots (see Sect. 13.3.3) or intact root systems. With intact root systems, several methods have been used which may be differentiated according to whether nutrients are supplied to the whole root system or only to specific root regions, whether mineral nutrients or tracers are supplied, whether nutrient or tracer depletion is measured in the root medium or whether tracer accumulation is in whole plants or root segments (Fig. 13.5). With intact root systems, nutrient uptake into various root regions is most commonly assessed by localizing nutrient supply to specific root zones, e.g. by injection of tracers into various soil zones, by applications of nutrient-containing agar on specific root zones, or by forcing the roots into compartmented containers.

Injection of Tracers into Various Soil Zones. The variation in time and space of the relative activity of roots for nutrient uptake can be assessed by injecting tracers into the soil at various dates and depths or distances from the plant, and subsequent measurement of tracer depletion from soil or accumulation within the plant. For tracer injection, holes are usually drilled into the soil with an auger. The application of the tracer (solid or solution) is made through pipes which are inserted into the holes. After tracer applications, the holes are refilled carefully with soil. Further details of tracer application are described by Böhm (1979).

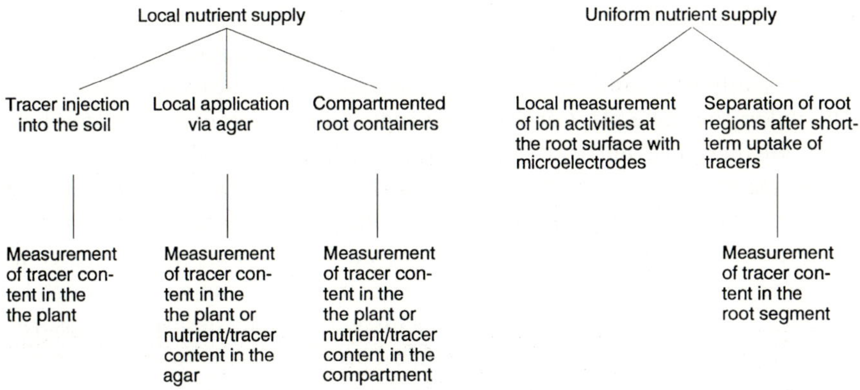

Fig. 13.5. Methods to assess the variation of nutrient uptake within an intact root system

Various tracers have been used, including rare chemical tracers such as Li, Sr, Rb and Br (Fitter 1986; Benjamin et al. 1996), stable isotopes e.g. ^{15}N (McKane and Grigal 1990), or radioisotopes e.g. ^{32}P (Kurien et al. 1992). As mineral nutrients are absorbed by uptake systems which are specific to each nutrient, studies with rare chemical tracers may give an indication of the distribution of living roots rather than of the ability of the roots for uptake of essential minerals. Even with isotopes, the quantitative assessment of the ability of specific root regions for nutrient uptake is tied to several prerequisites: (1) supply of the tracer to the uptake sites is not limited, (2) supply of the tracer is confined to the specific root region, and (3) the specific activity of the tracer at the uptake sites is uniform in the various root regions which are tested (Pearson 1974). To meet these requirements, the amount of applied tracers should be sufficiently high (possibly a slow-release tracer source), the uptake period should be sufficiently short, the tracer should be biologically, chemically and physically inert under the specific soil conditions (e.g. potassium in a non-fixing soil), and the nutrient/tracer application should at least transiently equilibrate differences between various root regions in the rhizospheric concentrations of native nutrients (resulting e.g. from differences between root regions in root activity). From this listing, which is certainly not complete, it becomes obvious that tracer injection techniques under field conditions may at best lead to semi-quantitative answers concerning the ability of different root regions to take up nutrients.

Local Application via Nutrient-Containing Agar. If the roots are readily accessible, e.g. roots growing along a removable plexiglas cover either in rhizoboxes or in the field (see Chap. 8), nutrients or tracers can be applied carefully directly to specific root segments via agar. The uptake ability of the root segment may be assessed by measuring tracer uptake into the plant (Ernst et al. 1989; Brady et al. 1993) or nutrient depletion from the agar (Reidenbach and Horst 1995; Box 13.5). Generally, measuring nutrient depletion from the medium surrounding specific regions of the root system allows the assessment of the uptake ability of different root regions within one plant, and measurements can be repeated in the same plant or even root segment e.g. in course of its development. Furthermore, nutrient concentration in the rooting medium can be a parameter which responds much more sensitively to nutrient uptake than the nutrient concentration in the plant, particularly if nutrients are supplied to small root segments in large plants, but the method is indirect. The change in nutrient concentration in the rooting medium may result not only from uptake by roots but also from physical (gaseous losses, leaching, desorption, adsorption), chemical (precipitation) and biological processes (mineralization, immobilization) in the medium. To assess the importance of the latter processes, which should be of minor importance in nutrient-containing agar, the change in nutrient concentration in the rooting medium should be tested under the experimental conditions in the absence of roots.

Compartmented Containers. The ability of different root zones to take up nutrients may be assessed by placing roots in compartmented containers containing soil or nutrient solution. Various substances have been used to seal the root segments in the compartments, including vaseline (Moritsugu et al. 1993), silicone grease (Siebrecht et al. 1995) and silicon putty (Häussling et al. 1988). Uptake of different root segments can be estimated by measuring the nutrient depletion from the individual compartments (Häussling et al. 1988), or by supply of tracers to one compartment and measuring tracer accumulation in the plant (Drew and Saker 1986).

Various experimental sets have been used (Fig. 13.6) which differ in technical expenditure and thus suitability for field measurements. Marschner et al. (1991) investigated the gradients in nutrient uptake ability along individual roots of spruce in a 60-year-old stand. The roots were gently uncovered from the upper soil horizon and placed into a simple compartmented container with nutrient solution (Fig. 13.6A), in which ion uptake rates from the individual compartments were determined over a 3-day-period from the decrease in solution volume and ion concentration. As a control, solution volume and ion concentration were also measured in a compartment containing no roots. This non-destructive field technique is suitable even for large trees and measurements can be repeated within the same plant or root segment. Under laboratory conditions, root segments can be placed in plastic tubes or compartments in which the root segment is supplied with a continuous flow of nutrient solution during the uptake period, allowing the measurement of tracer accumulation in the plant at constant external concentrations. In the experimental set up shown in Fig. 13.6B, the two compartments adjacent to the root segment under study were supplied with tracer-free solution to flush away tracer ions diffusing through the cortex and thus to prevent contamination of the outer solution with tracers. The detail with which variation of uptake along the root axis or with increasing root age can be studied depends on the length of the segment used for labeling. Harrison-Murray and Clarkson (1973) used small labeling chambers (3 mm diameter) which could be fitted into the gaps between the lateral roots. Thus, nutrient uptake of the main root could be studied with increasing root age without the confounding effect of young laterals (Clarkson 1996).

Separation of Root Regions After Short Periods of Uptake. The variation of nutrient uptake within an intact root system may also be assessed after short exposure of the entire root system to labelled nutrients followed by rapid separation of the various root regions (Canning and Cramer 1958; Yoneyama et al.

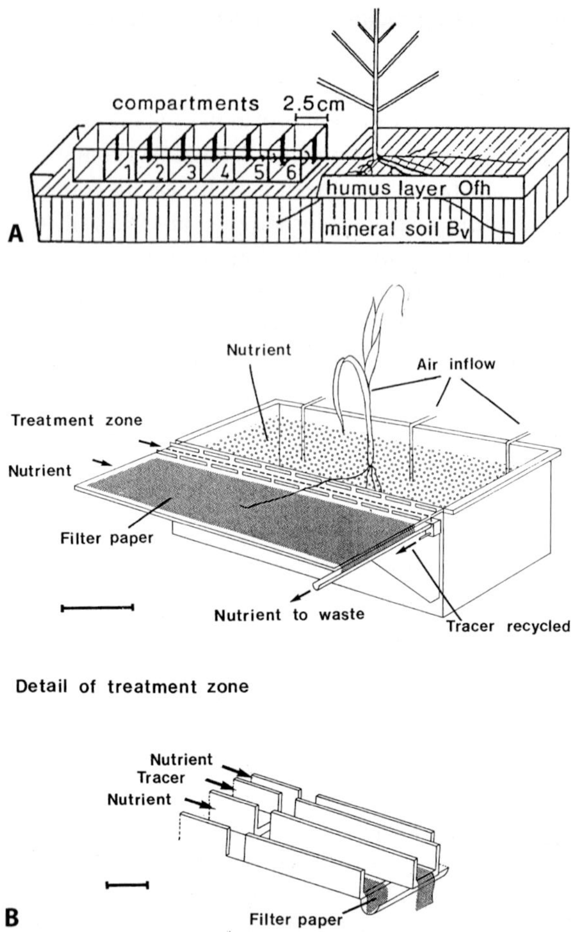

Fig. 13.6. Examples for compartmented containers; for further details see text. **A** Simple compartmented container used for field studies (Häussling et al. 1988, reprinted with permission from Gustav Fischer Verlag). **B** Apparatus for measurement of tracer uptake under controlled environmental conditions. (Drew and Saker 1986; with permission of Oxford University Press)

1975; Lazof et al. 1992; Box 13.5). Exposure to labelled solutions must be sufficiently short to minimize translocation among root regions. Results from short-term time courses of uptake by intact roots indicate that radial ion transport across the root to the xylem can occur within a few minutes (Siddiqi et al. 1991; Lazof et al. 1992), and considerable interregional transport is likely to occur with exposure to labelled nutrients for 30 min or more. In view of the brief exposure of the roots to labelled nutrients (tracers), the ratio of apoplastic to symplastic tracers is high. To remove tracers from the apoplast before analysis, careful rinsing of the roots is necessary, e.g. with a solution containing non-labelled nutrients (see Sect. 13.3.3).

Measurement of Ion Activities at the Root Surface with Microelectrodes. The variation of nutrient uptake along the root axes can also be assessed by mea-

BOX 13.5. Experimental Protocols for Measurement of Nutrient Uptake Along the Root

Measurement of Nutrient Depletion from the External Medium

Using the agar technique, Reidenbach and Horst (1995) measured the ability of different root regions for NO_3^- uptake in maize growing in large rhizoboxes up to maturity. To minimize the disturbance of roots directly before the uptake measurement, non-branched apical regions of roots growing along the removable plexiglass front cover were underlaid with gauze (4 × 4 cm, 40 μm pore width) which allowed nutrient and water transfer from the soil to the root. One day before measurement, the gauze was exchanged with waterproof plastic. The next day a small block of agarose (0.8% weight/volume, 2 × 1 × 0.6 cm, 1 mM KNO_3, 1 mM $CaSO_4$, pH 5) was put on the underlaid root segments and covered with plastic to prevent desiccation. After 8 h, the NO_3^- concentration in the agar blocks was determined by dissolving the blocks for 10 min in water at 90 °C, followed by centrifugation and measurement of NO_3^- concentration in the filtered supernatant with an autoanalyser. The root segments were photographed before agar application and after removal, and the mean root length during the uptake period was estimated from the photographs.

Measurement of Short-term Accumulation Within Root Segments of Intact Plants

Lazof et al. (1992) examined the transport characteristics of six different root regions by measuring the change in accumulation rates during short time courses (<15 min) of $^{15}NO_3^-$ supply to the whole root system. The roots were harvested and separated into regions after 1, 3, 6 or 15 min of ^{15}N supply. The rate of ^{15}N accumulation during the first minute was taken as an estimate of unidirectional influx. The 15-min accumulation rate was taken as the steady state accumulation rate which is the result of both influx and efflux from the external solution, and translocation to or import from other root regions. Within 15 min, accumulation rates of ^{15}N increased in the apical root regions indicating import from other root regions, and decreased in basal root regions indicating translocation to root tip and shoot. The accumulation rates of ^{15}N in the whole plant were constant during the 15 min of ^{15}N supply indicating that efflux was negligible. The results on translocation to and import from other root regions were confirmed with pulse chase experiments where the change in ^{15}N concentrations in different root regions were measured after ^{15}N supply for 6 min followed by a 12 min chase period with a supply of non-labelled NO_3^-.

suring ion activities in the unstirred layer immediately external to the root surface using ion-selective electrodes (Newman et al. 1987; Henriksen et al. 1992). From the measurements of ion activity at several positions in the diffusion boundary layer net fluxes of ions into the root can be estimated (Henriksen et al. 1990, 1992). With this method ion fluxes can be estimated with high resolution in time and space, and with minimal disturbance of the root system. But, the method is only suitable for plants growing in nutrient solution.

13.3.3 Nutrient Uptake into Excised Roots

Principles. The ability of roots to take up nutrients may be estimated from isolated root segments by measuring either nutrient depletion from solutions (Cruz et al. 1995), or nutrient accumulation in the roots (Högberg et al. 1995). As uptake studies with isolated roots have to be confined to a short time duration (see below), the roots are usually supplied with tracers to increase the sensitivity of measurements. For uptake studies roots may be sampled from plants growing in the field (Jackson et al. 1990), or in pots with soil (Matzner and Richards 1996) or nutrient solution (Cruz et al. 1995).

Methodology. In principle, the method involves the following steps: (1) adaptation of the roots to the temperature and pH at which uptake will be measured, (2) transfer to the uptake solution containing the tracer, (3) termination of uptake and removal of tracer from the root surface and free space of the cortex by rinsing the roots in cold (e.g. 4 °C) washing solution, and (4) measurement of tracer content in the roots, or in the uptake and washing solutions. To facilitate the transfer of roots from one solution to another, the roots may be enclosed in "tea bags" (Epstein et al. 1963) or "swimfeeders" (Harrison et al. 1978). Instead of transferring the roots from one solution to another, the solutions may be exchanged e.g. via vacuum withdrawal (Fox et al. 1996). For steps (1) to (3), the solutions must be aerated, and it is recommended to add Ca (final concentration about 0.2 to 0.5 mM) to maintain membrane integrity. The pH may be buffered during the uptake period, e.g. with 10 mM MES, pH 5.5 (Léon et al. 1995) or 5 mM Mes-Tris, pH 5.0 (Fox et al. 1996). The washing solutions (step 3) may contain the non-labelled form of the nutrient being studied at concentrations much higher (10–100 fold) than that in the uptake solutions. Alternatively, the washing solutions may contain $CaCl_2$ for the exchange of ions. Cations like Cu which are tightly bound to exchange sites in the free space may be exchanged with $Pb(NO_3)_2$ (Harrison et al. 1979). The adequacy of the desorption regime must be checked by measuring desorption (tracer content of the washing solution) in relation to the duration or frequency of rinsing. In chilling sensitive plants like cucumber or maize, a rapid decrease in solution temperature during transfer of the roots from the uptake to the washing solution may not only stop nutrient uptake into the root cells but cause substantial efflux from the cells into the washing solution (Minorsky and Spanswick 1989). Therefore, rapid changes in root temperature during washing should be avoided. To test the nutrient loss from the symplast to the solution, nutrient efflux during the washing procedure can be measured in roots from plants which were fed with labelled nutrients during preculture (Lazof et al. 1992).

Root Characteristics Assessed. With excised roots, their ability to take up nutrients may be assessed under the specific experimental conditions provided in the uptake solution (e.g. pH, temperature, nutrient concentration). Because the experimental conditions in the uptake solution and the length of the uptake period can be exactly regulated, the kinetics of net uptake or influx (see Sect. 13.3.6) may also be measured (Chapin et al. 1986). If root samples are taken from specific regions along the root axes, uptake characteristics can be related to root age or distance from the root tip (Cruz et al. 1995). The method has also been used to assess the uptake ability of roots subjected to drought (Matzner and Richards 1996) or a non-homogenous supply of nutrients in the field (Jackson et al. 1990).

Problems. Measuring nutrient uptake in excised roots suffers from two severe problems: (1) root injuries occuring during handling of the roots and (2) physiological changes associated with isolation of roots or root segments from the rest of the plant. Both factors impair nutrient uptake. Therefore, this method is appropriate for comparing relative differences between treatments in the ability of roots to take up nutrients rather than absolute values or actual in situ uptake rates. The extent of root injuries, e.g. loss of fine lateral roots and root hairs or damage of root epidermal cells which are of particular importance for nutrient uptake from soil (Marschner 1995), varies with the experimental system. It may be expected that injuries are more severe in root samples washed from soil cores taken from the field than from plants grown in nutrient solution. But, even gently cutting the roots or transferring roots from one solution to another may decrease nutrient uptake (Bloom and Sukrapanna 1990), presumably because of a transient decrease in the transmembrane electrochemical potential gradient (see Sect. 13.3.1). Pretreatment of the roots e.g. for 30 min in a solution containing Ca (see above, step 1) may lead to a recovery of the roots from this transient decrease in uptake ability. The effect of root injuries on nutrient uptake can be roughly estimated from comparison of the measured data on uptake with data in the literature obtained from intact plants.

Isolation of roots from the rest of the plant may decrease nutrient uptake within hours because of the interruption of both import of sugars and other "shoot signals" via the phloem, and export of nutrients via the xylem (see Sect.

13.3.1). The effect of these factors on nutrient uptake increases with time from excision, and is dependent on the specific nutrient (Bloom and Caldwell 1988). In order to find out the period after which root isolation leads to a marked depression of uptake, it is therefore necessary to make a time study of nutrient uptake with increasing time from excision.

13.3.4 Collection of Xylem Exudate

Principle. Nutrients are translocated from the roots to the shoot in the xylem by massflow with water. In transpiring plants, the massflow in the xylem vessels is driven by the water loss from the plants to the atmosphere, which occurs mainly via the open stomata. Under conditions of low or no transpiration, e.g. during the night or before leaf emergence in spring, or in decapitated plants, the massflow is driven by "root pressure". The permeability of plant membranes to water is much higher than to ions. Therefore, roots behave as osmometers, and "root pressure" may develop as a consequence of metabolically driven nutrient uptake into the roots and release into the xylem vessels, creating a difference in the osmotic potential between the external solution and the xylem sap (Marschner 1995). The amount of nutrients exuded with the xylem sap from detopped plants or individual roots can be taken as parameter to estimate the actual delivery of nutrients from the roots to the shoot. For the estimation of the amount of nutrients exuded, the nutrient concentration in the xylem sap, and the amount of xylem sap which is exuded per unit of time have to be measured.

Collection of Xylem Sap. In field-grown plants bleeding sap has been obtained from the stem cut just above the ground (Engels et al. 1994), or from individual nodal roots (Canny and McCully 1988) and laterals (Jeschke and Pate 1995) after careful uncovering of their distal end from the soil while the rest of the roots were still undisturbed in the soil. To collect the xylem sap a sloping cut can be made and the sap running down to the lower edge of the cut sampled with a micropipette (Canny and McCully 1988). For quantitative analysis of xylem sap, a sleeve of silicone rubber tubing is fitted over the cut surface and sealed at the base e.g. with silicone grease. To reduce bacterial growth during exudation, the cut surface can be cleaned with a paper towel and moistened with a drop of Micropur.

The duration of the collection period should be as short as possible to collect enough xylem sap for nutrient analysis, because translocation of nutrients from the roots to the shoot is increasingly affected by interruption of phloem supply from the shoot with carbohydrates and other "shoot signals" (see Sect. 13.3.1). Exudate sampled during the first 5–10 min after decapitation is often discarded for technical reasons (time needed to fit and seal the rubber tubing), and also because it may contain abnormally high concentrations of nutrients and hormones due to mechanical stress on the stem tissue during cutting or contamination with the contents of severed cells from the cut surface (Else et al. 1994).

The instantaneous content of the xylem vessels of field-grown roots or root segments may be sampled by applying a mild vacuum or pressure to excavated roots (Canny and McCully 1988; Jeschke and Pate 1995). To protect the root segments from drying they should be wrapped e.g. in a damp paper towel. In maize nodal roots to which reduced pressure was applied with a syringe, each 1 cm of bare root yielded about 1 mg of xylem sap (Canny and McCully 1988).

In plants growing in pots or nutrient solution, root xylem sap may be obtained by applying (1) a positive hydrostatic pressure to the whole root system (Else et al. 1994) or (2) a negative hydrostatic pressure (suction) to the cut surface to simulate transpiration (Gil de Carrasco et al. 1994). To pressurize root systems (method 1) the culture vessels including the roots are placed in Scholander-bomb-like vessels in which the pressure can be raised with compressed air (see Fig. 13.7). By varying the hydrostatic pressure, the sap flow rates can be adjusted to those measured in intact plants. Also roots (including the pot) of intact plants can be placed in pressure chambers allowing collection of xylem sap exuding from the cut leaf tips (Passioura and Munns 1984; Jeschke et al. 1996).

Assessment. Collection and analysis of xylem exudate allows an estimation of the actual root activity under field conditions. Compared with the method using the Williams formula (see Sect. 13.3.2.1) the resolution in time of the estimate for root activity is substantially improved. The disadvantages of the method are similar to those for the Williams formula, namely difficulties of estimating the

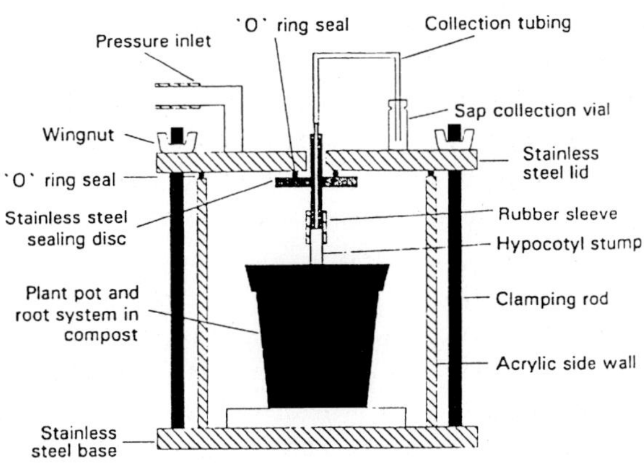

Fig. 13.7. Diagram of vessel used to pressurize root systems to obtain xylem sap. To avoid damage to the 20 mm hypocotyl stump, pressure sealing was made with an "O" ring between the lid and a sealing disc rather than against the hypocotyl itself (Else et al. 1994; with permission of Oxford University Press)

root size in the field, and problems in differentiating between root characteristics associated directly with nutrient uptake and translocation to the shoot and those modifying the nutrient availability in the rhizosphere. Furthermore, several factors have to be considered in the interpretation of xylem exudate analysis. Xylem sap contains shoot-derived nutrients cycled from the shoot to the roots in the phloem and reloaded to the xylem again in the roots (Marschner 1995). Translocation of solutes in the exudate of decapitated plants may substantially differ from that of intact plants because of not only the interruption of the supply of "shoot signals" (see Sect. 13.3.1) but also the differences in water flux rates. However, in contrast to the translocation rates of hormones, those of most nutrients seem to be not very sensitive to the rate of water flux in the xylem either in intact (Tanner and Beevers 1990) or decapitated plants (Else et al. 1995).

13.3.5 Contribution of Mycorrhiza to Nutrient Acquisition

General. Mycorrhizas are symbiotic associations between plant roots and soil fungi. In annual plants and also in fruit trees, the most common type of association is formed by vesicular-arbuscular mycorrhizal (VAM) fungi (also called arbuscular mycorrhizal [AM] fungi by some authors) of the order *Glomales* (*Zygomycotina*). Roots colonized by VAM fungi are characterized by internal fungal structures (hyphae, and often arbuscules and vesicles) and by external hyphae growing away from the root (Fig. 13.8). Spores are formed in the soil on

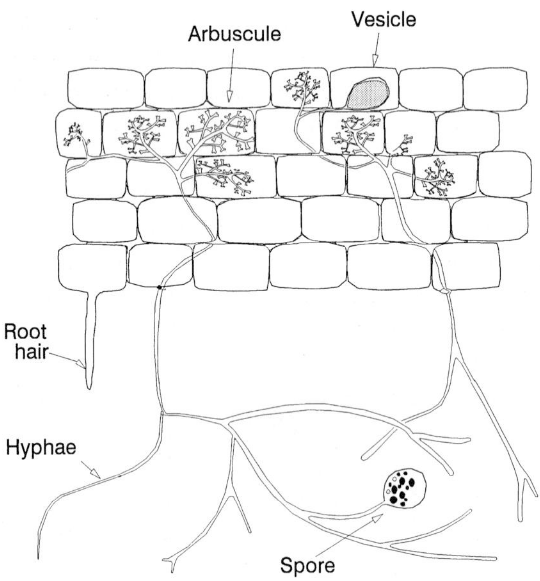

Fig. 13.8. Schematic presentation of internal and external structures of root colonisation with a vesicular-arbuscular mycorrhizal fungus

the external mycelium. The external hyphae are well suited to absorb mineral elements from root-distant soil and transport them to the plant.

Vesicular arbuscular mycorrhizal fungi occur in most terrestrial ecosystems, in agricultural fields, and can colonize most crop and pasture plants. Mycorrhizal fungi have probably been associated with roots since the occurrence of vascular land plants more than 300 million years ago (Taylor et al. 1995; Simon 1996) and only a few plant species (for example, lupins or some *Brassica* species) have developed a resistance against mycorrhizal colonization. In addition to a contribution to plant nutrient uptake, mycorrhizal fungi also modify the carbon economy, the water use, and the pathogen susceptibility of plants. In particular, non-nutritional effects of mycorrhizal fungi in disease resistance have recently received much interest (St-Arnaud et al. 1997). A summary of the current view on the role of mycorrhizas in plant growth is given by Smith and Read (1997) and Marschner (1995).

In this chapter, only a small part of the methodology used in mycorrhiza research is described. For a more detailed description, the reader is referred to, for example Schenck (1982), Sieverding (1991), Norris et al. (1992), and Brundrett et al. (1994).

Methods for Determining Mycorrhizal Colonization. It is not possible to see root colonization with VAM fungi without staining and microscopic observation. However, the staining process does not require expensive instrumentation. For assessment of mycorrhizal colonization, roots must be separated from soil and the tissue cleared with 10% potassium hydroxide. After washing and acidification of the root sample, fungal structures in the roots are stained by Trypan blue, acid fuchsin, chlorazole black or other compounds (Brundrett et al. 1984; Grace and Stribley 1991). A staining of active fungal parts only is also possible (Schaffer and Peterson 1993). After staining, in most cases the percentage of the total root length colonised by mycorrhizal fungi is determined (Giovanetti and Mosse 1980).

Spores of mycorrhizal fungi can be isolated from soil by a wet sieving and decanting technique (Gerdemann and Nicolson 1963). Also, the length of the external mycorrhizal mycelium in the soil can be determined after washing the hyphae from soil and subsequent staining (Hamel et al. 1990). In a recent development, fatty acid profiles are used to determine mycorrhizal biomass in the soil and within roots (Olsson et al. 1997). In the future, DNA-based methods will be used to characterise the distribution of different fungal populations in the soil and in the roots.

Methods for Determining the Effect of Mycorrhizal Fungi on Nutrient Uptake. Because of the ubiquity of mycorrhizal colonisation in most plant species, usually non-mycorrhizal individuals of these species cannot be found in fields or natural environments. The colonisation of roots by VAM fungi can be reduced by use of pesticides, but side-effects of these pesticides on other micro-organisms and on nutrient availability in soil have to be accounted for (Weber et al. 1993).

Most of the current knowledge on the contribution of mycorrhiza to plant nutrient uptake and growth is derived from pot experiments with partially sterilised (mycorrhiza-free) soils, soil mixtures or other growing substrates. Heating a soil to 70–80 °C for one day in most cases eradicates all mycorrhizal propagules. In appropriate treatments, mycorrhizal colonization of roots is then re-established by using fungal inoculum. Such mycorrhizal inoculum can be produced by isolating spores from soil and subsequent multiplication in a plant pot culture, but is also commercially available. In the experiments, nutrient uptake, water use, biomass production and other properties of mycorrhizal plants are compared with those of non-mycorrhizal plants, and the difference is attributed to mycorrhizal colonization.

Up to now, it has not been possible to grow mycorrhizal fungi on nutrient media without a host plant. Thus, little is known about the physiological requirements of different VAM fungal isolates. The use of in vitro cultures of mycorrhizal fungi with Ri T-DNA-transformed plant roots (Chabot et al. 1992) will be a means of investigating functions of mycorrhizal hyphae and roots independent of shoot influences.

To specifically study the extraradical mycelium, nets with a fine mesh size of 10–30 μm (required mesh size depends on minimum diameter of roots of plant species grown in the experiment) can be used to spatially separate root- and hyphae-growing zones in soil (George et al. 1995). Such meshes can be obtained from commercial suppliers of bolting cloth. The method requires the construction of pots where a net can be used to separate the growth substrate. With even less investment, nutrient depletion in root-distant soil can be studied in conventional pot experiments with the help of a soil tube accessible to hyphae only (Fig. 13.9).

When soil is sieved and stored, the indigenous mycorrhizal flora may become inactive. Also, mycorrhizal fungi rarely become established on nutrient solutions, except when the special requirements of the symbiosis are taken into account (Hawkins and George 1997). Thus, up to now many studies in root physiology are inadvertently carried out with non-mycorrhizal root systems, although field-grown roots are usually colonized by mycorrhizal fungi.

13.3.6 Measurement of Influx, Efflux, and Uptake Kinetics

Estimation of Influx. Net uptake of mineral nutrients into the roots is the result of simultaneous influx i.e. nutrient flux from the external solution into the cells,

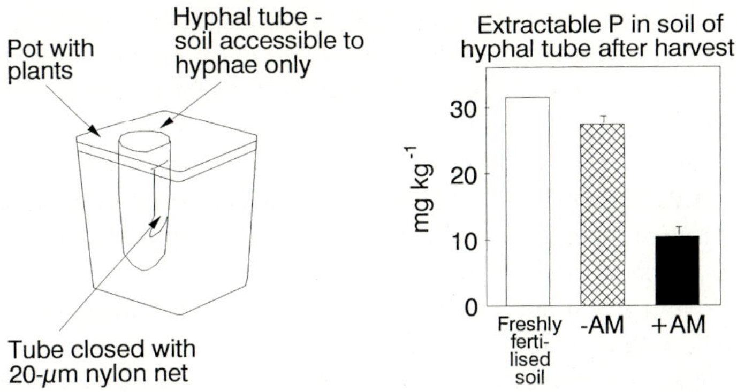

Fig. 13.9. Concentration of extractable phosphorus in the soil of a hyphal tube at the onset of the experiment (freshly fertilized soil) and after 40 days of growth of non-mycorrhizal (–AM) and mycorrhizal (+AM) ryegrass plants (*right side*). Uptake of phosphorus from the soil tube by mycorrhizal hyphae is indicated by depletion of extractable phosphorus in the soil enclosed by a 20μm nylon net (*left side*; E. George et al. unpubl.)

and efflux i.e. nutrient flux from the cells into the external solution. For separate estimation of influx, accumulation of labelled nutrients (tracers) in the plants or depletion from the external solution may be measured in time durations short enough to exclude substantial efflux of labelled nutrients into the external solution. Because efflux of labelled nutrients from the cytoplasmic pool of the roots to the external solution will lead to a time-dependent underestimation of influx, the measurements have to be done well within the time it takes to fill the cytoplasmic pools (Clarkson et al. 1996). Typical periods for the measurement of tracer accumulation in plant tissue which take into consideration the probable turnover times of cytoplasmic pools are 5 min for $^{45}Ca^{2+}$ (White et al. 1992), 5–15 min for ^{15}N-nitrate and -ammonium (Muller et al. 1995; Clarkson et al. 1996), and 10 to 25 min for ^{32}P-phosphate, ^{33}P-phosphate, ^{86}Rb, and ^{35}S-sulfate (Elliott et al. 1984; Siddiqi and Glass 1987; Clarkson et al. 1989; Lee 1993).

The lower time limits of influx studies are given by the sensitivity of measurements of tracer accumulation in the plant tissue or tracer depletion from the external solution. This is no problem with radioactive tracers which can be measured very sensitively against a very low background in the plant tissue, but possibly is with ^{15}N, particularly if the kinetic parameters of influx are to be determined, because this requires measurement of influx from low external concentrations (see below). The discrimination of ^{15}N absorbed from the uptake solution from that already in the tissue should be improved if the natural "background" is reduced by supply of ^{15}N-depleted N sources during preculture. However, it has been suggested that the trade-off for improved discrimination would be decreased precision, because the ^{15}N "background" of the ^{15}N-depleted plants may become more variable. This would result from differences in the contribution of N sources with natural ^{15}N abundance to the N supply of the plant, such as seed N reserves, NH_3 or NO_x absorbed via the stomata, or symbiotically fixed N_2 (Clarkson et al. 1996). For laboratories sufficiently close to a cyclotron, the radioactive isotope ^{13}N provides a sensitive tracer for investigating fluxes of nitrogen in plant material over short periods (Lee and Drew 1986; Kronzucker et al. 1995; Clarkson et al. 1996; see also Chap. 12).

It is obvious that influx studies over short, well-defined periods of time can only be done with roots to which the tracers are supplied in a nutrient solution. The procedure for removal of tracers from the apoplast at the end of the influx study is described in Section 13.3.3.

Estimation of Efflux. Efflux of nutrients from the roots to the external solution may be estimated either as the difference between influx and net uptake (Macduff and Jackson 1992; Lee 1993), or directly by measurement of nutrient release from the roots into the external solution (Lee and Clarkson 1986; Aslam et al. 1994). For the latter method the roots are transferred from an uptake solution which contains the nutrient or tracer to be studied to an efflux solution which is identical to the uptake solution except for lack of the specific nutrient or tracer (Aslam et al. 1994). Possibly, efflux is affected by the nutrient concentrations in the external solution, and related to the concurrently occuring influx. To study efflux from roots into which concurrently influx takes place, the plants are loaded with a tracer for time periods long enough to ensure isotopic equilibrium, or near equilibrium, even in the slowest-exchanging compartments under investigation. Preloading periods reported are variable, e.g. 15 h for loading of ^{15}N into soybean seedlings (Muller et al. 1995), or 2 days to label the P pools of maize with ^{32}P (Elliott et al. 1984). After preloading and short rinsing of the roots to remove apoplastic tracers (see Sect. 13.3.3) plants are transferred to efflux solutions which are identical in composition to the preloading solutions except that the nutrient being studied is supplied in its natural form (non radioactive, natural mass), and tracer accumulation is measured in this efflux solution. As tracers released from the roots by efflux may be recaptured by the roots again, leading to an underestimation of efflux, the periods for efflux should be short (10–20 min) and the volume of the efflux solution large. However, tracer efflux is not linear with time even within short time periods after transfer of plants to the efflux solution (Lee and Clarkson 1986; Muller et al. 1995). This is because the cytoplasmic tracer pool which is the immediate source for efflux is continuously diluted e.g. by nutrient influx and tracer fluxes to the external solution and the xylem. This becomes a problem particularly for nitrate and ammonium where the cytoplasmic pool size in the roots is small (about 0.25 to 1.25 $\mu mol\, g^{-1}$ root fresh weight, calculated on the basis of the cytoplasm occupying 5% of the total root volume and nitrate concentrations in the cytoplasm ranging between 5 and 25 mM) and influx is high (about 10 to 20 μmoles h^{-1} g^{-1} root fresh weight; Muller et al. 1995). In maize, influx and efflux of P in roots of intact plants have been simultaneously measured by preloading of the plants with ^{32}P for 48 h. Then the plants were transferred to a solution containing ^{33}P for 10 min to measure ^{32}P accumulation in the external solution (efflux) and ^{33}P accumulation in the plants (influx; Elliott et al. 1984).

Estimation of Uptake Kinetics. As a rule, ion uptake by plant roots shows Michaelis-Menten kinetics at least in the concentration range below 1 mM that is typical for soil solutions (Fig. 13.10). Thus, the potential ability of roots for ion uptake may be characterized by I_{max}, K_m, and c_{min} (see Sect. 13.3.1). Methods to determine the transport kinetics of nutrients by plant roots involve time-course recording of ion/tracer depletion from the external solution (Claassen and Barber 1974; Drew et al. 1984) or tracer accumulation in the plant (Siddiqi et al. 1990). As the rate of tracer accumulation in the plants has to be measured at different external concentrations, this method needs a separate set of plants for each external concentration to be tested. For the estimation of V_{max} and K_m of enzymatic reactions it has been recommended to measure enzyme activities (corresponding to influx) at least at 10 different concentrations ranging from a factor 10 smaller to a factor 10 higher than K_m (Bisswanger 1994).

In contrast to tracer accumulation in the plants, nutrient/tracer depletion from the external solution can be measured with one set of plants over the whole range from saturating concentrations to c_{min}. The use of tracers either for the measurements of depletion from the external solution or accumulation in the plants, allows the estimation of I_{max} and K_m for the unidirectional influx from the external solution into the roots, if the measurements at each discrete external concentration are confined to a time period short enough to minimize

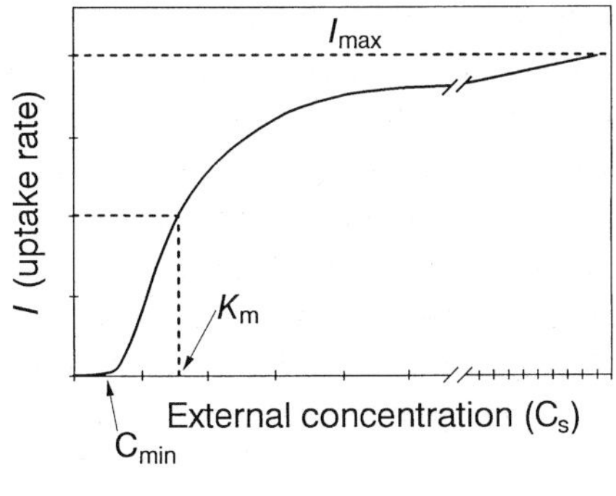

Fig. 13.10. Schematic presentation of the relationships between net uptake rates (net influx = influx − efflux) of ions and their external concentrations; c_{min} = net uptake zero (influx = efflux); K_m = Michaelis-Menten constant (external ion concentration allowing half maximal net uptake)

tracer and nutrient efflux (see above). It is obvious that for the estimation of kinetic parameters for influx several sets of plants have to be used regardless of whether tracer depletion from the external solution or tracer accumulation in the plant is measured.

Measurement of Rate of Depletion of the External Solution. Depletion of the external solution by roots can be measured at several discrete concentrations ranging from saturating concentrations to minimum concentrations by transferring the plants from one solution to another. In each solution normally a 20–25% decrease in ion concentration is allowed so that a change in concentration can be reliably detected. This means that only one data point is generated over a 20–25% change in concentration, while the uptake rate could have been continually changing throughout that interval (Goyal and Huffaker 1986). Therefore, ideally uptake kinetics should be determined by measuring nutrient depletion continuously over the whole concentration range. In this case, the volume of the vessels used for the upake study should be coordinated with the root size of the experimental plants (number of plants and root fresh weight per plant) to reach c_{min} in reasonable time (6–10 h). This helps to avoid adaptational responses of the roots to the diminishing external concentrations such as decreases in c_{min} and K_m. Drew et al. (1984) suggest volume ratios of root: solution between 1:10 and 1:40.

In the first minutes after transfer of the roots to the "depletion solution", a large efflux may occur, particularly if the plants were precultured at high nutrient concentrations. Therefore, before the depletion rate is measured, the plants may be transferred for a few hours to a nutrient solution deficient in the nutrient to be studied (Drew et al. 1984). After transferring the plants to the depletion solution, samples are taken at regular intervals for tracer or nutrient analysis. Goyal and Huffaker (1986) designed a fully automated microcomputer-based system for monitoring nitrate, nitrite and ammonium depletion from an uptake solution. In this system samples of the nutrient solution taken over time intervals of 2.5 min were directly injected into an HPLC, allowing measurement of these nutrients in real-time. Alternatively, samples can also be taken using a peristaltic pump (Swiader and Freiji 1996) or by hand. Samples are collected until 30–60 min after the nutrient concentration in the depletion solution has reached a minimum. Vigorous aeration of the nutrient solution is recommended to promote mixing. Water losses from the nutrient solution caused by evapotranspiration and sampling can be replaced by adding water according to volume changes of the solution in the pots or weight changes of the whole experimental unit (pots, solution, plants). Alternatively, water losses are determined from the difference in weight between start and end of the uptake study. Then the water losses as well as the nutrient losses from sampling can be considered for the calculations of uptake rates.

At the end of the experiment the plants are harvested for determination of the root parameter on which uptake rates are to be based (e.g. weight, length, surface area). When depletion is measured over time periods long enough to have an appreciable root growth, root size (R) at time (t) can be calculated using the expression

$$R = R_0 e^{kt},$$

where R_0 is the root size (weight, length, surface area) at the start of the experiment and k is a constant describing the rate of root size increase with time (Claassen and Barber 1974).

Calculations. The net influx (In) between two sampling dates is calculated from the volume of the nutrient solution (V), the decrease in its concentration per unit time ($\triangle C/\triangle t$) and the root size (R e.g. root fresh weight or root length or root surface area) according to the following formula:

$$\text{In} = -V/R\ \triangle C/\triangle t.$$

To calculate the kinetic parameters for net uptake, the experimental data are fitted to the Michaelis-Menten equation as described by Claassen and Barber (1974):

$$\text{In} = [I_{\max} \times C/(K_m + C)] - C_{\min},$$

where In is the net influx, expressed as μmol h^{-1} (unit root)$^{-1}$, $I_{\max}$ is the maximum net influx at saturating concentrations, C is the nutrient concentration in the nutrient solution (μmol l^{-1}) and K_m the Michaelis constant. $C_{\min}$ may be directly obtained from the final concentration reached in the "depletion solution" (see above). $I_{\max}$ and K_m may be estimated by the least-squares method, or after linear transformation of the Michaelis-Menten equation, allowing the results to be plotted as points on a straight line (Cornish-Bowden 1995; see Table 13.1).

Due to experimental errors, frequent sampling is associated with large variability in uptake rates if calculated for an interval of two subsequent samplings. To overcome this problem, the data obtained from the depletion experiment can be fit to a series of cubic equations using a cubic spline computer program (Swiader and Freiji 1996). The equation of the cubic spline is differentiated to obtain In as a function of time. $I_{\max}$ is calculated from the negative slopes of the linear portion of each curve. K_m is calculated from the derivative equations of the cubic spline by computing the solution concentration at which $I_{\max}/2$ occurs (Swiader and Freiji 1996). Further methods for the calculation of the kinetic parameters are described by Claassen and Barber (1974) and Drew et al. (1984).

Table 13.1. Linear transformations of the Michaelis-Menten equation

Name of the plot	Y	X	Intercepts with		Slope of the line
			X-axis	Y-axis	
Eadie-Hofstee	In	In/C	I_{max}/K_m	I_{max}	$-K_m$
Lineweaver-Burk	1/In	1/C	$-1/K_m$	$1/I_{max}$	K_m/I_{max}
Hanes	C/In	C	$-K_m$	K_m/I_{max}	$1/I_{max}$

13.3.7 Nutrient Analysis

The methods of sampling, handling and analyzing plant tissue for mineral nutrients are well described in handbooks of plant analysis (Jones and Case 1990).

Sample Preparation. If the plant samples are contaminated with dust or pesticide residues, they should be washed e.g. with a weak (0.1–0.3%) detergent solution, followed by rinsing in deionized water. Only fresh plant samples should be washed and the washing procedure should be done quickly and gently, to avoid leaching of mineral nutrients from the plant tissue. Washing is always recommended if the tissue is to be analyzed for Fe, Al or Si (Jones and Case 1990).

After washing, the samples are dried at 65 °C to constant weight to stop biological activity of the tissue which can reduce the plant mass by respiration. Drying at temperatures much lower than 65 °C, e.g. 50 °C, may leave too much moisture in the tissue and lead to enzymatic degradation of the tissue during storage. Drying at temperatures much higher than 65 °C, e.g. 105 °C, may lead to thermal decomposition of the tissue, particularly if ground tissue samples are re-dried before nutrient analysis (Steyn 1959).

If only a subsample of the collected plant tissue is used for nutrient analysis, the dried material has to be ground. The purpose of grinding is to obtain subsamples of uniform composition. The particle size which is necessary to obtain uniformity is mainly dependent on the size of the subsample needed for analysis. Grinding with a commercial coffee mill (particle size reduction to pass a 20-mesh screen) may be sufficient if 500 mg aliquots of plant tissue are analyzed, whereas fine pulverization of the tissue is necessary if only 3–10 mg aliquots are analyzed e.g. for ^{15}N in a mass-spectrometer. During grinding, the plant tissue may be contaminated mainly with Al, Cu, Fe or Zn (Jones and Case 1990). So, when these elements have to be analyzed, the tissue should be ground with an agate mill (Steyn 1959). The ground material should be dried again for

24 h at 65 °C to remove any moisture added during grinding, and then stored e.g. in sealed bottles until analysis.

Chemical Analysis. The dried and ground material may be directly used for analysis of S in automated sulfur analyzers and N in automated Dumas instruments or by Kjeldahl digestion. Methods for the determination of other elements involve destruction of the tissue's organic component to convert the elements to a soluble form for analysis. There are essentially two decomposition procedures, wet acid digestion and high temperature dry oxidation, frequently referred to as wet and dry ashing, respectively (for a detailed discussion of wet and dry ashing see Jones and Case 1990). Wet ashing is not recommended for B, since this element can be partially lost by volatilization during wet oxidation (Feldman 1961).

For dry ashing, aliquots of the sample are weighed into commercially available ashing vessels or small glass bottles in which contamination of the sample e.g. by B or Na should be excluded. The plant tissue is ashed for 6–8 h in a muffle furnace at 450–500 °C. Complete oxidation of the plant tissue is indicated by the "white" appearance of the remaining ash. If a clean white ash is not obtained after 8 h, re-ashing for 2 h after addition of ashing aids such as HNO_3 or 10% H_2SO_4 is necessary. Thereafter, the ash is dissolved in dilute acid (HCl or HNO_3). The clear solution is ready for elemental assay, with or without further dilution.

Wet ashing is the destruction of organic matter by high temperature acid digestion. The common acids used are H_2SO_4, HNO_3, and $HClO_4$, usually in some combination of two or all three. The wet oxidation is frequently done in commercially available sealed digestion tubes inserted into ports of a temperature-controlled digestion block. Alternatively, plant tissue can also be digested in a mixture of HNO_3 and 30% H_2O_2 heated by ultraviolet radiation (Ogner 1983) or in a microwave oven (White and Douthit 1985).

Instruments for Analysis. Colorimetric methods involving measurement in spectrophotometers are used for B (Basson et al. 1969) and P (Gericke and Kurmies 1952; Murphy and Riley 1962). The most commonly used methods for determination of the other elements involve flame emission spectrometry, atomic absorption spectrometry, and plasma spectrometry (Table 13.2). Ions may be measured by HPLC with anion and cation exchange columns (Weiß 1991; Table 13.2).

13.4 Concluding Remarks

In future, the development of more precise analytical techniques in the laboratory will allow collection of smaller samples, and thus, sampling of roots or

Table 13.2. Instruments for the measurement of mineral nutrients

Instrument	Mineral nutrient
Automated sulfur analyzer	S
Automated Dumas instrument	N
Spectrophotometer	P, Mo
Flame photometer	Ca, K, Na, Cu, Fe, Mn
Atomic absorption spectrometer	Ca, Mg, Cu, Fe, Mn, Zn
Inductively coupled plasma emission spectrometer (ICP)	P, K, Ca, Mg, Cu, Fe, Mn, Zn, B
HPLC (ion chromatography)	SO_4^{2-}, HPO_4^{2-}, NO_3^-, MoO_4^{2-}, cations

compounds in the immediate vicinity of roots with higher resolution in space and time. The critical step in obtaining information about nutrient acquisition capacity of roots will remain sampling itself. There is no generally accepted and recommended method for measuring root capacity for nutrient acquisition. Artificial growing conditions e.g. in rhizoboxes or along root windows may allow sampling and in situ measurements of root activity with minimal disturbance of the roots during sampling, but possibly do not exactly reflect root behaviour under natural conditions. Natural growing conditions often require strong disturbance of the root/soil system prior to or during measurement of acquisition capacity, or give little resolution of root activity in time and space. Thus, the best way to collect data on root activity will strongly depend on the specific research question to be solved.

Further Reading

Brundrett MC, Melville L, Peterson RL (1994) Practical methods in mycorrhizal research. Mycologue Publications, Waterloo, Canada

Clarkson DT (1996) Root structure and sites of ion uptake. In: Waisel Y, Eshel A, Kafkafi U (eds) Plant roots: the hidden half. 2nd edn. Marcel Dekker, New York, pp 483–510

Grayston SJ, Vaughan D, Jones D (1996) Rhizosphere carbon flow in trees, in comparison with annual plants: the importance of root exudation and its impact on microbial activity and nutrient availability. Appl Soil Ecol 5: 29–56

Jones JB Jr, Case VW (1990) Sampling, handling and analyzing plant tissue samples. In: Westerman RL (ed) Soil testing and plant analysis. Soil Science Society of America, Madison, pp 389–427

Marschner H (1995) Mineral nutrition of higher plants. Academic Press, London

References

Alef K (1995) Sterilization of soil and inhibition of microbial activity. In: Alef K, Nannipieri P (eds) Methods in applied soil microbiology and biochemistry. Academic Press, London, pp 52–54

Amann C, Amberger A (1989) Phosphorus efficiency of buckwheat (*Fagopyrum esculentum*). Z Pflanzenernähr Bodenkd 152: 181–189

Armstrong J, Armstrong W, Beckett PM (1992) Venturi- and humidity-induced pressure flows enhance rhizome aeration and rhizosphere oxidation. New Phytol 120: 197–207

Aslam M, Travis RL, Huffaker RC (1994) Stimulation of nitrate and nitrite efflux by ammonium in barley (*Hordeum vulgare* L.) seedlings. Plant Physiol 106: 1293–1301

Azaizeh HA, Marschner H, Römheld V, Wittenmayer L (1995) Effects of vesicular-arbuscular fungus and other soil microorganisms on growth, mineral nutrient acquisition and root exudation of soil-grown maize plants. Mycorrhiza 5: 321–327

Barber DA, Gunn KB (1974) The effect of mechanical forces on the exudation of organic substances by the roots of cereal plants grown under sterile conditions. New Phytol 73: 39–45

Barber SA (1984) Soil nutrient bioavailability. A mechanistic approach. John Wiley, New York

Barber SA, Ozanne PG (1970) Autoradiographic evidence for the differential effect of four plant species in altering the Ca content of the rhizosphere soil. Soil Sci Soc Am Proc 34: 635–637

Bartlett EM, Lewis DH (1973) Surface phosphatase activity of mycorrhizal roots of beech. Soil Biol Biochem 5: 249–257

Basson WD, Bohne RG, Stanton DA (1969) An automated procedure for the determination of boron in plant tissue. Analyst 94: 1135–1141

Benjamin LR, Peach L, van Woerden IC, Oppelaar A (1996) A technique to estimate the radial extent of active mineral absorption by individual plants in carrot stands. J Exp Bot 47: 687–692

Bhat KKS, Nye PH (1973) Diffusion of phosphate to the plant roots in soil. Quantitative autoradiography of the depletion zone. Plant Soil 38: 161–175

Bieleski RL, Johnson PN (1972) The external location of phosphatase activity in phosphorus-deficient *Spirodela oligorrhiza*. Aust J Biol Sci 25: 707–720

Bienfait HF, Bino RJ, van der Bliek AM, Duivenvoorden JR, Fontaine JM (1983) Characterization of ferric reducing activity in roots of Fe-deficient *Phaseolus vulgaris*. Physiol Plant 59: 196–202

Bisswanger H (1994) Enzymkinetik. Theorie und Methoden. VCH-Verlag Weinheim

Bloom AJ, Caldwell RM (1988) Root excision decreases nutrient absorption and gas fluxes. Plant Physiol 87: 794–796

Bloom AJ, Sukrapanna SS (1990) Effects of exposure to ammonium and transplant shock upon the induction of nitrate absorption. Plant Physiol 94: 85–90

Boero G, Thien S (1979) Phosphatase activity and phosphorus availability in the rhizosphere of corn roots. In: Harley JL, Russell S (eds) Soil-root interface. Academic Press, London

Boeuf-Tremblay V, Plantureux S, Guckert A (1995) Influence of mechanical impedance on root exudation of maize seedlings at two developmental stages. Plant Soil 172: 279–287

Böhm W (1979) Methods of studying root systems. Springer, Berlin Heidelberg New York

Brady DJ, Gregory PJ, Fillery IRP (1993) The contribution of different regions of the seminal roots of wheat to uptake of nitrate from soil. Plant Soil 155/156: 155–158

Brown JC, Ambler JE (1974) Iron stress response in tomato (*Lycopersicon esculentum*). 1. Sites of Fe reduction, absorption and transport. Physiol Plant 31: 221–224

Brundrett MC, Piché Y, Peterson RL (1984) A new method for observing the morphology of vesicular-arbuscular mycorrhizae. Can J Bot 62: 2128–2134

Brundrett MC, Melville L, Peterson RL (1994) Practical methods in mycorrhizal research. Mycologue Publications, Waterloo, Canada

Cairney JWG, Ashford AE (1989) Reducing activity at the root surface in *Eucalyptus pilularis-Pisolithus tinctorius* ectomycorrhizas. Aust J Plant Physiol 16: 99–105

Cakmak I, Marschner H (1988) Increase in membrane permeability and exudation in roots of zinc deficient plants. J Plant Physiol 132: 356–361

Canning RE, Kramer PJ (1958) Salt absorption and accumulation in various regions of roots. Am J Bot 45: 378–382

Canny MJ, McCully ME (1988) The xylem sap of maize roots: its collection, composition and formation. Aust J Plant Physiol 15: 557–566

Chabot S, Bécard G, Piché Y (1992) Life cycle of *Glomus intraradix* in root organ culture. Mycologia 84: 315–321

Chang CW, Bandurski RS (1964) Exocellular enzymes of corn roots. Plant Physiol 39: 60–64

Chapin FS III, Van Cleve K (1989) Approaches to studying nutrient uptake, use and loss in plants. In: Pearcy RW, Ehleringer JR, Mooney HA, Rundel PW (eds) Plant physiological ecology. Field methods and instrumentation. Chapman and Hall, London, pp 185–207

Chapin FS III, Van Cleeve K, Tyron PR (1986) Relationship of ion absorption to growth rate in taiga trees. Oecologia 69: 238–242

Claassen N, Barber SA (1974) A method for characterizing the relation between nutrient concentration and flux into roots of intact plants. Plant Physiol 54: 564–568

Claassen N, Jungk A (1982) Kaliumdynamik im wurzelnahen Boden in Beziehung zur Kaliumaufnahme von Maispflanzen. Z Pflanzenernähr Bodenkd 145: 513–525

Claassen N, Hendriks L, Jungk A (1981) Erfassung der Mineralstoffverteilung im wurzelnahen Boden durch Autoradiographie. Z Pflanzenernähr Bodenkd 144: 306–316

Clarkson DT (1985) Factors affecting mineral nutrient acquisition by plants. Annu Rev Plant Physiol 36: 77–115

Clarkson DT (1996) Root structure and sites of ion uptake. In: Waisel Y, Eshel A, Kafkafi U (eds) Plant roots: the hidden half, 2nd edn. Marcel Dekker, New York, pp 483–470

Clarkson DT, Saker LR, Purves JV (1989) Depression of nitrate and ammonium transport in barley plants with diminished sulphate status. Evidence of co-regulation of nitrogen and sulphate intake. J Exp Bot 40: 953–963

Clarkson DT, Gojon A, Saker LR, Wiersema PK, Purves JV, Tillard P, Arnold GM, Paans AJM, Vaalburg W, Stulen I (1996) Nitrate and ammonium influxes in soybean (*Glycine max*) roots: direct comparison of ^{13}N and ^{15}N tracing. Plant Cell Environ 19: 859–868

Cornish-Bowden A (1995) Analysis of enzyme kinetic data. Oxford University Press, Oxford

Cruz C, Lips SH, Martins-Louçao MA (1995) Uptake regions of inorganic nitrogen in roots of carob seedlings. Physiol Plant 95: 167–175

Darrah PR (1996) Rhizodeposition under ambient and elevated CO_2 levels. Plant Soil 187: 265–275

Deane-Drummond CE (1990) Biochemical and biophysical aspects of nitrate uptake and its regulation. In: YP Abrol (ed) Nitrogen in higher plants. Research Studies Press, Taunton, UK, 37 pp

Delhaize E, Ryan PR (1995) Aluminum toxicity and tolerance in plants. Plant Physiol 107: 315–321

Delhaize E, Ryan PR, Randall PJ (1993) Aluminum tolerance in wheat (*Triticum aestivum* L.). II. Aluminum-stimulated excretion of malic acid from root apices. Plant Physiol 103: 695–702

Dinkelaker B, Marschner H (1992) In vivo demonstration of acid phosphatase activity in the rhizosphere of soil-grown plants. Plant Soil 144: 199–205

Dinkelaker B, Römheld V, Marschner H (1989) Citric acid excretion and precipitation of calcium citrate in the rhizosphere of white lupin (*Lupinus albus* L.). Plant Cell Environ 12: 285–292

Dinkelaker B, Hahn G, Römheld V, Wolf GA, Marschner H (1993a) Non-destructive methods for demonstrating chemical changes in the rhizosphere I. Description of methods. Plant Soil 155/156: 67–70

Dinkelaker B, Hahn G, Marschner H (1993b) Non-destructive methods for demonstrating chemical changes in the rhizosphere II. Application of methods. Plant Soil 155/156: 71–74

Dinkelaker B, Hengeler C, Neumann G, Eltrop L, Marschner H (1996) Root exudates and mobilization of nutrients. In: Rennenberg H, Eschrich W (eds) Trees – contributions to modern tree physiology. SPB Academic Publishing, Amsterdam, pp 3–14

Dinkelaker B, Hengeler C, Neumann G, Eltrop L, Marschner H (1997) Root exudates and mobilization of nutrients. In: Rennenberg H, Eschrich W, Ziegler H (eds) Trees – contributions to modern tree physiology, Backhuys, Leiden, The Netherlands, pp 441–451

Drew MC, Saker LR (1975) Nutrient supply and the growth of the seminal root system in barley. II. Localized compensatory increases in lateral root growth and rates of nitrate uptake when nitrate supply is restricted to only part of the root system. J Exp Bot 29: 79–90

Drew MC, Saker LR (1986) Ion transport to the xylem in aerenchymatous roots of *Zea mays* L. J Exp Bot 37: 22–33

Drew MC, Saker LR, Barber SA, Jenkins W (1984) Changes in the kinetics of phosphate and potassium absorption in nutrient-deficient barley roots measured by a solution-depletion technique. Planta 160: 490–499

Elliott GC, Lynch J, Läuchli A (1984) Influx and efflux of P in roots of intact maize plants. Plant Physiol 76: 336–341

Else MA, Davies WJ, Whitford PN, Hall KC, Jackson MB (1994) Concentrations of abscisic acid and other solutes in xylem sap from root systems of tomato and castor-oil plants are distorted by wounding and variable sap flow rates. J Exp Bot 45: 317–323

Else MA, Hall KC, Arnold GM, Davies WJ, Jackson MB (1995) Export of abscisic acid, 1-aminocyclopropane-1-carboxylic acid, phosphate, and nitrate from roots to shoots of flooded tomato plants. Accounting for effects of xylem sap flow rate on concentration and delivery. Plant Physiol 107: 377–384

Engels C, Marschner H (1992) Root to shoot translocation of macronutrients in relation to shoot demand in maize (*Zea mays* L.) grown at different root zone temperatures. Z Pflanzenernähr Bodenkd 155: 121–128

Engels C, Buerkert B, Marschner H (1994) Nitrogen and sugar concentrations in the xylem exudate of field-grown maize at different growth stages and levels of nitrogen fertilization. Eur J Agron 3: 197–204

Epstein E, Schmid WE, Rains DW (1963) Significance and technique of short-term experiments on solute absorption by plant tissue. Plant Cell Physiol 4: 79–84

Ernst M, Römheld V, Marschner H (1989) Estimation of phosphorus uptake capacity by different zones of the primary root of soil-grown maize (*Zea mays* L.). Z Pflanzenernähr Bodenkd 152: 21–25

Farr E, Vaidynathan V, Nye PH (1969) Measurement of ionic concentration gradients in soil near roots. Soil Sci 107: 385–391

Feldman C (1961) Evaporation of boron from acid solutions and residues. Anal Chem 33: 1916–1920

Felipe MR, Pozuelo JM, Cintas AM (1979) Acid phosphatase localization at the surface of young corn roots. Agrochimica 23: 143–150

Fitter AH (1986) Spatial and temporal patterns of root activity in a species-rich alluvial grassland. Oecologia 69: 594–599

Flessa H, Fischer WR (1992) Redoxprozesse in der Rhizosphäre von Land- und Sumpfpflanzen. Z Pflanzenernähr Bodenkd 155: 373–378

Fox TC, Shaff JE, Grusak MA, Norvell WA, Chen Y, Chaney RL, Kochian LV (1996) Direct measurement of ^{59}Fe-labeled Fe^{2+} influx in roots of pea using a chelator buffer system to control free Fe^{2+} in solution. Plant Physiol 111: 93–100

Gahoonia TS (1993) Influence of root-induced pH on the solubility of soil aluminium in the rhizosphere. Plant Soil 149: 289–291

Gahoonia TS, Nielsen NE (1991) A method to study rhizosphere processes in thin soil layers of different proximity to roots. Plant Soil 135: 143–148

Gahoonia TS, Nielsen NE (1992) Control of pH at soil-root interface. Plant Soil 140: 49–54

Gahoonia TS, Claassen N, Jungk A (1992) Mobilization of phosphorus in different soils by ryegrass supplied with ammonium and nitrate. Plant Soil 143: 241–248

George E, Marschner H, Jakobsen I (1995) Role of arbuscular mycorrhizal fungi in uptake of phosphorus and nitrogen from soil. Crit Rev Biotechnol 15: 257–270

George E, Gorgus E, Schmeisser A, Marschner H (1996) A method to measure nutrient uptake from soil by mycorrhizal hyphae. In: Azcon-Aguilar C, Barea JM. Mycorrhizas in integrated systems from genes to plant development. Office for Official Publications of the European Community, Luxembourg, pp 535–538

Gerdemann JW, Nicolson TH (1963) Spores of mycorrhizal *Endogone* species extracted from soil by wet-sieving and decanting. Trans Br Mycol Soc 46: 235–244

Gericke S, Kurmies B (1952) Die colorimetrische Phosphorsäurebestimmung mit Ammonium-Vanadat-Molybdat und ihre Anwendung in der Pflanzenanalyse. Z Pflanzenernähr Düng Bodenkd 59: 235–247

Gerke J, Römer W, Jungk A (1994) The excretion of citric and malic acid by proteoid roots of *Lupinus albus* L.: effects on soil solution concentrations of phosphate, iron, and aluminium in the proteoid rhizosphere samples of an oxisol and a luvisol. Z Pflanzenernähr Bodenkd 157: 289–294

Gil de Carrasco C, Guzman M, Lorente FA, Urrestarazu M (1994) Xylem sap extraction: a method. Commun Soil Sci Plant Anal 25: 1829–1839

Giovanetti M, Mosse B (1980) An evaluation of techniques for measuring vesicular-arbuscular mycorrhizal infection in roots. New Phytol 84: 489–500

Glass ADM, Shaff JE, Kochian LV (1992) Studies of the uptake of nitrate in barley IV. Electrophysiology. Plant Physiol 93: 456–463

Gollany HT, Schumacher TE (1993) Combined use of colorimetric and microelectrode methods for evaluating rhizosphere pH. Plant Soil 154: 151–159

Göttlein A, Hell U, Blasek R (1996) A system for microscale tensiometry and lysimetry. Geoderma 69: 147–156

Goyal SS, Huffaker RC (1986) A novel approach and a fully automated microcomputer-based system to study kinetics of NO_3^-, NO_2^-, and NH_4^+ transport simultaneously by intact wheat seedlings. Plant Cell Environ 9: 209–215

Grace C, Stribley DP (1991) A safer procedure for routine staining of vesicular-arbuscular mycorrhizal fungi. Mycol Res 95: 1160–1162

Grayston SJ, Vaughan D, Jones D (1996) Rhizosphere carbon flow in trees, in comparison with annual plants: the importance of root exudation and its impact on microbial activity and nutrient availability. Appl Soil Ecol 5: 29–56

Grierson PF (1992) Organic acids in the rhizosphere of *Banksia integrifolia* L. Plant Soil 144: 259–265

Grzebisz W, Floris J, van Noordwijk M (1989) Loss of dry matter and cell contents from fibrous roots of sugar beet due to sampling, storage and washing. Plant Soil 113: 53–57

Hamel C, Fyles H, Smith DL (1990) Measurement of development of endomycorrhizal mycelium using three different vital stains. New Phytol 115: 297–302

Harrison SJ, Lepp NW, Phipps DA (1978) Uptake of copper by excised roots. I. A modified experimental technique for measuring ion uptake by excised roots, and its application in determining uptake characteristics of "free" copper ions in excised *Hordeum* roots. Z Pflanzenphysiol 90: 443–450

Harrison SJ, Lepp NW, Phipps DA (1979) Uptake of copper by excised roots. II. Copper desorption from the free space. Z Pflanzenphysiol 94: 27–34

Harrison-Murray RS, Clarkson DT (1973) Relationships between structural development and the absorption of ions by the root system of *Cucurbita pepo*. Planta 114: 1–16

Häussling M, Leisen E, Marschner H, Römheld V (1985) An improved method for non-destructive measurements of the pH at the root-soil interface (rhizosphere). J Plant Physiol 117: 371–375

Häussling M, Jorns CA, Lehmbecker G, Hecht-Buchholz Ch, Marschner H (1988) Ion and water uptake in relation to root development in Norway spruce [*Picea abies* (L.) Karst.]. J Plant Physiol 133: 486–491

Hawkins H-J, George E (1997) Hydroponic culture of the mycorrhizal fungus *Glomus messeae* with *Linum usitatissimum* L., *Sorghum bicolor* L. and *Triticum aestivum* L. Plant Soil 196: 143–149

Headley AD, Callaghan TV, Lee JA (1985) The phosphorus economy of the evergreen tundra plant, *Lycopodium annotinum*. Oikos 45: 235–245

Helal HM, Sauerbeck D (1983) A method to study turnover processes in soil layers of different proximity to roots. Soil Biol Biochem 15: 223–225

Hendriks L, Jungk A (1981) Erfassung der Mineralstoffverteilung in Wurzelnähe durch getrennte Analyse von Rhizo- und Restboden. Z Pflanzenernähr Bodenkd 144: 276–282

Henriksen GH, Bloom AJ, Spanswick RM (1990) Measurement of net fluxes of ammonium and nitrate at the surface of barley roots using ion-selective microelectrodes. Plant Physiol 93: 271–280

Henriksen GH, Raman DR, Walker LP, Spanswick RM (1992) Measurement of net fluxes of ammonium and nitrate at the surface of barley roots using ion-selective microelectrodes. II. Patterns of uptake along the root axis and evaluation of the microelectrode flux estimation technique. Plant Physiol 99: 734–747

Hether NH, Olsen RA, Jackson LL (1984) Chemical identification of iron reductants exuded by plant roots. J Plant Nutr 7: 667–676

Hodge A, Grayston SI, Ord BG (1996) A novel method for characterisation and quantification of plant root exudates. Plant Soil 184: 97–104

Hoffland E, Findenegg GR, Nelemans JA (1989) Solubilization of rock phosphate by rape. II. Local root exudation of organic acids as a response to P-starvation. Plant Soil 113: 161–165

Högberg P, Jensen P, Näsholm T, Ohlsson H (1995) Uptake of ^{24}Mg by excised pine roots: a preliminary study. Plant Soil 172: 323–326

Horst WJ, Wagner A, Marschner H (1982) Mucilage protects root meristems from aluminium injury. Z Pflanzenphysiol 105: 435–444

Horst WJ, Asher CJ, Cakmak I, Szulkiewicz P, Wissemeier AH (1992) Short-term responses of soybean roots to aluminium. J Plant Physiol 140: 174–178

Hülster A, Marschner H (1994) PCDD/PCDF-Transfer in Zuchini und Tomaten. Veröff PAÖ 8: 579–589

Jackson RB, Manwaring JH, Caldwell MM (1990) Rapid physiological adjustment of roots to localized soil enrichment. Nature 344: 58–60

Jaillard B, Ruiz L, Arvieu IC (1996) pH mapping in transparent gel using color videodensitometry. Plant Soil 183: 85–95

Janzen HH (1990) Deposition of nitrogen into rhizosphere by wheat root. Soil Biol Biochem 22: 1155–1160

Jeschke WD, Pate JS (1995) Mineral nutrition and transport in xylem and phloem of *Banksia prionotes* (Proteaceae), a tree with dimorphic root morphology. J Exp Bot 46: 895–905

Jeschke WD, Klagges S, Bhatti AS (1996) Collection and composition of xylem sap and root structure in two halophytic species. Plant Soil 172: 97–106

Johnson LF, Curl EA (1972) Control of soil environment. In: Johnson LF, Curl EA (eds) Methods for research on the ecology of soil-borne plant pathogens. Burgess Publishing Company, Minneapolis, pp 82–91

Johnson JF, Allan DL, Vance CP, Weiblen G (1996) Root carbon dioxide fixation by phosphorus-deficient *Lupinus albus*. Plant Physiol 112: 19–30

Jones DL, Darrah PR (1992) Re-sorption of organic compounds by roots of *Zea mays* L. and its consequences in the rhizosphere. I. Resorption of ^{14}C labelled glucose, mannose and citric acid. Plant Soil 143: 259–266

Jones DL, Darrah PR (1993) Re-sorption of organic compounds by roots of *Zea mays* L. and its consequences in the rhizosphere II. Plant Soil 153: 47–59

Jones DL, Darrah PR (1994) Amino-acid influx at the soil-root interface of *Zea mays* L. and its implications in the rhizosphere. Plant Soil 163: 1–12

Jones DL, Darrah PR (1995) Influx and efflux of organic acids across the soil-root interface of *Zea mays* L. and its implications in rhizosphere C flow. Plant Soil 173: 103–109

Jones JB Jr, Case VW (1990) Sampling, handling, and analyzing plant tissue samples. In: Westerman RL (ed) Soil testing and plant analysis. Soil Science Society of America, Madison, pp 389–427

Kape R, Wex K, Parniske M, Görge E, Wetzel A, Werner D (1992) Legume root metabolites and VA-mycorrhiza development. J Plant Physiol 141: 54–60

Karpov EA, Potapov NG (1975) Reducing activity of the root surfaces in corn plants in connection with its differentiation and absorptive capacity. Fiziol Rast 22: 298–304

Keith H, Oades IM, Martin IK (1986) Input of carbon to soil from wheat plants. Soil Biol Biochem 18: 445–449

Kraus M, Fusseder A, Beck E (1987) In situ determination of the phosphate gradient around a root by radioautography of frozen soil sections. Plant Soil 97: 407–418

Kronzucker HJ, Siddiqi MY, Glass ADM (1995) Compartmentation and flux characteristics of ammonium in spruce. Planta 196: 691–698

Kuchenbuch R, Jungk A (1982) A method for determining concentration profiles at soil-root interface by thin slicing rhizosphere soil. Plant Soil 68: 391–394

Kuchenbuch R, Jungk A (1984) Wirkung der Kaliumdüngung auf die Kaliumverfügbarkeit in der Rhizosphäre von Raps. Z Pflanzenernähr Bodenkd 147: 435–448

Kurien S, Goswami AM, Deb DL (1992) Root activity of two citrus rootstocks assessed using radiotracer techniques. J Hortic Sci 67: 87–94

Lavy TL, Barber SA (1964) Movement of molybdenum in soil and its effect on availability to the plants. Soil Sci Soc Am Proc 28: 93–97

Lazof DB, Rufty TW Jr, Redinbaugh MG (1992) Localization of nitrate absorption and translocation within morphological regions of the corn root. Plant Physiol 100: 1251–1258

Lee RB (1993) Control of net uptake of nutrients by regulation of influx in barley plants recovering from nutrient deficiency. Ann Bot 72: 225–230

Lee RB, Clarkson DT (1986) Nitrogen-13 studies of nitrate fluxes in barley roots. I. Compartmental analysis from measurements of ^{13}N efflux. J Exp Bot 37: 1753–1767

Lee RB, Drew MC (1986) Nitrogen-13 studies of nitrate fluxes in barley roots. II. Effect of plant N-status on the kinetic parameters of nitrate influx. J Exp Bot 37: 1768–1779

Leiser AT (1968) A mucilaginous root sheat in *Ericaceae*. Am J Bot 55: 391–398

Léon M, Lainé P, Ourry A, Boucaud J (1995) Increased uptake of native soil nitrogen by roots of *Lolium multiflorum* Lam. after nitrogen fertilization is explained by a stimulation of the uptake process itself. Plant Soil 173: 197–203

Li XL, George E, Marschner H (1991) Phosphorus depletion and pH decrease at the root-soil and hyphae-soil interfaces of VA mycorrhizal white clover fertilized with ammonium. New Phytol 119: 397–404

Lipton DS, Blanchar RW, Blevins DG (1987) Citrate, malate, and succinate concentration in exudates from P-sufficient and P-stressed *Medicago sativa* L. seedlings. Plant Physiol 85: 315–317

Maathuis FJM, Sanders D (1996) Mechanisms of potassium absorption by higher plant roots. Physiol Plant 96: 158–168

Macduff JH, Jackson SB (1992) Influx and efflux of nitrate and ammonium in Italian ryegrass and white clover roots: comparisons between effects of darkness and defoliation. J Exp Bot 43: 525–535

Marschner H (1995) Mineral nutrition of higher plants. Academic Press, London

Marschner H, Richter C (1973) Akkumulation und Translokation von K^+, Na^+ und Ca^{2+} bei Angebot zu einzelnen Wurzelzonen von Maiskeimpflanzen. Z Pflanzenernähr Bodenkd 135: 1–15

Marschner H, Römheld V (1983) In vivo measurement of root-induced pH changes at the soil-root interface: effect of plant species and nitrogen source. Z Pflanzenphysiol 111: 241–251

Marschner H, Römheld V, Ossenberg-Neuhaus H (1982) Rapid method for measuring changes in pH and reducing processes along roots of intact plants. Z Pflanzenphysiol 105: 407–416

Marschner H, Römheld V, Kissel M (1986) Different strategies in higher plants in mobilization and uptake of iron. J Plant Nutr 9: 695–713

Marschner H, Römheld V, Kissel M (1987) Localization of phytosiderophore release and iron uptake along intact barley roots. Physiol Plant 71: 157–162

Marschner H, Häussling M, George E (1991) Ammonium and nitrate uptake rates and rhizosphere pH in non-mycorrhizal roots of Norway spruce [*Picea abies* (L.) Karst.]. Trees 5: 14–21

Matzner SL, Richards JH (1996) Sagebrush (*Artemisia tridentata* Nutt.) roots maintain nutrient uptake capacity under water stress. J Exp Bot 47: 1045–1056

McKane RB, Grigal DF (1990) Spatiotemporal differences in ^{15}N uptake and the organization of an old-field plant community. Ecology 71: 1126–1132

Meharg AA, Kilham K (1991) A novel method of quantifying root exudation in the presence of soil microflora. Plant Soil 133: 111–116

Meharg AA, Kilham K (1995) Loss of exudates from the roots of perennial ryegrass inoculated with a range of micro-organisms. Plant Soil 170: 345–349

Miller AJ, Smith SJ (1996) Nitrate transport and compartmentation in cereal root cells. J Exp Bot 47: 843–854

Miller DM (1981) Pressure-flow characteristics of the roots of *Zea mays*. Plant Soil 63: 15–18

Minorsky PV, Spanswick RM (1989) Electrophysiological evidence for a role for calcium in temperature sensing by roots of cucumber seedlings. Plant Cell Environ 12: 137–143

Miyasaka SC, Buta JG, Howell RK, Foy CD (1991) Mechanism of aluminum tolerance in snapbeans. Root exudation of citric acid. Plant Physiol 96: 737–743

Morel JL, Mench M, Guckert A (1986) Measurement of Pb^{+}, Cu^{2+} and Cd^{2+} binding with mucilage exudates from maize (*Zea mays* L.) roots. Biol Fertil Soils 2: 29–34

Moritsugu M, Shibasaka M, Kawasaki T (1993) Where is the most important and efficient site for absorption and translocation of cations in excised barley roots? Soil Sci Plant Nutr 39: 299–307

Muller B, Tillard P, Touraine B (1995) Nitrate fluxes in soybean seedling roots and their response to amino acids: an approach using ^{15}N. Plant Cell Environ 18: 1267–1279

Murphy J, Riley JP (1962) A modified single solution method for the determination of phosphate in natural water. Anal Chim Acta 27: 31–36

Neumann G, Dinkelaker B, Marschner H (1995) Kurzzeitige Abgabe organischer Säuren aus Proteoidwurzeln von *Hakea undulata* (*Proteaceae*). In: Merbach W (ed) Pflanzliche Stoffaufnahme und mikrobielle Wechselwirkungen in der Rhizosphäre. BG Teubner, Stuttgart, pp 129–136

Newman IA, Kochian LV, Grusak MA, Lucas WJ (1987) Fluxes of H^{+} and K^{+} in corn roots: characterization and stoichiometries using microelectrodes. Plant Physiol 84: 1177–1184

Norris JR, Read DJ, Varma AK (1992) Methods in microbiology, vol 24. Academic Press, London

Ogner G (1983) Digestion of plants and organic soils using nitric acid, hydrogen peroxide and UV radiation. Commun Soil Sci Plant Anal 14: 936–943

Ohwaki Y, Hirata H (1990) Phosphorus absorption by chickpea (*Cicer arietinum*) as affected by VA mycorrhiza and carboxylic acids in root exudates. In: van Beusichem ML (ed) Plant nutrition – physiology and applications. Kluwer, Amsterdam, pp 171–177

Ohwaki Y, Hirata H (1992) Differences in carboxylic acid exudation among P-starved leguminous crops in relation to carboxylic acid contents in plant tissues and phospholipid levels in roots. Soil Sci Plant Nutr 38: 235–243

Olsson PA, Bååth E, Jakobson I (1997) Phosphorus effects on the mycelium and storage structures of an arbuscular mycorrhizal fungus as studied in the soil and roots by analysis of fatty acid signatures. Appl Environ Microbiol 63: 3531–3538

Papavizas GC, Davey CB (1961) Extent and nature of the rhizosphere of lupinus. Plant Soil 14: 215–236

Passioura JB (1972) Quantitative autoradiography in the presence of crossfire. In: Lüttge U (ed) Microradioautography and Electron Probe Analysis, Springer, Berlin Heidelberg New York, pp 49–59

Passioura JB, Munns R (1984) Hydraulic resistance of plants. II. Effects of rooting medium, and time of day, in barley and lupin. Aust J Plant Physiol 11: 341–350

Pearson RW (1974) Significance of rooting pattern to crop production and some problems of root research. In: Carson EW (ed) The Plant Root and Its Environment. The University Press of Virginia, Charlottesville, pp 247–270

Pellet DM, Grunes DL, Kochian LV (1995) Organic acid exudation as an aluminum-tolerance mechanism in maize (*Zea mays* L.). Planta 196: 788–795

Petersen W, Böttger M (1991) Contribution of organic acids to the acidification of the rhizosphere of maize seedlings. Plant Soil 132: 159–163

Powlson DS, Jenkinson DS (1976a) The effects of biocidal treatments on metabolism in soil. I. Fumigation with chloroform. Soil Biol Biochem 8: 167–177

Powlson DS, Jenkinson DS (1976b) The effects of biocidal treatments on metabolism in soil. II. Gamma irradiation, autoclaving. Soil Biol Biochem 8: 179–188

Prikryl Z, Vancura V (1980) Root exudates of plants. VI. Wheat exudation as dependent on growth, concentration gradient of exudates and the presence of bacteria. Plant Soil 57: 69–83

Reidenbach G, Horst WJ (1995) Bedeutung verschiedener Wurzelzonen für die Nitrataufnahmerate bei Mais (*Zea mays* L.). VDLUFA-Schriftenr 40: 121–124

Reining E, Merbach W, Knof G (1995) ^{15}N distribution in wheat and chemical fractionation of root-borne ^{15}N in the soil. Isotopes Environ Health Stud 31: 345–349

Riley D, Barber SA (1971) Effect of ammonium and nitrate fertilization on phosphorus uptake as related to root induced pH changes at the root-soil interface. Soil Sci Soc Am Proc 35: 301–306

Römheld V (1991) The role of phytosiderophores in acquisition of iron and other micronutrients in graminaceous species: an ecological approach. Plant Soil 130: 127–134

Rufty TW Jr, Thomas RF, Remmler JL, Campbell WH, Volk RJ (1986) Intercellular localization of nitrate reductase in roots. Plant Physiol 82: 675–680

Sakai H, Tadano T (1993) Characteristics of response of acid phosphatase secreted by the roots of several crops to various conditions in the growth media. Soil Sci Plant Nutr 39: 437–444

Sanders FE (1971) Effect of root and soil properties on the uptake of nutrients by competing roots. D Phil Thesis, Oxford, England

Schaffer GF, Peterson RL (1993) Modifications to clearing methods used in combination with vital staining of roots colonized with vesicular-arbuscular mycorrhizal fungi. Mycorrhiza 4: 29–35

Schaller G, Fischer WR (1985) pH-Änderungen in der Rhizosphäre von Mais und Erdnußwurzeln. Z Pflanzenernähr Bodenkd 148: 306–320

Schenk NC (1982) Methods and principles of mycorrhizal research. The American Phytopathological Society, St Paul, Minnesota

Schönwitz R, Ziegler H (1982) Exudation of water soluble vitamins and some carbohydrates by intact roots of maize seedlings (*Zea mays* L.) into a mineral nutrient solution. Z Pflanzenphysiol 107: 7–14

Schwab SM, Menge JA, Leonard RT (1983) Quantitative and qualitative effects of phosphorus on extracts and exudates of sudangrass roots in relation to vesicular-arbuscular mycorrhiza formation. Plant Physiol 73: 761–765

Seggewiss B, Jungk A (1988) Einfluss der Kaliumdynamik im wurzelnahen Boden auf die Magnesiumaufnahme von Pflanzen. Z Pflanzenernähr Bodenkd 151: 91–96

Shepherd T, Davies HV (1994a) Effect of exogenous amino acids, glucose and citric acid on the patterns of short-term accumulation and loss of amino acids in the root-zone of sand-cultured forage rape (*Brassica napus* L.). Plant Soil 158: 111–118

Shepherd T, Davies HV (1994b) Patterns of short-term amino acid accumulation and loss in the root-zone of liquid cultured forage rape (*Brassica napus* L.). Plant Soil 158: 99–109

Siddiqi MY, Glass ADM (1987) Regulation of K^+ influx in barley: evidence for a direct control of influx by K^+ concentration of root cells. J Exp Bot 38: 935–947

Siddiqi MY, Glass ADM, Ruth TJ, Rufty TW Jr (1990) Studies of the uptake of nitrate in barley. I. Kinetics of $^{13}NO_3^-$ influx. Plant Physiol 93: 1426–1432

Siddiqi MY, Glass ADM, Ruth TJ (1991) Studies of the uptake of nitrate in barley. III. Compartmentation of NO_3^-. J Exp Bot 42: 1455–1463

Siebrecht S, Mäck G, Tischner R (1995) Function and contribution of the root tip in the induction of NO_3^- uptake along the barley root axis. J Exp Bot 46: 1669–1676

Sieverding E (1991) Vesicular-arbuscular mycorrhiza management in tropical agriculture. Deutsche Gesellschaft für Technische Zusammenarbeit (GTZ), Eschborn, Germany

Sihna BK, Singh NT (1976) Salt distribution around roots of wheat under different transpiration rates. Plant Soil 44: 141–147

Simon L (1996) Phylogeny of the *Glomales* – deciphering the past to understand the present. New Phytol 133: 95–101

Smart DR, Ferro A, Ritchie K, Bugbee BG (1995) On the use of antibiotics to reduce rhizoplane microbial populations in root physiology and ecology investigations. Physiol Plant 95: 533–540

Smiley RW (1974) Rhizosphere pH as influenced by plants, soils and nitrogen fertilizers. Soil Sci Soc Am Proc 38: 795–799

Smith SE, Read DJ (1997) Mycorrhizal symbiosis, 2nd edn. Academic Press, San Diego

Starkey RL (1931) Some influences of the development of higher plants upon the micro-organisms in the soil: IV. Influences of proximity to roots on abundance and activity of micro-organisms. Soil Sci 32: 367–393

St-Arnaud M, Hamel C, Vimard B, Caron M, Fortin JA (1997) Inhibition of *Fusarium oxysporum* f. sp. dianthi in the non-VAM species *Dianthus caryophyllus* by co-culture with *Tagetes patula* companion plants colonized by *Glomus intraradices*. Can J Bot 75: 998–1005

Steyn WJA (1959) Leaf analysis. Errors involved in the preparative phase. J Agric Food Chem 7: 344–348

Swiader JM, Freiji FG (1996) Characterizing nitrate uptake in lettuce using very-sensitive ion chromatography. J Plant Nutr 19: 15–27

Tadano T, Sakai H (1991) Secretion of acid phosphatase by the roots of several crop species under phosphorus-deficient conditions. Soil Sci Plant Nutr 37: 129–140

Tagaki S, Nomoto K, Takemoto T (1984) Physiological aspect of muginieic acid, a possible phytosiderophore of graminaceous plants. J Plant Nutr 7: 469–477

Tang C-S, Young C-C (1982) Collection and identification of allelopathic compounds from the undisturbed root system of Bigalte Limpograss (*Hemarthria altissima*). Plant Physiol 69: 155–160

Tanner W, Beevers H (1990) Does transpiration have an essential function in long-distance transport in plants? Plant Cell Environ 13: 745–750

Tarafdar JC, Jungk A (1987) Phosphatase activity in the rhizosphere and its relation to the depletion of soil organic phosphorus. Biol Fertil Soils 3: 199–204

Taylor TN, Remy W, Hass H, Kerp H (1995) Fossil arbuscular mycorrhizae from the early Devonian. Mycologia 87: 560–573

Trolldenier G (1988) Visualisation of oxidizing power of rice roots and of possible participation of bacteria in iron deposition. Z Pflanzenernähr Bodenkd 151: 117–121

Uren NC (1981) Chemical reduction of an insoluble higher oxide of manganese by plant roots. J Plant Nutr 4: 64–71

Uren NC, Reisenauer HM (1988) The role of root exudates in nutrient aquisition. Adv Plant Nutr 3: 79–114

Van Vuuren MMI, Robinson D, Griffiths BS (1996) Nutrient inflow and root proliferation during the exploitation of a temporally and spatially discrete source of nitrogen in soil. Plant Soil 178: 185–192

Vaughan V, Cheshire MV, Ord BG (1994) Exudation of peroxidase from roots of *Festuca rubra* and its effects on exuded phenolic acids. Plant Soil 160: 153–155

von Wirén N, Mori S, Marschner H, Römheld V (1994) Iron inefficiency in maize mutant Ys1 (*Zea mays* L. cv Yellow-Stripe) is caused by a defect in uptake of iron phytosiderophores. Plant Physiol 106: 71–77

von Wirén N, Römheld V, Shioiri T, Marschner H (1995) Competition between micro-organisms and roots of barley and sorghum for iron accumulated in the root apoplasm. New Phytol 130: 511–521

Weber E, Saxena MC, George E, Marschner H (1993) Effect of vesicular-arbuscular mycorrhiza on vegetative growth and harvest index of chickpea grown in northern Syria. Field Crops Res 32: 115–128

Weiß J (1991) Ionenchromatographie, 2nd edn. VCH Verlagsgesellschaft, Weinheim

White PJ, Banfield J, Diaz M (1992) Unidirectional Ca^{2+} fluxes in roots of rye (*Secale cereale* L.). A comparison of excised roots with roots of intact plants. J Exp Bot 43: 1061–1074

White RT Jr, Douthit GE (1985) Use of microwave oven and nitric acid-hydrogen peroxide digestion to prepare botanical materials for elemental analysis. J/Assoc Off Anal Chem 68: 766–769

Wilkinson HF, Loneragen JF, Quirk JP (1968a) Calcium supply to plant roots. Science 161: 1245–1246

Wilkinson HF, Loneragen JF, Quirk JP (1968b) The movement of zinc to plant roots. Soil Sci Soc Am Proc 32: 831–833

Williams RF (1948) The effects of phosphorus supply on the rates of intake of phosphorus and nitrogen and upon certain aspects of phosphorus metabolism in gramineous plants. Aust J Sci Res B1: 333–361

Yoneyama T, Komamura K, Kumazawa K (1975) Nitrogen transport in intact corn roots. Soil Sci Plant Nutr 21: 371–377

CHAPTER 14

Water Uptake

J.E. Fernández[1], B.E. Clothier[2], and M. van Noordwijk[3]

[1] Instituto de Recursos Naturales y Agrobiología de Sevilla, Avenida de Reina Mercedes, No. 10, Apartado 1052, 41080 Sevilla, Spain

[2] Environment and Risk Management Group, HortResearch, PB11-030, 5301 Palmerston North, New Zealand

[3] ICRAF, JI CIFOR, PO Box 161, Situ Gede, Sindang Barang Bogor 16680, 16001 Bogor, Indonesia

CONTENTS

A.L. Smit et al. (Eds.), Root Methods

14.1 Introduction

Uptake of water by the root system is critical for plant functioning, as it balances aboveground water losses by transpiration and facilitates movement towards the roots of nutrients and other chemicals. The coupled processes in the soil-plant-atmosphere continuum are controlled at the plant, the cellular, or even molecular level (Kramer and Boyer 1995; Steudle and Peterson 1998). A basic understanding is needed of the process of root water uptake, combining soil physical and plant physiological perspectives in models of plant behaviour. This can be related to the water status in their environment to determine accurate plot-level soil water balances, evaluate plant adaptation to drought, and analyse below-ground competition in mixed vegetation systems.

At the most basic level, the soil-plant-atmosphere continuum concept (Huber 1924; Gradmann 1928; Van den Honert 1948) assumes steady-state flow and constant resistance. Obviously this is an oversimplification (Kramer and Boyer 1995), but it serves to remind us of the interrelationships between the soil, plant, and atmospheric factors that have to be taken into account when determining plant water status. Any analytical description, or model, of the soil water balance needs to consider many processes that act simultaneously across a number of different time and spatial scales. These processes include (Fig. 14.1):

1. Water entry into the soil by either rainfall or irrigation and its separation into either infiltration or surface run-off. This process of water entry into the

Fig. 14.1. Soil water balance for a unit of soil with vertical as well as lateral inputs and outputs

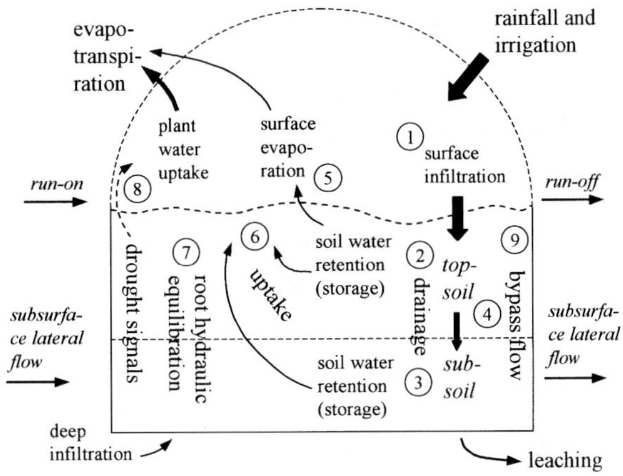

root zone operates on a seconds-to-minutes time scale throughout the depth of the root zone, say centimetres to many metres.

2. Infiltration through and drainage from the cascade of soil layers, and/or via bypass flow through macropores. This internal movement operates on a seconds-to-hours time scale across distances from millimetres to metres.
3. Evaporation from surface soil which operates on an hours-to-days time scale across the scale of centimeters.
4. Water uptake across the spatially distributed root system, which operates on a hour-to-days time scale up to a spatial scale of many (at least up to 50 in some trees) metres. Uptake normally occurs during daytime when the stomata are open, although capacitance recharge of the plant itself can result in nocturnal uptake.
5. Root uptake and discharge from and into soil ("hydraulic lift") can occur through hydrostatic equilibration across the root systems on an hour-to-days time scale, but mostly at night when plant transpiration is negligible. The spatial scale is that of the entire root system.
6. Root uptake strategies in response to hormonal signals can occur locally on an hours-to-weeks time scale.

Better understanding of root uptake can thus be seen to demand knowledge of the roots themselves, as well as information about the coupled biophysical processes operating in the soil and the atmosphere.

When the water balance of a plant, or tree, is calculated over a whole growing season, total water uptake will often be constrained by the amount of water freely available in the soil. The traditional definition of "available" water on the basis of a water retention curve as the difference between "field capacity" and "wilting point" may be adequate for the topsoil, but in the deeper layers of the root zone, not all of this "available" water might actually be accessed by

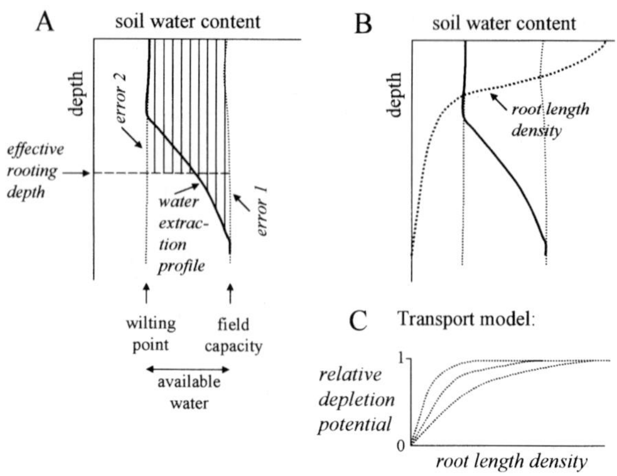

Fig. 14.2. A Definition of "effective rooting depth" for water uptake and "available water" (McKeague et al. 1984) In this pragmatic definition error 1 cancels error 2, so that the total amount of water available for the plant per unit soil area can be estimated from the difference in volumetric soil water content between field capacity and wilting point, times the "effective rooting depth"; **B** and **C** provide alternative concepts to estimate total water uptake on the basis of the complete root system and relative water depletion efficiency as a function on root length density (de Willigen and van Noordwijk 1987)

plants either in quantity, or at the rate at which it is needed. As illustrated in Fig. 14.2, a simple way around this conundrum is via the definition of an "effective rooting depth". This then makes up for the difficulties in defining "available" water because the "effective rooting depth" is defined in such a way that the overestimate of possible water uptake above the "effective" rooting depth equals the underestimate due to lack of complete water uptake below this depth. This definition (McKeague et al. 1984) couples and may appear to render useful both the concepts of "available water" and "effective rooting depth", even though it is not very satisfactory from a rigorous scientific viewpoint. Thus the challenge is to quantify better the form of the entire root system, and to predict better the biophysical functioning of roots in their quest to take up water and chemicals. To achieve this link between form and function, we need better descriptions of the root system, and improved comprehension of the chain of resistances to water flow in the soil-root-shoot-leaf-atmosphere pathway.

Researchers of root system form and function, being aware of the importance of understanding the processes of water uptake, have developed a considerable number of techniques to improve the measurement and modelling of water uptake (Fig. 14.3). The objective of this chapter is to give a comprehensive and detailed description of the most effective of these techniques. We

Table 14.1. Spatial resolution of the methods used for assessing water uptake by roots

Measured variable	Resolution 1–100 μm	0.1–10 mm	0.01–1 m
Water content, θ	?	Gamma densiometry	Gravimetry Neutron scattering NMR TDR FD
Soil water potential, Ψ_s	?	Microtensiometer Psychrometer	Tensiometer Resistance blocks
Root water potential, Ψ_r and resistance, $1/L_p$	Pressure probe	Psychrometer	Pressure chamber
Flow within roots, J_v	–	–	Sap flow techiques Split-rootbox

first discuss methods for measuring water uptake by roots that can be detected via changes in the soil's water content or water potential. We then describe methods for determination of water fluxes in plants, as well as methods to assess the sources of water absorbed by roots, and determination of the resistance to flow and uptake. We conclude with methods to quantify root-shoot communication via plant hormones. These methods all differ in their temporal and spatial resolution, and thus they are intended for use for different purposes (Table 14.1).

14.2 Evaluation of Water Uptake from Changes in Soil Water Status

A variety of methods can be used to determine changes in both soil water content and potential for identifying areas of the soil where roots are actively taking up water. Especially interesting are those suitable for continuously monitoring changes in soil water content with the required resolution in both space and time, and with minimum disturbance to soil conditions.

14.2.1 Methods for Determining Soil Water Content

The Gravimetric Method. The water content of soils can be determined from the moist and dry weights of soil samples. To obtain the moist weight, samples

Fig. 14.3. An apple tree whose root zone has been isolated by a 3 × 3 m trench. This isolated tree can also be covered by a ground level rainout shelter. The tree is instrumented with heat-pulse sensors of sap flow measurement in both the trunk and diametrically opposite structural roots. The soil is instrumented with TDR probes, minirhizotron tubes, and tensiometers. (Courtesy of Peter Spring, HortResearch, New Zealand)

are weighed at the time of sampling, or later if they are kept in sealed containers in a cold environment. To determine the dry weight, the sample is dried to a constant weight in an oven at 105 °C. The water content per unit soil dry weight or gravimetric wetness is the ratio of the weight loss by drying, to the weight of the dried sample. If bulk density of the soil is known or measured, volumetric soil water content can be calculated by multiplying the gravimetric reading by the bulk density.

The method does not require expensive instrumentation. However, it is destructive and time consuming. Errors can be made at every step – sampling, transporting, drying and weighing. Recommendations about the use of the technique can be found elsewhere (Hillel 1980; American Society of Agronomy

1986). Errors can be reduced by increasing the size and number of samples, but this requires more time and effort, and eventually disturbs the experimental plot.

The Neutron Scattering Method. This is one of the methods most widely used for determining soil water content. It requires the use of a neutron probe, an instrument with a source of fast neutrons and a detector of slow neutrons (Gardner and Kirkham 1952). When the probe is lowered into an access tube inserted into the soil, the fast neutrons that are emitted collide with various atomic nuclei in the surrounding medium. Through repeated collisions the neutrons are deflected and "scattered", and lose kinetic energy. Their speed decreases to the characteristic speed for particles at the ambient temperature. Those slow neutrons are said to be thermalised. The nuclei of hydrogen, being of similar size, are the most effective in thermalising the fast neutrons, and so the flux of slow neutrons scattered by the soil is proportional to the soil water content. The source of fast neutrons is obtained by mixing a radioactive emitter of alpha particles (radium or americium) with beryllium. The flux of slow neutrons is converted to a count rate by an electronic scaler.

The neutron scattering method is less time consuming and not as laborious as the gravimetric method. It is less destructive, and repeated measurements can be made at the same locations, directly surrounding the access tubes. The main disadvantages are the high cost of the neutron probe and the emission of neutron and gamma radiation. In addition, the method is not suitable close to the soil surface zone due to the escape of fast neutrons through the surface. The effective volume of measurement is a sphere which varies in size with soil water content. It ranges from a radius of less than 10 cm in a wet soil, to more than 25 cm in a dry soil (Stone et al. 1955; Holmes 1956). The neutron probe has to be calibrated, usually by making many measurements of soil moisture with the neutron and the gravimetric methods simultaneously. Once calibrated, it should work on different soil types, except in unusual soils, like very heavy clays or in soils with a high boron content. When changing soils, one or two readings can be obtained gravimetrically to see if points fall on the calibration line, to find out whether a new calibration is required. Different soil layers may require different calibration lines. A good example of the use of the neutron probe to infer root uptake is the work by Rambal (1984). He used the technique for 7 consecutive years to examine seasonal water extraction to 4.5 m in a *Quercus coccifera* evergreen scrub system. Different calibration curves were required for soils with different rock distributions.

Gamma Densiometry. The theory and practice of gamma densiometry were documented by van Bavel (1959) and Gardner (1986). Briefly, when a source of gamma radiation is placed in the soil close to a detector, the amount of radiation passing through the soil decreases in proportion to soil water content, if bulk density is constant. Simultaneous measurements of bulk density and soil moisture changes can be made with dual-source gamma-ray scanners, in which both ^{137}Cs and ^{241}Am are used (Gardner 1965). Small changes in soil water content can be detected with spatial resolution and precision that is difficult to attain using other methods (Baker and van Bavel 1986, 1988). For example, the effective depth resolution of measurement is about 1 cm, much better than the neutron probe (Table 14.1). The access tubes for the source and detector probes must be strictly parallel, which make installation difficult in certain soils. The technique is likely to be used less in future, as non-nuclear methods improve.

Dielectric Methods. Over the last decade a revolution has occurred in the technology of soil water measurement. The water content and electrolyte concentration of a soil can be accurately determined from the measurement of its dielectric properties. Two methods can be used, one with measurement in the frequency domain between 30-3000 MHz (Schmugge et al. 1980), and the other in the time domain at frequencies above a GHz (Topp and Davis 1985a). These methods, based on the measurement of the electrical capacitance of the soil, depend on the fact that water has a much higher dielectric constant than either air or the dry constituents of soil. The dielectric constant for free water is about 80, and for soil-bound water varies between about 4 and 80, while for dry soil it is about 5 and for air, 1.

14.2.1.1 Time Domain Reflectometry (TDR)

For the time domain method, the most popular instrument uses time domain reflectometry (TDR). The principle of the technique is well explained by Topp and Davis (1985a). TDR determinations are based on the propagation of high frequency ($\approx$GHz) electromagnetic waves through the soil. The propagation constants for electromagnetic waves in the soil depend on the soil properties, especially its water content and electrical conductivity. In the TDR technique, a voltage pulse is propagated along a transmission line which terminates with a wave guide inserted in the soil. Wave guides are composed of two or three parallel metal rods. The signal is reflected from the end of the wave guide and returns to the TDR receiver. The propagation velocity and the amplitude of the reflected signal are then recorded. Because the propagation velocity depends on the dielectric constant of the medium, it is proportional to volumetric water content, becoming smaller as water content increases (Fig. 14.4). The determination of water content is essentially independent of soil texture, temperature and salt content. However, Herkelrath et al. (1991) found that calibrations run on soil cores did not compare well with published curves

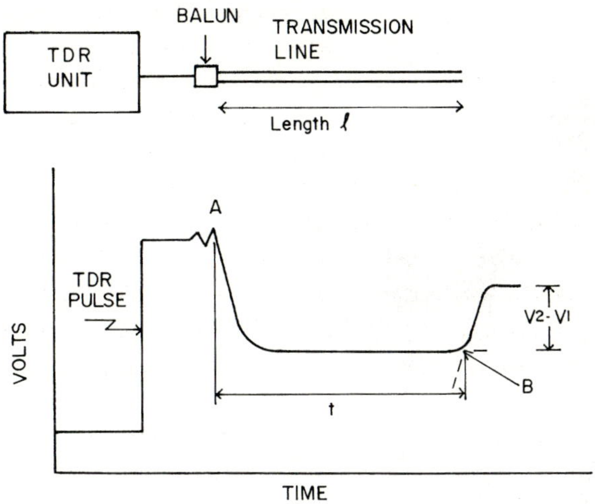

Fig. 14.4. Essential components for measurement of soil water content by TDR (*top*) and idealised output trace. Electrical conductivity is calculated from the signal travel time (t) between the beginning (*A*) and the end (*B*) of the soil transmission line, and from the reflected signal amplitude (v2-v1). (From: Topp and Davis 1985b)

developed for mineral soils, probably because of high soil organic matter content. The authors did not explain why organic soils shift the calibration. Measurements in dry cracking soils may also be difficult, due to the poor contact between probes and soil.

The TDR technique has the same order of accuracy and reproducibility as nuclear methods, and usually does not require site-specific calibration (Wraith and Baker 1991). Several wave guides can be multiplexed and continuously monitored by an unattended TDR unit (Baker and Allmaras 1990; Heimovaara and Bouten 1990). This makes TDR a powerful technique for studying root water uptake. Wraith and Baker (1991) used an automated TDR system to study water uptake and reported high spatial and temporal resolution with high precision; standard deviations in water-content measurements were between 0.0008 and 0.001 $m^3\,m^{-3}$. Clothier and Green (1994) used 22 TDR probes around a kiwifruit vine to monitor the pattern of root water uptake (Fig. 14.5). The near surface preference of roots for wetted soil can be seen in the lower figure. They also used a bank of tensiometers and sap flow measurements in the stem of the tree, in an example of the combined use of several techniques to understand and quantify the spatial and temporal pattern of water uptake by roots.

Further developments of the TDR technique will continue to improve the precision and accuracy of measurement and reduce cost.

14.2.1.2 Frequency Domain Method (FD)

The first dielectric soil water content sensors that worked with frequency domain (FD) were developed in the 1930s (Ferguson 1933; Smith-Rose 1933; Lück 1964). Modern instruments (Turski and Malicki 1974; Wobschall 1978;

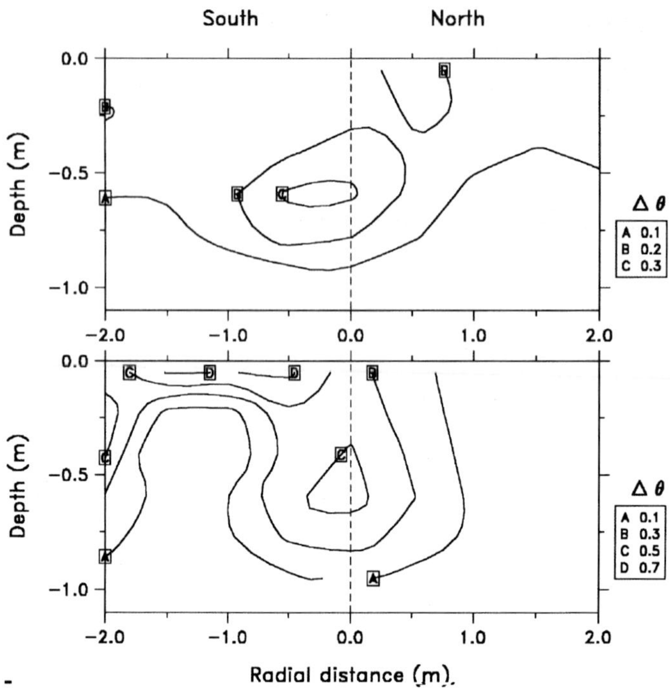

Fig. 14.5. An example of the spatial resolution of measurements by TDR where the ordinate is soil depth and the abscissa is the radial distance from the kiwifruit vine. The *upper part* shows the average daily change in soil water content around a kiwifruit vine over a 4-week period; the *lower part* gives results for a 2-week period following irrigation of the south side (*left side*); the rate of water extraction $\Delta\theta/\Delta t$ is given in $m^3 m^{-3} day^{-1}$. (From Clothier and Green 1994)

Hilhorst et al. 1992) use improvements that make the technique an accurate method of measuring soil water content, suitable for automatic measurements in situ (Hilhorst and Dirksen 1994; Dirksen and Hilhorst 1994). The FD method involves the measurement of capacitance and conductivity at a fixed frequency. Most commercially available FD sensors have an operating frequency between 10 and 150 MHz. A sinusoidal current is fed through an impedance formed by two electrodes with the soil as the dielectric. From the amplitude of the voltage developed across the electrodes and the phase angle between current and voltage, the dielectric properties of the soil can be calculated. Electrodes can be plates, rods or rings. The electronic equipment associated with the FD technique is less expensive than that of TDR, and they are easy to operate. Calibration, however, is soil specific and non linear (Bell et al. 1987). A description of different instruments using capacitance probes and a comparative study with the TDR method can be found in Paltineanu and Starr (1997) and Robinson et al. (1998).

Measurements of the dielectric properties of soil in the time domain and the frequency domain were compared by Hilhorst and Dirksen (1994). Despite the advantages attributed by the authors to the FD technique, TDR instruments have become more widely used.

14.2.2 Methods for Measuring Soil Water Potential

The measurement of water potential both in the soil and in roots is used for determining the state of water in both systems. It is the thermodynamic potential of water that provides the continuous and causal link in the transport chain from water in the soil, to the plant and to the atmosphere.

The main components of the total soil water potential are the matric potential, the osmotic potential and the gravitational potential (Kramer and Boyer 1995). The matric potential results from the capillary and adsorptive forces imposed on water by the soil matrix. The osmotic potential is due to the presence of solutes in the soil water. The gravitational potential of the soil water is determined by the elevation of a given point relative to a reference level, usually the soil surface or the water table. For measuring matric potential – the main component of the total water potential in the field – resistance blocks and tensiometers are used. Total soil water potential can be measured with thermocouple psychrometers. Hydraulic gradients can also be determined, with several tensiometers placed at different depths, so that Darcy's Law can be used to compute fluxes of water in the soil.

Electrical Resistance Blocks. Blocks of porous materials, such as gypsum, nylon or fiberglass, containing a pair of electrodes, can be used to measure matric potential (Bouyoucos and Mick 1940; Colman and Hendrix 1949). The underlying principle relies on the fact that the electrical resistance of many materials depends on water content. The resistance of a porous block in soil decreases as the quantity of soil water absorbed by the block increases. At equilibrium, the water content of the block is dependent on the matric potential of surrounding soil, which is therefore proportional to the resistance of the block. Porous blocks have to be calibrated per individual soils. Measurements are made by an ohmmeter connected to the electrodes of the resistance block.

Blocks made of gypsum behave better than blocks made of inert materials, since they are less sensitive to variations in the salinity of the soil solution. However, the solubility of gypsum makes the blocks eventually deteriorate in the soil. Gypsum blocks are more responsive in the dry range, whereas porous nylon blocks, because of their larger pore sizes, are more sensitive in the wet range of soil moisture variation (Hillel 1980). Porous blocks can be considered a complementary tool to tensiometers, since they work better in soils drier than –0.05 MPa. Reasonably accurate measurements can be obtained at matric potentials as low as –2.0 to –3.0 MPa. Goltz et al. (1981) described electronic improvements to make resistance blocks useful at water potentials as high as –0.03 MPa. Although the porous blocks are not very accurate, they are frequently used to schedule irrigation, partly because they can be used for automatic recording of in situ changes in soil suction (Schlub and Maine 1979).

Tensiometers and Microtensiometers. A tensiometer consists of a water-filled container with porous cup, usually ceramic, which is inserted into the soil. The water container is connected to a manometer. The de-aired water within the cup equilibrates with the soil water through the pores of the ceramic walls. Soil water, usually under suction, draws water out of the tensiometer to achieve equilibrium, causing a drop in pressure inside the container which is indicated by the manometer. The manometer can be a water, or mercury-filled U tube, a vacuum gauge or an electrical transducer.

Tensiometers may be used at water potentials between 0 and –0.08 MPa. Higher suction causes air entry through the cup. For dry areas, as near the soil surface, however, potentials can often go below –0.08 MPa. The size of the ceramic cup will determine the sphere of influence within which the soil matric potential is measured. Several tensiometers are needed for characterising the distribution of moisture in the soil. Maintenance and measurement operations are not greatly time-consuming, and little training is required.

Vetterlein et al. (1993) described microtensiometers for measuring variations in soil matric potential in the vicinity of a living root system. They used ceramic capillaries with a length of 5 mm, a diameter of 1 mm and a wall thickness of 0.3 mm. The ceramic cells were melted into a glass tube, which was connected to a pressure transducer via a plexiglass body. Measurements were taken every 36 min throughout a 14-day period. The microtensiometers were operational at matric potentials between 0 and –0.08 MPa, with a standard error <5% below –0.008 MPa. Temporal resolution was in the range of seconds.

Soil Psychrometers. Soil psychrometers are thermocouple psychrometers housed in a ceramic cup (Fig. 14.6), which is embedded in the soil. The thermocouple is used to measure the relative humidity inside the cup, which, at equilibrium, is equal to that of the soil atmosphere. This is accomplished by using the Peltier effect to condense water on the thermocouple junction within the ceramic cup. Relative humidity is then determined, in the psychrometric mode of operation, from the wet-bulb depression of the junction, or, in dew point mode, from its dew point temperature (see Sect. 14.6.1). Soil psychrome-

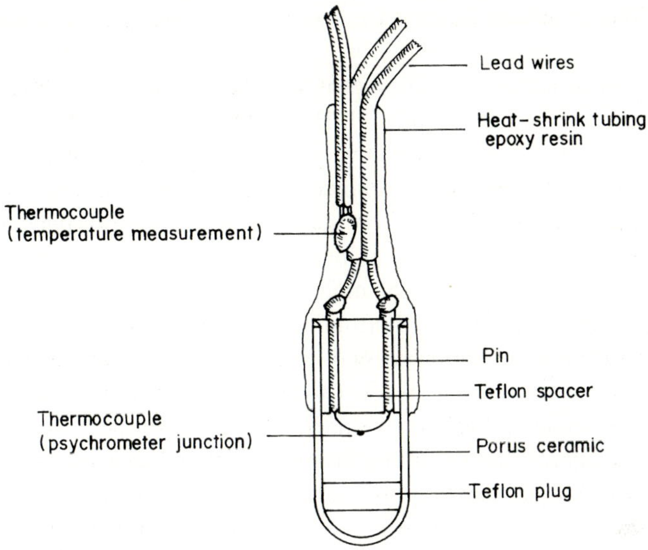

Fig. 14.6. Structure of a typical soil psychrometer. (From Pearcy et al. 1989)

ters can be used when the potential is below the tensiometer range, but they are not accurate above –0.2 MPa.

Thermocouple psychrometers have been used to measure soil water potential by many authors. Savage et al. (1987) described the use, calibration, installation and maintenance of commercial soil psychrometers, both in disturbed and undisturbed soil. They listed precautions necessary for accurate measurement of soil water potential in the laboratory, glasshouse and lysimeters. Thus, direct sun on the aluminium covered lead wires of the soil psychrometers has to be avoided to achieve an acceptable value of zero offset. In addition, the screen cage cover cannot be in contact with soil, in order to reduce thermal gradients within the psychrometer to acceptable levels. Differences in temperature between soil and instrument, as well as within the psychrometer circuitry itself, can greatly affect the accuracy and precision of measured water potentials. Brown and Chambers (1987) discussed the magnitude of these errors and analysed the influence on results of temperature gradients between psychrometer and soil. Fonteyn et al. (1987) and Schlesinger et al. (1987) used soil psychrometers in the field. They found a good correlation between soil water potential and predawn plant xylem sap potential.

Soil psychrometers are easily installed, but should be placed parallel to the soil surface to minimise thermal gradients. They retain calibration characteristics for several years, but salt deposition in saline soils can seriously degrade their performance.

14.2.3 The Nuclear Magnetic Resonance (NMR) Technique

The principles and other features of proton nuclear magnetic resonance (NMR) imaging are described in Chapter 11. Here we show main applications to the study of water distribution and transport in plant root systems, a field in which the technique has been widely used since the 1980s (Bottomley et al. 1986, 1993; Brown et al. 1986). The technique is ideal for determining water distribution around and within roots. The signal intensity in NMR images is most sensitive to protons with a high degree of molecular-level mobility such as those of hydrogen in water (Rogers and Bottomley 1987). Bottomley et al. (1986) were able to analyse uptake and movement of water within *Vicia faba* plants growing in pots with different potting media. They were able to monitor the process of plant wilt and recovery by using a $CuSO_4{\cdot}5H_2O$ solution with Cu^{2+} as a tracer (Cu^{2+} is paramagnetic) and water application alone to a plant growing in a plot with dry soil. NMR imaging has been used by Omasa et al. (1985) for monitoring changes in root and soil water content. MacFall et al. (1990, 1991) used it for observing the development of a water depletion zone around suberized pine roots.

The major advantage of the technique is that it is non-destructive, as it employs non-ionising radiation that penetrates the soil media. This has no known effect on plants (Bottomley et al. 1986). The technique is sensitive to the presence of ferromagnetic particles occurring naturally in the soil, which can disturb the magnetic field and thereby distort the NMR information. Rogers and Bottomley (1987) investigated the range of potential applications of NMR imaging in root systems and assessed the impact of soil magnetic properties and water content on the results from the use of the technique. They found that the NMR technique was not useful in soils with ferromagnetic particle contents greater than about 4% by weight. The equipment is expensive and requires highly trained technicians. The technique cannot provide images of large root systems.

14.2.4 Applications to the Study of Water Movement Between Soil Layers via the Root System

Water efflux from roots into surrounding soil can occur under certain conditions. Water will move out of roots that are permeable to water when the water potential of the soil falls below the water potential within the root. Hydraulic continuity via root systems can lead to transfers of water between soil layers, on the basis of water potential and resistance. If the subsoil is wet and the surface layers are dry, this process is called "hydraulic lift" (Baker and van Bavel 1986; Caldwell and Richards 1989; Xu and Bland 1993). The reverse process, transfers from wet surface layers to dry subsoil is possible as well and has recently been observed in Machakos (Kenya; Smith et al. 2000). Although the total quantities involved in these water transfers may be relatively small, it can be important in the competition between shallow and deep rooted plants. Hydraulic lift can re-wet nutrient-rich dry topsoil layers and thus facilitate nutrient uptake. The reverse process, deep water storage by deep rooted plants after moderate rainfall which only infiltrates into the topsoil, can increase their overall resource capture vis-a-vis shallow rooted plants.

Water efflux from roots is a phenomenon that needs to be taken into account when considering survival and competition within a plant community. Thus, hydraulic lift can help the plant to withstand a period of drought. It can even benefit neighbouring plants with shallow root systems. On the other hand, rapid storage of water in subsoil around the deep roots of trees, beyond the wetting front, may give trees a competitive edge over rather shallow rooted plants.

Measurement of Water Efflux and Hydraulic Lift. Water efflux and hydraulic lift can be determined by detailed measurement of local changes in soil water conditions. Vetterlein and Marschner (1993) used microtensiometers to measure the transfer of water to upper soil layers from lower layers by hydraulic lift. The microtensiometer technique was sensitive enough to detect water release by roots in the top soil. Vetterlein and Marschner (1993) mentioned other attempts to measure hydraulic lift by using gamma densiometry or psychrometers (Mooney et al. 1980; Dirksen and Raats 1985; Baker and van Bavel 1988; Saab and Sharp 1989). Dawson (1993) used standard water relations methods and stable hydrogen isotope analysis to determine the magnitude and radial extent of hydraulic lift by sugar maple trees. He observed that ground water taken up by the trees was redeposited at night into the upper 35 cm of soil, benefiting neighbouring plants. Direct measurements of the direction and rate of water movement in roots based on heat-pulse methods will be discussed below.

Split-Root Boxes. Water movements within the root system can be also studied using *split-root* boxes. In these experiments, the root zones of plants are divided by barriers so that water can move from one side to another only through root systems which bridge the partitions. Different conditions can be imposed on each side. Split-root experiments were used by Zhang and Kirkham (1995a) to determine the effect on crop behaviour of having half of the root system under water and nutrient stress. The authors explain a way of splitting the root systems of grain sorghum and sunflower. Baker and van Bavel (1986) worked with *Cynodon* growing in a split-root box. After creating a large difference in soil water potential between the two sides of the root system, they observed noc-

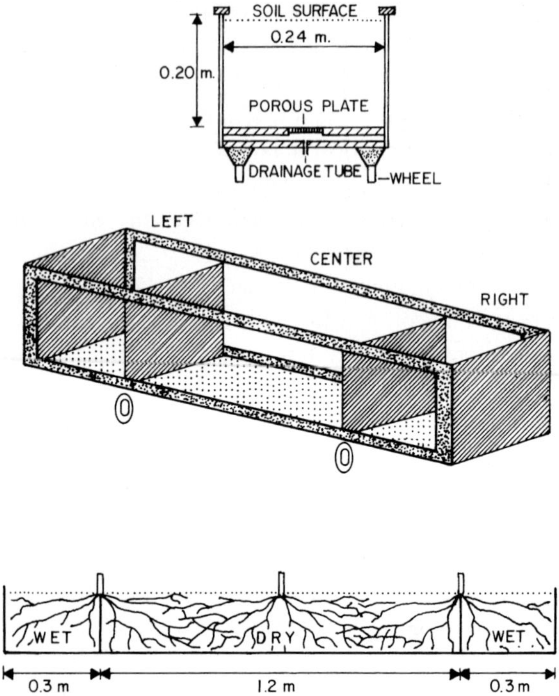

Fig. 14.7. A split-root box for quantifying hydraulic equilibration of soil via plant roots from wet to dry compartments. (From Baker and van Bavel 1986)

turnal transfer through the plant of significant amounts of water from one side to the other. This was detected by frequent measurements of volumetric water content in the box with a gamma probe. Overnight movements of water from wet to dry soil in a split-box (Fig. 14.7) were also observed by Baker and van Bavel (1988) when the potential gradient between dry and wetted compartments reached 1.0 MPa.

14.3 Evaluation of Water Uptake from Water Movement in Plants

Some 60 years ago Huber (1932) used heat as a qualitative tracer of sap flow in plants, by monitoring the downstream temperature rise following application of a short pulse of heat. He observed the rate of ascent of sap. Three thermal tracer techniques have been developed to quantify the volumetric rise of sap; the heat pulse method, the heat probe, and that of heat balance technique. These methods have been applied to a wide range of woody and herbaceous plants, both in the stem and in roots (van Bavel 1992; Swanson and Perttu 1994; Zhang a

b

c

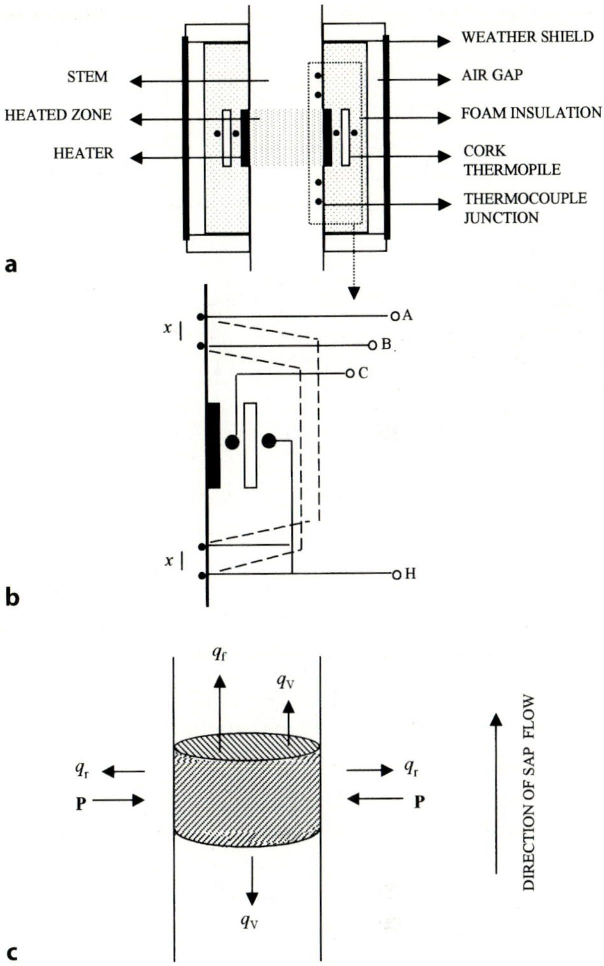

Fig. 14.8. Principles of heat balance and heat pulse technique for recording water flow through roots and stems. (From Smith and Allen 1996). **a** Vertical cross section through a stem heat balance gauge. The *dashed box* in **a** is expanded in **b** to show wiring details of thermocouples in the gauge; copper wires are shown as *solid lines* and constantan wires as *dotted lines*. For the determination of sap flow, the temperature differentials ΔT_a, ΔT_b and ΔT_r are obtained from measurements of the voltages across *AH*, *BH* and *CH*, respectively. **c** The heat balance of the length of stem heated by the gauge, where *P* is heat applied to the stem, q_V is the rate of vertical heat loss by conduction, q_r is radial heat loss by conduction and q_f is heat uptake by the sap stream. **d** Configuration of a single set of heat-pulse probes implanted radially into a stem of radius *R* at the cambium and *h* at the heartwood boundary. The upstream temperature sensor is installed at a distance X_u below the heater and the downstream sensor at a distance X_d above the heater. For calculating the heat-pulse velocity, the system measures the time delay for the temperatures at points X_u and X_d to become equal

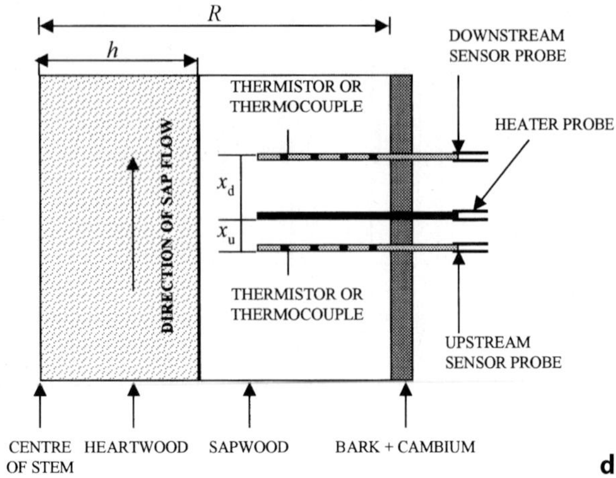

Fig. 14.8. (*cont.*)

and Kirkham 1995b). Smith and Allen (1996) reviewed principles and applications of both the heat balance and heat pulse methods (Fig. 14.8). Scaling up from measurements of individual plants to that of a whole stand over a growing season requires special attention.

Heat-Pulse. In the heat pulse technique, which is best adapted to woody plants, a line-source heater is inserted radially into the stem, or root (Fig. 14.9). Every so often, say every 15–30 minutes, a short (1–2 s) pulse of heat is applied by the heater. Temperature sensors can be located both upstream and downstream of the heater, as in the case of the compensation method (Swanson and Whitfield 1981); or simply on the downstream side (Cohen et al. 1981). Solution of the heat flow equation (Marshall 1958), with appropriate wound factors (Green and Clothier 1988), allows the local velocity of sap flow to be inferred from the measured time-of-travel of the heat pulse. Sap flow rate is then obtained by integration of sap velocity over the cross-sectional area of the stem.

Heat Probe. In the probe technique of Granier (1985) the dissipation of heat from a heater probe is empirically related to sap flow rates. Two probes are inserted radially into the stem or root of a woody plant; one probe is both a heater and temperature sensor, whereas the other measures the reference temperature upstream of the heater. From the regularly measured rise in temperature whilst the heater is supplied with a constant current, an empirical relation provides the time-varying trace in the local sap velocity.

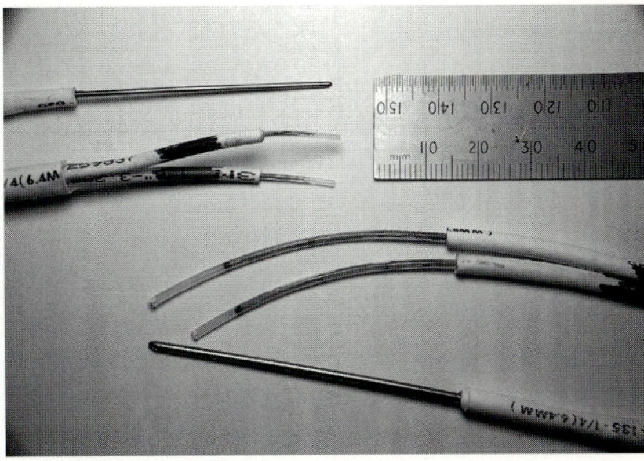

Fig. 14.9. Heat-pulse sensors for both measurement in the trunk using thermistors, and a miniaturised version for use in roots which uses thermocouples. Also shown is the stainless steel heater probe. (Courtesy of Dr. Brent Clothier, HortResearch, New Zealand)

Heat Balance. Unlike the two previous techniques, wherein a temperature change is related to a sap velocity, the heat-balance technique relies on measuring all-but-one of the components of the thermal balance of a stem or root section. The basic equation is:

$$q_i = q_v + q_r + q_f, \qquad (14.1)$$

where q_i is the heat input to the stem, q_v is the vertical heat loss by conduction in the stem, q_r is the radial heat loss and q_f is the heat uptake by the moving sap stream, the only term of the equation which cannot be measured directly; it must be found by difference.

In closing the thermal balance, the amount of heat carried away by the ascending sap can be used to determine the volumetric efflux of sap from the stem, or root section. This technique is well adapted to herbaceous plants or fruit rachilla (Sakuratani 1981), as well as for larger trunks of woody species using a different configuration (Cermák et al. 1973).

Over the last 25 years, these three techniques have been extensively applied in the trunks and stems of a wide variety of plants. These data have provided many insights into the aerial mechanisms controlling the amount of water transpired by plants, and also into the mechanisms of plant hydraulic functioning. More recently these techniques have been applied in roots so that the subterranean functioning of plants has become observable in more detail.

Sap-Flow Measurements in Individual Roots. Green and Clothier (1991) miniaturised their heat-pulse equipment so that they could observe sap flow in 25-mm-diameter roots of kiwifruit. Their results (Green and Clothier 1995) provided a clearer picture of how vine roots that have dried down the soil can respond preferentially to localised wetting (Fig. 14.10). In Fig. 14.10, The rise of sap flow in R1 3–4 days after wetting on DOY 84 was thought to have resulted

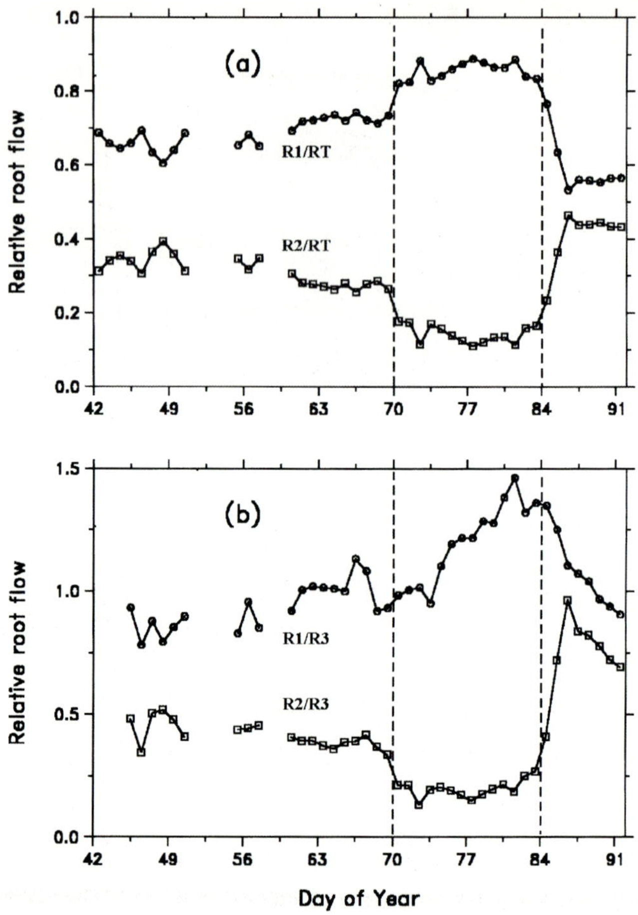

Fig. 14.10. Sap flow in kiwifruit roots measured with the heat-pulse technique: **a** relative sap flow in two roots, R1 and R2, relative to their total, R_T. $R1/R_T$ is the *top line*, and $R2/R_T$ the *lower*. In **b** the root flow is expressed as a ratio to that measured in a control vine R3. Here R1/R3 is the *top line*, and R2/R3 the *lower*. On day 70 the soil surrounding R1 only was wetted and on day 84 the entire root zone was wetted. (From Green and Clothier 1995)

from growth of new roots, as sap flow in R1 was higher, after rewetting, than prior to the dry down.

For apples, Green, Clothier, McLeod (1996) reported quite a different response to localised wetting (Fig. 14.11). The sap flow in R1 shows an immediate and sustained response to the localised wetting of DOY 47. The rapidity of the response, and the subsequently unchanging ratio to stem flow would be the response expected from Gardner's (1960) hydraulic model of water uptake.

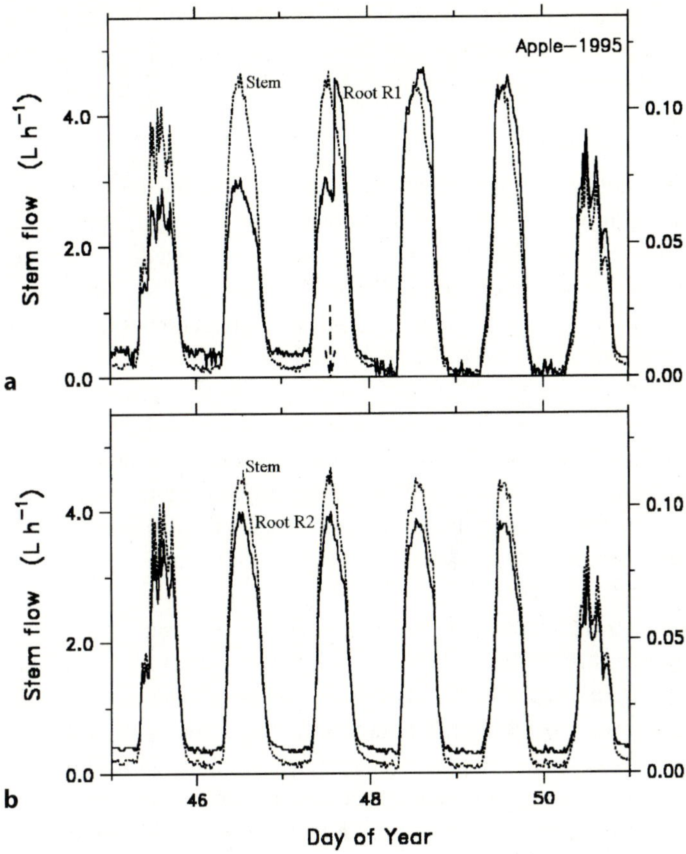

Fig. 14.11. Response of sap flow in two apple roots to localised wetting around R1 at the time indicated by the *arrow* in the *top* diagram (from Green et al. 1996). **a** The *dotted line* is the sap flow measured in the trunk and the ordinate scale is on the *left*. The sap flow in root R1 is given as the *solid line* and the flow can be seen to rise immediately following the irrigation on Day 47. The ordinate for root flow is on the *right*. **b** Again stem sap flow is given as the *dotted line*, and here the sap flow in root R2 is given as the *solid line*

In the case of apples, the root form is one of high occupancy in the explored zone, and so it would appear that with such a root system form, apple uptake functioning does not rely on root growth. Kiwifruit root system form, in comparison, has low occupancy in the explored zone, so that uptake functioning relies in part on the growth of new roots, much as Nobel et al. (1990) found with desert succulents.

Using the heat-probe technique, Cabibel and Do (1991) observed the diurnal trace of sap flow in apple roots that were either drawing water from a zone of localised wetting, or from drier soil. They found sap flow in the dry root tended to be out of phase with that in the trunk such that flow actually peaked at night, and they attributed this to be a result of nocturnal recharge of the tree's capacitance.

Using the heat-balance technique on the roots of the African tree *Croton melagocarpus*, Ong and Khan (1993) were able to examine, by branch pruning and root amputation, the ability of the root system to function when the soil's water is depleted, or when the root system is damaged. A commercially available brand of heat balance equipment (Dynamax Inc., Houston, Texas, USA) was used to monitor sap flow in 20–30 mm diameter roots of *Eucalyptus grandis* (Lightbody et al. 1994). By sequential root severance, they noted that other remaining roots were able to increase their flow to accommodate the flow required by the tree. Thus flow in individual roots and shoots of perennials can be quite decoupled, whereas this might not be the case in annuals.

For the sclerophyllous olive, Moreno et al. (1996) used the heat pulse technique to observe the inability of near-surface roots to recover immediately from emboli caused by water stress, despite a thorough but single wetting of the soil around the once very-dry root.

By extension into roots of heat tracer techniques of sap flow monitoring, new observations are being made of the link between root form and functioning (Fig. 14.12). This includes details of root response to localised wetting, the nocturnal recharge of a plant's capacitance, the respective roles of near-surface roots and roots deeper in the profile, the impact of emboli, and the fraction of the plant's uptake functioning that can be carried out by just some fraction of the root system. Further use of heat-tracer techniques will be useful in providing modellers with data that have hitherto been unavailable. It is important to realise that water flow in any part of a stem or root can be in all directions, and deep roots may pump water out of subsoil during droughts, but may also leak water taken up after rainfall has rewetted the surface layers.

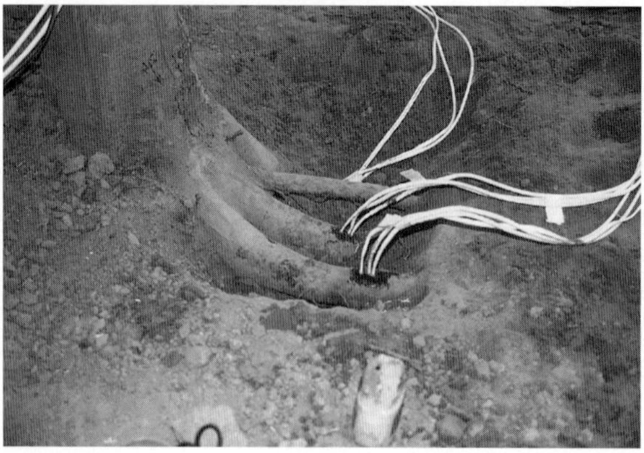

Fig. 14.12. A set of three instrumented apple tree roots. Each of the roots dips away from the trunk at a different angle, so that comparison between the measured flows can be used to indicate the depthwise uptake strategy by the root system. (Courtesy of Dr. Brent Clothier, HortResearch, New Zealand)

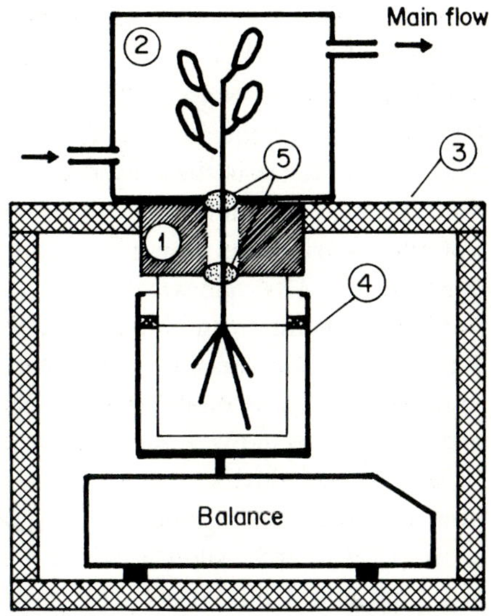

Fig. 14.13. Schematic design of water-budget meter: *1* control unit with attached inner and outer containers mounted on a balance; *2* gas exchange cuvette for determining CO_2 exchange and transpiration; *3* steel cage connecting control unit with platform of the balance; *4* outer container on the pan of the balance with oil seal to prevent evaporation to ambient air; *5* air-tight seals fixing the plant to the control unit and gas exchange cuvette; wind and radiation shields not shown. (From Flach et al. 1995)

The Water-Budget-Meter. For detailed laboratory-scale experiments, Flach et al. (1995) used a improved potometric water-budget-meter (WBM; Fig. 14.13), based on a system described by Eller et al. (1993). Plants are mounted in a gas exchange cuvette where CO_2 exchange and transpiration can be measured. The roots are bathed in water in a container on a balance which is used to determine water uptake. Water uptake is the weight change of the container. They reported an error of less than ±3% for water uptake, if a time period of more than 1 day was considered. For shorter periods (6–24 h) the error was larger, but still less than ±6%. This improved WBM is able to quantify water uptake of a whole plant during time periods of several weeks. It will be a useful instrument for detailed and controlled studies of the processes controlling root water uptake.

14.4 Measuring Stable Isotopes in Plants to Assess Water Uptake from Different Depths

The principles underlying the use of stable isotopes for understanding of water uptake processes are explained in Chapter 12. In this section we only mention illustrative applications of these techniques that can be used for assessing the sources of water absorbed by roots.

Deuterium. The isotopic composition of a sample is usually expressed in terms of delta units relative to some standard (Friedman and O'Neil 1978):

$$\delta X_{std} = (R_{sam}/R_{std} - 1)1000, \tag{14.2}$$

where δX_{std} is the isotopic ratio in delta units relative to the standard, and R_{sam} and R_{std} are the isotope abundance ratios of the sample and standard respectively. Water is the only hydrogen source for a plant under the natural conditions of rainfall. Water absorbed by roots can either come from precipitation during the growing season, or from stored soil water, or from the groundwater. The values of δD, the deuterium/hydrogen ratio of the water from those sources can be quite different. This difference arises from the fact that in temperate areas, with a continental climate, δD values of precipitation commonly show a seasonal cycle. High δD values occur in summer, with lower δD values in the winter (Dansgaard 1964; Gat 1980). Thus, the δD values of rain falling during the summer growing season are generally different from those of stored groundwater. Flanagan et al. (1992) used measurements of the D/H ratio in stem water to determine the relative uptake of summer precipitation by four co-occurring plant species in southern Utah. They found that it was possible to determine the uptake of summer precipitation by measuring the stable isotopic composition of stem water (Flanagan and Ehleringer 1991). Since no isotopic fractionation occurs during water uptake by plant roots (Dawson and Ehleringer 1991), the isotopic composition of water in roots and stems reflects that of water taken up by roots. White et al. (1985) carried out measurements of D/H ratios of tree sap to determine the source of water. They took sap samples from the trunk around noon, when sap flow rate was high. Wood samples were taken from the trunk using an increment borer and the sap drawn into capillary tubes. They demonstrated that the δD values of sap could be used to quantify the relative contributions of summertime rainfall and groundwater. Dawson and Pate (1996) were able to identify the sources of the water taken up by the lateral and tap roots belonging to several woody shrubs and trees. They did this by studying the seasonal changes in the δD of xylem water within different parts of the root system and lower shoots.

Dalton (1988) studied water extracted by the roots of tomato plants. He explained that water will be either H_2O, characterised by water when both hydrogen atoms are the isotope $^1H_1$, or HDO (D for deuterium) when one hydrogen atom is the isotope $^2H_1$. The vapour pressure of the lighter H_2O molecule is slightly greater than that of the DHO molecule. Therefore, vapour phase water in equilibrium with liquid phase water will have a slightly lower ratio HDO/H_2O than that of the liquid phase. If there is no vapour component to root water extraction, then the proportion of HDO in the residual soil water will remain constant. This experimental technique and isotopic ratio measurements were shown to be useful for determining the vapour component of water uptake by the plant roots.

The $^{18}O/^{16}O$ Ratio. Measurements of $^{18}O/^{16}O$ ratios in samples of water drawn from the xylem can also be used for determining whether the water absorbed by roots is coming from either surface or deep waters. Allison and Hughes (1983) analysed isotope ratios of ^{18}O and D in soil profiles down to 15 m depth. This was carried out in areas of *Eucalyptus*, and *Eucalyptus* areas that had been converted to wheat. During the rainy season, the soil surface of the cropped land showed an increase of the isotopic ratio, as compared with the measurements taken during summer. This indicates a degree of infiltration of the precipitation that possesses a higher δD and $\delta^{18}O$. Such an increase was not observed in the soils with *Eucalyptus*, indicating negligible infiltration since most of the precipitation has been intercepted by the tree canopy.

The $^{18}O/^{16}O$ composition of water at equilibrium is known. The $^{18}O/^{16}O$ composition of water can be determined by equilibration with a small volume of CO_2, followed by repeated cryogenic distillation to free the CO_2 which has reached isotopic equilibrium (Epstein and Mayeda 1953; Compston and Epstein 1958). For very small samples, the $^{18}O/^{16}O$ composition can be found by a reaction with guanidine hydrochloride to produce CO_2 (Dugan et al. 1985; Wong et al. 1987).

Smith et al. (1996, 1997) found a clear difference in the $^{18}O/^{16}O$ isotopic ratio between late and early rainfall. In the early season rain, the rain results from the northward advance of the inter tropical convergence zone (ITCZ), and during the later part of the growing season in the Sahelian region of Africa, the rain is due to the southward return of the ITCZ. This isotopic discrimination can be used to distinguish between surface and deep water uptake by the roots of Sahelian trees.

14.5 Measuring the Ability of Roots to Take Up Water

Water uptake by roots can be assessed with the methods described so far. In this section, we refer to methods for studying the ability of roots to take up water, and the role of root-shoot signalling mechanisms which lead to stomatal closure and, ultimately, to water uptake.

14.5.1 Soil and Root Hydraulic Conductivity

The water status of a plant is largely determined by the balance of transpiration and water uptake from the soil. Water uptake depends on the soil water status, root distribution and activity in the soil, and the hydraulic properties along the pathway of water movement. The four main components of the soil-root-shoot pathway are: (1) the soil, (2) the root-soil interface, (3) entrance to the xylem in the root, and (4) the axial pathway from root to shoot in the xylem (Huang and Nobel 1994). The conductance of this pathway is not independent of rate of flow or soil water content, and the relative importance of the four main components varies. Changes in the conductivity of the soil-root-shoot pathway can be due to:

1. Reduction in soil water conductivity as the soil dries out,
2. Changes in continuity of the soil-root pathway by air gaps and reduced root-soil contact,
3. Changes in osmotic potential outside of the root,
4. Changes in true permeability of the cell walls along the pathway,
5. Changes in osmotic potential inside the root,
6. Changes in the continuity of the axial pathway by embolism.

Different perspectives exist on the relative importance of these processes (Sands et al. 1982; Weatherley 1982; Passioura 1988; Rüdinger et al. 1994; Stirzaker and Passioura 1996), but there is now general agreement that past conclusions on the importance of (4) ignored a number of alternative explanations. When the soil's water content is high, water movement from the soil to the shoots depends mainly on the hydraulic conductivity of the root itself (3–5 and possibly 6; Blizzard and Boyer 1980; Passioura 1988). As the soil dries, the conductance of the soil-root interface (2) can become limiting for water movement, due to root shrinkage (Taylor and Willatt 1983; Fernández et al. 1992), but possibly also because of increases in the osmotic potential of the soil solution (3; Stirzaker and Passioura 1996). At lower soil water contents, the soil conductivity (1) becomes the primary limitation on water movement in the root-soil pathway. See Box 14.1 for the components related to the permeability of roots.

A range of techniques is available for determining L_p, L_r and K_h, usually based on measuring the volume of water flowing through a root that is exposed to a known difference in water potential, by varying hydrostatic or osmotic pressure (Huang and Nobel 1994). L_p has been determined for individual roots by measuring J_v induced by applying a partial vacuum to the distal end of an excised root (Nobel et al. 1990), but also by applying external pressure to the whole root system. Typical results for L_p, L_r and K_h, as measured by different authors in different species are reviewed by Huang and Nobel (1994). They show how hydraulic conductivity changes with plant stage and environmental conditions. From all the techniques, the root pressure probe technique seems to be the most widely accepted method to determine L_p, L_r and K_h (Sect. 14.5.1).

The relative importance of the axial and radial component of L_p varies with plant species and the developmental stage of the root (Huang and Nobel 1994). Rowse and Goodman (1981) studied water movement in broad bean roots. They analysed the relative effects of axial resistance, calculated from the xylem dimensions, and radial resistance calculated using a potometer developed to

BOX 14.1. The Permeability of Roots

Water from the surrounding soil moves radially into the root and axially along the xylem. Therefore, the permeability of roots to water depends on the root hydraulic conductivity, L_p ($m\,s^{-1}\,Pa^{-1}$) which can be defined as (Moreshet and Huck 1991; Nobel 1991):

$$L_p = J_v / (\Psi_{surface} - \Psi_{xylem}), \tag{14.3}$$

where J_v ($m^3\,m^{-2}\,s^{-1}$) is the volumetric flux density of water into the root, $\psi_{surface}$ (Pa) is the water potential at the root surface, and ψ_{xylem} is the water potential of the root xylem. Also,

$$L_p = Q_v / (\Delta P\, A), \tag{14.4}$$

where A (m^2) is the root surface area, Q_V is the volumetric flow rate of water ($m^3\,s^{-1}$), and ΔP (Pa) is the pressure drop along the root. L_p can be partitioned into two components: radial conductivity (L_r) for water movement into the roots and up to the xylem, and axial conductance (K_h) for water movement along the root xylem (Dryden and van Alfen 1983; Nobel et al. 1990). For a root segment of length L (m), K_h ($m^4\,s^{-1}\,Pa^{-1}$) can be expressed as:

$$K_h = (L\, Q_v) / \Delta P. \tag{14.5}$$

The radial conductivity L_r ($m\,s^{-1}\,Pa^{-1}$) averaged over the root segment can be calculated by incorporating L_p, K_h, L, and the root radius R_0 (m) into the model of Landsberg and Fowkes (1978):

$$L_r = (L_p\, \alpha\, L) / \tanh(\alpha\, L), \tag{14.6}$$

where α (m^{-1}) equals $(2\,\pi\, R_0\, L_r / K_h)^{1/2}$, which represents the inverse of the length along the root xylem between which the pressure drops to a half (Landsberg and Fowkes 1978; Frensch and Steudle 1989; Alm et al. 1992).

measure the rates of water uptake by small portions of an intact root system. They found that axial resistance was not likely to be important. This conclusion may apply to most dicotyledonous plants with secondary development of xylem transport capacity; in monocots without the capacity to adjust xylem dimensions, however, axial resistance may dominate L_p in well-branched root systems. The continuity of the axial pathway may be dramatically reduced in all plants, however, by embolism.

Sperry and Tyree (1990) observed that embolism caused by water stress reduced hydraulic conductivity in roots of *Picea rubens*. The appearance of air embolism in the xylem, as the soil dries, caused a marked reduction of K_h. The

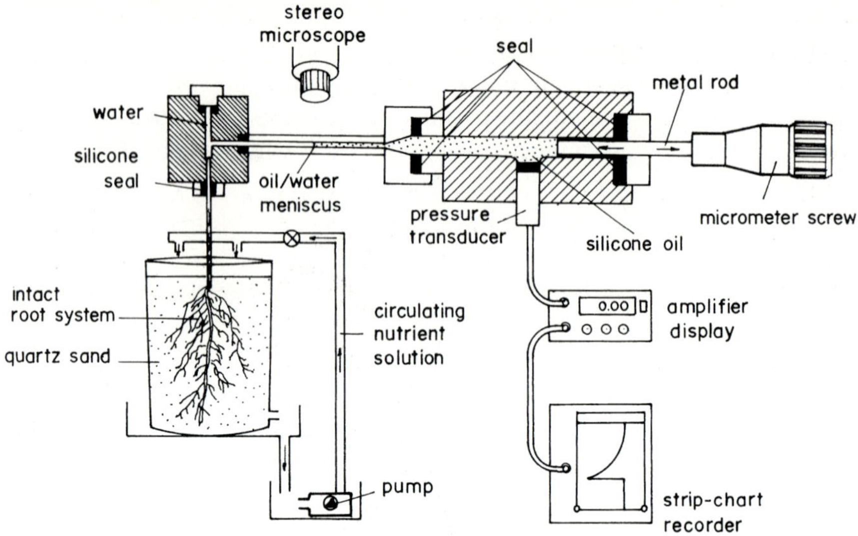

Fig. 14.14. Device for measuring root pressure and root pressure relaxation on spruce roots; root pressure was monitored by a pressure transducer connected to an amplifier and display; movements of the oil water meniscus are recorded in a measuring capillary (stereo microscope); hydrostatic pressure relaxations are measured by moving the micrometer screw rapidly into and out of the pressure probe. (From Rüdinger et al. 1994)

extent of air embolism can be determined by the ratio of axial conductance measured on segments before, and after, pressurisation in solution to remove the emboli (Sperry et al. 1994; Magnani and Borghetti 1995). Xylem at the junctions between lateral roots and main roots, and between main roots and stems, differs from that elsewhere in the root system. Embolism occurs more easily at the junctions.

The Pressure Probe Techniques. The cell pressure probe technique has been adapted for use on excised roots (Steudle and Jeschke 1983). The root pressure probe is described by Hallgren et al. (1994), Melchior and Steudle (1994), and Steudle (1994). Root pressure is measured using a small manometer tightly fixed to the cut end of excised roots (Fig. 14.14). Excised roots exude xylem sap because of the active accumulation of solutes in the xylem. When exudation is stopped by attaching a "root manometer", the system reaches a root pressure which can be recorded continuously. Half of the interior of the root pressure probe is filled with silicone oil and the other half with water, so a meniscus is formed. The position of the meniscus is monitored with the aid of a stereomicroscope. Volume changes can be induced in the measuring system, being responses of root pressure monitored. Measurements begin when a steady root pressure is attained. Changes in pressure with time can be converted into water flows (Melchior and Steudle 1993). Hydraulic conductivity can be calculated from analysis of the pressure/flow relations (Peterson and Steudle 1993; Peterson et al. 1993). Rüdinger et al. (1994) used the root pressure probe to determine the hydraulic and osmotic properties of a spruce root.

Root pressures can be measured on root tips, segments, or whole root systems. In a single root tip, measurements are usually possible for 1 to 2 days. In root systems, measurements have been performed for up to 10 days (Steudle 1994). Measurements on excised roots may not, however, be representative of those of intact plants.

Pressure in individual xylem vessels of intact plants can be directly measured by the *xylem pressure probe*, described in detail by Balling and Zimmermann (1990) and Zimmerman et al. (1994). Zhu et al. (1995) used this technique to measure hydraulic properties in roots of maize. This technique allowed, for the first time in intact plants, the direct measurement of diurnal and seasonal changes in xylem pressure, plus xylem flow and, in principle, the solute composition of the xylem sap at the single vessel level. The xylem pressure probe is filled with "denucleated" water and incorporates a pressure transducer. A microcapillary probe is advanced slowly through xylem tissue until the transducer registers a pressure below atmospheric. In order to verify that the probe is present in a vessel, the microcapillary is loaded within a dye solution prior to insertion, so that the penetrated vessel can be seen in cross-sections after the removal of the probe (Benkert et al. 1991). Model experiments have shown that the probe can accurately read pressure values down to at least about −1 MPa (Balling et al. 1988; Zimmermann et al. 1994).

Zimmermann et al. (1995) stated that single measurements with the xylem pressure probe indicate that the xylem tension in the leaves of intact, transpiring plants is often much smaller than that predicted for transpiration-driven water ascent through continuous water columns. Studies with the xylem pressure probe, together with other evidence, indicates that different mechanisms need to be considered to explain water movement within plants.

14.15.2 Resistance to Flow

Principle. The relation between applied hydrostatic pressure and flow out of decapitated root systems is non-linear (Fig. 14.15). This non-linearity was initially interpreted as a change in the inherent conductance of roots with increasing flow rates. Fiscus (1975) and Dalton et al. (1975) showed that the non-linearity can be explained by the interaction of osmotic and hydrostatic pressures, at a constant hydraulic conductance of the roots. In fact, the osmotic

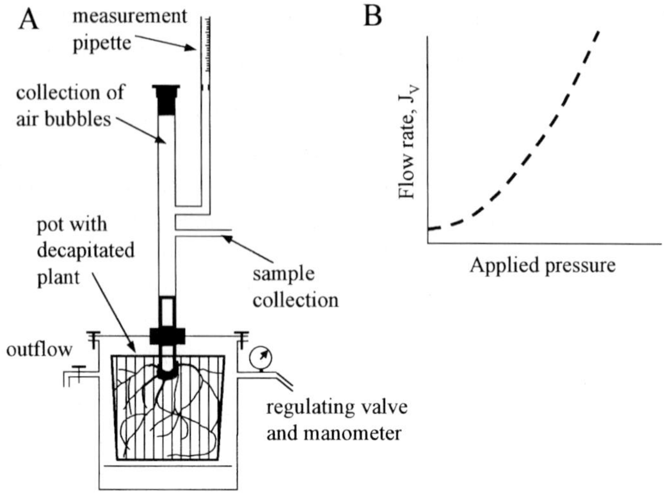

Fig. 14.15. A Pressure bomb design to measure the rate of water flow through excised root systems at different external hydrostatic pressures. **B** Typical result, showing non-linear relationship of flow rate and applied pressure. (From de Willigen and van Noordwijk 1987)

pressure component changes with changing flow rates, depending on the solute reflection coefficient σ_r of the membrane. The models were initially based on a single membrane in a well-stirred solution, but can be extended to a membrane in an unstirred solution with salt accumulation in front of the membrane, counteracted by diffusion away from the root surface (Fiscus 1977; Passioura 1984; de Willigen and van Noordwijk 1987). Recently, Steudle and Peterson (1998) described the composite transport model, to explain how external factors may influence water passage across roots, and how plants optimize water uptake according to demands from the shoot.

Methods. Rieger (1995) measured the hydraulic conductivity of the roots of fruit trees using the technique of Rieger and Motisi (1990). The canopy of each plant was enclosed in a gas exchange chamber, with the root container and one basal side shoot left outside the chamber. The container was then placed in 1 cm distilled water to ensure that the soil water potential was virtually zero. Leaves on the basal shoot outside the chamber were wrapped in parafilm and foil, and allowed to equilibrate with the water potential of the stem base. At steady-state, the water potential gradient across the root system was assumed equal in magnitude, but opposite in sign, to the water potential of the wrapped leaf. Total water flow through the plant was measured by pumping chamber air through a desiccant column at a rate that maintained a constant dewpoint. A regression equation was developed for water potential gradient (y) versus water flow on a root length basis (x) from the two points obtained for each plant, assuming linearity. The y-intercept of the regression gives the minimum gradient required for flow through the root system (Passioura and Munns 1984; Rieger and Motisi 1990), and the reciprocal of the slope gives L_p. With this technique, L_p measurements are time consuming.

Measurements of the flow rate through decapitated root systems can be made for single roots or whole root systems in permeable pots, by enclosing the roots in a pressure chamber with the stem attached to a pipette (Fig. 14.15). At low applied pressures (around 0.1 MPa) the osmotic component is not negligible and samples of the exudate can be collected for analysis. At high applied pressures (around 0.5 MPa) the osmotic component is negligible, but physical damage to the roots may occur. If root systems are used which were grown under incomplete aeration and have created cortical air spaces in response to this, a substantial volume of air may bubble out of the root, as the cut end of the stem includes the cortex. The air bubbles can disturb the measurement of flow rates, but their escape from the root can also cause physical damage to the root. A sequence of low and high pressures should be used, including zero-pressures to check for root leakiness under the pressure of the water column in the pipette.

The method can be applied with large potted plants, provided the dimensions of the pressure chamber are suitable. For roots in their natural habitat, however, a reversal of the direction of flow is needed. Tyree et al. (1994, 1995) developed a novel method of measuring hydraulic conductance of tree root systems using a high-pressure flow meter in the laboratory and field. They measured the amount of water flowing into root systems through a cut end on the proximal side (either close to the tree stem or at some distance away) at a range of pressures. By rapidly testing a range of pressures, total water flow during the measurements is small and therefore osmotic adjustments or changes in soil water content around the roots are probably negligible. By pressuring the roots internally, root-soil contact may be improved. At pressures above 0.1 MPa a linear response of flow J_v to applied pressure ΔP can be obtained, from which L_p for the whole root system can be derived if total root surface area is known. The latter will need destructive measurement on the root after completing measurements, and/or the use of allometric equations relating tree root diameter to total root surface area. When the method is applied to dead roots very high flow rates can be obtained, so a check on the integrity of the root system is needed, especially for "outliers".

14.5.3 Root-Shoot Communication

Tardieu and Davies (1993) consider that the root-shoot signalling mechanism acting over the stomatal conductance of plants in drying soil is better explained by considering the integrated effect of hydraulic and chemical signals, and not only chemical signals or water relations alone. They studied the ABA concentration in the xylem sap. ABA was synthesized by dehydrating the roots (Cornish and Zeevaart 1985). This synthesis can take place in different root tissues (Hartung and Davies 1991). Synthesis is roughly proportional to the root water potential. Therefore, the "root message" in drying soil should be linked to the water status of the root system. The authors mention different hypothesis for root messaging.

Ternesi et al. (1994) analysed root-shoot signalling in sunflower. They collected the sap of container-grown sunflower plants by cutting the shoot of the plant at 2.0 cm above the beginning of the root, and placing a latex tube on the top. Sap was collected in test tubes during 6 h and immediately frozen and stored at –20 °C until required. The ABA content in collected sap was measured by using an ABA-specific monoclonal antibody (Ephyscience EP120, Phytoscience, France) using the protocol suggested by the manufacturer. Results suggested the synthesis of a chemical signal in roots of plants under mechanical stress – a seven times higher ABA concentration in confined roots than in roots of control plants –, which can induce the inhibition of plant growth.

Jackson (1994) mentioned the difficulties of correct measurement of the amounts of hormones transported from root to shoot by the transpiration stream. The problem is that sap flow rates have a marked diluting influence on hormone concentration. If the sap taken for analysis flows at a slower rate than in the intact plants, which is usual, estimates of concentration present in the xylem sap of intact plants will be in error. Thus, variable sap flows mean that concentration alone is not a reliable guide to the amounts moving into the shoot. What is needed is a delivery rate per unit of time. Such a value is obtained easily by multiplying concentration with the sap flow rate. The most appropriate concentration measurements will be those made in sap flowing from the roots at rates as close as possible to the transpiration rate. Jackson (1994) mentioned that in pot-grown plants, the latter can be obtained gravimetrically over a few minutes immediately before the shoot is removed, or by using similar whole plants set aside for this purpose. One alternative approach has been to expel small amounts of sap from cut veins of leaves after pressurizing the roots of intact plants (Schurr et al. 1992). A second alternative method has been to sample the first few drops of sap issuing from the cut surface of the detopped root system (Zhang and Davies 1990), though the samples can be contaminated with ABA, sucrose, and other substances as a consequence of compression forces exerted on the stem when tubing is applied to collect the sap (Else et al. 1994).

Else et al. (1995) mentioned the difficulties in determining whether root signals are hydraulic or chemical. They observed that ABA, ethylene, and the ethylene precursor ACC increased in shoots of flooded plants. Trejo et al. (1995) determined ABA concentration in leaves of *Phaseolus acutifolius*. Sampled leaves were frozen in liquid nitrogen and freeze-dried. Bulk leaf ABA concentration was determined using a radioimmunoassay. The monoclonal antibody AFR MAC 62 was used, which is specific for (+)-ABA. Samples were extracted using different ratios of extraction (leaf dry weight: solvent) depending on the concentration of the ABA fed (1:65–1:150), and further dilutions were made when found necessary.

14.6 Measurement of Plant Water Status

Measurements of the plant water status at a certain time could give us valuable information for estimating water uptake rates by the roots. Techniques for measuring stem diameter changes are also discussed in this section, as a more direct way to estimate water uptake.

14.6.1 Root Water Potential

Root water potential can be indirectly calculated (Jones 1983) from measurements of leaf water potential made using the pressure-chamber technique (Scholander et al. 1965). For more direct determination, thermocouple psychrometers, the sucker method or the root pressure probe can be used. However, these are all techniques that, by virtue of their delicate nature, are limited to use in the laboratory.

The Psychrometer Technique. Thermocouple psychrometers have been widely used for measuring root water potential. See Box 14.2 for details on the techniques. In situ psychrometers were developed in the 1970s for leaves (Boyer 1972; Campbell and Campbell 1974), stems (Michel 1977) and roots (Fiscus 1972). The last is easy to install on roots of various diameters, just by changing sleeves and bottom caps. By placing soil psychrometers close to root psychrometers, the potential gradient from the bulk soil to the root surface can be estimated.

Errors associated with the use of thermocouple psychrometers are caused by vapour pressure disequilibria; the resistance to vapour diffusion from the sample to the chamber atmosphere; thermal gradients and instability caused by fluctuations in environmental temperature; respiratory heat production from the sample; and by solar heating (Koide et al. 1989). Changes in sample water potential can also result from excision of the sample and growth during the equilibration period. This causes a reduction in water potential since the turgor

BOX 14.2. Thermocouple Psychrometers

After placing a tissue sample in an enclosed psychrometer chamber held at a constant temperature, the water in the sample equilibrates with the chamber atmosphere. The value of the sample water potential can then be determined by measuring the equilibrium relative humidity (h) of this atmosphere (Turner 1981):

$$\psi = \frac{RT\ 1n(h)}{V_W}, \tag{14.7}$$

where R (J K^{-1} mol^{-1}) is the universal gas constant, T is the temperature (K) and V_w (m^3 mol^{-1}) is the partial molar volume of water. Since a water potential of −2.0 MPa represents a wet bulb depression of about 0.25 °C, facilities for accurate temperature control and measurement have to be available. Slow drifts in temperature can be tolerated, so long as the reference temperature is known at the time when the sample is measured. In practice, the sealed chambers are usually placed in an accurate water bath in a constant temperature room controlled 1 or 2 °C below that of the water bath. There are two main techniques to measure h in the psychrometer chamber. In the first, the psychrometric mode, pure water is allowed to evaporate into the chamber atmosphere from the surface of a thermocouple junction. The resulting wet-bulb depression at the thermocouple junction is related to h by

$$h = \frac{e_W - a\,\gamma(T - T_W)}{e}, \tag{14.8}$$

where a (Pa) is the atmospheric pressure, γ (K^{-1}) is the psychrometric constant, T the dry-bulb temperature of the psychrometer, T_W is the wet-bulb temperature, e (Pa) is the saturation vapour pressure at T, and e_W (Pa) is the saturation vapour pressure at T_W (List 1968). A procedure based on the Peltier effect is commonly used for placement of the pure water on the thermocouple junction. A cooling current reduces the temperature of the junction below the dew point of the chamber atmosphere, so water is condensed on it (Spanner 1951).

With the second technique, called dewpoint hygrometry, h is determined by measuring inside the psychrometer chamber the dew-point temperature (Neumann and Thurtell 1972; Campbell et al. 1973). Dew-point hygrometers hold the thermocouple junction at the dewpoint temperature using a pulsed cooling current. An advantage of this technique is that the dew-point depression is larger than the wet-bulb depression; the thermocouple signal is therefore larger. In addition, the signal is stable, as there is no net movement of water at the dew point, so that vapour equilibrium in the chamber is not disturbed by measurement.

pressure will decline in the absence of a water supply. To minimise errors, psychrometers must be kept clean and the sample surface area per unit chamber volume should me maximised. Errors due to thermal gradients and instability can be reduced by assuring good contact between the sample and the chamber and by using chamber materials with a high thermal conductivity. Errors due to changes in water potential resulting from the excision and growth of the sample in the chamber can be reduced by working at low temperatures or with mature tissue.

Brown and Collins (1980) noted errors caused by temperature gradients after using different types of psychrometers to measure water potential, and they developed a new screen-caged psychrometer. Oosterhuis (1987) described a method to measure components of the root water potential in cotton plants grown in pots. He used screen-caged thermocouple psychrometers mounted in sample chambers. Water potential was determined after a 4-h equilibration time in a 25 °C isothermal water bath. Osmotic potential was measured by freezing the sample-chamber in liquid nitrogen for 3 min, and then allowing it to thaw at room temperature for 30 min, before reequilibration in the water bath for 4 h. Pressure potential was obtained by subtraction.

Simonneau and Habib (1991) measured root water potential on small root samples from peach trees with psychrometers. For calibration, they used strips of filter paper soaked in standard sodium chloride solutions (Brown and Collins 1980); calibration curves were then obtained for temperatures between 20 and 35 °C. Schildwacht (1988) avoided errors resulting from the nutrient solution adhering to roots by centrifuging the root sample in wire mesh on a plastic tube before transferring it to the psychrometer chamber. Bennet and Cortes (1985), among others, stated that the volume of sample has to be large enough to limit the effect of water adsorption by the chamber.

Yamauchi et al. (1995) used a modification of the method proposed by Fiscus (1972; Fig. 14.16) to measure xylem water potential in cotton taproots. They constructed teflon chambers so that screen-caged thermocouple psychrometers could be attached to the taproot at points 0.4 and 0.6 m apart. A screen-caged thermocouple psychrometer was buried near each root psychrometer to measure the soil water potential. Subsequent changes in root and soil water potential were measured. Temperature fluctuations were minimized by isolation of the system. Psychrometer readings were corrected for any observed temperature difference from the psychrometer calibration temperature at the time of the reading. Axial resistance was also measured by root psychrometers attached to roots growing in a device that ensured all the water flowed only through the taproot to which the psychrometer was attached.

The Pressure-Chamber Technique (Sucker Shoots). When leaves are not transpiring, the pressure chamber (Scholander et al. 1965) can be used to determine

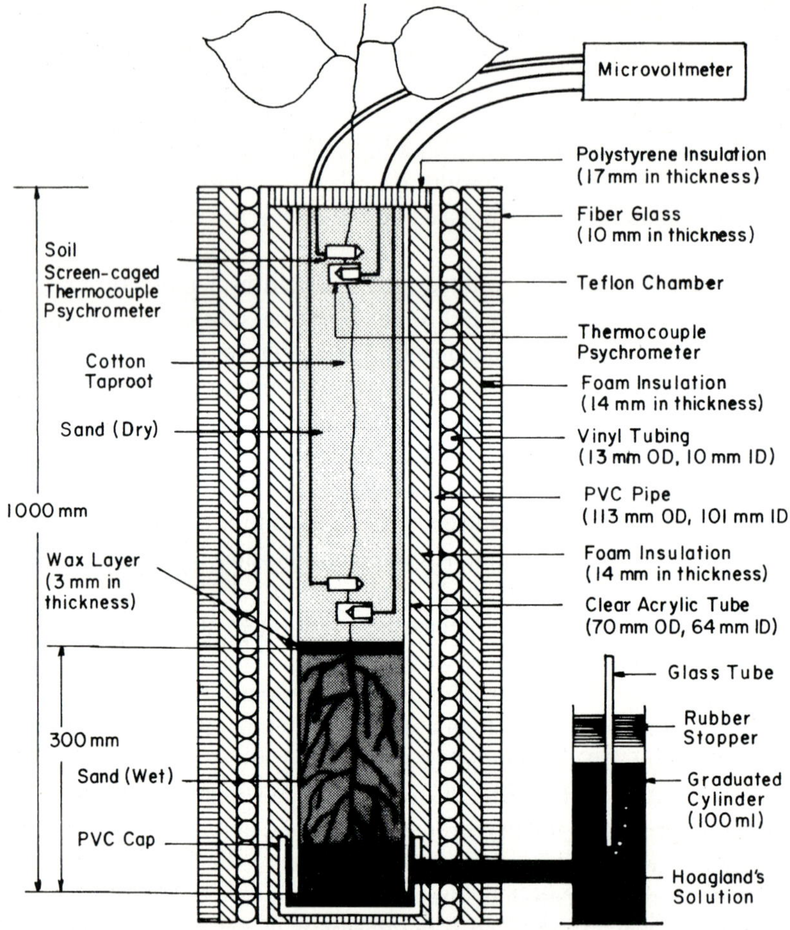

Fig. 14.16. Apparatus for measuring root axial resistance to water flow. (From Yamauchi et al. 1995)

root water potential. Simonneau and Habib (1991) developed the sucker shoot method. They used sucker shoots of young peach trees (Fig. 14.17) to estimate the water potential at the root-sucker branching point. They enclosed the sucker shoots in plastic bags and aluminium foil to prevent transpiration and measured the water potential of sucker shoots with a pressure chamber. As there is no flow in the suckers, the measured water potential should be equal to the water potential in the root xylem at the branching point.

Turner (1981, 1987) detailed the main precautions for measuring water potential with a pressure chamber. Thus, in rapidly transpiring leaves, measurements with the pressure chamber have to be taken as soon as possible after

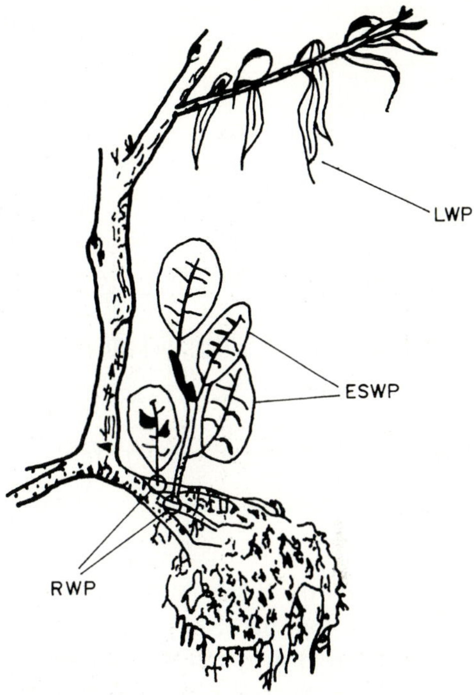

Fig. 14.17. Water potential in trees at different positions; *LWP* leaf water potential, *RWP* root water potential, *ESWP* enclosed sucker water potential. (From Simonneau and Habib 1991)

excision (not later than 10–30 s); the leaf petiole should not be recut; the length of the petiole left outside of the pressure vessel should be as short as possible; gas must not escape from the pressure vessel during the measurement process; pressurization in the vessel has to be slower than 0.005 MPa s^{-1}. The pressure chamber is robust and inexpensive. Time required for each measurement varies among species. The technique is, however, destructive and, in some species, the determination of the end point is complicated by resin exudation from ducts in the xylem or fluid exudation from the pith and cortex (Ritchie and Hinckley 1975).

14.6.2 Stem Diameter Changes

Water stress in a plant can lead to shrinking and swelling of the stem. Measurement of stem diameter variations and the analysis of the relation with the water status of the plant have been made in both herbaceous and woody plants (Molz and Klepper 1973, in cotton; So et al. 1979, in soybean; Brough et al. 1986, in apple trees; Castel 1994, in citrus trees). Molz and Keppler (1972) found that the relationship between stem diameter and leaf water potential had significant hysteresis, mainly due to the time lag in transmittance of water potential

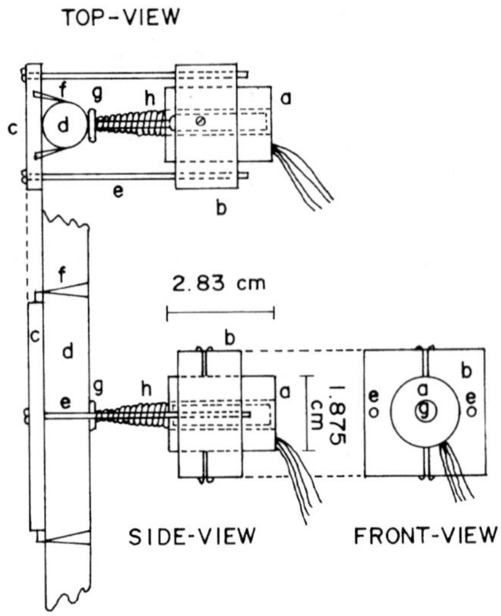

Fig. 14.18. Design (top, side and front view) of linear variable difference transformer (LVDT) attached to a plant stem to record changes in stem diameter; *a* LVDT, *b* perspex holder, *c* perspex backing plate, *d* plant stem, *e* machine screw for adjusting mechanical zero, *f* rubber bands, *g* LVDT-core, *h* light metal spring. (From So et al. 1979)

changes from the xylem into the phloem and its surrounding cortical tissues. Using an analogy between a tensiometer-soil system, and a linear variable differential transformer (LVDT – other authors use the word 'transducer' instead of 'transformer') plant system, So et al. (1979) developed a dynamic method to correct for the time lag observed when using stem diameter fluctuations to predict changes in plant water potential. They used a LVDT system continuously to monitor changes in stem diameter. The holders were similar to that used by Klepper et al. (1971; Fig. 14.18). They reported small errors due to wind fluctuations. LVDT holders with a low expansion coefficient have also been described by Huguet (1985) and Li et al. (1989). The holder and sensor system can reach a precision of 5 μm. The system can be mounted on a stem or branch. Outputs can be electronically recorded and stored in a data logger.

Simmoneau et al. (1993) used a LVDT system in roots of a 4-year-old peach tree growing in pots supplied with nutrient solution. The sensor and holder was of the type described by Huguet (1985). They observed that the diurnal evolution in diameter of the terminal shoot, main branch, trunk and main root followed a similar pattern, close to changes in climatic conditions.

14.7 Summary and Conclusions

The biophysical processes involved in the movement of water through soil to roots, hydraulic lift (and the reverse process), and root-shoot communication, provide many challenges for the researcher. A considerable variety of techniques have been developed for such studies, with different spatial and temporal resolution. Sophisticated technological methods for evaluation of water movement, and water distribution around and within roots, e.g. the NMR technique and the use of isotopes, are now at the disposal of researchers. Also, new electronic devices and innovative sensors are now capable of working in the soil, in the vicinity of roots, and are miniaturised and automated to provide the researcher with powerful tools for studying the biophysical functioning of the root system. Using such high-resolution tools, however, will not necessarily contribute to the understanding of water balances on ecologically relevant time scales, unless such study is accompanied by a better understanding of the range of internal regulation and feedback processes operating in the plant as a whole. The functionality of plant responses to temporary or local droughts depends on the chances it provides for long-term survival and reproduction, rather than in maximizing short-term productivity. The tools that the researcher chooses will thus continue to depend on the goals of the study, and the way detailed knowledge contributes to the overall objectives.

References

Allison GB, Hughes MW (1983) The use of natural tracers as indicators of soil-water movement in a temperate semi-arid region. J Hydrol 60: 157–173

Alm DM, Cavelier J, Nobel PS (1992) A finite-element model of radial and axial conductivities for individual roots: development and validation for two dessert succulents. Ann Bot 69: 87–92

American Society of Agronomy – Soil Science Division (1986) Methods of soil analysis, Part 1, Physical and mineralogical methods, Agronomy Monograph No 9, 2nd edn., American Society of Agronomy Dnc., Madison, Wisconsin

Baker JM, Allmaras RR (1990) System for automating and multiplexing soil moisture measurement by time-domain reflectometry. Soil Sci Soc Am J 54: 1–6

Baker JM, van Bavel CHM (1986) Resistance of plant roots to water loss. Agron J 78: 641–644

Baker JM, van Bavel CHM (1988) Water transfer through cotton plants connecting soil regions of differing water potential. Agron J 80: 993–997

Balling A, Zimmermann U (1990) Comparative measurements of the xylem pressure of *Nicotiana* plants by means of the pressure bomb and pressure probe. Planta 182: 325–338

Balling A, Zimmermann U, Büchner KH, Lange LO (1988) Direct measurement of negative pressure in artificial-biological systems. Naturwissenschaften 75: 409–411

Bell JP, Dean TJ, Hodnett MG (1987) Soil moisture measurement by an improved capacitance technique, part II. Field techniques, evaluation and calibration. J Hydrol 93: 79–90

Benkert R, Balling A, Zimmermann U (1991) Direct measurements of pressure and flow in the xylem vessels of *Nicotiana tobacum* and their dependence on flow resistance and the transpiration rate. Bot Acta 104: 423–432

Bennet JM, Cortes PM (1985) Errors in measuring water potentials of small samples resulting from water adsorption by thermocouple psychrometer chambers. Plant Physiol 79: 184–188

Blizzard WE, Boyer JS (1980) Comparative resistance of the soil and the plant to water transport. Plant Physiol 66: 809–814

Bottomley PA, Rogers HH, Foster TH (1986) NMR imaging shows water distribution and transport in plant root systems in situ. Proc Natl Acad Sci *USA* 83: 87–89

Bottomley PA, Rogers HH, Prior SA (1993) NMR imaging of root water distribution in intact *Vicia faba* L. plants in elevated atmospheric CO_2. Plant Cell Environ 16: 335–338

Bouyoucus GJ, Mick AH (1940) An electrical resistance method for the continuous measurement of soil moisture under field conditions. Mich Agric Exp Stn Techn Bull 172, pp 1–38

Boyer JS (1972) Use of isopiestic technique in thermocouple psychrometry. II. Construction. In: Brown RW, van Haveren BP (eds) Psychrometry in water relations research. Utah Agricultural Experimental Station, Utah State University, pp 98–102

Brough D, Jones H, Grace J (1986) Diurnal changes in water content of the stems of apple trees, as influenced by irrigation. Plant Cell Environ 9: 1–7

Brown JM, Johnson GA, Kramer PJ (1986) In vivo magnetic resonance microscopy of changing water content in *Pelargonium hortorum* roots. Plant Physiol 82: 1158–1160

Brown RW, Chambers JC (1987) Measurements of in situ water potential with thermocouple psychrometers: a critical evaluation. Proceedings of the International Conference on Measurement of Soil and Plant Water Status. July 6–10, Utah. Logan, USA 125–136

Brown WR, Collins JM (1980) A screen-caged thermocouple psychrometer and calibration chamber for measurements of plant and soil water potential. Agron J 72: 851–854

Cabibel B, Do F (1991) Mesures thermiques des flux de sève et comportement hydrique des arbres. II. Evolution dans le temps des flux de sève et comportement hydrique des arbres en présence ou non d'une irrigation localisée. Agronomie 11: 757–766

Caldwell MM, Richards JH (1989) Hydraulic lift: water effux from upper roots improves effectiveness of water uptake by deep roots. Oecologia 79: 1–5

Campbell GS, Campbell MD (1974) Evaluation of a thermocouple hygrometer for measuring leaf water potential in situ. Agron J 66: 24–27

Campbell EC, Campbell GS, Barlow WK (1973) A dewpoint hygrometer for water potential measurement. Agric Meteorol 12: 113–121

Castel JR (1994) Response of young clementine citrus trees to drip irrigation. I. Irrigation amount and number of drippers. J Hortic Sci 69: 481–489

Cermák J, Dernl J, Penka M (1973) A new method of sap flow rate determination in trees. Biol Plant 15: 171–178

Clothier BE, Green SR (1994) Root zone processes and the efficient use of irrigation water. Agric Water Manage 25: 1–12

Cohen Y, Fuchs M, Green GC (1981) Improvement of the heat pulse method for determining sap flow in trees. Plant Cell Environ 4: 391–397

Colman EA, Hendrix TM (1949) Fiberglass electrical soil moisture instrument. Soil Sci 67: 425–438

Compston W, Epstein S (1958) A method for the preparation of carbon dioxide from water vapor for oxygen isotope analysis. Trans Am Geophys Union 39: 511–512

Cornish K, Zeevaart JAD (1985) Abscisic acid accumulation by roots of *Xanthium strumarium* L. and *Lycopersicon esculentum* Mill. in relation to water stress. Plant Physiol 79: 653–658

Dalton FN (1988) Plant root water extraction studies using stable isotopes. Plant Soil 111: 217–221

Dalton FN, Raats PAC, Gardner WR (1975) Simultaneous uptake of water and solutes by plant roots. Agron J 67: 334–339

Dansgaard W (1964) Stable isotopes in precipitation. Tellus 16: 399–401

Dawson TE (1993) Hydraulic lift and water use by plants: implications for water balance, performance and plant-plant interactions. Oecologia 95: 565–574

Dawson TE, Ehleringer JR (1991) Streamside trees that do not use stream water. Nature 350: 335–337

Dawson TE, Pate JS (1996) Seasonal water uptake and movement in root systems of Australian phraeatophytic plants of dimorphic root morphology: a stable isotope investigation. Oecologia 107: 13–20

de Willigen P, van Noordwijk M (1987) Roots, plant production and nutrient use efficiency. Doct Thesis Wageningen Agricultural University, Wageningen, the Netherlands

Dirksen C, Hilhorst MA (1994) Calibration of a new frequency domain sensor for soil water content and bulk electrical conductivity. Proc Symp on TDR in Environmental, Infrastructure and Mining Applications. September 7–9, 1994 Evanston, Illinois

Dirksen C, Raats PAC (1985) Water uptake and release by alfalfa roots. Agron J 77: 621–626

Dryden P, van Alfen NK (1983) Use of the pressure bomb for hydraulic conductance study. J Exp Bot 34: 523–528

Dugan JP, Borthwick J, Harmon RS, Gagnier MA, Glahn JE, Kinsel EP, MacLeod S, Viglino JA, Hess JW (1985) Guanidine hydrochloride method for determination of water oxygen isotope ratios and the oxygen-18 fractionation between carbon dioxide and water at 25 °C. Anal Chem 57: 1734–1736

Eller BM, Egli A, Flach BM-T (1993) Influence of transpiration and photosynthetic pathway on water uptake of *Cotyledon orbiculata* (CAM) and *Othonna opima* (C_3). Bot Helv 103: 207–221

Else MA, Davies WJ, Whitford PN, Hall KC, Jackson MB (1994) Concentrations of abscisic acid and other solutes in xylem sap from root systems of tomato and castor-oil plants are distorted by wounding and variable sap flow rates. J Exp Bot 45: 317–323

Else MA, Davies WJ, Malone M, Jackson MJ (1995) A negative hydraulic message from oxygen-deficient roots of tomato plants? Plant Physiol 109: 1017–1024

Epstein S, Mayeda T (1953) Variation of ^{18}O content of waters from natural sources. Geochim Cosmochim Acta 4: 213–224

Ferguson JG (1933) Classification of bridge methods of measuring impedances. Bell Syst Technol J 12: 452–459

Fernández JE, Moreno F, Martín-Aranda J, Fereres E (1992) Olive-tree root dynamics under different soil water regimes. Agric Mediterr 122: 225–235

Fiscus EL (1972) In situ measurement of root-water potential. Plant Physiol 50: 191–193

Fiscus EL (1975) The interactions between osmotic- and pressure-induced water flow in plant roots. Plant Physiol 55: 917–922

Fiscus EL (1977) Effects of coupled solute and water flow in plant roots with special reference to Brouwer's experiment. Plant Physiol 55: 917–922

Flach BM-T, Eller BM, Egli A (1995) Transpiration and water uptake of *Senecio medley-woodii* and *Aloe jucunda* under changing environmental conditions: measurements with a potometric water-budget-meter. J Exp Bot 291: 1615–1624

Flanagan LB, Ehleringer JR (1991) Stable isotope composition of stem and leaf water: applications to the study of plat water use. Funct Ecol 5: 270–277

Flanagan LB, Ehleringer JR, Marshall JD (1992) Differential uptake of summer precipitation among co-occurring trees and shrubs in a pinyon woodland. Plant Cell juniper Environ 15: 831–836

Fonteyn PJ, Schlesinger WH, Marion GM (1987) Accuracy of soil thermocouple hygrometer measurements in desert ecosystems. Ecology 68: 1121–1124

Frensch J, Steudle E (1989) Axial and radial hydraulic resistance in roots of maize (*Zea mays* L.). Plant Physiol 91: 719–726

Friedman I, O'Neil JR (1978) Isotopes in nature. In: Wedepohl KH (ed) Handbook of geochemistry. Springer, Berlin Heidelbery New York 1B1–1B8

Gardner WR (1960) Dynamic aspects of water and availability to plants. Soil Sci 89: 63–73

Gardner WH (1965) Water content. In: American Society of Agronomy, Methods of soil analysis, Monograph 9. Klute, Madison, Wisconsin pp 92–127

Gardner WH (1986) Water content. In: American Society of Agronomy, Methods of soil analysis. Part1, 2nd edn. Klute, Madison, Wisconsin pp 43–544

Gardner WR, Kirkham D (1952) Determination of soil moisture by neutron scattering. Soil Sci 73: 391–401

Gat JR (1980) The isotopes of hydrogen and oxygen in precipitation. In: Fritz P, Fontes Ch (eds) Handbook of environmental isotope geochemistry, vol 1. Elsevier, Amsterdam, pp 24–48

Goltz SM, Benoit G, Schimmelpfennig H (1981) New circuitry for measuring soil water matric potential with moisture blocks. Agric Meteorol 24: 25–82

Gradmann H (1928) Untersuchungen über die Wasserverhältnisse des Bodens als Grundlage des Pflanzenwachstums. Jahrb Wiss Bot 69: 1–100

Granier A (1985) Une nouvelle méthode pour la mesure du flux de sève brute dans le tronc des arbres. Ann Sci For 42: 193–200

Green SR, Clothier BE (1988) Water use of kiwifruit vines and apple trees by the heat-pulse technique. J Exp Bot 39: 115–123

Green SR, Clothier BE (1991) Heat pulse measurement of sap flow in trees and vines. In: van Bavel CHM (ed) 1992 Symposium on sap flow measurements. Collected Summaries of papers at the 83rd ASA meeting, Denver CO, Oct27–Nov2 1992 37pp. Dynamax Inc, Houston, Texas. 37 pp

Green SR, Clothier BE (1995) Root water uptake by kiwifruit vines following partial wetting of the root zone. Plant Soil 173: 317–328

Green SR, Clothier BE, McLeod DJ (1996) The preferential uptake response of sap flow in apple roots due to localised irrigation. Agric Water Manage 33: 63–78

Hallgren SW, Rüdinger M, Steudle E (1994) Root hydraulic properties of spruce measured with the pressure probe. Plant Soil 167: 91–98

Hartung W, Davies WJ (1991) Drought-induced changes in physiology and ABA. In: Davies WJ, Lones HG (eds) Abcisic acid, physiology and biochemistry. Bios Scientific Publishers, Oxford, pp 63–79

Heimovaara TJ, Bouten W (1990) A computer-controlled 36-channel time domain reflectometry system for monitoring soil water contents. Water Resour Res 26: 2311–2316

Herkelrath WN, Hamburg SO, Murphy F (1991) Automatic, real-time monitoring of soil moisture in a remote field area with time domain reflectometry. Water Resour Res 27: 857–864

Hilhorst MA, Dirksen (1994) Dielectric water content sensors: time domain versus frequency domain. Proc Symp on TDR in Environmental, Infrastructure and Mining Applications. September 7–9, 1994 Evanston, Illinois

Hilhorst MA, Groenwold J, Groot JF (1992) Water content measurements in soil and rockwool substrates: dielectric sensors for automatic in situ measurements. Sensors in Horticulture. Acta Hortic 304: 209–218

Hillel D (1980) Fundamentals of soil physics. Academic Press, New York

Holmes JW (1956) Calibration and field use of the neutron scattering method of measuring soil water content. Aust J Appl Sci 7: 45–58

Huang B, Nobel PS (1994) Root hydraulic conductivity and its components, with emphasis on desert succulents. Agron J 86: 767–774

Huber B (1924) Die Beurteilung des Wasserhaushaltes der Pflanze. Jahrb Wiss Bot 64: 1–120

Huber B (1932) Beobachtung und Messung pflanzlicher Saftströme. Ber Dtsch Bot Ges 50: 89–109

Huguet J (1985) Appréciation de l'état hydrique d'une plante à partir des variations micrométriques de la dimension des fruits ou des tiges au cours de la journée. Agronomie 5: 733–741

Jackson MB (1994) Root-to-shoot communication in flooded plants: involvement of abscisic acid, ethylene, and 1-aminocyclopropane-1-carboxylic acid. Agron J 86: 775–782

Jones HG (1983) Estimation of an effective soil water potential at the root surface of transpiring plants. Plant Cell Environ 6: 671–674

Klepper B, Browning VD, Taylor HM (1971) Stem diameter as a measure of plant water status. Plant Physiol 46: 683–685

Koide RT, Robichaux RH, Morse SR, Smith CM (1989) Plant water status, hydraulic resistance and capacitance. In: Pearcy RW, Ehleringer J, Mooney HA, Rundel PW (eds) Plant physiological ecology. Field methods and instrumentation. Chapman and Hall, London, pp 161–183

Kramer PJ, Boyer JS (1995) Water relations of plants and soils. Academic Press, London

Landsberg JJ, Fowkes ND (1978) Water movement through plant roots. Ann Bot (Lond) 42: 493–508

Li S, Hughet J, Bussi C (1989) Irrigation scheduling in a mature peach orchard using tensiometers and dendrometers. Irrig Drainage Syst 3: 1–12

Lightbody KE, Savage MJ, Graham ADN (1994) In situ measurement of the sap flow rate in lateral roots and stems of *Eucalyptus grandis* under conditions of marginality, using a steady state heat balance technique. J S Afr Soc Hortic Sci 4: 1–7

List RJ (1968) Smithsonian meteorological tables, 6th revised edn. Smithsonian Institution Press, Washington, DC

Lück W (1964) Physikalische Grundlagen und Begriff der Feuchtemessung: Gebergeräte für Gasfeuchte und Materialfeuchte. VDI – Bildungswerkhandbuch. BW 1421, S 1–2, and BW 1429, S 1–7

MacFall JS, Johnson GA, Kramer PJ (1990) Observation of a water-depletion region surrounding loblolly pine roots by magnetic resonance imaging. Proc Natl Acad Sci USA 87: 1203–1207

MacFall JS, Johson GA, Kramer PJ (1991) Comparative water uptake by roots of different ages in seedlings of loblolly pine (*Pinus taeda* L.). New Phytol 119: 551–560

Magnani F, Borghetti M (1995) Interpretation of seasonal changes of xylem embolism and plant hydraulic resistance in *Fagus sylvatica*. Plant Cell Environ 18: 689–696

Marshall DC (1958) Measurement of sap flow in conifers by heat transport. Plant Physiol 33: 385–396

McKeague JA, Eilers RG, Thomasson A, Reeve MJ, Bouma J, Grossman RB, Favrot JC, Renger M, Strebel O (1984) Tentative assessment of soil survey approaches to the characterisation and interpretation of air-water properties of soils. Geoderma 34: 69–100

Melchior W, Steudle E (1993) Water transport in onion (*Allium cepa* L.) roots. Changes of axial and radial hydraulic conductivities during root development. Plant Physiol 101: 1305–1315

Melchior W, Steudle E (1994) Hydrostatic and osmotic hydraulic conductivities of onion (*Allium cepa* L.) roots. In: Baluska F, Gasparikova O, Ciamporova M (eds) Structure and function of roots. Kluwer, Dordrecht

Michel BE (1977) A miniature stem thermocouple hygrometer. Plant Physiol 60: 645–647
Molz FJ, Klepper B (1972) Radial propagation of water potential in stems. Agron J 64: 469–473
Molz FJ, Klepper B (1973) On the mechanism of water-stress-induced stem deformation. Agron J 65: 304–306
Mooney HA, Gulman SL, Rundel PW, Ehleringer J (1980) Further observations on the water relations of *Prosopis tamarugo* of the northern Atamama desert. Oecologia 44: 177–180
Moreno F, Fernández JE, Clothier BE, Green SR (1996) Transpiration and root uptake by olive trees. Plant Soil 184: 85–96
Moreshet S, Huck MG (1991) Dynamics of water permeability, In: Waisel V, Eshel A, Kafkafi U (eds) Plant roots: the hidden half. Marcel Dekker, New York pp 605–623
Neumann HH, Thurtell GW (1972) A Peltier-cooled thermocouple dewpoint hygrometer for in situ measurement of water potentials. In: Brown RW, van Haveren BP (eds) Psychrometry in water relations research. Utah Agricultural Experiment Station, Utah State University, Utah, pp 103–112
Nobel PS (1991) Physicochemical and environmental plant physiology. Academic Press, San Diego
Nobel PS, Schulte PJ, North GB (1990) Water influx characteristics and hydraulic conductivity for roots of *Agave deserti* Engelm. J Exp Bot 41: 409–415
Omasa K, Onoe M, Yamada H (1985) NMR imaging for measuring root systems and soil water content. Environ Control Biol 23: 99–102
Ong CK, Khan AAH (1993) The direct measurement of water uptake by individual tree roots. Agrofor Today Oct–Dec 2–5
Oosterhuis DM (1987) A technique to measure the components of root water potential using screen-caged thermocouple psychrometers. Plant Soil 103: 285–228
Paltineanu IC, Starr JL (1997) Real-time soil water dynamics using multisensor capacitance probes: laboratory calibration. Soil Sci Soc Am J 61: 1576–1585
Passioura JB (1984) Hydraulic resistance of plants. I. Constant or variable? Aust J Plant Physiol 11: 333–339
Passioura JB (1988) Water transport in and to roots. Annu Rev of Plant Physiol 39: 245–265
Passioura JB, Munns R (1984) Hydraulic resistance of plants. II. Effects of rooting medium, and time of day, in barley and lupin. Aust J Plant Physiol 11: 341–350
Pearcy RW, Ehleringer J, Mooney HA, Rundel PW (eds) (1989) Plant physiological ecology. Chapman and Hall, New York
Peterson CA, Steudle E (1993) Lateral hydraulic conductivity of early metaxylem vessels in *Zea mays* L. roots. Planta 189: 288–297
Peterson CA, Murrmann M, Steudle E (1993) Location of the major barriers to water and ion movement in young roots of *Zea mays* L. Planta 190: 127–136
Rambal S (1984) Water balance and pattern of root water uptake by a *Quercus coccifera* evergreen scrub. Oecologia 62: 18–25
Rieger M (1995) Offsetting effects of reduced root hydraulic conductivity and osmotic adjustment following drought. Tree Physiol 15: 379–385
Rieger M, Motisi A (1990) Estimation of root hydrulic conductivity on intact peach and citrus rootstocks. HortScience 25: 1631–1634
Ritchie GA, Hinckley TM (1975) The pressure chamber as an instrument for ecological research. Adv Ecol Res 9: 165–254
Robinson DA, Gardner CMK, Evans J, Cooper JD, Hodnett MG, Bell JP (1998) The dielectric calibration of capacitance probes for soil hydrology using an oscillation frequency response model. Hydrol Earth Syst Sci 2: 83–92

Rogers HH, Bottomley PA (1987) In situ nuclear magnetic resonance imaging of roots: influence of soil type, ferromagnetic particle content, and soil water. Agron J 79: 957–965

Rowse HR, Goodman D (1981) Axial resistance to water movement in broad bean (*Vicia faba*) roots. J Exp Bot 128: 591–598

Rüdinger M, Hallgren SW, Steudle E, Schulze E-D (1994) Hydraulic and osmotic properties of spruce roots. J Exp Bot 279: 1413–1425

Saab IN, Sharp RE (1989) Nonhydraulic signals from maize roots in drying soils: inhibition of leaf elongation but not stomatal conductance. Planta 179: 466–474

Sakuratani T (1981) A heat balance method for measuring water flow rate in the stem of intact plants. J Agric Meteorol 40: 273–277

Sands R, Fiscus EL, Reid CPP (1982) Hydraulic properties of pine and bean roots with varying degrees of suberation. Vascular differentiation and mycorrhizal infection. Aust J Plant Physiol 9: 959–69

Savage MJ, Ritchie JT, Khuvutlu IN (1987) Soil hygrometers for obtaining water potential. Proc Int Conf on Measurement of Soil and Plant Water Status. July 6–10, Utah. Logan, USA, pp 119–124

Schildwacht PM (1988) Changes in the osmotic potential of the root as a factor in the decrease in the root-shoot ratio of *Zea mays* plants under water stress. Plant Soil 111: 271–275

Schlesinger WH, Fonteyn PJ, Marion GM (1987) Soil moisture content and plant transpiration in the Chihuahuan Desert of New Mexico. J Arid Environ 12: 119–126

Schlub RL, Maine JW (1979) Portable recorder for the continuous monitoring of soil moisture resistance blocks. J Agric Engin Res 24: 319–323

Schmugge TJ, Jackson TJ, McKim HL (1980) Survey of methods for soil moisture determination. Water Resour Res 16: 961–979

Scholander PF, Hammel HT, Bradstreet ED, Hemmingsen EA (1965) Sap pressure in vascular plants. Science 148: 339–113

Schurr U, Gollan T, Sculze E-D (1992) Stomatal responses to soil drying in relation to changes in the xylem sap composition of *Helianthus annuus*: II. Stomatal sensitivity to abscisic acid imported from xylem sap. Plant Cell Environ 15: 551–560

Simonneau T, Habib R (1991) The use of tree root suckers to estimate root water potential. Plant Cell Environ 14: 585–591

Simonneau T, Habib R, Goutouly JP, Huguet (1993) Diurnal changes is stem diameter depend upon variations in water content: direct evidence in peach trees. J Exp Bot 260: 615–621

Smith DM, Allen SJ (1996) Measurement of sap flow in plant stems. J Exp Bot 47: 1833–1844

Smith DM, Jarvis PG, Odongo JCW (1996) Water use by Sahelian windbreak trees. In: Buerkert B, Allison BE, von Oppen M (eds) Wind erosion in West Africa: the problem and its control. Margaf Verlag, Weikersheim, pp 115–128

Smith DM, Jarvis PG, Odongo JCW (1997) Sources of water used by trees and millet in Sahelian windbreak systems. J Hydrol 198: 140–153

Smith DM, Jackson NA, Roberts JM (2000) A new direction in hydraulic lift: can tree roots siphon water downwards? Agrofor Forum 8 (1): 23–26

Smith-Rose RL (1933) The electrical properties of soils for alternating currents at radio frequencies. Proc R Soc Lond 140: 359–377

So HB, Reicosky DC, Taylor HM (1979) Utility of stem diameter changes as predictors of plant canopy water potential. Agron J 71: 707–713

Spanner DC (1951) The Peltier effect and its use in the measurement of suction pressure. J Exp Bot 2: 145–168

Sperry JS, Tyree MT (1990) Water-stress-induced xylem embolism in three species of conifers. Plant Cell Environ 13: 427–436

Sperry JS, Nichols KL, Sullivan JEM, Eastlack SE (1994) Xylem embolism in ring-porous, diffuse-porous, and coniferous trees of northern Utah and interior Alaska. Ecology 75: 1736–1752

Steudle E (1994) Water transport across roots. Plant Soil 167: 79–90

Steudle E, Jeschke WD (1983) Water transport in barley roots. Planta 158: 237–248

Steudle E, Peterson CA (1998) How does water get through roots? J Exp Bot 49: 775–788

Stirzaker RJ, Passioura JB (1996) The water relations of the root-soil interface. Plant Cell Environ 19: 201–208

Stone JF, Kirkham D, Read AA (1955) Soil moisture determination by a portable neutron scattering moisture meter. Soil Sci Soc Am Proc 19: 419–423

Swanson RH, Perttu AAH (1994) Significant historical developments in thermal methods for measuring sap flow in trees. Agric For Meteorol 72: 113–132

Swanson RH, Whitfield DWA (1981) A numerical analysis of heat pulse velocity and practice. J Exp Bot 32: 221–239

Tardieu F, Davies WJ (1993) Integration of hydraulic and chemical signalling in the control of stomatal conductance and water status of droughted plants. Plant Cell Environ 16: 341–349

Taylor HM, Willatt T (1983) Shrinkage of soybean roots. Agron J 75: 818–820

Ternesi M, Andrade AP, Jorrin J, Benlloch M (1994) Root-shoot signalling in sunflower plant with confined root systems. Plant Soil 166: 31–36

Topp GC, Davis JL (1985a) Time-domain reflectometry (TDR) and its application to irrigation scheduling. Adv Irrig 3: 107–127

Topp GC, Davis JL (1985b) Measurement of soil water content using time domain reflectometry (TDR): a field evaluation. Soil Sci Soc Am J 49: 19–24

Trejo CL, Clephan AL, Davies WJ (1995) How do stomata read abscisic acid signals? Plant Physiol 109: 803–811

Turner NC (1981) Techniques and experimental approaches for the measurement of plant water status. Plant Soil 58: 339–66

Turner NC (1987) The use of the pressure chamber in studies of plant water status. Proc. Int. Conf. Measurement Soil Plant Water Status 2: 13–24

Turski R, Malicki M (1974) A precise laboratory meter of a dielectric constant of soil of a different moisture. Pol J Soil Sci 7: 71–79

Tyree MT, Yang S, Cruizat P, Sinclair B (1994) Novel method of measuring conductivity of tree root systems and interpretation using AMAIZED. Plant Physiol 104: 189–199

Tyree ME, Patino S, Bennink J, Alexander J (1995) Dynamic measurements of hydraulic conductance using a high-pressure flow meter in the laboratory and field. J Exp Bot 46: 83–94

van Bavel CHM (1959) Densitometry by gamma transmission. Soil Sci 87: 50–58

van Bavel CHM (ed) (1992) Symposium on sap flow measurement. Collected Summaries of papers at the 83rd ASA meeting, Denver CO, Oct27–Nov2 1992. Dynamax, Houston, Texas 37pp

van den Honert TH (1948) Water transport in plants as a catenary process. Discuss Faraday Soc 3: 146–153

Vetterlein D, Marschner H (1993) Use of microtensiometer technique to study hydraulic lift in a sandy soil planted pearl millet (*Pennisetum americanum* L. Leeke). Plant Soil 149: 275–282

Vetterlein D, Marschner H, Horn R (1993) Microtensiometer technique for in situ measurement of soil matric potential and root water extraction from a sandy soil. Plant Soil 149: 263–273

Weatherley PE (1982) Water and flow in roots. In: Lange OL, Nobel PS, Osmond CB, Ziegler H (eds) Encyclopedia of plant physiology. Springer, Berlin Heidelbery New York, Vol 12B, pp 79–109

White JWC, Cook ER, Lawrence JR, Broecker WS (1985) The D/H ratios of sap in trees: Implications for water sources and tree ring D/H ratios. Geochim Cosmochim Acta 49: 237–246

Wobschall D (1978) A frequency shift dielectric soil moisture sensor. IEEE Trans Geosci Electron 15: 112–118

Wong WW, Lee LS, Klein PD (1987) Oxygen isotope ratio measurements on carbon dioxide generated by reaction of microliter quantities of biological fluids with guanidine hydrochloride. Anal Chem 59: 690–693

Wraith JM, Baker JM (1991) High-resolution measurement of root water uptake using automated time-domain reflectometry. Soil Sci Am J 55: 928–932

Xu X, Bland WL (1993) The short life and replant problems of deciduous fruit trees. Agron J 85: 384–388

Yamauchi A, Taylor HM, Upchurch DR, McMichael BL (1995) Axial resistance to water flow of intact cotton taproots. Agron J 87: 439–445

Zhang J, Davies WJ (1990) Does ABA in the xylem control the rate of leaf growth in soil-dried maize and sunflower plants? J Exp Bot 41: 1125–1132

Zhang J, Kirkham MB (1995a). Water relations of water-stressed, split-root C_4 (*Sorghum bicolor*, Poaceae) and C_3 (*Helianthus annuus*, Asteraceae) plants. Am J Bot 82: 1220–1229

Zhang J, Kirkham MB (1995b) Sap flow in a dicotyledon (sunflower) and a monocotyledon (sorghum) by the heat-balance method. Agron J 87: 1106–1114

Zhu JJ, Zimmermann U, Thürmer F, Haase A (1995) Xylem pressure response in maize roots subjected to osmotic stress: determination of radial reflection coefficients by use of the xylem pressure probe. Plant Cell Environ 18: 906–912

Zimmermann U, Meinzer FC, Benker R, Zhu JJ, Schneider H, Goldstein G, Kuchenbrod E, Haase A (1994) Xylem water transport: is the available evidence consistent with the cohesion theory? Plant Cell Environ 17: 1169–1181

Zimmermann U, Meinzer F, Bentrup F-W (1995) How does water ascend in tall trees and other vascular plants? Ann Bot 76: 545–551

CHAPTER 15

Modelling Water and Nutrient Uptake

P. de Willigen[1], N.E. Nielsen[2], N. Claassen[3], and A.M. Castrignanò[4]

[1] Alterra, Green World Research, Postbus 47, 6700 AA Wageningen, The Netherlands
[2] The Royal Veterinary and Agricultural University, 40 Thorvaldsensvej, 1871 Frederiksberg C, Copenhagen, Denmark
[3] Georg-August University, von Siebold-Str. 6, 37075 Göttingen, Germany
[4] Istituto Sperimentale Agronomico, Via C. Ulsiani 5, 70125 Bari, Italy

CONTENTS

A.L. Smit et al. (Eds.), Root Methods

15.1 Introduction

Supplying plants with inorganic nutrients and water is a major function of roots (Nye and Tinker 1977; Nielsen 1983; Barber 1984; Jungk and Claassen 1989; Gahoonia and Nielsen 1991, 1992a,b). It is still a major challenge to agricultural research to quantitatively predict production of a crop as influenced by uptake of nutrients and water. Models can be of great assistance to understand how different factors and their interactions influence crop functioning. A model is per definition a simplification of part of reality. To quote Van Wijk (1963): "Generally it is necessary to introduce drastic simplifications in an agricultural problem to make it amenable to a mathematical analysis. Objections are sometimes raised that such simplifications restrict or even preclude the applicability.... However this need not be the case. The very purpose of the mathematical physical treatment is to isolate the essential characteristics of the problem and to abstract them from the less essential ones".

Nye (1992a,b,c) mentioned the major advantages of modelling: "a) The act of constructing a model ... clarifies thinking, b) the overall model serves to identify gaps in knowledge or understanding, c) the model provides a framework for discussion and argument, and serves to identify precisely the points at issue, d) it helps to decide research priorities by highlighting the most worthwhile experiments." This chapter gives an overview of different models in use and discusses the advantages and disadvantages of the different approaches, at the scale of a single root and that of a crop root system. Further detail can be found in several books and review papers dealing with uptake of nutrients (Nye and Tinker 1977; Barber 1984; Rengel 1999) and of water (Molz 1981; Jones and Ritchie 1991). In this chapter we emphasise the similarity of the transport problems involved in water and nutrient uptake, and discuss these in an analogous fashion. We first begin with the fundamentals of transport to and uptake by a single root.

15.2 Single Root Models

15.2.1 Basic Transport Equation

The equation of continuity for a substance in a porous region reads:

$$\frac{\partial S}{\partial T} = -\nabla . F + U, \tag{15.1}$$

where S is the bulk density of the substance [$L.T^{-3}$], F is the flux [$M.L^{-2}.T^{-1}$], ∇ is the gradient operator [L^{-1}], T is time [T], and U is the production or consumption [$M.L^{-3}.T^{-1}$]. The flux of the substances under consideration can, at least within the scope of this book, be assumed to consist of two components: a convective component and a diffusive component:

$$F = F_C + F_D. \tag{15.2}$$

The convective flux is due to movement of the fluid which contains the substance, brought about by e.g. gradients of total pressure. The moving fluid carries with it the substance it contains:

$$F_C = VC, \tag{15.3}$$

where V is the volume flux of the fluid [$L.T^{-1}$], and C is the concentration of the substance in the fluid [$M.L^{-3}$]. The diffusive flux is assumed here to be given by Fick's first law, i.e. proportional to the gradient of the concentration:

$$F_D = -D\nabla C, \tag{15.4}$$

where D is the diffusion coefficient [$L^2.T^{-1}$]. Substitution of Eqs. (15.2), (15.3), and (15.4) into (15.1) results in:

$$\frac{\partial S}{\partial T} = -\nabla \cdot VC + \nabla \cdot D\nabla C + U, \tag{15.5}$$

and this then is the general equation that is used to describe transport in the soil to the root. It assumes a particular form where water or solutes are concerned. The same symbols C, S, U and D (with the exception of diffusivity D_w) will be used. From the context it will be clear whether they refer to water or solutes.

15.2.2 Geometry and Boundary Conditions

Though the earlier descriptions of nutrient transport and uptake by a root considered a single root in an infinite medium (Bouldin 1961; Nye 1966), most single root models consider a single cylindrical root confined in an isolated soil cylinder. The single root is assumed to be representative of a set of roots all having the same dimensions and operating under the same conditions. In order to solve the partial differential equation (15.5) boundary conditions have to be formulated. Two boundary conditions have to be formulated, one at the outer boundary of the medium and one at the inner boundary, the surface of the root.

For a root in an infinite medium the boundary condition at infinity is that of finite bulk density of the substance. For a root confined within a soil cylinder, with radius R_1 the condition at the outer boundary of the latter is that of vanishing flux:

$$R = R_1, \quad \boldsymbol{F}_C + \boldsymbol{F}_D = 0 \tag{15.6}$$

At the inner boundary, i.e. at the root surface where $R = R_0$, the boundary condition is usually that of prescribed flux:

$$R = R_0, \quad \boldsymbol{F}_C + \boldsymbol{F}_D = \boldsymbol{F}_0, \text{ and} \tag{15.7}$$

sometimes that of given bulk density:

$$R = R_0, S = S_0 \tag{15.8}$$

where both $\boldsymbol{F}_0$ and S_0 can be functions of the spatial co-ordinates, time, and in the case of flux of the concentration at R_0.

15.2.3 Initial Condition

The initial distribution of the bulk density in the soil cylinder is generally given by:

$$T = 0, \quad S = S_i \tag{15.9}$$

where the initial bulk density can be a function of the spatial co-ordinates.

15.2.4 Nutrients

15.2.4.1 Transport Equation

The total bulk density of nutrient in inorganic form is composed of the amount in solution, which is mobile, and that in the solid phase, which is considered immobile. The latter consists of a pool Q which exchanges at a fast rate with the pool in solution and a pool from which transfer to the solution is so slow that in that can be ignored for our purposes:

$$S = Q + \theta C. \tag{15.10}$$

It will be assumed here that at any time there exists a unique functional relation – the adsorption isotherm – between Q and C, symbolised by:

$$Q = f(C). \tag{15.11}$$

Substitution of (15.11) and (15.10) into (15.5) yields an equation with only C as the dependent variable:

$$[f'(C)+\theta]\frac{\partial C}{\partial T} = \nabla \cdot D\nabla C - \nabla \cdot VC + U. \tag{15.12}$$

In the following we will generally neglect the production/consumption term U. In Eq. (15.12) the role of the soil chemical properties is given by the derivative of the adsorption isotherm with respect to concentration as shown in the left hand side of Eq. (15.12). The soil physical properties and conditions manifest themselves via their influence on the transport parameters V and D. Water is both the carrier of solutes – reflected in the role of the flux V – and the medium through which diffusion takes place. The diffusion coefficient in soil is very much dependent on water content and can be calculated as (Nye and Tinker 1977):

$$D = D_0 f_l \theta, \tag{15.13}$$

where D_0 is the diffusion coefficient in water, and f_l the impedance factor, a function of water content.

Equation (15.12) assumes special forms, some examples of which follow, depending on subsequent assumptions. An obvious and widely used assumption is that of the cylindrical form of the root, which leads to the choice of expressing the gradient and divergence in cylindrical co-ordinates. Moreover, when tangential and vertical gradients are assumed negligible, Eq. (15.12) assumes the form:

$$[f'(C)+\theta]\frac{\partial C}{\partial T} = \frac{1}{R}\frac{\partial}{R}DR\frac{\partial C}{\partial R} - \frac{1}{R}\frac{\partial R V_R C}{\partial R}. \tag{15.14}$$

When transport is dominated by diffusion and the mass flow component can be neglected, Eq. (15.14) simplifies to:

$$[f'(C)+\theta]\frac{\partial C}{\partial T} = \frac{1}{R}\frac{\partial}{R}DR\frac{\partial C}{\partial R}. \tag{15.15}$$

When the root has contact with the soil solution only over a part of its radial surface, tangential gradients will develop, and when transport is by diffusion alone, Eq. (15.12) becomes:

$$[f'(C)+\theta]\frac{\partial C}{\partial T} = \frac{1}{R}\frac{\partial}{\partial R}DR\frac{\partial C}{\partial R} + \frac{1}{R^2}\frac{\partial}{\partial \psi}\frac{\partial DC}{\partial \psi}, \tag{15.16}$$

where ψ is the tangential co-ordinate.

With some special experimental techniques it is possible to obtain conditions where transport is one-dimensional in a rectangular geometry (Dunham and Nye 1973; Dunham and Nye 1974; Gahoonia et al. 1994). In such conditions the appropriate transport equation reads:

$$[f'(C)+\theta]\frac{\partial C}{\partial T}=\frac{\partial}{\partial X}D\frac{\partial C}{\partial X}-\frac{\partial V_X C}{\partial X}. \tag{15.17}$$

When the adsorption isotherm is linear then $f(C) = K_a.C$, and $f'(C)$ is constant. Provided D is independent of C, Eqs. (15.15), (15.16) and (15.17) are linear and can be solved analytically by classical mathematical techniques. If the adsorption isotherm is non-linear, the equations generally have to be solved by numerical methods.

15.2.4.2 Boundary Conditions

In the case of an infinite medium, the outer boundary condition states that the concentration converges to a finite value:

$$R \to \infty, \quad C \to C_i. \tag{15.18}$$

When the root is confined within a soil cylinder, the boundary condition at the outer boundary of the soil cylinder is that of vanishing flux Eq. (15.6), in cylindrical co-ordinates:

$$R = R_1, \quad -D\frac{\partial C}{\partial R}+VC=0. \tag{15.19}$$

The boundary condition at the root surface usually involves some assumptions about physiological properties of the root. Bouldin (1961) and later the school of Nye assumed a relationship between uptake rate and concentration at the root surface. Nye and Tinker (1977, p. 116) state: "It is most usual to consider root uptake in relation to solution concentration at the surface as the boundary condition". The earlier models (Bouldin 1961; Nye 1966) assumed uptake to be proportional to the concentration at the root's surface:

$$R = R_0, \quad -D\frac{\partial C}{\partial R}+VC=\alpha C, \tag{15.20}$$

where α is the root absorbing power [$L.T^{-1}$]. Later models used Michaelis-Menten kinetics, sometimes with some modification, to describe the relation between concentration and uptake (Nye and Marriott 1969):

$$R = R_0, \quad -D\frac{\partial C}{\partial R}+VC=U_{max}\frac{C}{K_m+C}, \tag{15.21}$$

where U_{max} is the maximum uptake rate, occurring at very high values of C, and K_m is the Michaelis-Menten constant.

De Willigen and van Noordwijk (1978, 1987) took as a starting point the demand of the plant, which generally is a function of time, but in the case of crops with closed canopies is constant during a considerable fraction of the growing season. They assumed that, as long as the concentration at the root surface exceeded a certain minimum value, uptake equalled demand. When the minimum value is reached, uptake is such that the minimum value C_{lim} of the concentration is maintained:

$$R = R_0, \quad \text{if } C \geq C_{lim}, \quad D\frac{\partial C}{\partial R} = \frac{A}{2\pi H L_{rv} R_0}, \quad \text{else } C = C_{lim}. \tag{15.22}$$

15.2.4.3 *Analytical Solutions and Approximations for Simple Geometry*

Generally, analytical solutions can only be found for linear conditions, i.e. a linear adsorption isotherm and the diffusion coefficient and flow of water being independent of the concentration.

The exact solutions are generally quite complicated, consisting of infinite series of special functions, therefore simpler approximations are used. These can be employed in macroscopic uptake models. Baldwin et al. (1973) proposed to describe the concentration-distance profile around a root, taking up with a rate proportional to the concentration at its surface, by a sequence of steady-state solutions. In this approach, at each time the average concentration in the soil cylinder – when transport is by diffusion alone – is given by (for convenience of notation dimensionless variables are used, Table 15.1 gives the meaning of the symbols used.):

$$\bar{c} = \exp\left(\frac{-2pt}{B\rho^2\left\{1 - \frac{p}{2} + p\frac{\rho^2}{\rho^2 - 1}\ln\rho\right\}}\right). \tag{15.23}$$

The uptake rate can then be calculated as:

$$\omega = \frac{p}{1 - \frac{p}{2} + p\frac{\rho^2}{\rho^2 - 1}\ln\rho}\bar{c}. \tag{15.24}$$

The approximate cumulative uptake calculated with Eqs. (15.23) and (15.24) was shown to give good agreement with the uptake calculated by a numerical model (Nye and Mariott 1969) especially for $p < 5$. Taking the limit of Eq. (15.24) as $p \to \infty$ yields the solution for zero concentration at the root surface. Moreover, if one can assume $\rho >> 1$, as usually is the case, Eq. (15.24) simplifies to:

Table 15.1. List of symbols

Symbol	Name	Dimension	Dimensionless symbol
T	Time	T	$t = DT/R_0^2$
R	Radial co-ordinate	L	$r = R/R_0$
R_0	Root radius	L	–
R_1	Soil cylinder radius	L	ρ
Δz	Root length (layer thickness)	L	η
A	Plant demand for nutrients	$M.L^{-2}.T^{-1}$	$\omega = \rho^2\, B/(2\phi\eta)$
S_i	Initial amount of nutrient	$M.L^{-3}$	–
ϕ	Supply/demand parameter	–	$\phi = D\, S_i/(AR_0)$
C	Concentration	$M.L^{-3}$	c
C_i	Initial concentration	$M.L^{-3}$	–
$B\ (= K_a + \Theta)$	Buffer capacity	–	$\beta = (K_a + \Theta)/\Theta$
w	Water uptake rate	$L^3.T^{-1}$	–
V	Flux of water	$L.T^{-1}$	$v = W/(2\pi HD)$
			$\mu = E/\{2\pi HD(\rho^2 1)\}$
L_{rv}	Root density	L^{-2}	–
K_a	Adsorption constant	–	–
Θ	Water content	–	–
D	Diffusion coefficient	$L^2.T^{-1}$	–
α	Root absorbing power	$L.T^{-1}$	$p = \alpha R_0/D$
K_m	Michaelis-Menten constant	$M.L^{-3}$	$\kappa = K_m/C_i$

$$\omega = \frac{1}{\ln\rho - \frac{1}{2}}\bar{c}. \tag{15.25}$$

Claassen (1989) used the same approach as Baldwin et al. (1973), but used Michaelis-Menten kinetics. The uptake rate can now be given as:

$$\omega = \frac{\bar{c} - c_0}{\ln\rho - \frac{1}{2}}, \tag{15.26}$$

where c_0 is the concentration at the root surface, which itself is a non-linear function of the average concentration:

$$c_0 = \frac{\bar{c} - \kappa - \kappa p f(\rho)}{2} + \sqrt{\frac{[\bar{c} - \kappa - \kappa p f(\rho)]^2}{4} + \kappa\bar{c}} \tag{15.27}$$

where $f(\rho)$ is defined as: $f(\rho) = \ln\rho - 1/2$. In the case of zero-sink uptake, i.e. when $c_0 = 0$, Eq. (15.26) is identical to Eq. (15.25).

De Willigen and van Noordwijk (1987) described the development of the concentration distribution as a sequence of steady rate situations, in accordance with the boundary condition of zero concentration at the root surface. Now the average concentration is given by:

$$\bar{c} = \exp\left(\frac{-2t}{B\left\{\frac{1-3\rho^2}{4} + \frac{\rho^4}{\rho^2-1}\ln\rho\right\}}\right), \tag{15.28}$$

and the uptake rate by:

$$\omega = \frac{\rho^2-1}{\frac{1-3\rho^2}{4} + \frac{\rho^4}{\rho^2-1}\ln\rho}\bar{c}. \tag{15.29}$$

When $\rho >> 1$ Eq. (15.29) simplifies to:

$$\omega = \frac{1}{\ln\rho - \frac{3}{4}}\bar{c}. \tag{15.30}$$

The cumulative uptake can be calculated in both cases with:

$$\Omega = 1 - \bar{c}. \tag{15.31}$$

Both Eqs. (15.24) and (15.29) give the uptake rate as a function of the average concentration, the buffer power does not play a role in this relation. De Willigen (1990) showed that the zero-sink uptake rate can be approximated by Eq. (15.29) even if the adsorption isotherm is non-linear.

Similar approximations as Eqs. (15.24) and (15.29) apply in cases of transport by mass flow playing a role where the replenishment of water takes place at the outer boundary of the soil cylinder (Cushman 1980; De Willigen and van Noordwijk 1994a,b).

Up to now it has been assumed that the root over its total (active) surface had direct contact with either the soil solution or the soil solid phase. However, inspection of root-soil contact in the field, both macroscopically, and microscopically by observing thin sections (Altemüller and Haag 1983), reveals that complete contact between the root and the soil liquid and solid phases may be the exception rather than the rule, especially for heavy soils. De Willigen and van Noordwijk (1987) derived a steady rate solution of the distribution of concentration around a root of which only a fraction of its radial surface partakes in uptake. From their solution an approximate zero-sink uptake rate can be derived as:

$$\omega = \frac{\rho^2-1}{\frac{1-3\rho^2}{4} + \frac{\rho^4}{\rho^2-1}\ln\rho + \frac{2(\rho^2-1)}{\psi_1}\sum_{k=1}^{\infty}\frac{\rho^{2k}+1}{\rho^{2k}-1}\frac{\sin k\psi_1}{k^2}}\bar{c}. \tag{15.32}$$

15.2.4.4 Examples and Applications

Availability. The concepts of soil fertility and nutrient availability have shifted from essentially static approaches to ones in which the transport of solute in the soil-root system plays an important role, a tale well told by Nye and Tinker (1977).

Bray (1954) was among the first to point out the significance of the mobility of nutrients in the soil with respect to plant nutrition. Barber (1962) formulated this more precisely when he distinguished two mechanisms of transport towards the plant root: transport by mass flow and transport by diffusion, concepts which are now generally accepted (Nye and Tinker 1977). How much of the available amount will actually be taken up by the plant at the required rate depends on the rate itself, the path length of transport (which is a function of root length density), the transport possibilities in the soil, and the rate of depletion by sinks other than the plant root. The available amount (in the definition of de Willigen and van Noordwijk 1987) is that part of the total amount present in the root zone which can be taken up by the crop when transport through soil is not limiting. Within the root zone during the uptake period many processes can occur which render part of the originally available pool, at least temporarily, unavailable (chemical fixation, microbial immobilisation), or transform former unavailable fractions into a readily available forms (mineralization, release from minerals). When relevant, considerations about crop uptake have to take into account these amounts released or fixed.

Because of its finite root density and because transport rates in soil are finite, the plant can take up only a fraction of the available pool. The amount that can be taken up at the required rate, will be indicated as the *unconstrained uptake capacity*. Then the *total uptake capacity* is that amount which will be taken up in a certain period, e.g. a growing season, with a rate less than or equal to the required rate. The difference between total uptake capacity and available pool is the amount remaining in the soil due to transport limitations. De Willigen and van Noordwijk (1987) applied the steady rate solution to estimate the amount of a nutrient which is unconstrained and available, for $\rho >> 1$, it is given by:

$$Q = S_i - \frac{A(K_a + \Theta)}{D\Delta z} \frac{R_1^2}{2} \left\{ \ln \rho - \frac{3}{4} \right\}. \tag{15.33}$$

The unconstrained uptake capacity thus equals the available amount S_i diminished by a term representing the amount which cannot be taken up at the required rate. The unconstrained uptake capacity equals the available amount if root length density and/or the diffusion coefficient becomes infinite, i.e. when transport is not limited. This can be expressed as a fraction of the available amount, the fractional depletion:

$$F_d = 1 - \frac{\Theta\beta\rho^2}{2(\rho^2-1)} \frac{AR_0^2}{DS_i\Delta z} \left\{ \frac{1-3\rho^2}{4} + \frac{\rho^4 \ln\rho}{\rho^2-1} \right\}. \tag{15.34}$$

Figure 15.1 shows the fractional depletion, i.e. the fraction of the available nutrient which can be taken up with the required rate. For the parameters used it shows that in absence of adsorption, e.g. in case of nitrate, a rather sparse root system can take up almost all of the available nutrient.

Application to Pot Experiments. The single root models are especially suitable for analysing laboratory and pot experiments. A good example of the use of a model with simple geometry is given by the study of Gahoonia et al. (1994) concerning the transport of phosphorus to the rhizosphere as influenced by soil moisture. By employing a special technique in pot experiments, the effects of water content on transport of P and water could be studied separately, i.e. isolated from effects of water availability on root growth and physiology. The experimental set up was such that transport in one direction in a planar geometry and zero-sink uptake could be assumed. The adsorption isotherm of P was taken to consist of two linear branches, so that for each branch separately an analytical solution could be used. There were two treatments comprising a (constant) moisture content of 0.14 and 0.20 respectively leading to diffusion

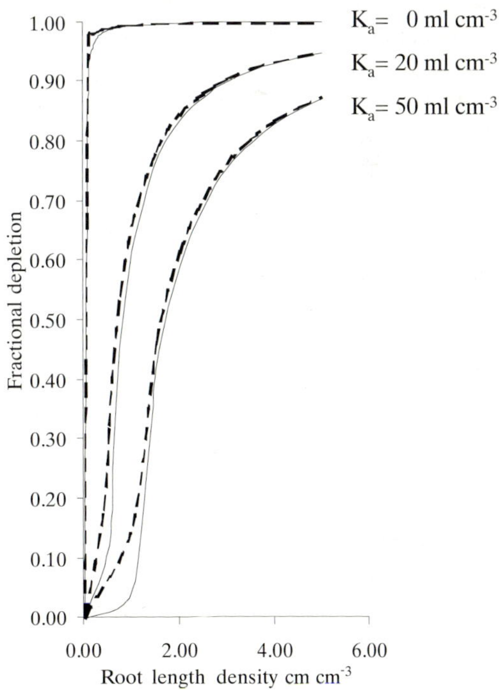

Fig. 15.1. Fractional depletion as a function of root length density, mass flow and adsorption strength. Parameters: uptake rate 2 kg $ha^{-1}.day^{-1}$, available amount 400 kg ha^{-1}, diffusion coefficient 0.1 $cm^2 day^{-1}$. *Interrupted lines*: transpiration 0.5 cm day^{-1}, *uninterrupted lines* no transpiration, transport by diffusion only. The numbers at the *curves* give the values of the adsorption constant

coefficients differing by a factor of 4. Figure 15.2 a and b show the good agreement between theory and experiment, i.e. the differences in distribution of (available) P around the root could, to a great extent, be explained by the differences in diffusion coefficients, which in their turn are caused by difference in water content, so confirming the validity of the transport theory.

Claassen (1989) added to the usual form of the transport equation (15.14) a sink term to account for the uptake of root hairs. The strength of the sink was calculated by using an equation like (15.26), where now R_0 and R_1 denote the radius of the root hair and the half-distance between root hairs respectively. The model was put to the test by calculating P-uptake by a number of species in pot experiments. Taking uptake by root hairs into account considerably improved the agreement between measured and calculated uptake (Fig. 15.3a and b).

15.2.5 Water

15.2.5.1 Transport Equation

In temporate regions the soil solution under conditions of interest is very dilute [in the order of 0.01 M (Bolt and Bruggenwert 1979)], accordingly, gradients of the concentration of water in the soil solution will be very small, and can be neglected for our purposes, i.e. in the case of water one can assume [cf. Eq. (15.4)]:

$$\frac{\partial S}{\partial T} = -\nabla . F_C . \tag{15.35}$$

The convective flux is given by Darcy's law which states that the flux of the soil solution is proportional to the gradient of total potential. The total potential is the sum of various component potentials, but as we deal with horizontal transport only, we will only take into account matric potential, reflecting the forces by which the water is held by the soil matrix. The treatment below is taken, with some modifications, from Stroosnijder (1976). Darcy's law reads:

$$V = -K_w \nabla P_T = -K_w \nabla P_m, \tag{15.36}$$

where K_w is the hydraulic conductivity $[M^{-1}.L^3.T]$, P_T the total potential $[M.L^{-1}.T^{-2})]$, and P_m the matric potential. It is convenient and commonly done, to express the potential in terms of length of water column. The flux equation then becomes:

$$V = -K \nabla h, \tag{15.37}$$

where h is the pressure head [L], and K the hydraulic conductivity $[L.T^{-1}]$. Substitution of Eq. (15.3) and Eq. (15.37) into Eq. (15.35) results in:

Fig. 15.2a,b. Concentration of inorganic P in soil solution (*points* observed, *lines* calculated) at varying distance from a root mat of maize, at two soil moisture levels. (After Gahoonia et al. 1994)

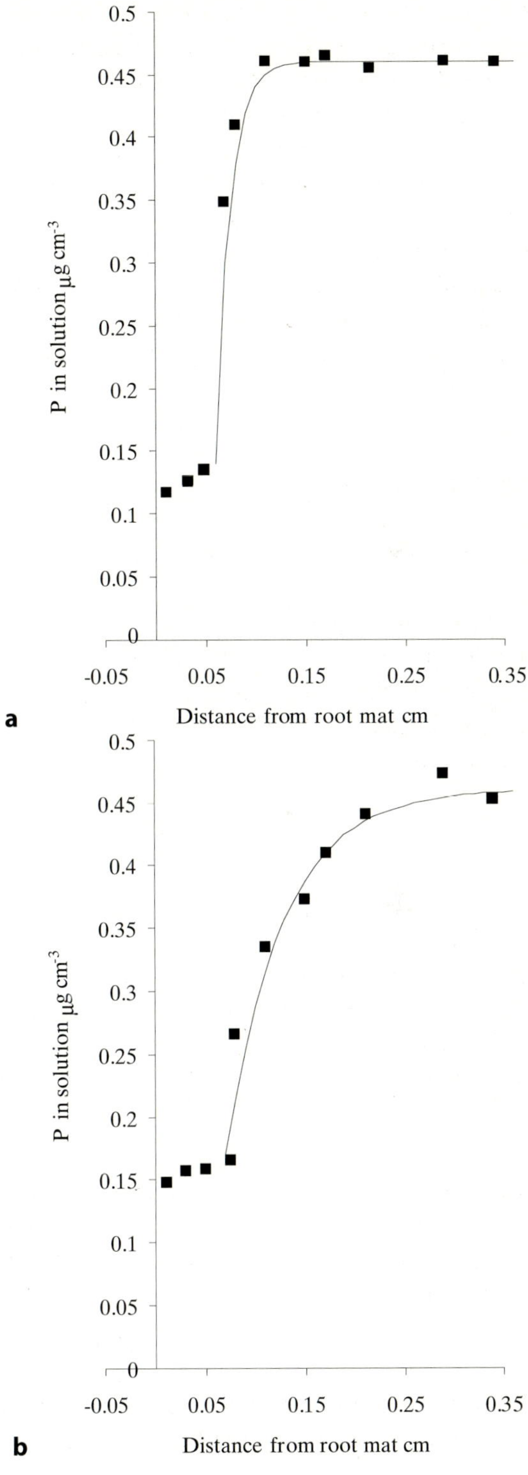

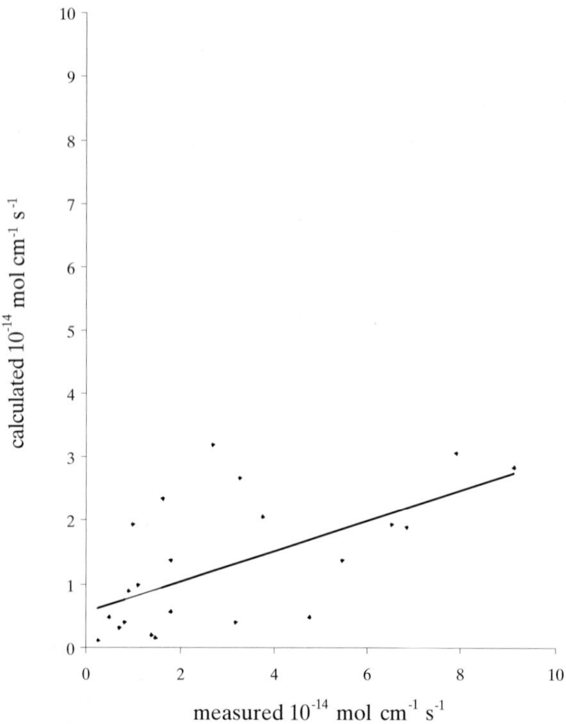

a

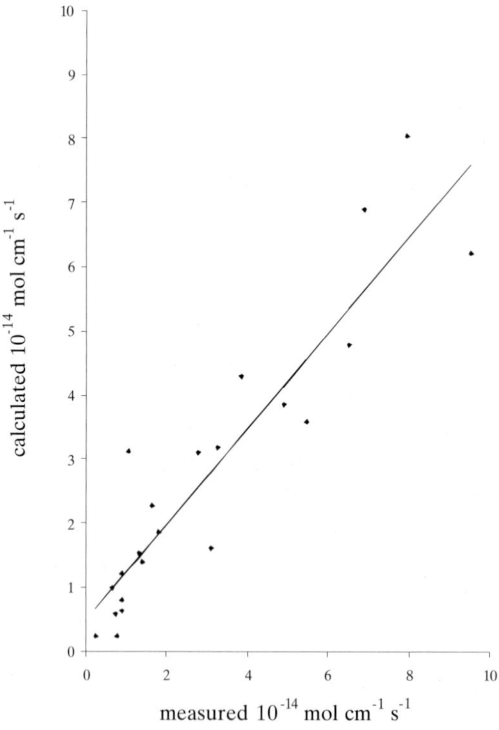

b

Fig. 15.3. Comparison of measured and calculated influx of P for seven plant species. **a** Calculations without root hairs. **b** Calculations taking the presence of root hairs into account. (After Claassen 1989)

$$\frac{\partial S}{\partial T} = \nabla \cdot KC\nabla h. \tag{15.38}$$

As $S = \theta C$, where θ is the volumetric water content of the soil and since C can be taken to be constant because of the low concentration of solutes, one finally obtains:

$$\frac{\partial \theta}{\partial T} = \nabla K \nabla h. \tag{15.39}$$

To solve Eq. (15.39) the relation between P_m (or H_p) and θ should be known. This relation is usually called the water retention curve. Now the water diffusivity $D_w(\theta)$ can be defined as:

$$D_w = -K\frac{dh}{d\theta}. \tag{15.40}$$

Substitution of Eq. (15.40) into Eq. (15.39) leads to:

$$\frac{\partial \theta}{\partial T} = \nabla D_w \nabla \theta. \tag{15.41}$$

In cylindrical co-ordinates with neglect of tangential and vertical gradients term one obtains:

$$\frac{\partial \theta}{\partial T} = \frac{1}{R}\frac{\partial}{\partial R} R D_w \frac{\partial \theta}{\partial R}. \tag{15.42}$$

Because of the non-linear relations between D_w and θ, and K and θ, a solution of Eq. (15.41) can usually be found only by numerical methods, though in special cases analytical solutions can be found (e.g. Knight and Philip 1974).

15.2.5.2 Boundary Conditions

The same type of boundary conditions as with nutrient uptake apply, viz. given flux or water content. The former is mostly used, where the flux at the root surface is related to root length density and transpiration.

But contrary to uptake of nutrients, uptake of water is usually assumed to be a passive process, a consequence of gradients in water potential in the path: bulk soil-rhizosphere-root-leaf-atmosphere. The models dealt with here pertain to a part of the path viz. to the transport of water from bulk soil to the rhizosphere and from there into the root. This leads to a boundary condition of the form (Campbell 1985; De Willigen 1990):

$$R = R_0, \quad D_w \frac{\partial \theta}{\partial R} = \frac{F_1}{2\pi R_0}, \tag{15.43}$$

where R_0 is the root radius (L), and F_1 is the uptake rate per unit length of root ($L^2.T^{-1}$). The uptake rate is assumed to be proportional to the difference in pressure between water in the rhizosphere and in the roots (the root water potential):

$$F_1 = K_P(h_{R_0} - h_P), \tag{15.44}$$

where h_{R_0} is the pressure head of the rhizosphere (L), h_P is the pressure head in the root (L), and K_P is the conductance of the root ($L.T^{-1}$). Equation (15.44) thus describes the flow across the root surface. On the other hand, F_1 is also equal to the transpiration rate per unit root length:

$$F_1 = \frac{E}{\Delta z L_{rv}}, \tag{15.45}$$

where E is the actual transpiration rate ($L.T^{-1}$), and Δz (L) is the layer thickness (root length). Equation (15.45) describes the flow of water from bulk soil to the root surface. From Eqs. (15.44) and (15.45) it follows that

$$E = L_{rv}\Delta z K_P(h_{R_0} - h_P). \tag{15.46}$$

The actual transpiration rate in its turn may depend on the root water potential:

$$E = f(h_P)E_{pot}, \tag{15.47}$$

where E_{pot} is the potential transpiration rate, as determined by atmospheric evaporative demand ($L.T^{-1}$) and $f(h_P)$ denotes the reduction of potential transpiration due to root water potential. Campbell (1991) proposed the function:

$$f(h_P) = \frac{1}{1 + \left(\frac{h_P}{h_{P,1/2}}\right)^a} \tag{15.48}$$

where $h_{P,1/2}$ is the leaf water potential where $E = T_{pot}/2$, and the exponent a is a species dependent constant.

When the level of dissolved salts in the soil solution becomes appreciable, the difference in total soil moisture potential (including the osmotic as well as the matric potential) between the soil and the root-water potential must be taken into account. A detailed description of water and solute transport from soil to root must consider both the differential hydraulic conductance and the differential osmotic permeability of the various tissues and flow paths, with the osmotic factor being particularly important at the endodermis.

Because the membranes of the roots exclude most solutes from the stream of solution passing through them, there will be a build-up of solutes at these membranes and hence the osmotic pressure at the membranes will be greater than that of the external solution. Localised draw-downs of both matric and osmotic components of soil moisture potential around the roots can be important, particularly in the case of young plants and/or if the irrigation water or the soil is saline. Water and solute transport through root systems or segments of roots is more precisely expressed in the plant physiological literature by the relationship:

$$F_1 = K_P(\Delta P - \sigma\Delta\pi), \tag{15.49}$$

which derives from irreversible thermodynamics and is routinely used for describing the flow of water across an individual semipermeable membrane (Dainty 1976). ΔP is the drop in hydrostatic pressure across the membrane; σ is the reflection coefficient, which takes into account the permeability of the membrane to a given solute; and $\Delta\pi$ is the rise in osmotic pressure, across the membrane. Equation (15.49) applies, strictly, only to a well-defined membrane with uniform properties.

15.2.5.3 Analytical Solutions and Approximations

Contrary to the flux equation for solutes, which for our purposes can be assumed to be linear, i.e. the diffusion coefficient and the flux of water are independent of the solute concentration, the flux equations for soil water are highly non-linear, the conductivity decreasing strongly with decreasing pressure head and water content. Next to the non-linearity of the flux equation there is a non-linear relation between the pressure head and the water content. Due to these non-linearities, the transport equation generally has to be solved numerically. However, some analytical approximations are mentioned in the literature and will be discussed here.

Van den Honert (1948) introduced the attractive analogue of a electrical current flowing through a series of resistances. For the analogy to apply, a steady-state situation should exist. Gardner (1960) described the drying of soil, due to root water uptake, as a sequence of steady-states.

For a homogeneous soil and horizontal transport, the non-linearity of the flux can be removed by introducing the matric flux potential (Raats 1970), defined by:

$$\Phi = \int_{h_{ref}}^{h} K dh = \int_{\theta_{ref}}^{\theta} D_w d\theta, \tag{15.50}$$

where Φ is the matric flux potential ($L^2.T^{-1}$), h_{ref} is a reference value of the pressure head and the corresponding water content. This transforms Eq. (15.37) into:

$$V = -\nabla\Phi. \tag{15.51}$$

Next to the non-linearity of the flux equation there is a non-linear relation between the pressure head and the water content. By assuming a steady-rate situation, i.e. supposing $\partial\theta/\partial t$ = constant, Passioura (1980) obtained an analytical expression relating uptake of water to root length density and difference in matric flux potential in the bulk soil and at the root surface:

$$\Phi_{R1} - \Phi_{R_0} = \frac{E}{2\pi L_{rv} H}\left(\frac{\rho^2}{\rho^2-1}\ln\rho - \frac{\rho^2-1}{2(\rho^2-1)}\right). \tag{15.52}$$

De Willigen and van Noordwijk (1987) have shown the distribution of water content around the root calculated using Eq. (15.52) is in excellent agreement with that calculated using a model, which solves Eq. (15.39) numerically. From Eq. (15.52), the water flow from the bulk soil towards the root surface can be expressed as a function of the average value of the matric flux potential in the bulk soil and in the rhizosphere:

$$F_1 = \frac{\rho^2 - 1}{\frac{1-3\rho^2}{4} + \frac{\rho^4}{\rho^2-1}\ln\rho^2}[\overline{\Phi} - \Phi_{R_0}]. \tag{15.53}$$

15.2.5.4 Examples and Applications

Availability. As in the case of nutrients, the following definition of availability is used: the available amount of water in the root zone is that amount of water which can be taken up at the rate required to fulfil the demand of the plant for water, if transport in the soil/root system were not limiting uptake, i.e. if the diffusivity and root conductance were infinitely large. An infinitely small gradient of pressure head would then suffice to transport water to the root surface and from there into the root at the required rate. Suppose that the transpiration of the plant is reduced when the pressure head of water in the plant equals a certain value $h_{P,c}$. In the absence of any gradient, the pressure head and water content are the same everywhere in the soil, and the uptake of water by such a root system can proceed unhampered until the value of the pressure head in the rhizosphere equals $h_{P,c}$ with a corresponding water content of $\theta_{P,c}$. The amount of available water W (L) in a soil layer with given water content $\theta_i > \theta_{P,c}$ and thickness Δz (L) can thus be given as:

$$W = \frac{2\pi\Delta z \int_{R_0}^{R_1} R(\theta_i - \theta_{P,c})dR}{\pi R_1^2} = \pi\Delta z L_{rv}(\theta_i - \theta_{P,c})(R_1^2 - R_0^2). \tag{15.54}$$

Availability in this definition thus depends on the crop only through the plant-specific value of $h_{P,c}$. In reality, of course, the diffusivity and the root conductance have finite values. This means that when the root water potential equals $h_{P,c}$, the pressure head in the rhizosphere equals (cf. Eqs. (15.44) and (15.45)):

$$h_{R_0} = h_{P,c} + \frac{E_{pot}}{\Delta z L_{rv} K_P}. \tag{15.55}$$

Moreover, there will be a gradient in the water content around the root. The unconstrained available water W_u which can be taken up at the required rate, taking into account the transport possibilities of the soil/root system, is defined by:

$$W_u = 2\pi L_{rv}\Delta z \int_{R_0}^{R_1} R(\theta_i - \theta)dR = W - 2\pi L_{rv}\Delta z \int_{R_0}^{R_1} R(\theta - \theta_{P,c})dR. \tag{15.56}$$

The integral on the right hand side of Eq. (15.56) can be evaluated using Eq. (15.52), where Φ_{R_0} is evaluated with (15.50) at $h = h_{R0}$, with h_{R0} given by Eq. (15.55). The fractional depletion of the available water realised in the period of unconstrained uptake is:

$$F_{D,W} = \frac{W_u}{W} = 1 - \frac{2\int_{R_0}^{R_1} R(\theta - \theta_{P,c})dR}{(\theta_1 - \theta_{P,c})(R_1^2 - R_0^2)}. \tag{15.57}$$

Figure 15.4 shows the fractional depletion as a function of root length density for a Dutch clay soil (clayb10, Wösten 1987) with $h_{P,c} = -2$ MPa, corresponding to an amount of available water of 3.37. 10^{-2} m, or a maximum uptake period of 6.74 days, E_{pot} being 5.10^{-3} m day^{-1}. The unconstrained availability and uptake periods are zero, when the pressure head in the rhizosphere equals the value corresponding to field capacity, with the parameter values used here at a root length density of 2.51 10^3 m m^{-3}. If root length density is only slightly higher, the unconstrained fractional depletion increases fast, e.g. to 0.4 and 0.7 at root length densities of 3.10^3 m m^{-3} and 6.10^3 m m^{-3} respectively. Above a root length density of 0.5 10^3 m m^{-3} further increase in root length density causes a limited increase of the fractional depletion, i.e. from 0.9 at $L_{rv} = 0.5$ cm cm^{-3} to 0.98 at $L_{rv} = 5\ 10^3$ m m^{-3}. This means that doubling the total root length within a layer, e.g. increasing the root length density by a factor two, is much less favourable than maintaining the same average root length density and doubling the root length, i.e. in the case discussed here distributing the total root length over 0.4 rather than over 0.2 m.

Root and Soil Resistance. The single root uptake model can be used to asses the relative importance of soil resistance with respect to root resistance. The root resistance is calculated as:

$$\omega_P = \frac{h_{R_0} - h_P}{E}, \tag{15.58}$$

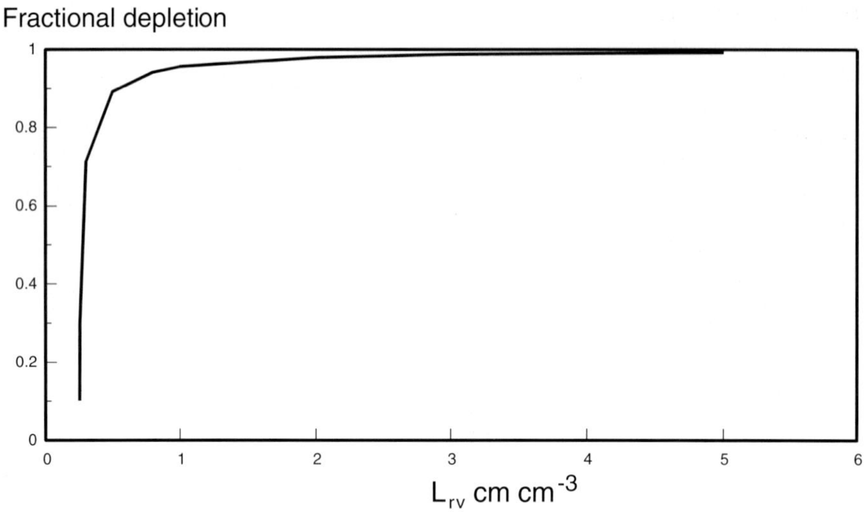

Fig. 15.4. Fractional depletion as a function of root length density for a clay soil. Critical root water potential −2 MPa, potential transpiration rate 5.10^{-3} m day^{-1}

where ω_P is the resistance of the root (d), so:

$$\omega_P = \frac{1}{K_P L_{rv} \Delta z}. \tag{15.59}$$

The soil resistance is, in analogy to Eq. (15.58), equal to the difference between pressure head in the bulk soil, $\bar{h}$, and that found in the rhizosphere, h_{R0}, divided by the transpiration:

$$\omega_s = \frac{\bar{h} - h_{R_0}}{E}. \tag{15.60}$$

Figure 15.5 shows the plant and soil resistance as a function of the average pressure head (given as pF), for a root length density of 10^3 m m^{-3}, and transpiration of 5 10^{-3} m day^{-1}, for a clay, sand, and peat soil in the Netherlands (Wösten 1987), and a root conductance of 23.10^{-4} m d^{-1}, the highest value found in the literature (De Willigen and van Noordwijk 1987). Clearly, even with this high value of root conductance, plant resistance exceeds the soil resistance for pF-values up to 3 (sand) or even 4.5 (clay). This is an illustration of the conclusion of Molz (1981): "... it [the root resistance] seems certain to dominate in the upper 75% of the water content under normal rooting conditions".

Herckelrath et al. (1977) were the first to quantitatively evaluate the effects of limited root-soil contact for water uptake. The uptake rate in their approach is multiplied with the water content relative to the water content at saturation.

Fig. 15.5. Soil and plant resistance as a function of average pF-value [=log(h)] of the soil. Parameters: root radius 0.01 cm, root length 20 cm, root length density 1 cm cm^{-3}, root conductance 23.10^{-6} cm day^{-1}

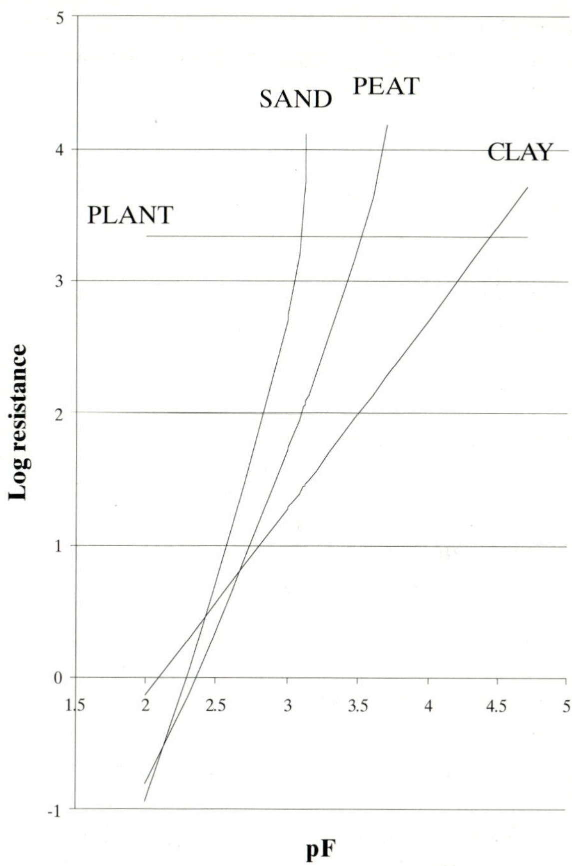

The same procedure has been used by Hansen et al. (1991) and Bouten (1992). The approximation in Eq. (15.32) can be used to estimate the extra resistance in the soil due to incomplete contact, again the matric flux potential plays an analogous role to the concentration. For the clay soil and the parameters of Fig. 15.5, the point where the soil resistance exceeds the root resistance shifts from 2.77 MPa in case of complete contact to 1.57 MPa when contact is 10%.

15.3 Modelling Uptake by a Crop Root System

15.3.1 Nutrients

For a root system consisting of regularly distributed roots confined to a single layer of uniform initial nutrient concentration, upscaling from a single root model is rather trivial, the uptake rate per unit soil surface simply is the product of the root length density and the uptake rate of a single root. Upscaling is considerably more difficult, even for parallel identical roots confined to a single soil layer, when roots or nutrient are not regularly distributed. Barley (1970) proposed constructing a polygon around each root, consisting of the locus of points in the soil nearer to that root than to any other. Such a construction is called the Dirichlet tessellation (Green and Sibson 1977). Barley then considered the uptake from a cylinder with the same area as the polygon. Uptake rate was assumed to be proportional to the concentration at the root surface. He found small effects of non-regular root distribution. Baldwin et al. (1973), using an electrical analogue, found considerably larger effects. De Willigen and van Noordwijk (1987) constructed a Dirichlet tessellation for root maps made in the field, and calculated the potential uptake from equivalent cylinders, as Barley had done, but also took into account the eccentric position of the root in such a cylinder. They found no reduction in uptake for mobile solutes like nitrate, but a reduction of 75% for a nutrient with an adsorption constant of 20.

Rappoldt (1992) presented a highly interesting theory on the use of an equivalent model system to which the relevant diffusion equation can be applied. The model system consists of spherical, cylindrical or plane aggregates, which do not represent the individual aggregates of the soil, however. The length scales in the real and model system are the same. A model of diffusion in soil aggregates is evaluated by solving a differential equation in the model system. The overall result for the soil is then found as a weighted sum of the results for the various length scales (cf. Sect. 7.3.7 on the root effectivity ratio R_{per}).

Claassen and Barber (1976) presented a model of a growing root system, where it was assumed that the roots do not interfere [boundary condition, Eq. (15.18)], i.e. it deals with a single root, the length of which is increasing in time. Barber and Cushman (1981) extended this model to one where the boundary condition Eq. (15.19) was used, but where root length density apparently stayed constant.

Hoffland et al. (1990) simulated phosphate uptake for a regularly distributed root system where roots behave as zero-sinks, increasing in root length density in time. With increasing density the radius of the soil cylinder allocated to each root decreases and the concentration gradients in the soil volumes surrounding the root are truncated. Newly grown roots sample the soil with the largest concentration. The model was successfully applied to a pot experiment with rape. Later (Hoffland 1991) this was extended to include exudation of organic acids by the root system in order to explain the unexpectedly high uptake when rock phosphate was applied.

Geelhoed et al. (1997) extended this model with a sink term for zero sink uptake by root hairs. The zero sink influx term was taken from de Willigen and van Noordwijk [1994b, cf. Eq. (15.29)]. The model appeared to give good results for P-uptake of maize plants grown in a medium of quartz sand coated with goethite, at least when the level of adsorbed phosphate was high (Fig. 15.6). Low level uptake was severely underestimated, due to the large decrease in pH in the course of the experiment, which was not incorporated into the model.

Most crop growth models contain a module which deals with uptake by the root system. Van Keulen et al. (1975) used a numerical model to calculate the uptake possibilities of a root system. They came to the conclusion that in case of non-adsorbed nutrients, like nitrate, under normal conditions availability is almost 100%, even at low root length densities (see also Fig. 15.1). Based on such calculations, in many crop growth models (Greenwood et al. 1985; van Keulen and Seligman 1987; Rijtema and Kroes 1991) dealing with uptake of nitrate the (distribution of) root length density does not play a role, the only relevant parameter being the vertical extent of the root system.

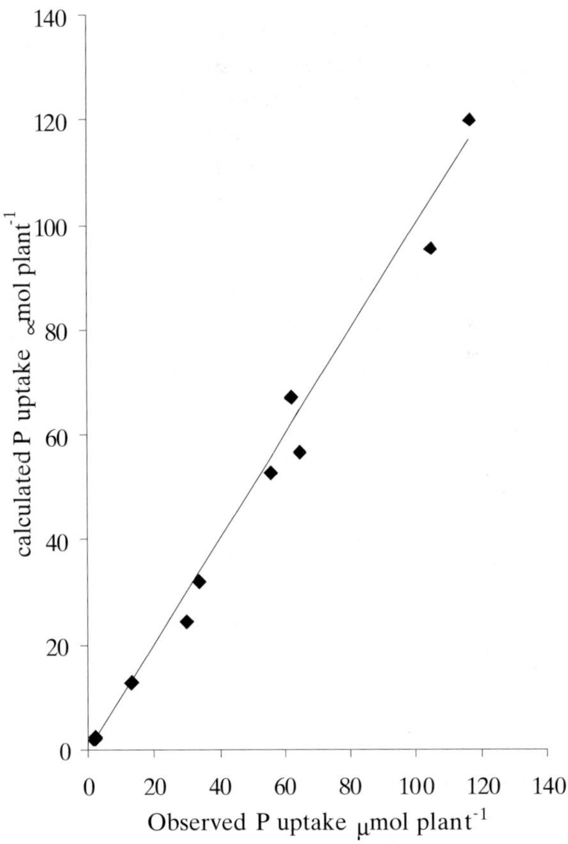

Fig. 15.6. Observed and calculated P uptake by maize plants. Uptake calculated with zero sink Eq. (15.29). (After Geelhoed 1998)

In the models or modules where distribution of root length density is taken into account, the root zone is divided into layers or compartments within which the roots are assumed to be distributed regularly (e.g. Jones and Kiniry 1986; Hansen et al. 1991; De Willigen and van Noordwijk 1995). For calculation of the uptake rate the steady-state, steady-rate and zero-sink approximations are used.

15.3.1.1 Examples and Applications

Stationary Root Systems. The assumption of a stationary root system in one layer of soil is acceptable for phosphate uptake by a full-grown arable crop. Due to the low mobility of phosphate ions, transport of P from the plough layer to the deeper layers can be ignored, and uptake will mainly take place from the plough layer (the upper 20 to 25 cm of the soil profile, Prummel 1957). The phosphate recommendation for arable crops in the Netherlands is based on the so-called Pw-value of the plough layer. This value gives the amount of phosphate, expressed in mg $P_2O_5 l^{-1}$ soil, soluble in water at an extraction ratio of 60 volumes of water to 1 volume of soil (van der Paauw 1971). It was shown by de Willigen and van Noordwijk (1978) and van Noordwijk et al. (1990) that the Pw-value can be calculated with reasonable accuracy on the basis of 24 h P-adsorption isotherms and the total amount of labile P as measured by anion exchange resins or iron-hydroxide-impregnated filter paper.

Van Noordwijk et al. (1990) used Eq. (15.29) to calculate the minimum required phosphate supply of the soil for maximum growth of a number of crops and two groups of soils, and compared these with experimentally established values for required phosphate supply. For each crop, representative values for daily P demand and root length density were used. Although the agreement between theoretical and experimental results was not 1 : 1, the correlation between the two appeared considerable. This at least indicates that the P level in the soil necessary to meet the phosphate requirement of crops can be estimated using Eq. (15.29).

Growing Root Systems. Three examples of models of uptake by a growing root system will be treated here. These models are in fact modules and as such a part of a model on crop growth, in which above ground growth and soil processes like mineralization, leaching, etc. are also dealt with.

The first example is the nutrient uptake module of the Danish model DAISY (Hansen et al. 1990, 1991). In this model the root depth and root density profile are calculated on a daily basis and are used as input variables in the models for water and nutrient uptake.

The nitrogen uptake model is based on the concept of a potential nitrogen demand simulated by the crop model, and the plant availability of soil nitrogen, i.e. the rate by which nitrogen can be made available at the root surfaces. The transport of nitrogen from the bulk soil to the root surfaces is based on the assumption that each root exploits an average effective volume of soil which is a cylinder around each root. The radius of this cylinder corresponds to the average half distance between the assumed parallel roots. The nitrogen transfer to the root surface takes place by mass flow and diffusion. It is assumed that the concentration distance profile develops in time in a stepwise manner and, at each time step, approximates to a steady state profile. In the present model it is assumed that nitrogen uptake equals the nitrogen flux towards the root surface. Based on these assumptions, the flux I of nitrogen towards the root surface is calculated as:

$$I = \begin{cases} 2\pi D(\overline{C} - C_r)\left[\dfrac{\rho^2 \ln\rho}{\rho^2 - 1} - \dfrac{1}{2}\right] & \nu = 0 \\ q_r \dfrac{(\rho^2 - 1)\overline{C} - C_r \ln\rho^2}{(\rho^2 - 1) - \ln\rho^2} & \nu = 2, \\ q_r \dfrac{(\rho^2 - 1)(1-\nu)\overline{C} - (\rho^{2-2\nu} - 1)C_r}{(\rho^2 - 1)(1-\nu) - (\rho^{2-2\nu} - 1)} & \text{else} \end{cases}$$

where q_r is the radial water flow rate ($L^3.L^{-1}.T^{-1}$), and C_r is the concentration in the soil solution at the root surface ($M.L^{-3}$).

If the uptake is limited by the availability of nitrogen, the concentration at the root surface is assumed to be equal to zero and hence the root will act as a zero sink. In this case total uptake of nitrogen is calculated by integrating the flux over the entire root system. In the case of ample nitrogen supply, the total nitrogen uptake is determined by the potential nitrogen demand. Then total uptake is distributed over the entire root zone by assuming a common concentration to exist along the root surfaces of the entire root system. Soil layers in which the concentration is less than the common concentration are assumed not to contribute to nitrogen uptake. The calculations are performed for both ammonium and nitrate. It is assumed that ammonium is taken up by the plant roots in preference to nitrate. The mobility of the ammonium in soil is considered less than that of nitrate due to adsorption of ammonium to soil colloids which is described by an adsorption-desorption isotherm. The vertical movement of nitrogen is modelled by means of a numerical solution of the convection-dispersion equation for ammonium as well as for nitrate. The source sink term in the convection dispersion equation integrates the transformation processes for ammonium as well as for nitrate.

A second example pertains to the uptake module developed by De Willigen and van Noordwijk (1987, 1995) applied in the crop growth model of CP-BKF (Cultures Pluviales – Burkina Faso, AB DLO Haren). The potential uptake rate where the root is supposed to act as a zero-sink is first calculated based on Eq. (15.29), if the required uptake is less than the potential, the actual uptake rate equals the required uptake rate, otherwise it equals the potential uptake rate. Furthermore, in the module it is supposed, as indeed is often found, that roots in favourable positions can compensate for roots in less favourable positions by increasing their uptake rates (De Jager 1985). It is therefore assumed that the uptake rate of roots in a certain layer depends on the uptake potential of roots in other layers. The calculation procedure implies that roots growing under favourable conditions will compensate as much as possible for roots growing under less favourable conditions. It is thus assumed that information about the necessary behaviour, as far as uptake is concerned, is instantaneously available throughout the complete root system. The procedure described above can also be applied when roots within a horizontal layer are not distributed regularly. The layer then is divided over a number of compartments within which root distribution can be assumed to be regular.

The third example stems from the much used CERES-model, in this case the CERES-Maize (Jones and Kinity 1986). Here the potential uptake of nitrate from a certain layer is calculated as:

$$U_i = L_{v,i} * u_{max} * \alpha_w^2 * \Delta z_i * 100 * F_{av}, \qquad (15.61)$$

where U is uptake in kg ha^{-1} day^{-1}, $L_{v,i}$ is the root length density in cm cm^{-3}, α_w is a reduction factor depending on relative water content, Δz_i is the thickness of the layer in cm, the maximum uptake rate per unit root length (mg cm^{-1} d^{-1}) and F_{av} an availability factor depending on the nitrate content ($N_{NO3,i}$ in mg kg^{-1}):

$$F_{av,i} = 1 - \exp(-0.03 * N_{NO3,i}). \qquad (15.62)$$

The factor 100 is for conversion to kg ha^{-1} day^{-1}. Total uptake is calculated by summing the uptake rates from the individual layers. Then a nitrate uptake factor is computed as the ratio of demand and total potential uptake. If this factor exceeds 1, the actual uptake equals the potential uptake, otherwise the actual uptake is calculated as the potential uptake multiplied by the nitrate uptake factor.

For a situation as given in Table 15.2, and a plant demand of 4 kg ha^{-1} day^{-1}, the uptake distribution was calculated using the three models. Results are given in Fig. 15.7, where it is shown that the Ritchie formulation leads to a total uptake rate of less than half the demand, whereas the two other uptake models do satisfy demand completely or almost.

Table 15.2. Distribution of roots, nitrogen and water with depth as used for the calculations of distribution of nitrogen uptake by three different uptake models

Layer cm	Root length density $cm\ cm^{-3}$	Nitrogen content $kg\ ha^{-1}$	Water content –
0–20	1.	20	0.2
20–40	0.5	10	0.2
40–60	0.25	8	0.2
60–80	0.12	5	0.2
80–100	0.06	2	0.2

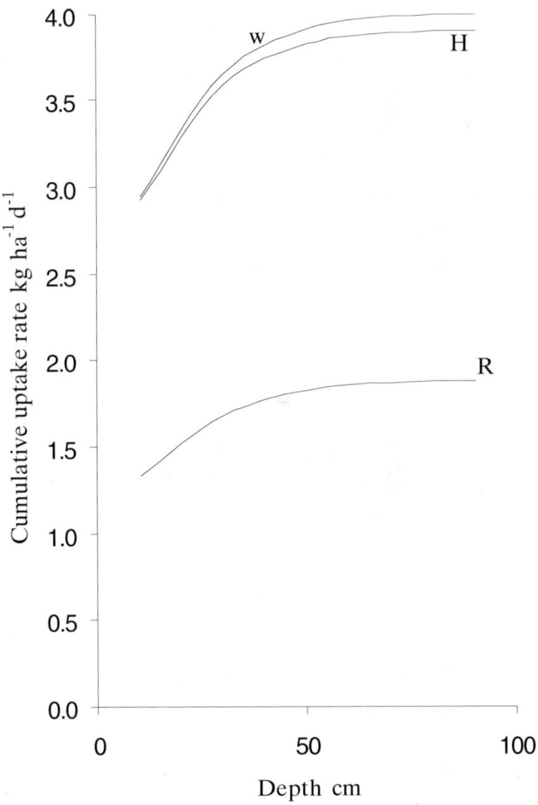

Fig. 15.7. Cumulative uptake rate with depth for the situation given in Table 15.2, as calculated with the uptake module of Hansen et al. (1991; indicated with *H*) the Ceres model (*R*) and that of de Willigen and van Noordwijk (1995) (*W*)

15.3.2 Water

Obviously, as with nutrients, upscaling from a single root to a set of regularly distributed parallel roots is simply a matter of multiplying the root length density by the uptake rate. Lafolie et al. (1991) used a numerical model to calculate uptake of water by a (stationary) root system consisting of parallel non-regularly distributed roots. A Dirichlet tessellation was constructed from root maps, and based on this a triangle finite element grid was generated.

As with nutrients, one can distinguish two types of uptake models. In the first group where root length density and root water potential do not play a role, uptake is determined by the depth of the root zone and the pressure head or water content of the soil layers (e.g. Feddes et al. 1978; van Keulen and Seligman 1987), the other is based upon steady-state or steady-rate approximations of the single root model (e.g. Nimah and Hanks 1973; Hillel et al. 1976; De Willigen and van Noordwijk 1995).

15.3.2.1 Examples and Applications

A first example is the water uptake module as used in the SWATRE model (Feddes et al. 1978). First, a potential water uptake rate is calculated, which is corrected for the pressure head in the soil. In its simplest form the potential water extraction rate (S_{max}) is assumed to be equally distributed over the root zone (Z_r):

$$S_{max} = \frac{E}{Z_r}. \tag{15.63}$$

Later (Bouten 1992), the potential uptake was distributed according to the relative distribution of the root length density:

$$S_{max,i} = \frac{L_{rv,i}}{\sum L_{rv,i}} \frac{E}{\Delta Z_i}. \tag{15.64}$$

The actual uptake rate (S) in a certain layer depends on the pressure head of that layer:

$$S = a(h) S_{max}. \tag{15.65}$$

De Willigen and van Noordwijk (1995) used a water uptake model (WATUP) based on an extension of the single root model treated in Section 15.2.5. Again, for a root system distributed over several soil layers (or compartments), they assumed that within each layer (compartment) the roots are regularly distributed and that again in each layer flow of water from the bulk soil to the rhizosphere equals that from the rhizosphere into the root:

$$[\bar{\Phi}_i - \Phi_{R_{0,i}}]\frac{\pi(\rho_i^2 - 1)}{G(\rho_i)} = K_P(h_{R_{0,i}} - h_P), \tag{15.66}$$

where the subscript i denotes the number of the layer. Also, the total uptake should equal transpiration:

$$\sum_i^N L_{rv,i}\Delta z_i K_P(h_{R_{0,i}} - h_P) = E_{act} = f(h_P)E_{pot}. \tag{15.67}$$

This yields N + 1 equations with N + 1 unknowns: the N values for $h_{Ro,i}$ and the root water potential h_P. Again, the equations have to be solved by iteration because of the non-linear relation between Φ and h, and the non-linear nature of $f(h_P)$. For a relatively wet soil the major resistance is found in the root (Fig. 15.5), and the gradients in the soil are accordingly very small, i.e. $\Phi_{Ro,i} \approx \bar{\Phi}_i$, and h_P is much larger than h_i. The uptake from a certain layer is then proportional to the relative root length density in that layer as in Eq. (15.64). So for relatively wet conditions, both models yield the same distribution of uptake.

The models are applied to results of an experiment of Sharp and Davies (1985). They measured root length and water uptake distribution of maize plants. After a preliminary growth period, half of the plants were watered daily, the other half were not watered at all. Figure 15.8a shows the results of the well-watered treatment. The results show that measured water uptake is more or less proportional to root length density as predicted by WATUP and SWATRE with a potential sink term as given in Eq. (15.64). The sink term in Eq. (15.63) leads to a uniform distribution of water uptake.

Figure 15.8b shows the uptake distribution for the treatment where no water was given during 18 days. Now the distribution of uptake is not determined by distribution of root length density but by distribution of pressure head, as presumably the resistance in the soil now exceeds that in the roots. The three models yielded more or less the same distribution as found in the experiment although the measured transpiration amounted to 0.93 cm day^{-1}, WATUP calculated 0.85 cm day^{-1}, and SWATRE, using Eq. (15.63), yielded 0.62 cm day^{-1}, and using Eq. (15.64) arrived at 0.49 cm day^{-1}.

15.4 Conclusions and Suggestions for Further Research

Though it can be said that we have come a long way in modelling uptake by root systems since the beginning in the early 1960s, a large number of uncertainties still exist. One of the difficulties is the validation of the uptake models as such in field situations. Especially in the case of uptake of nutrients, it is compulsory to consider other processes as well e.g. mineralization, dissolution,

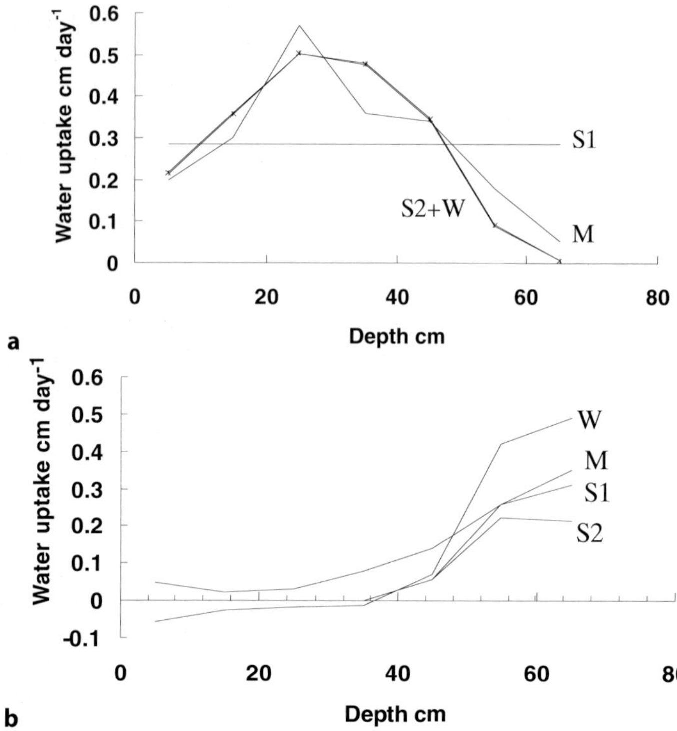

Fig. 15.8. a Distribution of water uptake with depth as observed (Sharp and Davies 1985, M), and calculated by SWATRE (Feddes et al. 1978, S1), modified SWATRE (Bouten 1992, S2) and WATUP (de Willigen and van Noordwijk 1995, W). Well-watered treatment. **b** As for **a** but during 18 days no water was given

aboveground growth, distribution of dry matter etc. Even when uptake is predicted satisfactorily this could be due to compensating errors. Moreover, differences between models when acting at a certain instant as shown in Section 15.3.1, may disappear when considered during longer periods due to feedback mechanisms, which makes it difficult to decide which is the "correct" one.

Many models perform reasonably when simulating uptake in a relative rich environment, and do much worse under poor conditions (De Willigen et al. 1998), this seems especially the case for uptake of phosphate. Amijee et al. (1991) state: "... in soils of low P concentration ... the predicted value always underestimates the observed value". This has to do with the many possible adaptations of the roots and the root system, apart from simply increasing its size and exploring new areas, which are not implemented in most models. Many P-starved plants exude organic acids, thereby influencing the adsorption properties of the soil with respect to phosphate in such a way that availability is increased. A few models (e.g. Hoffland 1991; Geelhoed 1998) have included these processes. Other processes triggered by root activity are temporal immobilisation of N and P by micro-organisms, and production and exudation of iron mobilising substances. Darrah (1993) gives a nice review of the relative importance of such processes. Future models should, where relevant, include more details of the processes in the rhizosphere and their interaction with root activity and functioning.

Another interesting area, but now at the scale of a root system rather than of a single root, is that of the influence of root architecture on efficiency of uptake. Fitter (1987) estimated this by combining his topological model with the depletion zone as defined by Baldwin and Nye (1974). Doussan et al. (1998a,b) coupled a model of root system architecture (Pagès et al. 1989) with laws describing flow of water into and within roots. They used the model to investigate the influence of conductance varying with time on the distribution of water potential in a maize root system. They found that considerable gradients in water potential could develop. It is to be expected that coupling of architectural models with appropriate uptake models will be one of the main activities in the field of modelling of uptake and functioning of root systems, but the influence of growth and decay of roots should also be dealt with.

On a still larger scale, there are models describing the competition between root systems of different species, which is of relevance for e.g. agroforestry systems or where interaction between crops and weeds is studied. Recently, two models investigating agroforestry systems have been published, Wimisa (Mayus 1998) and Wanulcas (Van Noordwijk and Lusiana 1998), the first one dealing with competition between trees and a crop for water, the second also dealing with competition for nutrients. Wimisa has been put to the test with reasonable results, Wanulcas is still at the development stage. Developing and validating such complicated models offer a major challenge to future research.

References

Abramowitz M, Stegun IA (1970) Handbook of mathematical functions. Dover Publications, New York

Altemüller HJ, Haag Th (1983) Mikroskopische Untersuchungen an Maiswurzeln im ungestörten Bodenverband. Kali-Briefe (Büntehof) 16 (6): 349–363

Amijee F, Barraclough PB, Tinker PB (1991) Modelling phosphorus uptake and utilisation by plants. In: Johansen C, Lee KK, Sahrawat KL (eds) Phosphorus nutrition of grain legumes in the semi-arid tropics, pp 63–77 ICRISAT, Pabanchever, Judia

Baldwin JP, Nye PH (1974) A model to calculate the uptake by a developing root system or root hair system of solutes with concentration variable diffusion coefficients. Plant Soil 40: 703–706

Baldwin JP, Nye PH, Tinker PB (1973) Uptake of solutes by multiple root systems from soil. III. A model for calculating the solute uptake by a randomly dispersed root system developing in a finite volume of soil. Plant Soil 38: 612–625

Barber SA (1962) A diffusion and mass flow concept of soil nutrient availability. Soil Sci 93: 39–48

Barber SA (1984) Soil nutrient bioavailability. A mechanistic approach. Wiley, New York

Barber SA, Cushman JH (1981) Nitrogen uptake model for agronomic crops. In: Iskandar IK (ed) Modelling wastewater renovation. John Wiley, New York, pp 382–410

Barley KP (1970) The configuration of the root system in relation to nutrient uptake. Adv Agron 22: 159–201

Bear J (1972) Dynamics of fluids in porous media. American Elsevier, New York, 764 pp

Bolt GH, Bruggenwert MGM (1979) Composition of the soil. In: Bolt GH, Bruggenwert MGM (eds) Soil chemistry A. Basic elements. Elsevier, Amsterdam, 1979, pp 1–11

Bouldin DR (1961) Mathematical description of diffusion processes in the soil-plant system. Soil Sci Soc Amer Proc 25: 476–480

Bouten W (1992) Monitoring and modelling forest hydrological processes in support of acidification research. Doct Thesis, University of Amsterdam, Amsterdam

Bray RH (1954) A nutrient mobility concept of soil plant relationships. Soil Sci 78: 9–22

Campbell GS (1985) Soil physics with BASIC. Elsevier, Amsterdam

Campbell GS (1991) Simulation of water uptake by plant roots. In: Hanks J, Ritchie JT (eds) Modeling plant and soil systems. Agronomy Monograph 31, American Society of Agronomy, Madison, Wisconsin, pp 273–285

Carlslaw HS, Jaeger JC (1959) Conduction of heat in solids. Oxford University Press, Oxford, 510 pp

Claassen N (1989) Die Aufnahme vom Nährstoffen aus dem Boden durch die höhere Pflanze als Ergebnis von Verfügbarkeit und Aneignungsvermögen. Habilitationsschrift, Georg-August Universität, Göttingen, 327 pp

Claassen N, Barber SA (1976) Simulation model for nutrient uptake from soil by a growing plant root system. Agron J 68: 961–964

Cushman JH (1979) An analytical solution to solute transport near root surfaces for low initial concentration: I. Equations development. Soil Sci Soc Am J 43: 1087–1090

Cushman JH (1980) Completion of the list of analytical solutions for nutrient transport to roots. 1. Exact linear models. Water Resour Res 16: 891–896

Dainty J (1976) Water relations of plant cells. In: Lüttge U, Pitman MG (eds) Transport in plants. II.A Cells. Encyclopaedia of Plant Physiology. New Series, vol 2. Springer, Berlin Heidelberg New York, pp 12–35

Darrah PR (1993) The rhizosphere and plant nutrition: a quantitative approach. In: Barrow NJ (ed) Plant nutrition from genetic engineering to field practice. Kluwer, Dordrecht, the Netherlands, pp 3–23

De Jager A (1985) Response of plants to localised nutrient supply. PhD Thesis, University of Utrecht, Utrecht, 137 pp

De Willigen P (1981) Mathematical analysis of diffusion and mass flow of solutes to a root assuming constant uptake. Inst Bodemvruchtbaarheid, Rapport 6–81, 56 pp

De Willigen P (1990) Calculation of uptake of nutrients and water by a root system. Nota 210, Institute for Soil Fertility, Haren, the Netherlands

De Willigen P, van Noordwijk M (1978) Calculations on phosphate uptake by a crop in relation to rooting density and P-status of the soil (In Dutch). Inst Bodemvruchtbaarheid Rapport 19–78, 63 pp

De Willigen P, van Noordwijk M (1987) Roots, plant production and nutrient use efficiency, PhD Thesis, Wageningen Agricultural University, Wageningen

De Willigen P, van Noordwijk M (1994a) Mass flow and diffusion of nutrients to a root with constant or zero-sink uptake. I. Constant uptake. Soil Sci 157: 162–170

De Willigen P, van Noordwijk M (1994b) Mass flow and diffusion of nutrients to a root with constant or zero-sink uptake. II. Zero-sink uptake. Soil Sci 157: 171–175

De Willigen P, van Noordwijk M (1995) Model for interactions between water and nutrient uptake. In: Kabat P, Marshall B, van den Broek BJ, Vos J, van Keulen H (eds) Modelling and parametrization of the soil-plant-atmosphere system, a comparison of potato growth models. Wageningen Press, Wageningen, pp 135–153

De Willigen P, Dijksterhuis G, Chardon W (1998) Dynamique et disponibilité du Phosphore dans les sols. In: Breman H, Sissiko K (eds) L'intensification agricole au Sahel. Karthala, Paris. pp 266–284

Doussan C, Vercambre G, Pagès L (1998a) Modelling of the hydraulic architecture of root systems: an integrated approach of water absorption – model description. Ann Bot 81: 213–223

Doussan C, Vercambre G, Pagès L (1998b) Modelling of the hydraulic architecture of root systems: an integrated approach of water absorption – distribution of axial and radial conductances in maize. Ann Bot 81: 225–232

Dunham RJ, Nye PH (1973) The influence of soil water content on the uptake of ions by roots. I. Soil water content gradients near a plane of onion roots. J Appl Ecol 10: 585–598

Dunham RJ, Nye PH (1974) The influence of soil water content on the uptake of ions by roots. II. Chloride uptake and concentration gradients in soil. J Appl Ecol 11: 581–596

Feddes RA, Kowalik PJ, Zaradny H (1978) Simulation of field water use and crop yield. Simulation monographs, PUDOC, Wageningen, 189 pp

Fiiter AH (1987) An architectural approach to the comparative ecology of plant root systems. New Phytol. (Suppl.) 106: 61–77

Gahoonia TS, Nielsen NE (1991) A method to study rhizosphere processes in thin soil layers of different proximity to roots. Plant Soil 135: 143–146

Gahoonia TS, Nielsen NE (1992a) Control of pH at soil root interface. Plant Soil 140: 49–54

Gahoonia TS, Nielsen NE (1992b) The effect of root-induced pH change on the depletion of inorganic and organic phosphate. Plant Soil 143: 185–191

Gahoonia TS, Raza S, Nielsen NE (1994) Phosphorus depletion in the rhizosphere as influenced by soil moisture. Plant Soil 159: 213–218

Gardner WR (1960) Dynamic aspects of water availability to plants. Soil Sci 89: 63–73

Geelhoed J (1998) Phosphateavailability in the soil-root ysytem: integration of oxide surface chemistry, transport and uptake. PhD Thesis, Wageningen Agricultural University, 177 pp

Geelhoed JS, Findenegg GR, van Riemsdijk WH (1997) Availability to plants of phosphate adsorbed on goethite: experiment and simulation. Eur J Soil Sci 48: 473–481

Green PJ, Sibson R (1977) Computing Dirichlet tessellations in the plane. Comput J 21: 168173

Greenwood DJ, Neeteson JJ, Draycott A (1985) Response of potatoes to N fertiliser. Plant Soil 85: 163–183

Hansen S, Jensen HE, Nielsen NE, Svendsen H (1990) DAISY: A Soil Plant System Model. Danish simulation model for transformation and transport of energy and matter in the soil plant athmosphere system. The national Agency for Environmental Protection, Copenhagen, 369 pp

Hansen S, Jensen HE, Nielsen NE, Svendsen H (1991) Simulation of nitrogen dynamics and biomass production in winter wheat using the Danish simulation model DAISY. Fert Res 27: 245–261

Herckelrath WM, Miller EE, Gardner WR (1977) Water uptake by plants. II The root contact model. Soil Sci Soc Am Proc 41: 1039–1043

Hillel D, Talpaz H, Van Keulen H (1976) A macroscopic-scale model of water uptake by a nonuniform root system and of water and salt movement in the soil profile. Soil Sci 121: 242–255

Hoffland E (1991) Mobilisation of rock phosphate by rape. PhD Thesis, Wageningen Agricultural University, Wageningen

Hoffland E, Bloemhof HS, Leffelaar PA, Findenegg GR, Nelemans JA (1990) Simulation of nutrient uptake by a growing root system considering increasing root density and interroot competition. Plant Soil 124: 149–155

Jones CA, Kinity JR (eds) (1986) CERES-maize: a Simulation model of maize growth and development. Texas A&M University Press, College Station, 194 pp

Jones JW, Ritchie JT (1991) Crop growth models. In: Hoffman GJ, Howell TA, Solomon KH (ed) Management of farm irrigation systems. ASAE, St. Joseph, MI. pp 63–89

Jungk A, Claassen N (1989) Availability in soil and acquisition by plants as the basis for phosphorus and potassium supply to plants. Z Pflanzenernaehr Bodenkd 152: 151–157

Knight J, Philip JR (1974) Exact Solutions in non linear diffusion. J Eng math 8: 219–227

Lafolie F, Bruckler L, Tardieu F (1991) Modelling root water potential and soil-root water transport: I. Model presentation. Soil Sci Soc Am J 55: 1203–1211

Mayus M (1998) Millet growth in windbreak-shielded fields in the Sahel, experiment and model. PhD Thesis, Wageningen Agricultural University, Wageningen

Molz FJ (1981) Models of water transport in the soil-plant system: a review. Water Resour Res 17: 1245–1260

Nielsen NE (1983) Plant parameters controlling the efficiency of nutrient uptake from soil. In: United Nations Economic Commission for Europe (ed) efficient use of fertilisers in agriculture. Developments in Plant and Soil Sciences, vol 10. Martinus Nijhoff, Dordrecht, the Netherlands, pp 199–217

Nimah A, Hanks RJ (1973) Model for estimating soil water, plant, and atmospheric interrelations. I Description and sensitivity. Soil Sci Soc Amer Proc 37: 522–527

Nye PH (1966) The effect of the nutrient intensity and buffering power of a soil, and the absorbing power, size and root hairs of a root, on nutrient absorption by diffusion. Plant Soil 25: 81–105

Nye PH (1992a) Towards the quantitative control of crop production and quality. I. The role of computer models in soil and plant research. J Plant Nutr 15 (6&7): 1131–1150

Nye PH (1992b) Towards the quantitative control of crop production and quality. II. The scientific basis for guiding fertiliser and management practice, particularly in poorer countries. J Plant Nutr 15 (6&7): 1151–1173

Nye PH (1992c) Towards the quantitative control of crop production and quality. III. Some recent developments in research into the root-soil interface. J Plant Nutr 15 (6&7): 1175–1192

Nye PH, Marriott FHC (1969) A theoretical study of the distribution of substances around roots resulting from simultaneous diffusion and mass flow. Plant Soil 30: 459–472

Nye PH, Tinker PB (1977) Solute movement in the soil-root system. Sudies in Ecology vol 4. Blackwell Scientific Publications, Oxford, 342 pp

Olsen SR, Kemper WD (1968) Movement of nutrients to plant roots. Adv Agron 20: 91–151

Pagès L, Jordan MO, Picard D (1989) Simulation of the three-dimensional architecture of the maize root system. Plant Soil 119: 147–154

Passioura JB (1980) The transport of water from soil to shoot in wheat seedlings. J Exp Bot 31: 333–345

Prummel J (1957) The significance of the thickness of the arable layer for the evaluation of soil analysis data. (In Dutch). Landbouwkd Tijdschr 69: 703–712

Raats PAC (1970) Steady infiltration from line sources and furrows. Soil Sci Am Proc 34: 709–714

Rappoldt C (1992) Diffusion in aggregated soil. PhD Thesis, Wageningen Agricultural University, 162 pp

Rengel Z (1999) Mineral nutrition of crops. Food Products Press, New York, 399 pp

Rijtema PE, Kroes JG (1991) Some results of nitrogen simulations with the model ANIMO. In: Groot JJR, de Willigen P, Verberne ELJ (eds) Nitrogen turnover in the soil-crop system. Kluwer, Dordrecht, pp 189–199

Sharp RE, Davies WJ (1985) Root growth and water uptake by maize plants in drying soil. J Exp Bot 36: 1441–1456

Stroosnijder L (1976) Infiltratie en herverdeling van water in grond. PUDOC, Wageningen, 213 pp

Turner NC (1974) Stomatal behaviour and water status of maize, sorghum and tobacco under field conditions. II At low water potential. Plant Physiol 53: 360–365

Van den Broek BJ (1992) Predicting times of irrigation with a simulation model. M Sc Thesis, University of Alberta, Alberta, 122 pp

Van den Honert TH (1948) Water transport in plants as a catenary process. Discuss Faraday Soc 3: 146–153

Van der Paauw F (1971) An effective water extraction method for the determination of plant-available soil phosphorus. Plant Soil 34: 467–481

Van Keulen H, Seligman NG (1987) Simulation of water use, nitrogen nutrition and growth of a spring wheat crop. PUDOC, Wageningen, Simulation Monographs, 310 pp

Van Keulen H, Seligman NH, Goudriaan J (1975) Availability of anions in the growth medium to roots of an actively growing plant. Neth J Agric Sci 23: 131–138

Van Noordwijk M, Lusiana B (1990) WaNulCAS 1.1. Background of a model of water nutrient and light capture in agroforestry systems. Working Document, International Centre for Research in Agroforestry (ICRAF), Bogor, Indonesia

Van Noordwijk M, de Willigen P, Ehlert PAI, Chardon WJ (1990) A simple model of P uptake as a possible basis for P fertiliser recommendations. Neth J Agric Sci 38: 317–322

Van Wijk WR (1963) Physiscs of plant environment. North-Holland publ. Co., Amsterdam, 382 pp.

Wösten JHM (1987) Beschrijving van de waterretentie- en doorlatendheidskarakteristieken uit de Staringreeks met analytische functies. Stiboka rapport 2019, Stiboka Wageningen, 53 pp

Youngs EG, Gardner WR (1963) A problem of diffusion in the infinite hollow cylinder. Soil Sci Soc Am Proc 27: 475–476

CHAPTER 16

Plant Anchorage

A.R. Ennos[1] and S. Pellerin[2]*

[1] The University of Manchester, School of Biological Sciences, 3.614 Stopford Building, Oxford Road, Manchester, M13 9PT, UK

[2] INRA, Domaine de la Grande Ferrade, B.P. 81, 71 Av. Ed. Bourleaux, 33883 Villenave d'Ornon Cédex, France

CONTENTS

16.1 Introduction

Together with water and nutrient absorption, a major function of roots is to anchor plants. This chapter is dedicated to methods used for studying root systems when considering this particular function.

* Principal author

A.L. Smit et al. (Eds.), Root Methods

Under natural conditions, plants may be subjected to two types of external forces which could cause failure to the anchorage: lateral or vertical pulling by a grazing herbivore, or lateral pushing by wind. The resistance of plants to uprooting influences the durability of grazed pastures, whereas resistance to lodging of arable crops such as wheat, maize or rape has consequences on yield (Carter and Hudelson 1988; Minami and Ujihara 1991), harvest quality (in the case of silage maize for instance) and time required for harvesting operations. Similarly, damage by windthrow may cause huge economic losses for forestry. Methods for evaluating root anchorage are therefore needed in these different contexts: resistance to uprooting of grassland species; resistance to lodging of arable crops; and resistance to windthrow of trees.

Resistance of grassland species to uprooting has been rarely studied. Devices for measuring forces exerted by grazing herbivores were, however, proposed by Hughes et al. (1991) and Laca et al. (1992). In contrast, lodging of arable crops and windthrow of trees have given rise to an abundant literature. Both lodging and windthrow are the consequence of very complex phenomena involving airflow above and within the canopy, the mechanics of plants under wind loading, and their anchorage characteristics (Baker 1995). Moreover, it has been shown that plants exhibit adaptive growth when subjected to mechanical stimulation, which modifies both shoot and root morphology (Stokes et al. 1995; Crook and Ennos 1996a; Goodman and Ennos 1996, 1997). Lodging or windthrow may happen to healthy plants, or to plants previously weakened either by remobilisation of carbohydrates from the stalk or roots, or by attacks of pathogens or pests. In some cases, stems or trunks may break or buckle, so that the problem is dependant on the mechanical properties of the stem or trunk. In most cases, however, plants are displaced from the vertical or uprooted without any stem damage occurring, so that the anchorage of the plant is clearly involved. In either case, resistance to lodging or windthrow involves the dynamic equilibrium between the aerial part, which is subject to wind, and the soil-root system, which anchors the plant.

The mechanics of trees, and to a lesser extent arable crops, under wind loading has given rise to numerous studies (Coutts and Grace 1995; Flesch and Grant 1992a,b). Models exist for calculating the transfer of wind momentum into canopies. The aerial part of individual plants is subjected to wind pressure and also creates a torque when displaced from the vertical because of its own weight. The characteristics of plants which are involved therefore include the area of the leaf and stem, their vertical distribution, and drag coefficient; the vertical distribution of their weight; and the mechanical properties of the stems. Plant root development, and the way in which roots and soil anchor plants has received less attention than what occurs above-ground. Pioneer studies have only relatively recently been carried out on trees and arable crops.

This chapter focuses on methods used for measuring plant anchorage. The reader must bear in mind, however, that most anchorage studies are only valuable if combined with investigations on aerial parts of the plant. Basic descriptions of the mechanics of plant anchorage are given in Section 16.2. Available methods for measuring anchorage components, or anchorage as a whole, are given in Sections 16.3 and 16.4.

16.2 The Mechanics of Anchorage

If we are to prevent catastrophic failure of crop plants it is clearly important to understand exactly how roots anchor plants in the soil. Unfortunately, the mechanics of anchorage are extremely complex, even when a plant is simply being pulled out of the ground. As we shall see, anchorage strength depends not only on the strength of individual roots, but also on their length, distribution and branching pattern, and also on the properties of the soil. Changes in each of these factors can affect not only the anchorage strength but also the pattern of failure itself.

16.2.1 Resistance to Uprooting

Roots resist uprooting in just the same way as tent pegs (Ennos 1989, 1990); friction transfers tension from the roots into the surrounding soil. Such a system can fail in several ways depending on the size and strength of the root and the strength of the soil. If a root is short and strong and the soil weak it will pull out of the ground. In contrast, a long, weak root in strong soil will break before the soil fails. In practise, unstrengthened roots more than a few millimetres in length would break long before being pulled out of the soil, and at very low forces.

Most climbing and procumbent plants which have to resist being uprooted have their anchorage systems improved over a single unstrengthened root in three ways (Ennos 1991a, 1993). First, the friction between root and soil is increased by possession of root hairs which are glued to soil particles. However, even then the rate at which tension can be transferred is ultimately limited by the shear strength of the soil. Hence in weak, wet soil friction is still transferred slowly. Second, their roots are strengthened towards their base where they are most highly stressed. Third, most of these plants produce fibrous root systems using large numbers of adventitious roots or a highly branched primary root to produce a root system with a large surface area. In these ways, therefore, the anchorage strength is maximised for a minimum investment in strengthening material.

In practise, the failure of real root systems occurs by a combination of three modes of failure (Ennos 1991a, 1993). When a plant is pulled up the soil fails in tension near the base of the stem and a ball of soil attached to the basal roots is broken off. Distal to this further failure occurs. Some of the longer or weaker roots fail in tension. The break is sometimes clean but often the inner stele pulls out from the cortex. However, the shorter or stronger ones are pulled out of the soil intact.

Exactly where failure occurs depends on the soil conditions. The shear strength of hard, dry soil is high so tension will be transferred rapidly from roots to soil. The roots will mostly snap near their base and only a small root ball will break free with the stem base. The uprooting resistance will be high and in extreme conditions will exceed the strength of the stem. In contrast, soft, wet soil has a low shear strength, so tension will be transferred only slowly into the soil and failure will occur much further away from the stem base. More roots will be pulled out intact from the soil together with a larger root ball, and the force required to uproot the plant will be much lower. This is why soil around plants should be dampened before transplanting them.

16.2.2 Resistance to Lodging and Windthrow

The mechanics of anchorage of self-supporting plants against lodging or windthrow is even more complex since the root system must transmit a torque into the soil rather than a simple upward force. There are many possible designs for such an anchorage system and the manner in which they fail is even more dependent on the properties of the soil in which the plant is growing. However, there is one fundamental aspect of the design of all these root systems; each contains at least one rigid element without which a torque could not be transmitted (Ennos and Fitter 1992; Ennos 1993). A plant with a fibrous root system would simply collapse if its stem was not propped up. The rigid element allows the soil to be loaded in one of two ways. In some systems an upward force is transmitted to the soil on the windward side of the stem and an equal downward force to the soil on the leeward side. Alternatively, upper soil layers can be loaded to the leeward side while lower soil layers are loaded by an equal windward force. Both patterns of forces produce a net torque which props up the plant and, if failure does occur, the mode will depend on the relative ability of the roots and soil to withstand these forces.

16.2.2.1 Systems Transmitting Vertical Forces to the Soil

At one extreme of the range of anchorage systems which transmit vertical forces to the soil are plants which have a broad, spreading root system with many horizontal lateral roots and some vertical sinkers. Such plants include many trees including spruce (Coutts 1983, 1986), larch (Crook and Ennos 1996b), poplar and many buttressed rainforest trees. In these spreading systems the resistance of the soil to downward movement of the laterals will be very high because of the large area of the root system and the high compressive strength of soil. For these reasons failure occurs mostly on the windward side, the plant being levered up about a hinge on the leeward side (Fig. 16.1A). Exactly what happens depends on the soil conditions, just as it does during uprooting of climbers. In wet soil a large root plate may be levered out of the ground together with many sinkers which are pulled out intact. In contrast, in drier, stronger soil a smaller root plate may be formed, windward roots may be broken and sinkers sheared off. In such systems three components of anchorage may be identified (Fig. 16.1A): the resistance of the leeward hinge to bending; the resistance of the windward side to uprooting and the weight of the root-soil plate.

At the other extreme of the range of anchorage types are plants which have a narrower root system with large numbers of nearly vertical roots. Such systems are found in the coronal root systems of winter wheat (Crook and Ennos 1993) and the prop root systems of maize (Ennos et al. 1993b). In such systems the resistance of the windward roots to being pulled out of the soil is higher than the combined resistance of the soil beneath the narrow root plate to being compressed and the leeward roots to buckling. As a result, the system will rotate about a *windward* hinge (Fig. 16.1B) and a cone of roots and attached soil will be levered *into* the ground. The size of the cone will depend on the soil strength, being large in weak wet soil and smaller in strong dry soil. There are only two possible components of anchorage: the resistance of the soil to compression and the buckling resistance of the leeward roots.

16.2.2.2 Systems Transmitting Horizontal Forces to the Soil

A final extreme form of anchorage system is the simple tap root system seen in many members of the Brassicaceae (Ennos and Fitter 1992) and some pioneer rainforest trees, such as members of the Euphorbiaceae (Crook et al. 1997). In such systems anchorage is almost entirely provided by the vertical root which acts just like a foundation pile. When such a plant is pushed over the tap root rotates about a point some way below the soil surface, the top moving to leeward and the bottom to windward (Fig. 16.1C). These movements are resisted by two components: the compressive resistance of the soil to lateral motion and the bending resistance of the tap root itself. Resistance of the lower regions of the tap root to uprooting may also play some slight role. As in the other systems the exact mode of failure will depend on the soil strength: soft, wet soil will fail deep into the ground and the plant will lean over permanently; in stronger, drier soil the tap root or stem are more likely to snap.

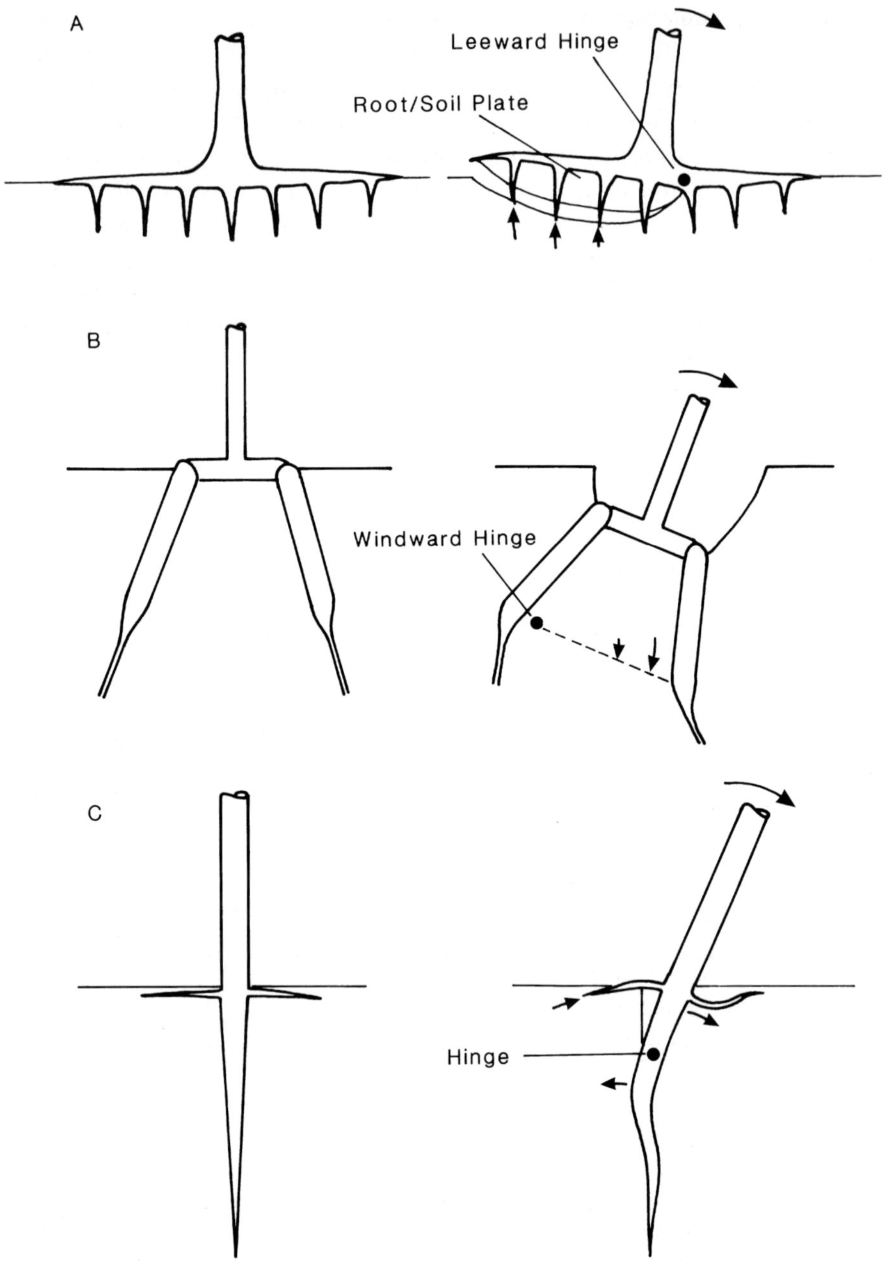

Fig. 16.1A–C

Understanding the behaviour of these three extreme types of anchorage systems is all very well but plants do not always behave in the textbook manner. Some species may have morphologies which are intermediate between two types. Sunflowers, for instance, have a large tap root as well as spreading laterals. In other species the pattern of failure may vary between varieties. In some varieties of maize, for example, the windward roots tend to shear off while in others the leeward roots can buckle above-ground. Similarly, spring wheat may fail in a way which is intermediate between winter wheat and trees, the plant rotating about a point directly beneath its stem (Ennos 1991b). The mode of failure may even depend on the conditions in which plants have been grown or on the soil moisture. Himalayan balsam, a dicot with a root system which looks like that of maize, only rarely fails in the same way (Ennos et al. 1993a); more often there is windward failure similar to that characteristic of many tree species.

16.3 Methods for Measuring Anchorage Components

16.3.1 Measuring the Relative Contributions of Anchorage Components

The first stages in determining the components of anchorage in a particular plant are to investigate the morphology of the root system and to determine how the anchorage fails. This is best achieved by cutting vertical trenches in the soil alongside the stem to expose a cross section of the root system (Coutts 1983). This has two advantages. It allows direct observation of the roots in situ and allows the movements of the roots and surrounding soil to be observed as the plant is pulled over. These movements may be better visualised by coating the sides of the trench with silver spray paint, which shows up cracks, and filming the process of uprooting using a video camera. Sounds of breaking also give clues about how failure is occurring.

Fig. 16.1. Failure modes due to horizontal forces in three types of root systems. In widely spreading root systems with sinker roots such as those possessed by trees and some herbaceous dicots (**A**) the system rotates up around a leeward hinge. Uprooting is resisted by three components: the stiffness of the hinge; the resistance to pullout of the windward sinker roots; and the weight of the root/soil plate. In the narrower systems seen in cereals (**B**) and other monocots, rotation occurs about a windward hinge. Lodging is resisted by two components: the resistance of the soil beneath the root/soil cone to compression; and the buckling resistance of the coronal roots. Simple tap root systems (**C**) are supported by the resistance of the soil on either side to compression. Rotation occurs directly beneath the stem

Once the pattern of failure has been observed, and the possible components of anchorage identified, their relative importance can be determined by carrying out a further series of tests.

In plate systems in particular it has proved possible to determine the relative size of the different components. This is done by comparing the torques required to pull over single plants both before and after particular components are destroyed. Typical procedures involve cutting of the leeward hinge and trenching around the root system (Coutts 1986; Ennos et al. 1993a). It is also possible to estimate the strength of the leeward hinge by summing the strengths of individual lateral roots on one side of the trunk against being bent upwards (Crook and Ennos 1996). In all the species that have so far been investigated the largest anchorage component is the uprooting resistance of windward roots.

In other systems it has proved much harder to work out the relative size of the different anchorage components. For some systems this is due to practical difficulties. In cone systems, for instance, it is hard to determine the importance of the buckling resistance of the basal roots because each buckles at a different depth and at a different distance from the centre of rotation (Crook and Ennos 1993). The importance of the resistance of soil to compression in such root systems can be determined, however, by rotating a model cone into the soil. For other anchorage systems the difficulty in determining the relative size of the anchorage components is intrinsic to the mechanics. The soil will only help prevent tap-rooted systems from lodging, for example, if the tap root has bending rigidity. It is therefore theoretically impossible to separate the components.

The most obvious technique to determine anchorage components, and one which has been extensively used by many early workers who investigated the lodging resistance of wheat and maize (Neenan and Spencer-Smith 1975), is to measure the force required to extract individual roots. Unfortunately this technique has very little value because roots are not pulled out of the ground when these species are pulled over. A further problem is that the failure which occurs when an individual root is pulled out is quite different from that which happens when the whole system fails and many roots pull out in a block.

In cases in which it is difficult to calculate the size of the components of anchorage it is perhaps best to simply measure the mechanical properties of the roots and the soil in which they are embedded. Measurement of root properties can provide much more information about the anchorage system and will allow much easier direct comparison between different varieties or species of plants.

16.3.2 Measuring Root Properties

Two sets of mechanical tests can be carried out on roots depending on their morphology and function. These are best performed on a universal loading frame that has been purpose built for materials testing (eg. Instron Corporation, Lloyd's).

16.3.2.1 *Tensile Tests*

Fibrous roots resist only tensile forces so their properties are best investigated by carrying out tensile tests, in which the upper regions of the roots are held in the clamps of a universal testing machine and stretched (Fig. 16.2A). The machine will produce a graph of force against extension (Fig. 16.2C) which can be used (see Box 16.1) to determine the root's mechanical properties: breaking strength, and the breaking stress and stiffness of the material from which it is made. The "overall" strength of the root system can finally be characterised as the sum of the basal strengths of all the roots.

BOX 16.1. Equations to Calculate Root Properties from Tensile Tests

The breaking strength of a root, F_{max}, is simply the greatest force it can withstand (see Fig. 16.2b). From the graph of force against extension the material properties of the material from which the root is made can also be calculated.

The breaking stress, σ_{max}, is the force per unit area the root can withstand. If the root is circular, breaking stress is given by the expression:

$$\sigma_{max} = 4F_{max} / \pi D^2, \qquad (16.1)$$

where D is the diameter of the root.

Finally, the material stiffness or Young's modulus, E, of the root, is the stress which would be required to double its length. This has units of newtons per square metre and is given by the expression:

$$E = \frac{dF}{dy}\frac{4L}{\pi D^2}, \qquad (16.2)$$

where dF/dy is the initial slope of the force/extension curve, and L is the initial length of the root between the clamps (Ennos 1989).

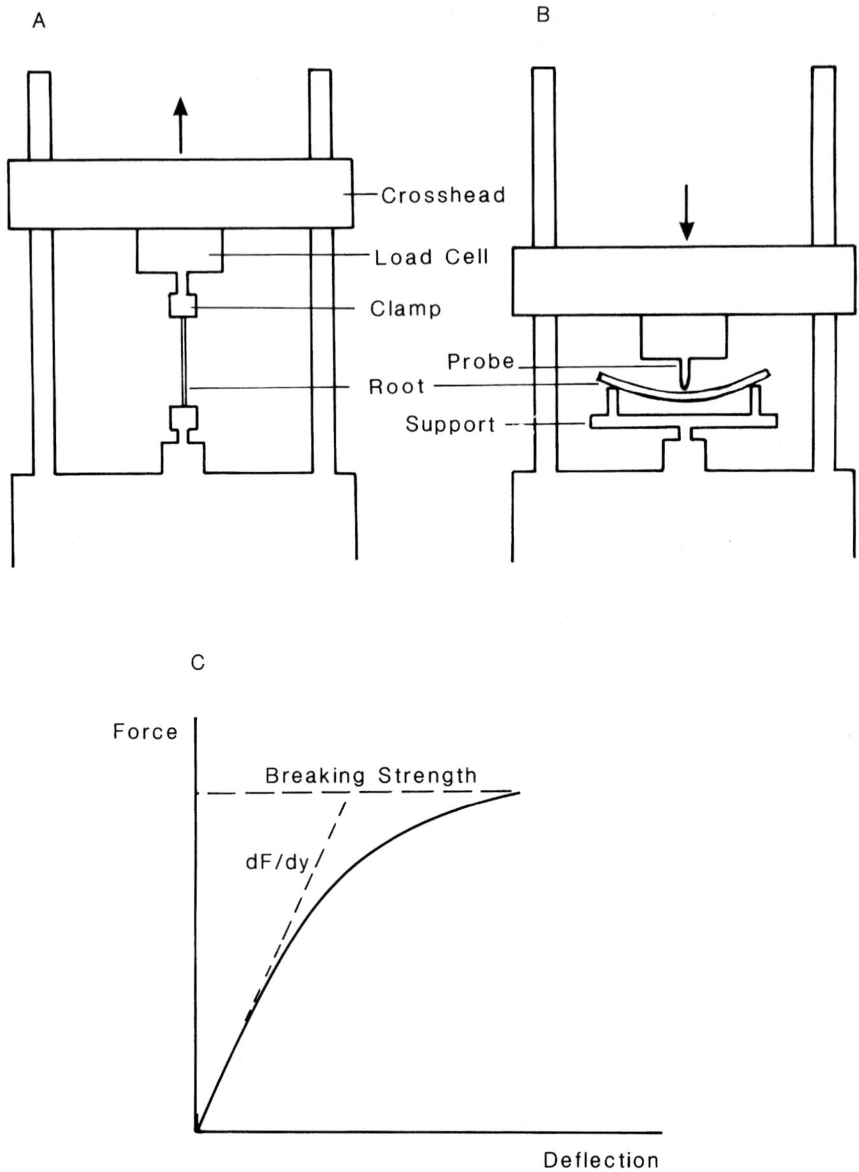

Fig. 16.2. Methods for determining the mechanical properties of roots using a universal testing machine. In tensile tests (**A**), fibrous roots are gripped by two clamps and are stretched by upward movement of the crosshead. The load cell measures the force required. In three point bending tests (**B**), a length of rigid root is balanced between two supports and is bent by a probe which is moved down and hits it midway between them. Both tests produce a force/deflection graph (**C**) which can be used to determine both the maximum bending moment a root can withstand, its rigidity, and the stiffness of the material of which it is composed

There are two main problems with tensile tests. The first is a theoretical problem; the strength of roots decreases with increasing distance from their base, so the results obtained may not correspond well with uprooting resistance. The second is practical; when stretched by the machine, roots tend to break at the clamps, at forces well below their breaking strength. Gluing the roots to the clamp using cyanoacrylate "superglues" may reduce this problem to an extent but when this is done the stele may simply slip out of the cortex at the clamps.

16.3.2.2 Bending Tests

The more rigid laterals and tap roots are broken more frequently in nature by being bent rather than by being stretched, so their properties are better investigated by carrying out bending tests. To study uniform bending forces on their own without additionally imparting shear to the roots it would be theoretically best to carry out four point bending tests, which are equivalent to holding a root section at its two ends and bending it. This provides a uniform bending load to the whole section. However, these four point bending tests are tricky to perform and roots are never, in fact, loaded this way. It is easier and just as valid to carry out three point bending tests. In such tests, the upper region of the root is rested between two supports and the root is bent by a probe which impacts on it half way between the supports (Fig. 16.2B). In these tests the centre of the section is subjected to the greatest bending loads, while the ends get no loading in bending. However, from the force/deflection curve (Fig. 16.2C) both the bending strength and rigidity of the root can be calculated, along with the stiffness, or Young's modulus, of the root material (Ennos et al. 1993b), but the calculations (see Box 16.2) are rather more complex than for tensile tests.

Bending tests have the advantage that they are very quick and easy to perform. Their disadvantages are that roots are seldom cylindrical, as the test assumes, but taper gradually. Hence the calculations of the strength, rigidity and Young's modulus are subject to some error. Despite this, three point bending tests have been extremely useful in comparing anchorage systems of different species (Ennos et al. 1993a,b), varieties of plants (Crook and Ennos 1994) and plants subjected to different treatments (Crook and Ennos 1994, 1995; Goodman and Ennos 1996). The rigidity and strength of the stem can also be measured for comparison using this technique. A simplified pocket device exists for field measurement of the material properties of woody tree roots (Mattheck et al. 1994). This measures both the lateral breaking stress and longitudinal compressive breaking stress of the wood (Stokes and Mattheck 1996). Though these properties are not the same as those calculated by the three point bending test, they are still useful because the breaking stress of wood tends to be proportional to its stiffness.

BOX 16.2. Equations to Calculate Root Properties from Bending Tests

The maximum bending moment a root can withstand, M_{max}, is a structural property of the whole organ. It has units newton metres and is given by the expression:

$$M_{max} = F_{max} L/4, \tag{16.3}$$

where F_{max} is the force required to break or bend the root past its failure point, and L is the length between the supports.

The rigidity of the root, R, which is also a structural property of the whole organ, is its resistance to being curved. It has units newton metres squared and is given by the expression:

$$R = \frac{dF}{dy}\frac{L^3}{48} \tag{16.4}$$

where dF/dy is the initial slope of the force/deflection curve.

The rigidity of the root is, in turn, the product of the stiffness (or Young's modulus) of the material, E, and I, the second moment of area of its cross section, so:

$$R = EI. \tag{16.5}$$

For a cylindrical root of diameter, D, I is given by the expression:

$$I = \pi D^4/64. \tag{16.6}$$

The stiffness of the material, E, which has units of newtons per metre squared, can therefore also be readily calculated for each root section by combining the last two equations to give:

$$E = 64R/\pi D^4. \tag{16.7}$$

Just as with the tensile tests, the sum of the bending strengths or rigidities of all the roots can then give a measure of the strength of the root system.

16.3.3 Measuring Soil Properties

As we have seen, the strength of a plant's anchorage system and even its uprooting behaviour depends greatly on the strength of the soil in which its roots are embedded. In turn this crucially depends on the water potential of the soil. It is therefore extremely important when performing lodging and uprooting tests to control soil properties, or at least be able to measure them. In this section we mention briefly some techniques used in measuring and controlling soil properties. The reader should refer to Chapter 3 of this book, and standard soil physics texts for further information (eg. Marshall and Holmes 1998; Smith and Mullins 1999).

Unfortunately, controlling soil water potential is extremely difficult, particularly in the field next to a transpiring plant. Perhaps the best that can be achieved in field conditions is to carry out tests when the soil is wet and anchorage failure is most likely to occur. This is best done by saturating the soil by heavily overwatering it before leaving it for over an hour to drain. In the laboratory, control of water potential is rather easier. Soil at field capacity can be achieved by saturating cores of soil, and allowing them to drain. A tension table can also be used to obtain a known matric potential between 0 and about −20 kPa. This range can be extended to about −80 kPa using ceramic tables with a vacuum pump attached. To do this the saturated cores are laid on filter paper which covers the tension table. The matric potential is adjusted by raising or lowering a water reservoir, and leaving the samples to equilibrate. The time taken for equilibration is greater for more negative matric potentials and taller soil samples. Equilibrium is achieved when the mass of the soil core remains constant with time, and may require a week or longer. Soil strength increases considerably in the range −50 to −1500 kPa, but it is unfortunately virtually impossible to evenly dry soil that contains a transpiring plant to such negative potentials.

No easy tests exist to reliably measure the tensile or compressive strengths of undisturbed soil in the field, as related to anchorage. Penetrometers measure the pressure required to push a metal cone through the soil (Chap. 3), involving compression of the soil around the probe tip. Penetrometer resistance involves a large component of frictional resistance, and a complex mode of failure around the probe tip that cannot be simply related to the failure of soil which occurs during anchorage tests.

The shear strength, τ, of soils is most easily measured using a simple vane shearmeter which can be modified to measure a wide range of soil strengths by changing the size of the vane. Soil is a variable material so as many tests as possible should be carried out close to the plants which are being examined; ten to twenty tests should give a good idea of the mean and variance of the soil strength. These tests should always be carried out on undisturbed soil as remoulding dramatically alters soil mechanical properties.

16.4 Methods for Evaluating Plant Anchorage Taken as a Whole

16.4.1 Direct Methods

16.4.1.1 Vertical Tests

Because the mechanics of plant anchorage is complex, and involves several root and soil characteristics, a convenient way to evaluate anchorage resistance is to measure it directly. The most commonly used method is to measure the vertical root-pulling resistance of the plant. Its basic principle is to measure the peak force required to pull the plant's root system from the soil. Practical problems may happen if stem breakage occurs instead of uprooting. This is most likely when the soil is very dry, so these conditions should be avoided. Simple experimental devices have been proposed which may be used for field studies (Nass and Zuber 1971; Rogers et al. 1976; O'Toole and Soemartono 1981; Peters et al. 1982; Jenison et al. 1981; Kevern and Hallauer 1983; Penny 1981). The equipment used includes a scale to measure force, a lever and a hand-operated clamp to grasp the plant, and a tripod or some other portable platform. The clamp, sash cord, or leather strap is placed around the base of the stalk and the plant is uprooted with a hand-powered winch, the force necessary being measured by a force transducer. This type of device has some limitations with respect to the standardisation of the test and the time required. More sophisticated devices have been proposed with a tractor or a skid-steer loader providing the required hydraulic power. These devices are less labour intensive and use a standardised root-pull measurement (Donovan et al. 1982; Beck et al. 1987). A device aimed at more accurate vertical tests performed on young plants in laboratory conditions was proposed by Ennos (1991b). In this latter case, the whole force/vertical displacement curve was recorded.

Significant correlations were found by some authors between this vertical test and the resistance to lodging of a range of maize genotypes (Fincher et al. 1985). Contradictory results were, however, found by other authors who found low or no correlations between this test and lodging resistance (Arihara and Crosbie 1982; Kevern and Hallauer 1983; Melchinger et al. 1986). One major advantage of this method is its simplicity, but the mode of failure is very different from what happens during natural lodging; the anchorage is tested vertically, whereas lodging involves resistance to a horizontal strain. Weak correlations between this test and resistance to lodging are therefore not surprising. Clearly, this method seems more adapted to studies of plant resistance to uprooting.

16.4.1.2 Horizontal Pushing or Pulling Tests

Another group of methods is based on the measurement of the horizontal pushing or pulling resistance of the plant. The basic principle of these methods is to measure the resistant torque of the plant versus its angle of displacement when it is horizontally pulled or pushed. Such a test was used in field or laboratory conditions on rice (Terashima et al. 1992, 1995), maize (Kushibiki 1979; Koinuma et al. 1990; Ennos et al. 1993), wheat (Ennos 1991b; Crook and Ennos 1993b; Crook et al. 1994) and sunflower (Ennos et al. 1993). Similar tests, using powerful winches and high tension force transducers, have been used to test the anchorage of trees (Fraser and Gardiner 1967; Coutts 1983, 1986; Stokes et al. 1995; Crook and Ennos 1996b; Crook et al. 1997). For laboratory trials, an accurate system was proposed by Ennos (1991b) using a modified tensile testing machine which could simultaneously push over the plant and measure the restoring anchorage moment supplied by its root system. In field conditions, many authors have used quite simple manual devices with only a force gauge and a ruler to measure angle displacement of the stem. A more sophisticated device was proposed recently by Fouéré et al. (1995; Fig. 16.3). The apparatus consists of a support, a force sensor, an angle sensor and a control head. It records a curve relating the resisting torque of the anchorage system to the inclination of the stem during an artificial pushing test. Data may be transferred to a computer using an RS232 serial transfer protocol. A typical response curve is given in Fig. 16.4.

Significant correlations were found between this horizontal test and resistance to lodging (Kushibiki 1979; Koinuma et al. 1990; Terashima et al. 1992; Fouéré et al. 1995). Although they do not exactly reproduce what happens during natural lodging, horizontal tests mimic lodging far better than vertical root-pull tests. The horizontal test may also be used to evaluate the relative contribution of the different components of the anchorage performing manipulations as proposed by Ennos et al. (1993a,b).

16.4.2 Indirect Methods

Many authors have tried to find correlations between important features of the root system and plant anchorage, measured with one of the previously mentioned methods, or resistance to lodging. Correlations with individual root characters are often weak because of the complexity of anchorage mechanics. The closest correlations were often found with characters which affect the size of the root-soil cone, such as the angle of spread of adventitious roots from higher internodes (Hébert et al. 1992). In wheat, the link between the diameter of the root-soil cone and anchorage resistance was theoretically and experi-

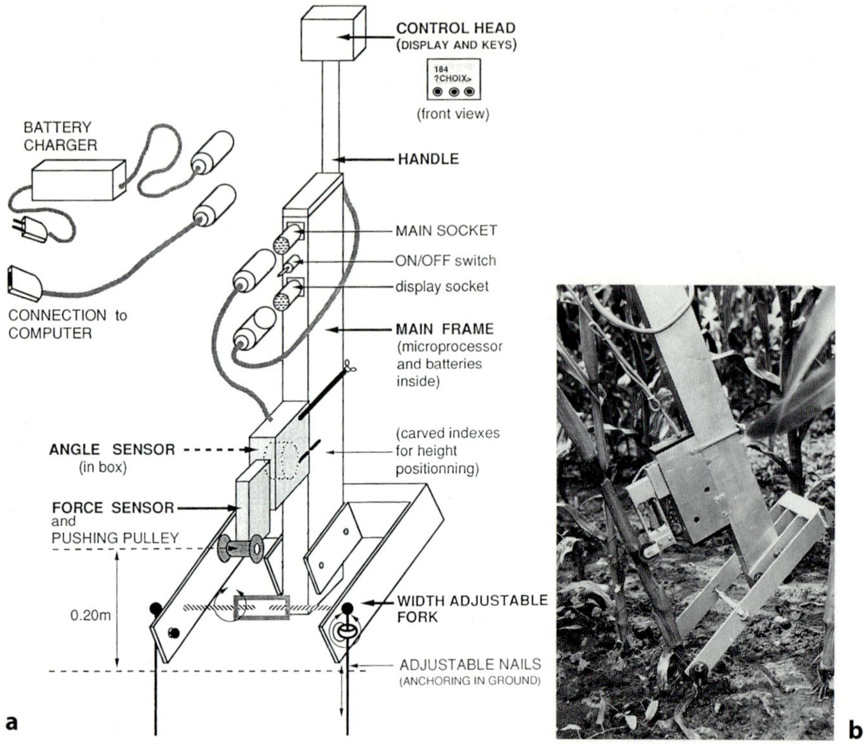

Fig. 16.3. General view of the portable electronic device for horizontal pushing test proposed by Fouéré et al. (1995). The free rotating nails are anchored in the ground, thus creating a horizontal rotation axis passing through the plant base and preventing back-sliding of the device during the pushing operation. During the pushing operation, the force sensor measures and transmits the force to the stalk. A freely rotating PVC pulley prevents any friction between the force sensor and the stalk. (Reproduced from Agronomy Journal)

mentally demonstrated by Crook and Ennos (1993). Obviously the search for root characters which are correlated with plant anchorage resistance should be based on a preliminary investigation of the anchorage mechanics of the species under consideration. Most promising developments are based on attempts to model root anchorage, which may help in identifying the most important root and soil characters for anchorage (Ennos 1991b; Crook and Ennos 1993).

16.5 Conclusion

A classical objective of people interested in plant anchorage is to compare the ability of different genotypes to withstand conditions which may cause lodging or windthrow. They have also investigated the relative performance of plants

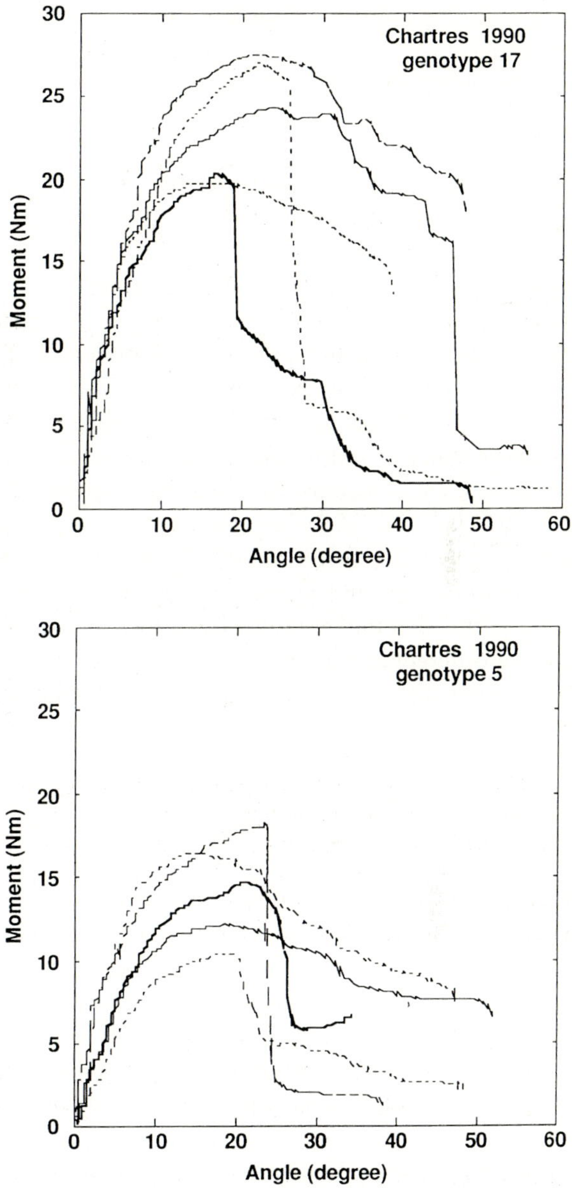

Fig. 16.4. Example of response curves for horizontal pushing resistance of maize (Fouéré et al. 1995; reproduced from Agronomy Journal)

subjected to different treatments, such as growing them at different densities, and in different soils. Static methods for evaluating anchorage as a whole, such as vertical uprooting or lateral pushing or pulling tests, have proved helpful and pragmatic tools in this context, even though they do not exactly mimic the natural challenge to anchorage when aerial parts are subjected to dynamic wind pressures. Until now, methods for measuring root mechanical properties have also been used in the context of comparison between species or genotypes.

Because the scientific investigation of anchorage is still in its early stages, there is, however, still a strong need for simply understanding how it works. Observation-based methods, which describe what happens when a plant is pushed or pulled, are still useful in this context. An important contribution may be expected from the use of strain gauges which record the distribution of mechanical strains in the stem and roots (Stokes et al. 1995; Crook and Ennos 1996b; Crook et al. 1997). A further step will be to model plant anchorage by simultaneously considering the root system architecture, the mechanical properties of the roots, and the mechanical properties of soils. Pioneer models have been proposed by Ennos (1991b), and Crook and Ennos (1993), but future models will probably have to be more sophisticated and make use of the critical state theory of soil mechanics, which up until now has mostly investigated the movement of narrow tines through topsoil. Such models, combined with experimental studies in which model roots are pulled through the soil, will help in identifying the most important root and soil characteristics for anchorage. Future work must also investigate the development of root systems over time, and the effect on the *secondary* growth of roots of the constriction imposed by soils of different strengths.

Finally, if the ultimate aim is to account for the ability of plants to withstand adverse climatic conditions, anchorage factors have to be combined with aerial factors. Although plant response to wind pressure and anchorage mechanics are both very complex, the most promising approaches for understanding plant resistance to lodging are those based on the simultaneous consideration of these two phenomena. As a plant develops and becomes taller, the mechanical demands imposed on the stems and root system increase. The equilibrium between aerial parts and anchorage has to be considered with a dynamic approach. Examples of such "safety factors" accounting for the equilibrium between shoot and root system have been proposed by Crook et al. (1994) for wheat, Duparque and Pellerin (1994) for maize, and Nielsen (1995) for trees. An even more sophisticated general approach has been developed by Baker (1995) and is being applied to the lodging of wheat.

Acknowledgements. The authors wish to thank Alexia Stokes and John Graham for helpful comments on the manuscript.

References

Anderson CJ, Coutts MP, Ritchie RM, Campbell DJ (1989) Root extraction force measurements for Sitkas pruce. Forestry 62: 127–137

Arihara J, Crosbie TM (1982) Relationships among seedling and mature root system traits of maize. Crop Sci 22: 1197–1202

Baker CJ (1995) The development of a theoretical model for the windthrow of plantsr. J Theor Biol 175: 355–372

Beck DL, Darrah LL, Zuber MS (1987) An improved technique for measuring resistance to root pulling in maize. Crop Sci 27: 356–358

Carter PR, Hudelson KD (1988) Influence of simulated wind lodging on corn growth and grain yield. J Prod Agric 1: 295–299

Coutts MP (1983) Root architecture and tree stability. Plant Soil 71: 171–188

Coutts MP (1986) Components of tree stability in Sitka spruce on peaty gley soil. Forestry 59: 173–197

Coutts MP, Grace J (1995) Wind and trees. Cambridge University Press, Cambridge

Crook MJ, Ennos AR (1993) The mechanics of root lodging in winter wheat *Triticum aestivum* L. J Exp Bot 44: 1219–1224

Crook MJ, Ennos AR (1994) Stem and root characteristics associated with lodging resistance in four winter wheat varieties. J Agric Sci 123: 167–174

Crook MJ, Ennos AR (1995) The effects of nitrogen and growth regulators on stem and root characteristics associated with lodging resistance in winter wheat. J Exp Bot 46: 931–938

Crook MJ, Ennos AR (1996a) Mechanical differences between free-standing and supported wheat plants *Triticum aestivum* L. Ann Bot 77: 197–202

Crook MJ, Ennos AR (1996b) The anchorage mechanics of mature larch *Larix europea x L. japonica*. J Exp Bot 47: 1509–1517

Crook MJ, Ennos AR, Sellers EK (1994) Structural development of the shoot and root systems of two winter wheat cultivars, *Triticum aestivum L.* J Exp Bot 45: 857–863

Crook MJ, Ennos AR, Banks JR (1997) The function of buttress roots: a comparative study of the anchorage systems of buttressed (*Aglaia* and *Nephelium ramboutan* species) and non-buttressed (*Mallotus wrayi*) tropical trees. J Exp Bot 48: 1703–1716

Donovan LS, Jui P, Kloek M, Nicholls CF (1982) An improved method of measuring root strength in corn (*Zea mays* L.). Can J Plant Sci 62: 223–227

Duparque A, Pellerin S (1994) An attempt to model the mechanism of root lodging resistance in maize. 3rd Congr European Society for Agronomy. Proceedings, 18–22 September 1994, Padova, Italy, pp 360–361

Ennos AR (1989) The mechanics of anchorage in seedlings of sunflowers *Helianthus annuus*. New Phytol 113: 185–192

Ennos AR (1990) The anchorage of leek seedlings: the effect of root length and soil strength. Ann Bot 65: 409–416

Ennos AR (1991a) The mechanics of anchorage in wheat *Triticum aestivum* L. I. The anchorage of wheat seedlings. J Exp Bot 42: 1601–1606

Ennos AR (1991b) The mechanics of anchorage in wheat *Triticum aestivum* L. II. Anchorage of mature wheat against lodging. J Exp Bot 42: 1607–1613

Ennos AR (1993) The scaling of root anchorage. J Theor Biol 161: 61–75

Ennos AR, Fitter AH (1992) Comparative functional morphology of the anchorage systems of annual dicots. Funct Ecol 6: 71–78

Ennos AR, Crook MJ, Grimshaw C (1993a) A comparative study of the anchorage systems of Himalayan balsam *Impatiens glandulifera* and mature sunflower. *Helianthus annuus*. J Exp Bot 44: 133–146

Ennos AR, Crook MJ, Grimshaw C (1993b) The anchorage mechanics of maize *Zea mays*. J Exp Bot 44: 147–153

Fincher RR, Darrah LL, et Zuber MS (1985) Root development in maize as measured by vertical pulling resistance. Maydica, 30: 383–394

Flesch TK, Grant RH (1992a) Corn motion in wind during senescence: I. Motion characteristics. Agron J 84: 742–747

Flesch TK, Grant RH (1992b) Corn motion in wind during senescence: II. Effect of dynamic plant characteristics. Agron J 84: 748–751

Fouéré A, Pellerin S, Duparque A (1995) A portable electronic device for evaluating root lodging resistance in maize. Agron J 87: 1020–1024

Fraser AI, Gardiner JBH (1967) Rooting and stability in Sitka spruce. Forestry Commission Bulletin 40. HMSO, London

Goodman AM, Ennos AR (1996) A comparative study of the response of roots and shoots of sunflower and maize to mechanical stimulation. J Exp Bot 47: 1499–1507

Goodman AM, Ennos AR (1997) The responses of field-grown sunflower and maize to mechanical support. Ann Bot 79: 703–711

Hébert Y, Barrière Y, Bertholeau JC (1992) Root lodging resistance in forage maize: genetic variability of root system and aerial part. Maydica 37: 173–183

Hughes TP, Sykes AR, Poppi DP, Hodgson J (1991) The influence of sward structure on peak bite force and bite weight in sheep. Proc N Z Soc Anim Prod 51: 153–158

Jenison JR, Shank DB, Penny LH (1981) Root characteristics of 44 maize inbreds evaluated in four environments. Crop Sci 21: 233–237

Kevern TC, Hallauer AR (1983) Relation of vertical root-pull resistance and flowering in maize. Crop Sci 23: 357–363

Koinuma K, Inoue Y, Kato A (1990) Evaluation of lodging resistance of maize (*Zea mays* L.) by the measurement of the horizontal pull resistance. Bull Nat Grassl Res Inst 43: 23–29

Kushibiki H (1979) A simple method of testing lodging resistance of maize. Bull Hokkaido Prefect Agric Exp Stn 42: 21–27

Laca EA, Ungar ED, Seligman NG, Ramey MR, Demment MW (1992) Am integrated methodology for studying short-term grazing behaviour of cattle. Grass forage sci 47: 81–90

Marshall TJ, Holmes JW (1998) Soil physics, 3rd edn. Cambridge University Press, Cambridge

Mattheck C, Bethge K, Zipse A (1994) Das Fractometer. Allg Forstztg 14: 790–791

Melchinger AE, Geiger HH, Schmidt GA (1986) Vertical root-pull resistance and its relationship to root lodging and forage traits in early maturing European inbred lines and F1 hybrids of maize. Maydica 31: 335–348

Minami M, Ujihara A (1991) Effect of lodging on dry matter production, grain yield and nutritional composition at different growth stages in maize. Jpn J Crop Sci 60: 107–115

Nass HG, Zuber MS (1971) Correlation of corn (*Zea mays* L.) roots early in development to mature root development. Crop Sci 11: 655–658

Neenan M, Spencer-Smith JL (1975) An analysis of the problem of lodging with particular reference to wheat and barley. J Agric Sci 85: 495–507

Nielsen CCN (1995) Recommendations for stabilisation of Norway spruce stands based on ecological survey. In: Coutts MP, Grace J (eds) Wind and trees. Cambridge University Press, Cambridge

O'Toole JC, Soemartono (1981) Evaluation of a simple technique for characterizing rice root systems in relation to drought resistance. Euphytica 30: 283–290

Pellerin S, Trendel R, Duparque A (1990) Relation entre quelques caractères morphologiques et la sensibilité à la verse en végétation du maïs (*Zea mays* L.). Agronomie 6: 439–446

Penny LH (1981) Vertical-pull resistance of maize inbreds and their test crosses. Crop Sci 21: 237–240

Peters DW, Shank DB, Nyquist WE (1982) Root-pulling resistance and its relationship to grain yield in F1 hybrids of maize. Crop Sci 22: 1112–1114

Rogers RR, Russell WA, Owens JC (1976) Evaluation of a vertical-pull technique in population improvement of maize for corn rootworm tolerance. Crop Sci 16: 591–594

Smith KA, Mullins CE (1999) Soil analysis: physical methods, 2nd edn. Marcel Dekker, New York

Stokes A, Mattheck C (1996) Variation of wood strength in tree roots. J Exp Bot 47: 693–699

Stokes A, Drexhage M, Heinze P, Guitard D (1995) Anchorage strength and distribution of strain in tree root systems. Colloque L'arbre, Montpellier, France

Terashima K, Akita S, Sakai N (1992) Ecophysiological characteristics related with lodging tolerance of rice in direct sowing cultivation. I: Comparison of the root lodging tolerance among cultivars by the measurement of pushing resistance. Jpn J Crop Sci 61: 380–387

Terashima K, Akita S, Sakai N (1995) Ecophysiological characteristics related with lodging tolerance of rice in direct sowing cultivation. III: Relationship between the characteristics of root distribution in the soil and lodging tolerance. Jpn J Crop Sci 64: 243–250

Further Reading

Ennos AR (1994) The biomechanics of root anchorage. Biomimetics 2: 129–137

Appendix: Suppliers

Company	Address	Chapter	Items
Bartz Technology Corporation	116-A East Yanonali Street Santa Barbara, CA 93101, USA tel.: +1 805 884 5156 fax: +1 805 884 5168 www. bartztechnology.com e-mail: bartz@silcom.com	8,9,10	Minirhizotron camera systems, digital image capture systems
BASF plc	Wembley Park Drive 151 HA9 8HQ Middlesex, UK	8	Styodur insulation material
BioRad Laboratories	Harbour Way South 1414 Richmond, CA 94804, USA	13	Ion exchange resins, material for solid phase extraction
Campbell Scientific Ltd.	Field Street 14–20 LE12 9AL Leicestershire, UK	3	Temperature sensors
Classen and Co.	Grevenweg 89 2000 Hamburg 26, Germany tel.: +49 40 254031	8	Endoscopes
Decagon	950 EN Nelson Court Pullman, WA 99163, USA tel.: +1 509 332-2756 fax: +1 509 332-5158 www.decagon.com e-mail: bryan@decagon.com	14	Plant water and osmotic potential, soil water potential
Delta-T Devices Ltd.	Low road 128 CB5 OEJ Cambridge, UK tel.: +44 1638 742922 fax: +44 168 743 155 www.delta-t.co.uk e-mail: sales@delta-t.co.uk	3,6,10,8	Temperature sensors, root washer, root image analysis system, minirhizotrons, desktop scanner systems for scanning root samples
DOW Europe S.A.	Bachtobelstrasse 3 8810 Horgen, Switzerland	8	Styrofoam insulation material
Dynamax Inc.	Fallstone 10808 Houston, TX 77099, USA fax: +1 713 564 5200	14	Sap flow
Easy Test Ltd.	P.O. box 24 Solarza 8b 20-815 Lublin 56, Poland	3	Redox probes, etc.

A.L. Smit et al. (Eds.), Root Methods

Company	Address	Chapter	Items
Eijkelkamp Agrisearch Equipment	Postbus 4 6987 ZG, Giesbeek, The Netherlands tel.:+31 313 631941 fax: +31 313 632167 www.eijkelkamp.com e-mail: eijkelkamp@eijkelkamp.com	6,14	Root augers, soil water content and matrix potential, (peat) profile samplers, penetrometers, plant physiological research equipment
ELE International Ltd.	Eastman Way HP2 7HB Herts, UK tel.: +44 1442 218 355 fax: +44 1442 252 474 www.eleint.co.uk. e-mail: ele@eleint.co.uk	3,16	Penetrometers, temperature sensors, shear meters
Environmental Measuring Systems	Turisticka 5 62100 Brno, Czech Republic tel.: +420 (0) 5 41225344 fax: +42 (0) 541 225 344 e-mail: jkucera-ems@iol.cz	14	Sap flow
Fluka Chemie	Messerschmittstrasse 17 89231 Neu-Ulm, Germany	13	Agarose (low endosmosis type), pH and redox indicators, ion exchange resins, material for solid phase extraction
Flurochem Ltd.	Wesley Street SK13 9RY Derbyshire, UK	3	Polyethylene glycol
Gillisons Variety Fabrication Inc.	3033 Benzie HWY Benzona, MI 496 16-9747,USA tel.: +1 231 882 5921/ fax: +1 231 882 5637 e-mail: gillison@gtii.com	6	Hydropneumatic root washer
Greenspan Technology Pty Ltd.	22 Palmerin Street Warwick QLD 4370, Australia tel.: +61 (0) 7 4660 1888 fax: +61 (0) 7 4660 1800 www.greenspan.com.au e-mail: admin@greenspan.com.au	14	Sap flow

Company	Address	Chapter	Items
Hawker De Havilland Victoria Ltd.	5108 Salisbury, Australia tel.: +61 8258 3422 fax: +61 8 281 2529	6	Comair root scanner
Hortresearch Environment & Risk Management Group	PB 11-030 Palmerston North New Zealand tel.: +64 (06) 356-8080 fax: +64 (06) 354 6731 www.hort.cri.nz e-mail: sgreen@hort.cri.nz	14	Sap flow
Ingold – Mettler Toledo	Siemensstrasse 9 61449 Steinbach, Germany	13	pH electrodes and microelectrodes
Institute of Hydrology	Crowmarsh Gifford OX10 8BB Oxfordshire, UK	14	Soil water content and matrix potential
Instron Ltd.	Coronation Road HP12 3SY Bucks, UK tel.: + 01494 464646 fax: + 01494 456123	16	Universal mechanical testing machines
Kiyowa	Hatfield Road 135 AL1 4LZ St Albans, UK tel.: + 01727 859 373 fax: + 01727 844272	16	Strain gauges, strain bridges
Machery & Nagel GmbH & Co. KG	Duren, Germany	13	Filter papers, membrane filters, phosphatase test papers, ion exchange resins, material for solid phase extraction
Mecmesin Ltd.	Lawson Hunt Industrial Park Horsham, West Sussex, UK tel.: + 01403 256276 fax: + 01403 240205 www.mecmesin.com e-mail: info@mecmesin.com	16	Force gauges, torque gauges, force testing, torque testing, force measurement, torque measurement, quality control, tension, compression, dynamometes, materials testing, software, universal testing medina, force transduces, torque transducers
Merck KG	64271 Darmstadt, Germany	13	Agar, ion exchange resins, material for solid phase extraction

Company	Address	Chapter	Items
Millipore Corporation	Maple Street 34 Milford, MA 01757, USA	13	Filter papers, membrane filters
Olympus Optical Co.	Wndenstrasse 14–16 2 Hamburg, Germany	8	Borescopes, fibroscopes
Regent Instruments Inc.	rue Blain 4040 G2B 5C3 Quebec, Canada tel.: +1(418) 840-1347 fax: +1(418) 840-1350 www.regent.qc.ca e-mail: sales@regent.qc.ca	8,10	Root image analysis system, desktop scanner systems for scanning root samples, root morphological measurement, root length measurement
Rhizosphere Research Products	Dolderstraat 62 6706 JG Wageningen, The Netherlands tel.: +31-317-410961 fax: +31-317-422415 www.rhizosphere.com e-mail: info@rhizosphere.com	13	Equipment for soil, rhizosphere and root research
Schott Microelectrodes Inc.	Rickingham Road 302 Lindindery, NH 03053, USA	13	pH electrodes and microelectrodes
Schott-Ruhrglas GmbH	Theodor Schmidt Str. 25 8589 Bayreuth, Germany	8	Duran borosilicate glass minirhizotrons
Sentek PTY Ltd.	King William Street 69 5067 Kent Town Australia fax: + (08) 362 8400	14	Soil water content and matrix potential
Sigma Chemie	Postbus 260 9470 Buchs, Switzerland	13	pH and redox indicators
Skye Instruments Ltd.	Ddole Industrial Estate Landrindod Wells LD1 6DF Powys UK fax: +44(0) 1597 824 812	14	Plant water and osmotic potential
Soilmoisture Equipment Corp.	801 S. Kellogg Ave. Goleta, CA93117, USA tel.: +1 805 9643525 fax: +1 805 683 2189 www.soilmoisture.com e-mail: websales@soilmoisture.com	14	Plant water and osmotic potential

Company	Address	Chapter	Items
Thermoplastic Processes Inc. TML (Tokyo Sokki Kenkyojo Co. Ltd. Japan)	Valley Road 1268 Stirling, NJ 07980, USA tel.: +1 201 6471000 fax: +1 201 753 6749	8	Clear tenite butyrate tubes and cylinders, clear polycarbonate tubing (minirhizotrons)
Transducers; agent UK/Ireland: Techni Measure	59 Alcester Road, Studley B80 7NJ Warwickshire, UK tel.: +44 1527 854103 fax: +44 1527 853267 e-mail: sales@techni-measure.co.uk	16	Force transducers, strain gauge-based transducers
Up GmbH	Ruffinstr. 12 80637 Munich, Germany fax: +49 (0) 89 1665 135	14	Sap flow
Zurich Bolting Cloth Mfg. Co. Ltd.	Ruschlikon 8803, Switzerland	13	Bolting cloth

Subject Index

Printing (Computer to Film): Saladruck, Berlin
Binding: H. Stürtz AG, Würzburg